CRISPR/CAS-MEDIATED GENOME EDITING IN PLANTS

Edited by

Deepu Pandita

Anu Pandita

APPLE ACADEMIC PRESS

First edition published 2023

Apple Academic Press Inc.
1265 Goldenrod Circle, NE,
Palm Bay, FL 32905 USA

760 Laurentian Drive, Unit 19,
Burlington, ON L7N 0A4, CANADA

CRC Press
6000 Broken Sound Parkway NW,
Suite 300, Boca Raton, FL 33487-2742 USA

4 Park Square, Milton Park,
Abingdon, Oxon, OX14 4RN UK

Library and Archives Canada Cataloguing in Publication

Title: CRISPR/Cas-mediated genome editing in plants / edited by Deepu Pandita, Anu Pandita.
Names: Pandita, Deepu, editor. | Pandita, Anu, editor.
Description: First edition. | Includes bibliographical references and index.
Identifiers: Canadiana (print) 20230133673 | Canadiana (ebook) 20230133770 | ISBN 9781774912126 (hardcover) | ISBN 9781774912133 (softcover) | ISBN 9781003331759 (ebook)
Subjects: LCSH: Food crops—Genetics. | LCSH: CRISPR (Genetics) | LCSH: CRISPR-associated protein 9. | LCSH: Gene editing. | LCSH: Plant genetics. | LCSH: Crop improvement.
Classification: LCC SB123.57 .C75 2023 | DDC 631.5/233—dc23

Library of Congress Cataloging-in-Publication Data

CIP data on file with US Library of Congress

ISBN: 978-1-77491-212-6 (hbk)
ISBN: 978-1-77491-213-3 (pbk)
ISBN: 978-1-00333-175-9 (ebk)

About the Editors

Deepu Pandita, MPhil
Senior Lecturer, Government Department of School Education, Jammu, Union Territory of Jammu and Kashmir, India

Deepu Pandita is working as a Senior Lecturer in the Government Department of School Education, Jammu, Union Territory of Jammu and Kashmir, India. Deepu Pandita has more than 20 years of teaching experience and has Masters in Botany from the University of Kashmir, Jammu and Kashmir, India, and Master of Philosophy in Biotechnology from the University of Jammu, Jammu and Kashmir, India. Deepu Pandita has a number of international and national courses to her credit. Deepu Pandita has qualified for fellowships, such as JRF-NET and SRF from the Council of Scientific and Industrial Research (CSIR), New Delhi India; Biotechnology Fellowship, Government Department of Science and Technology, Jammu & Kashmir, India; and IAS-INSA-NASI Summer Research Teacher Fellowship, India. Deepu Pandita has presented her research work at both the national and international conferences and was awarded Best Oral Presentation Award at an International Conference on Biotechnology for Better Tomorrow, a Women Researcher Award, and a Research Excellence Award from two professional associations in India. Deepu Pandita is a life member of various scientific societies. Deepu Pandita acts as a reviewer (18 journals), associate editor, and editor of several international journals. Deepu Pandita has published several editorials, book chapters (Springer, CRC, and Elsevier), reviews, and research articles in various journals of national and international repute, including *Cells, Frontiers in Plant Sciences, Journal of Fungi,* and *Frontiers in Physiology.* Deepu Pandia has currently several books in the process of production.

Anu Pandita, MSc
Senior Dietician, Vatsalya Clinic, New Delhi, India

Anu Pandita is working as a Senior Dietician at Vatsalya Clinic, E-5/11, Krishna Nagar, New Delhi, India. Previously she worked as a lecturer at Bee Enn College of Nursing, Talab Tillo, Jammu, India, and Dietician at Ahinsa Dham Bhagwan Mahavir Charitable Health Centre, New Delhi, India. Anu Pandita has done her MSc internship and a course in the Dietetics Department of PGI, Chandigarh, India. Anu Pandita conducted a case study at the Pediatric Gastroenterology Ward at Nehru Hospital, PGI, Chandigarh, India, on a patient suffering from chronic liver disease. She has a certificate course in Food & Nutrition as well. She has several training programs, refresher courses, and workshops to her credit. Anu Pandita is a life-time member of the Indian Dietetic Association and Indian Science Congress Association, Kolkata, India. Anu Pandita has published book chapters in Springer journal and a number of reviews and research articles in various journals of national and international repute, including *Cells, Journal of Fungi,* and *Frontiers in Physiology.* Anu Pandita also has various books in the process of production. Anu Pandita has presented her research work at both the national and international conferences.

Contents

Contributors

Abdel Rahman Al-Tawaha
Department of Biological Sciences, Al-Hussein Bin Talal University, Maan, Jordan

S. Ruth Assumi
Division of System Research and Engineering, ICAR Research Complex of NEH Region, Umiam, Meghalaya, India

Suman Bakshi
Nuclear Agriculture and Biotechnology Division, Bhabha Atomic Research Center, Mumbai – 400085, Maharashtra, India

Aditya Banerjee
Post-Graduate Department of Biotechnology, St. Xavier's College (Autonomous), 30, Mother Teresa Sarani, Kolkata – 700016, West Bengal, India

Pooja Bhadrecha
University Institute of Biotechnology, Chandigarh University, Mohali, Punjab, India

Lekha Bhagtaney
Intern, Caius Research Laboratory, St. Xavier's College (Autonomous), Mumbai – 400001, Maharashtra, India

Pradeep Bhati
Borlaug Institute for South Asia, Ladhowal, Ludhiana, Punjab – 141008, India

Shankar K. Bhujbal
Nuclear Agriculture and Biotechnology Division, Bhabha Atomic Research Center, Mumbai – 400085, Maharashtra, India

Protha Biswas
Department of Life Sciences, Presidency University, 86/1 College Street, Kolkata – 700073, West Bengal, India

Ercan Bursal
Department of Biochemistry, Mus Alparslan University, Turkey

Ambika Chaturvedi
Department of Botany, Faculty of Science, Dayalbagh Educational Institute, Dayalbagh, Agra – 282005, Uttar Pradesh, India

Ashraf Dar
Department of Biochemistry, University of Kashmir, Srinagar – 190006, Jammu and Kashmir, India

Fayaz Ahmad Dar
Department of Bioresources, Amar Singh College Campus, Cluster University, Srinagar – 190008, Jammu and Kashmir, India

Chinmaya Das
AICRP on Rapeseed-Mustard, Ranital (RRTTS, Ranital, OUAT), Bhubaneswar – 756111, Odisha, India

Abhijit Dey
Department of Life Sciences, Presidency University, 86/1 College Street, Kolkata – 700073, West Bengal, India

Heerah Eijaz
Department of Bioresources, University of Kashmir, Srinagar – 190006, Jammu and Kashmir, India

Vincent Ezin
Faculty of Agricultural Sciences, University of Aboemy-Calavi, Benin

Suman Ganger
Department of Microbiology, Vivekanand Education Society's College of Arts, Science and Commerce, Mumbai, Maharashtra, India

Mimosa Ghorai
Department of Life Sciences, Presidency University, 86/1 College Street, Kolkata – 700073, West Bengal, India

Arabinda Ghosh
Department of Botany, Guwahati University, Guwahati, Assam, India

Geetanjali Harale
Department of Biotechnology, Vivekanand Education Society's College of Arts, Science and Commerce, Mumbai, Maharashtra, India

Farhana Hassan
Department of Biotechnology, School of Biological Sciences, University of Kashmir, Hazratbal, Srinagar, Jammu and Kashmir – 190006, India

Muddasarul Hoda
Department of Biological Sciences, Aliah University, Kolkata, West Bengal, India

Sanjay J. Jambhulkar
Nuclear Agriculture and Biotechnology Division, Bhabha Atomic Research Center, Mumbai – 400085, Maharashtra, India

Prashant Kale
Dr. Panjabrao Deshmukh Krishi Veedyapeeth, Akola, Maharashtra – 444104, India

Ravindra Kale
Dr. Panjabrao Deshmukh Krishi Veedyapeeth, Akola, Maharashtra – 444104, India

Prashant Kawar
ICAR–Directorate of Floriculture Research, Shivaji Nagar, Pune, Maharashtra – 411005, India

Zulqurnain Khan
Institute of Plant Breeding and Biotechnology, MNS University of Agriculture, Multan, Pakistan

Firdous A. Khanday
Associate Professor, Signal Transduction Lab, Department of Biotechnology, School of Biological Sciences, Science Block, University of Kashmir, Hazratbal, Srinagar, Jammu and Kashmir – 190006, India

Bushra Hafeez Kiani
Department of Biological Sciences (Female Campus), International Islamic University, Islamabad – 44000, Pakistan

Uttam Kumar
Borlaug Institute for South Asia, Ladhowal, Ludhiana, Punjab – 141008, India

Parama Majumdar
Department of Botany, Hislop College, Nagpur, Maharashtra, India

Auqib Manzoor
Department of Biochemistry, University of Kashmir, Srinagar – 190006, Jammu and Kashmir, India

Sunidhi Mishra
Department of Horticulture, College of Agriculture, Korea, Indira Gandhi Agriculture University, Raipur – 497335, Chhattisgarh, India

Samapika Nandy
Department of Life Sciences, Presidency University, 86/1 College Street, Kolkata – 700073, West Bengal, India

Potshangbam Nongdam
Department of Biotechnology, Manipur University, Canchipur, Imphal, Manipur, India

Devendra Kumar Pandey
Department of Biotechnology, Lovely Professional University, Punjab, India

Saurabh Pandey
Department of Agriculture, Guru Nanak Dev University, Amritsar, Punjab – 143005, India

Anu Pandita
Dietician, Vatsalya Clinic, Krishna Nagar, New Delhi, India

Deepu Pandita
Government Department of School Education, Jammu, Jammu and Kashmir, India

Narsing Parlawar
Dr. Panjabrao Deshmukh Krishi Veedyapeeth, Akola, Maharashtra – 444104, India

Vashnavi Parma
Dr. Panjabrao Deshmukh Krishi Veedyapeeth, Akola, Maharashtra – 444104, India

Pratik Pusadkar
Dr. Panjabrao Deshmukh Krishi Veedyapeeth, Akola, Maharashtra – 444104, India

Archana Rai
Nuclear Agriculture and Biotechnology Division, Bhabha Atomic Research Center, Mumbai – 400085, Maharashtra, India

Rajiv Ranjan
Department of Botany, Faculty of Science, Dayalbagh Educational Institute, Dayalbagh, Agra – 282005, Uttar Pradesh, India

Aryadeep Roychoudhury
Post-Graduate Department of Biotechnology, St. Xavier's College (Autonomous), 30, Mother Teresa Sarani, Kolkata – 700016, West Bengal, India

Pooja Saraswat
Department of Botany, Faculty of Science, Dayalbagh Educational Institute, Dayalbagh, Agra – 282005, Uttar Pradesh, India

Amrina Shafi
Education Department, Government of J&K, Srinagar, Kashmir – 190002, Jammu and Kashmir, India

Humaira Shah
Department of Biochemistry, University of Kashmir, Srinagar – 190006, Jammu and Kashmir, India

Mahipal S. Shekhawat
Department of Plant Biology and Biotechnology, K.M. Center for Post Graduate Studies, Lawspet, Pondicherry, Tamil Nadu, India

Prashant Shingote
Vasantrao Naik College of Agricultural Biotechnology (Dr. PDKV, Akola), Waghapur Road, Yavatmal, Maharashtra – 445001, India

Joginder Singh
Department of Microbiology, School of Bioengineering and Biosciences, Lovely Professional University, Phagwara – 144411, Punjab, India

Shilpy Singh
Department of Biotechnology, Noida International University, Noida, Uttar Pradesh, India

Priya Sundarrajan
Associate Professor, Department of Life Science and Biochemistry and Director, Caius Research Laboratory, St. Xavier's College (Autonomous), Mumbai – 400001, Maharashtra, India

Shobha Surbhaiyya
Dr. Panjabrao Deshmukh Krishi Veedyapeeth, Akola, Maharashtra – 444104, India

Rachael C. Symonds
School of Biological and Environmental Sciences, Liverpool John Moores University, UK

Umer Majeed Wani
Department of Biotechnology, School of Biological Sciences, University of Kashmir, Hazratbal, Srinagar, Jammu and Kashmir – 190006, India

Dhiraj Wasule
Dr. Panjabrao Deshmukh Krishi Veedyapeeth, Akola, Maharashtra – 444104, India

Abbreviations

2AP	2-acetyl-1-pyrroline
AB	adenosine base
ABA	abscisic acid
ABEs	adenine base editors
AFP	anti-freezing protein gene
ALMT9	aluminum-activated malate transporter 9
ALS	acetolactate synthase
AM	arbuscular mycorrhizal
AT	abiotic stress tolerance
ATV	atroviolacea
BADH2	betaine aldehyde dehydrogenase 2
BCTV	beet curly top virus
BeYDV	bean yellow dwarf virus
bHLH	basic helix-loop-helix
BLP	bacterial lipoprotein
BSCTV	beet severe curly top virus
BSMV	barley stripe mosaic virus
BSV	banana streak virus
BZR1	brassinazole resistant 1
CAM	crassulacean acid metabolism
CaMV	cauliflower mosaic virus
CB	cytosine base
CBE	cytidine base editing
CBSV	cassava brown streak virus
CENH3	centromeric histone
CFIA	Canadian Food Inspection Agency
CH2	methane
CLCuAlV	cotton leaf curl Alabad virus
CLCuKoV	cotton leaf curl Kokhran virus
CLCuMuV	cotton leaf curl Multan virus
CLCuRaV	cotton leaf curl Rajasthan virus
CLCuV	cotton leaf curl virus
CLV-WUS	CLAVATA-WUSCHEL
ClYVV	clover yellow vein virus

CMV	cucumber mosaic virus
CO2	carbon dioxide
CP	coat protein
CPIP	CP-interacting protein
CRD	central repeat domain
CRISPR	clustered regularly interspaced short palindromic repeat
CRISPR/Cas9	clustered regularly interspaced short palindromic repeat/Cas9
CRISPR-BEST	CRISPR-base editing system
CRISPR-GA	CRISPR-genome analyzer
CRISPRi	CRISPR interference
crRNAs	CRISPR-derived RNAs
crRNP	CRISPR ribonucleoprotein
CVYV	cucumber vein yellowing virus
DAMPs	damage-associated molecular patterns
dCas9	dead Cas9
DCL	DICER-like
DEP1	dense and erect panicle 1
DFR	dihydroflavonol-4-reductase
DIPM	DspE-interacting proteins of malus
DR	direct repeat
DR	disease resistance
DREB/CBFs	Dehydration-Responsive Element Binding proteins/ C-repeat binding factors
DRT	donor repair template
DSB	double-strand break
dsDNA	double-stranded DNA
eBSV	endogenous banana streak virus
ECJ	European Court of Justice
ECQ	eating and cooking quality
EDR1	enhanced disease resistance
EMNs	engineered meganucleases
EMS	ethyl methane sulfonate
EPA	Environmental Protection Agency
EPSPS	5-enolpyruvylshikimate-3-phosphate synthase
ETI	Effector-triggered immunity
EU	European Union
FAD2	fatty acid desaturase 2
FAD2-1a	fatty acid desaturase 2-1a
FA-VOCs	fatty acid-derived volatile organic compounds

FDA	Food and Drug Administration
FHB	fusarium head blight
FISH	fluorescent in situ hybridization
GABA	gamma-aminobutyric acid
GAD	glutamate decarboxylase
GBSS	granule-bound starch synthase
GBSSI	granule-bound starch synthase I
GE	genetic engineering
GenEd	genome editing
GEEN	genome editing with engineered nucleases
GFP	green fluorescence protein
GM	genetic modification
GMO	genetically modified organism
GMP	genetically modified plants
GN1a	grain number 1a
GOLS	galactinol synthase
gRNA	guide RNA
GS3	grain size 3
GT	gene technology
GUS	glucuronidase
GW2	grain width 2
GWAS	genome-wide association study
HCV	hepatitis C virus
HD	homology-directed
HDR	homologous directed repair
HEs	homing endonucleases
HGTs	horizontal gene transfer
HMEJ	microhomology-mediated end joining
HR	homologous recombination
HSA1	heat-stress sensitive albino 1
HSNO	hazardous substances and new organisms
HT	herbicide tolerance
IMI	imidazolinone
INDELs	insertion and deletion mutations
IPA1	ideal plant architecture 1
IR	insect resistance
IR	intergenic region
IRRI	International Rice Research Institute
JA	jasmonic acid

LCY-B2	lycopene β-cyclase 2
LCYε	lycopene epsilon-cyclase
lncRNA	long noncoding RNA
MAPKs	mitogen-activated protein kinases
MAS	marker-assisted selection
MGEs	mobile genetic elements
MHLW	Ministry of Health, Labor, and Welfare
MNs	mega nucleases
MP	movement protein
MTL	matrilineal
nCBP-1	novel cap-binding protein-1
NGS	next-generation sequencing
NHEJ	non-homologous end joining
NI	nutritional improvement
NLRs	nucleotide-binding leucine-rich repeat receptor
NLS	nuclear localization signal
NUC	nuclease
NUE	nitrogen use efficiency
ODM	oligonucleotide-directed mutagenesis
OMM	oligonucleotide-mediated mutagenesis
OTEs	off-target effects
PA	polyamines
PACE	phage-assisted continuous evolution
PAM	protospacer adjacent motif
PEBV	pea early browning virus
PEG	polyethylene glycol
pegRNA	prime editing guide RNA
PGPM	plants growth promoting micro-organism
PI	protease inhibitors
PL	pectate lyase
PM	plant-microbe's
PMR4	powdery mildew resistance 4
PRR	phytophthora root rot
PRRs	pathogen recognition receptors
PRSV-W	papaya ringspot mosaic virus-W
psiRNAs	prokaryotic silencing RNAs
PSY1	phytoene synthase 1 gene
PTG	polycistronic tRNA-gRNA
PTI	PAMP-triggered immunity

PVA	potato virus A
PVY	potato virus Y
PYL1	pyrabactin resistance like 1
QPM	quality protein maize
QTL	quantitative trait locus
RCas9	RNA-targeted Cas9
RDT	recombinant DNA technology
REC	recognition
RFOs	raffinose family oligosaccharides
RGENs RNPs	RNA guided endonuclease ribonucleoproteins
RGR	ribozyme-gRNA-ribozyme
rin	ripening inhibitor
RMCE	recognition mediated cassette exchange
RNAi	RNA interference
RNP	ribonucleoprotein
ROS	reactive oxygen species
RSDR	regularly spaced direct repeats
RT	reverse transcriptase
RTSV	rice tungro spherical virus
RVD	repeat variable diresidues
SA	salicylic acid
SBEI	starch branching enzyme I
SDN	site-dirccted nuclease
Se	selenium
sgRNA	single-guide RNA
SlAGL6	SlAGAMOUS-LIKE 6
SlPelo	surveillance factor Pelota
SMs	secondary metabolites
SMV	soybean mosaic virus
SNPs	single nucleotide polymorphisms
SP	Streptococcus pyogenes
SRBSDV	southern rice black-streaked dwarf virus
sRNAs	small RNA molecules
SRSRs	short regularly spaced repeats
SSIVa	soluble starch synthase isoform
SSN	single specific nuclease
SSNs	site-specific nucleases
SYNV	Sonchus yellow net virus
TA	tartaric acid

TA	toxin-antitoxin
TALE	transcription activator-like effector
TALEN	transcription activator-like effector nucleases
TALE-R	TALE recombinases
TALE-TF	TALE transcription factors
TbCSV	tobacco curly shoot virus
T-DNA	Ti plasmid DNA
T-DNA	transfer DNA
TGW	thousand-grain weight
TIDE	tracking of indels by decomposition
TIRs	terminal inverted repeats
TMV	tobacco mosaic virus
TOM1	Target Of Myb1 Membrane Trafficking Protein
tracrRNA	trans-activating crRNAs
TRV	tobacco rattle virus
TuMV	turnip mosaic virus
TYLCCNV	tomato yellow leaf curl China virus
TYLCV	tomato yellow leaf curl virus
USDA	United States Department of Agriculture
VIT	vacuolar iron transporter
WDV	wheat dwarf virus
WSMV	wheat streak mosaic virus
ZFN	zinc finger nuclease
ZYMV	zucchini yellow mosaic virus

Preface

The Food and Agriculture Organization (FAO) of the United Nations reports that currently about 820 million people face hunger and fight for their survival. The rate will increase with the increase in population as well as with global warming and climate change. Also, the mutilation of food crop productivity is approximately 40% due to ecological stress. The poor plant traits and quality, in addition to biotic and abiotic factors, result in crop losses globally. So, the need of the hour is to design plant varieties with high yield, improved traits, and resistance to mitigate the detrimental impacts of biotic and abiotic stress conditions on food crops through clustered regularly interspaced short palindromic repeats (CRISPR)/CRISPR-associated nuclease (NUC) proteins. CRISPR/Cas-mediated genome editing (GE) is a robust, simple, precise, efficient, economical, and revolutionary toolbox that leads to improved plant traits with enhanced amount, quality, and resistance against various stresses. Today the eye of world-renowned scientists is on the toolbox of CRISPR/Cas to design CRISPR-edited crops to feed hungry mouths and for future sustainable agriculture.

The prime focus of this book is to deliver a potential podium for sharing recent knowledge and findings by distinguished researchers related to the toolbox of CRISPR/Cas.

Part I

Introduction to Genome Editing Tools

CHAPTER 1

Conventional Breeding, Genetically Modified Crops, and Genome Editing

ZULQURNAIN KHAN

Institute of Plant Breeding and Biotechnology,
MNS University of Agriculture, Multan, Pakistan,
E-mail: zulqurnain.khan@mnsuam.edu.pk

ABSTRACT

Over 50 years, after ensuring food security, the green revolution has exceeded its biological limits, reflected by the ongoing stagnation in crop yield and production. To meet the food demands of the growing population, plant breeders are facing big challenges. Breeding crops using new sophisticated approaches, such as marker-assisted selection (MAS), male sterility, genetic engineering (GE), and genome editing (GenEd). Crop growth and productivity is decreasing due to changing climatic conditions. So, using new techniques which can address these challenges is the need of the hour. Moreover, a new method of creation of desirable variations in the plants is very important to get more valuable traits in plants. The current chapter is going to summarize plant breeding, GE, and genome editing techniques along with their merits and demerits.

1.1 INTRODUCTION

Genetically modified (GM) crops are the crops whose genomes have been manipulated using genetic engineering (GE) methods. Genetic modification (GM) is aimed at enhancing existing characteristics or to introduce a new feature which does not naturally occur in that crop. Transgenic plants are those that have had particular regions of exogenous nucleic acid/gene sequence

CRISPR/Cas-Mediated Genome Editing in Plants, Deepu Pandita, Anu Pandita (Eds.)

inserted into their genome via transformation techniques. These techniques involve direct gene transfer or indirect gene transfer, such as *Agrobacterium*-mediated transformation (Griffiths et al., 2005). The foreign introduced gene, also referred to as a transgenic, might be from a different fungus, plant, virus, bacterium, or even animal species. Hence, the introduction of the technique of GM removed the primary barrier of traditionally applied plant breeding methods, which required having sexual compatibility among species in order to breed them. The natural capacity of *Agrobacterium* to reliably integrate Ti plasmid DNA (T-DNA) into the DNA of recipient cell of plant was identified in 1977 (Chilton et al., 1977); it led to the idea of using Ti plasmid as a vector to transfer genetic material into plant cells. This suggestion made a breakthrough in the creation of transgenic plants. After that, it was first shown that a particular gene sequence could be transmitted to plant cells via transformation protocol utilizing recombinant DNA (Herrera-Estrella et al., 1983). In the very same year, antibiotic-resistant petunia and tobacco were created in the first-ever successful attempt of making transgenic plants (Herrera-Estrella et al., 1983). It was then reported that bean 'phaseolin' gene was successfully expressed in sunflower (Murai et al., 1983). This was the first-ever demonstration of a gene expression even after being translocated to a taxonomically different family of angiosperms.

Later, other transgenic crops were approved for commercialization, including canola with improved composition of oil, Bt potatoes, Bt corn, Bt cotton, cotton with herbicide resistance, and soybeans with glyphosate-resistance, etc., (James, 1997). Transgenic crops that have been marketed have mostly used genetic components from microbes (Kumar et al., 2018). A total of more than 525 genetic transfer events in 32 crops have been commercialized to far (ISAAA database 2019). Maize has the most occurrences (238), followed by cotton with 61, potato with 49, Argentine canola with 42, soybean with 41, carnation with 19 genetic transfer events. Transgenic crop production has significantly enhanced global agricultural output during the last two decades. Research was made on a global scale, and it was reported in that research that gene editing has raised agricultural yields by a significant margin of 22% on average; it resulted in a 68% increase in farmer income (Klumper and Qaim, 2014). Even so, researchers have raised several concerns on crops containing foreign gene(s) due to the possibility of gene flow event among genetically engineered crops and their wild relatives. Other concerns include the possibility of transmission of antibiotic resistance genes to pathogenic bacteria and the health implications such as allergenicity and toxicity to humans. As a result of these factors, genetically modified crops have been met with a lack of popular approval in many regions of the globe, preventing their broad adoption. As a substitute to transgenes, two novel approaches – intra-genesis and cis-genesis, were developed to address concerns about foreign gene transfer. Both of these strategies use genetic material from the same or nearly similar species, i.e., from an interbreeding gene pool, to boost crop yield. Furthermore, the revolutionary technique of genome editing in recent times has allowed crop DNA to be modified with unparalleled simplicity, precision, and efficiency.

To a certain degree, novel editing technologies are based on site-specific nucleases (SSNs) such as zinc finger nucleases (ZFNs), transcription activator-like effector nucleases (TALENs), and the clustered regularly interspaced short palindromic repeats (CRISPR)/Cas system. The technologies have exceptionally lowered the uncertainty and low efficiency attributed to conventional methods of mutagenesis. These new genome editing technologies have the capability to resolve several of the regulatory complications connected with transgenics. As a result, they are poised to support the development of improved varieties through targeted genome editing. Despite the biosafety issues, transgenesis has been a preferred strategy for the quick production of better agricultural plants and the stacking of several beneficial features. The worldwide area on which transgenic crops are grown has expanded dramatically in the previous 22 years: In 1996, this area was 1.7 million ha only, and by 2018, it increased 113-fold to become an area of 191.7 million ha (ISAAA, 2018). Transgenic soybean occupied 95.9 million ha (50%), transgenic maize 58.9 million ha (31%), transgenic cotton 24.9 million ha (13%), transgenic canola 10.1 million ha (5%), and remaining other transgenic crops 1.9 million ha (1%) (ISAAA, 2018). As a result, the transgenic technique is recognized as the most rapidly adopted agricultural technology in contemporary agriculture. In 2017–2018, the worldwide market value of transgenic crops was USD 18.2 billion and over 17 million farmers in 26 countries grew these crops (ISAAA, 2018). Disease resistance (DR), abiotic stress tolerance (AT), herbicide tolerance (HT), nutritional improvement (NI), and insect resistance (IR) are the key features for which gm crops have indeed been created and authorized for commercialization. HT is by far the most prevalent, accounting for 47% of the total area covered by transgenic crops in 2017. Following HT, transgenic crops with stacked traits and IR trait account for around 41% and 12% of worldwide transgenic cultivated areas, respectively (Nazir et al., 2019). Less than 1% of the total area covered by transgenic crops is accounted for by the remaining features (ISAAA, 2017). The purpose of this review is to describe not only the existing status of commercially grown genetically modified crops but also the public concerns and biosafety concerns related to the transgenic crop production. The latest progress in GE tools for crops will also be discussed in this review. Moreover, we have also discussed the future scenario of transgenic crops created using genome editing tools (Kumar et al., 2020). A comparison of the three methods, viz; plant breeding, GE, and genome editing, of developing plant varieties is provided in Figure 1.1.

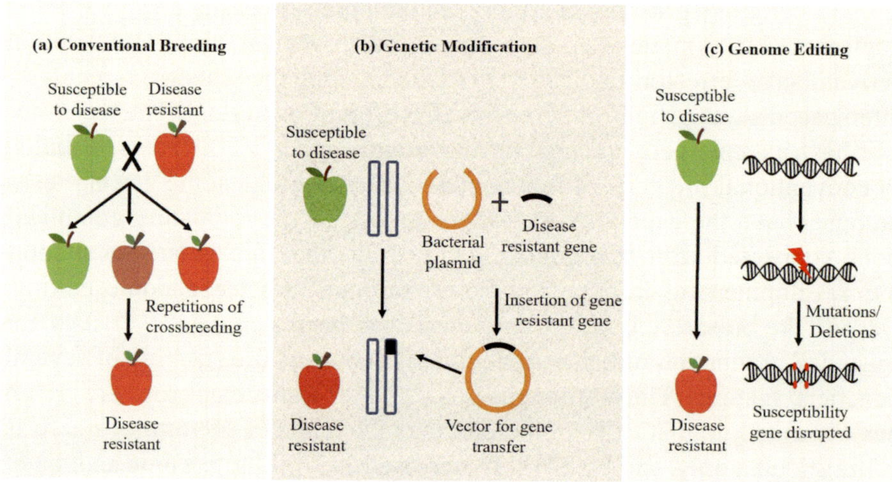

FIGURE 1.1 Comparison of conventional and modern plant breeding methods.
Source: Modified from Kato-Nitta, et al. 2019.

1.2 CONVENTIONAL PLANT BREEDING IN CROP IMPROVEMENT

Conventional Plant breeding (man-made evolution) is the method employed to form new varieties with genetic variability by manipulation and combination of exiting desirable variety. Thousands of years back, superior varieties developed by identifying superior genetic variants. This man-made evolution led to the use of useless weed plants into useful crops. Thus, plant breeding is an integral part of sustainable agriculture system and for the welfare of human.

Farmer selection of favored plant kinds of landraces or populations for subsequent sowing was the primary method of agricultural improvement, until the 19th century. In 1850, genetic improvement of the crop plant was laid in the writing of Darwin (variation of life) and the Mendel's law of inheritance. In 1900, scientist started working on genetics. The era of the plant breeding is divided into three categories on the basis of how species propagate themselves. One is the self-pollination of seed, the second is the cross-pollination, and the third one is the planting/grafting of the vegetative parts. The foundation of plant breeding lies in the genetic diversity of the population through intercrossing or mating selected germplasm with superior characteristics and selection of the superior varieties of the crop plants with desired traits (Borlaug, 1983).

In 1930, the manipulation of the hybrid vigor helped in the genetic improvement of the crop plant, i.e., the yield of the many major cereal crops enhanced gradually (Duvick, 1984). This also led to the phenomenal increase in many crop plants. Heterosis breeding showed remarkable results and more than 70 hybrid cultivars were produced (Jauhar, 2006). Between the 1960s–1970s, the advent of Green Revolution helped in the prevention of mass-scale starvation in Asia. The most remarkable feature of the 20th century was Green Revolution. The semi-dwarf varieties of rice and yields characterized with high resistance to lodging N responsiveness were introduced (Khush, 2001).

Where the world were facing malnourishment, there the conventional breeding proved to be useful not in the higher grain yield but also in the nutritional quality of crop plants. High-quality protein maize (QPM) was produced through conventional breeding and hybridization (Vasal et al., 2006). Hybrid sorghum rich in protein and lysine content were produced by a combination of varieties having high nutritional quality and grain yield. The development of QPM and hybrid sorghum proved an important scientific breakthrough in conventional breeding (Reddy et al., 2006). But the main problem with heterosis breeding was the maintenance of the heterozygosity. Hybrid seeds have had to be produced every year for farmers. The introduction of apomixis in heterosis breeding eliminates this problem and hybrids would enable them to maintain hybrid vigor of the crop plant through heterozygosity. Apomixis has been introduced in rice, pearl millet and maize but with limited success. However, the apomixis helped in the generation of superior hybrids with low cost of seed production cost (Koltunow and Tucker, 2003).

The hybrid science cytogenetics gain special attention of the researchers after the understanding of the chromosome behavior during meiosis thus scientist compile the art of cytology and genetics. Cytogenetics had a major impact on the development of genetic crop improvement and accelerating plant breeding programs. The tools of cytogenetics had greatly helped in the plant breeding by understanding the genomic relationship of crop plants specie. For example, haploidy technique helps in the development of homozygosity of crop throughout successive generations. The improvement of hybridization of crop plant with desired genes helped in the development of superior varieties. This type of plant breeding has helped to meet the demand of world population (Jauhar, 2006).

Chromosome pairing is key to the success of enhancement of crop improvement. As the closely related plant are rich in aggregation of resistance genes to pathogens and pests. Plant breeding programs are designed for pairing of parental chromosome of crop species in transferring genes to the cultivars. For example, the suppression of Ph1 pairing regulation in polyploidy wheat and oat for recombination of desirable alien chromatin by chromosome engineering (Sears, 1981).

Plant breeding is still evolving over the years. With the passage of time, the complexities of the genetic improvement are being arisen with. That is why, the adaptation of better method is needed because of the many plant diseases has become a threat for sustainability of agriculture system and for enough food supply. The genomic reconstruction through wide hybridization along with cytogenetic manipulation is a vital source for the introduction of beneficial characteristics into domesticated crop plants. Undoubtedly, traditional plant breeding has a huge impact on the agriculture system. But it also has some limitations like it can employ only between two plants that have the ability to sexually mate with each other (Table 1.1).

1.3 GM CROPS AND ITS CURRENT STATUS

The terminology genetically modified (GM) is used to describe the gene transfer across species in various steps, such as gene cloning followed by recombination of segments of DNA and finally the introduction of that gene into the host cell. The organisms (plants or animals) created in result of using this technique is called genetically modified organism (GMO). They are also known as transgenic, bioengineered, or even genetically engineered. Interestingly, 'genetically modified' could be an ambiguous word, since, as a matter of fact, almost all things being consumed today by us have been "genetically modified" over the course of time via human selection. This term has been used in this review since its usage has become common in the scientific community for describing recombinant DNA technology (RDT) applications. USDA has prohibited genetically modified seeds or any other genetically modified planting substance in organic farming. A detailed account of the genetically modified crops and the traits for which they were modified has been provided in Figure 1.2. Most modern genetically modified crops have been altered for IR and HT or both (Byrne, 2014).

TABLE 1.1 Conventional Plant Breeding in Crop Improvement

SL. No.	Technique	Crop	Trait	References
1.	Cisgenic	Potato (*Solanum tuberosum*)	Resistance to late blight disease by expression of R-genes, resistance to scab by expression of *HcVf2*	Vanblaere et al. (2011)
		Apple (*Malus domestica*)	Resistance to fungal disease by expressing *VVTL-1* and *NtpII*	–
		Grapevine (*Vitis vinifera*)	Multiple growth types by expressing growth genes	Han et al. (2011)
		Poplar (*Populous* sp.)	Enhance the phytase activity of grain by overexpression of *HvPAPhya*	Holme et al. (2012)
		Barley (*Hordeum vulgare*)	High efficiency of nitrogen utilization by expressing *gTIP2 and gGS1a*	Lutken et al. (2004)
		Durum wheat (*Triticum durum*)	Improved quality of baking by expression of *Lpvp1*	Gadaleta et al. (2008)
		Ryegrass (*Lolium* sp.)	Drought tolerance	Hanley (2008)
2.	Mutation breeding	Wheat (*Triticum aestivum*)	Resistance to drought and salinity	Mujjassim et al. (2019)
		Barely (*Hordeum vulgare*)	Enhance lodging resistance	Gómez et al. (2009)
		Peanut (*Arachis hypogaea*)	Early maturation of varieties	Kale et al. (2007)
		Chickpea (*Cicer arietinum*)	High yield and resistance to wilt	Kozgar (2009)
3.	Induced mutation breeding	Oat (*Avena sativa*)	Resistance to victoria blight	Konzak et al. (1984)
		Chickpea (*Cicer arietinum*)	Resistance to wilt and blight disease	Kharkwal (1996)
		Rice (*Oryza sativa*)	Resistance to YMV, sheath, and blight disease, bacterial leaf blight and *Helminthosporium*	Kharkwal (1996)
		Pearl millet (*Pennisetum glaucum*)	Resistance to *Pyricularia*	Navarro and Solazur (1995)
		Sorghum (*Sorghum bicolor*)	Resistance to charcoal rot, downy mildew, head molds	Kharkwal (1996)
		Mungbean (*Vigna radiate*)	Resistance Mungbean Yellow Mosaic Virus, *Macrophomina* blight and powdery mildew	Kharkwal (1996)
4.	Marker-assisted selection	Barley (*Hordeum vulgare*)	Resistance to barley yellow mosaic virus (BYMV) and stripe rust	Lema (2018)
		Rice (*Oryza sativa*)	Resistance to stem borer, blast disease, and bacterial blight.	–
		Wheat (*Triticum aestivum*)	Powdery mildew	–

TABLE 1.1 *(Continued)*

SL. No.	Technique	Crop	Trait	References
5.	QTL mapping	Rice (*Oryza sativa*)	Root penetration ability	Ray et al. (1996)
			Submergence	Nandi et al. (1997)
			Tolerance	Ni et al. (1998)
			Phosphorus deficiency	Fukuoka and Okuno (2001)
			Resistance	Tan et al. (2001)
			Blast resistance	Lanceras et al. (2004)
			Color, protein content, and milling quality.	Liu et al. (2008)
			Drought tolerance	Lou et al. (2009)
			Basal root thickness	Gomez et al. (2010)
			Grain quality	Tian et al. (2014)
			Drought resistance	Quarrie et al. (1994)
			Drought-induced abscisic acid production	Kulwal et al. (2005)
		Wheat (*Triticum aestivum*)	Pre-harvest sprouting tolerance	Castro et al. (2008)
			Growth response to stress hormones	Santra et al. (2008)
			Stripe rust resistance	Navakode et al. (2009)
			Aluminum toxicity	Vijayalakshmi et al. (2010)
			Senescence-related traits under high temperature	Agrama and Moussa (1996)
			Drought tolerance	Agrama et al. (1999)
		Maize (*Zea mays*)	Downy mildew resistance	Presterl et al. (2007)
			Early plant vigor in chilly environments	Li et al. (2008)
			Head smut resistance	–
			Reaction to phaeosphaeria leaf spot	Moreira et al. (2009)
			Leaf temperature responses to drought	Liu et al. (2011)

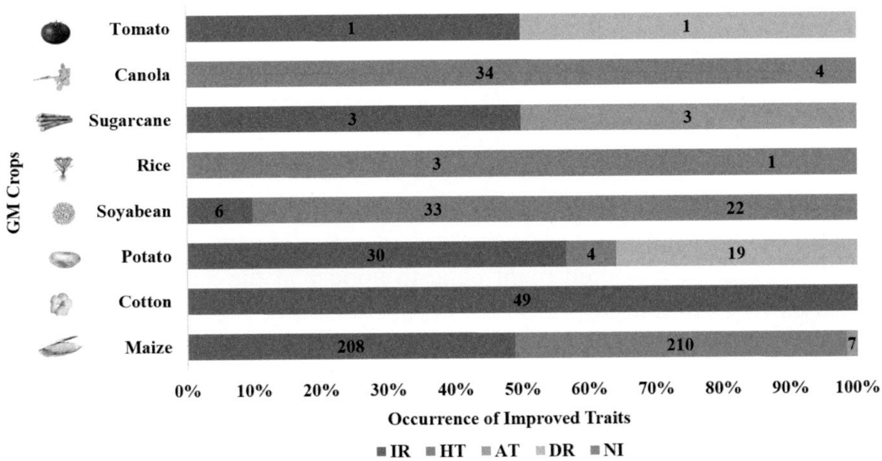

FIGURE 1.2 Commercialized transgenic (GM) crops with various improved traits viz., insect resistance (IR), herbicide tolerance (HT), nutritional improvement (NI), disease resistance (DR), and abiotic stress tolerance (AT). Numbers in bars represent numbers of commercialized transgenic events in a particular crop (Kumar et al., 2020).

1.4 METHODS OF DEVELOPING GM CROPS

When food is genetically altered, new genes are introduced or deleted through the methods of GE, which include *Agrobacterium*-mediated transfer, electroporation, and gene guns. In more recent times, CRISPR, and TALEN have provided considerably more accurate and easy editing tools.

Gene guns/biolistic direct molecules or radiations target genes. It is the most popular method for performing gene transfer. The copies of the gene of interest are bound to small gold/tungsten particles. These particles are then shot into host tissue/cell(s) under high pressure. The particles got accelerated and crossed the barriers of the cell wall and cell membrane to reach the nucleus. This technique has recently been implemented for just a wide range of plants, particularly monocots such as wheat and rice, for which conversion using *Agrobacterium tumefaciens* has become less effective. A benefit of this technique is that significant harm to the cellular tissue may occur (Qaim, 2018). There are many various kinds of GMs, each with its own associated side effects:

- *Agrobacterium tumefaciens*;
- Electroporation;
- Microinjection.

1.4.1 *AGROBACTERIUM TUMEFACIENS*

The presence of tumefaciens is also another typical strategy in inducing metamorphosis. The bacteria known as "agrobacteria" are natural plant parasites. Another way to modify genes is to use the inherent capacity of organisms to transmit genetic material to another.

The agrobacteria produce cell hybrids (they combine cells of two different plant species) and then inject those hybrid cells into the soil. This allows a large number of modified plant cells to appear at the surface of the soil (crown gall). That tumor development is encoded on circular DNA fragment (plasmid). Ascribing transfer of T-DNA to a random position in the genome to the DNA from *Agrobacterium*, when it infects a plant, is known as the T-DNA transfer mechanism. When it is utilized in genetic manipulation, the bacterial T-DNA is deleted as from bacterial genome and replaces with the host genome of choice. Bacteria are a vector in this technique. Using this strategy is particularly helpful when working with dicotyledonous plants, such as potatoes, tomatoes, and tobacco. However, this method was not a huge success in wheat and maize (Wang et al., 2017).

1.4.2 *ELECTROPORATION*

If the plant tissue does not possess cell walls, then electroporation is just the method of choice. The primary benefit of this method is that "DNA penetrates the plant cells via tiny holes that are created by brief bursts of electricity."

1.4.3 *MICROINJECTION*

It is used to introduce genetic material directly into cells by means of micro-injection. Supported by findings of contemporary complete profile of crop composition, agricultural scientists claim that crops improved using genome modification methods are less likely to be dangerous than other traditional approaches. The claim is backed by the fact that transformation procedures are very established, simple to work with, and understood genomes are easily reproduced. The mustard has served as a useful model organism. The new genes must be introduced into plants using a promoter unique to the location where the new gene is to be produced. For instance, in order to express a gene not in leaves but grains, an endosperm-specific promoter should be used in rice. The genes of the organism must have their codon use bias taken into consideration when they are coded (Table 1.2) (Qaim, 2018).

TABLE 1.2 Genetic Engineering Tools in Crop Improvement

SL. No.	Technique	Crop	Trait	References
1.	Transgenic approach	Banana (*Musca* sp.)	Resistance to bunchy top, *Fusarium wilt* and *Xanthomonas wilt*	Sridhar and Sathish, V. Chapter-7
			Tolerance to Sigatoka leaf spot	–
		Grapevine (*Vitis vinifera*)	Resistance to *Xylella fastidiosa*, powdery mildew	–
		Guava (*Psidium guajava*)	Increased anthocyanin and seedlessness	–
			Resistance to viral diseases, crown gall, fungal diseases	–
		Apple (*Malus domestica*)	Endochitinase gene against guava wilt, transformation of cold hardiness genes (CBF1, CBF2, and CBF3) and reduced polyphenol, oxidase, PQ-polyphenol oxidase, and ethylene	–
2.	*Agrobacterium*-mediated transformation	Cowpea (*Vigna unguculata*)	Disease resistance	Bello (2003)
		Field beans (*Pisum sativum*)	–	–
		Cotton (*Gysopium hirsutum*)	–	–
		Mango (*Mangifera indica*)	–	–
		Tobacco (*Nioctiana tobbacum*)	–	–
		Lime (*Citrus aurantifolia*)	Resistance to viral diseases	Fagoaga et al. (2006)
		Fragaria spp.	Modulation of fruit softening	Sridhar and Sathish, V. Chapter-7
3.	Microprojectile bombardment	Peach (*Prunus persica*)	Pest resistance	Sridhar and Sathish, V. Chapter-7
		Grape (*Vitis vinifera*)	–	–
4.	Electroporation	Mandarin (*Citrus reticulata*)	Diseases resistance	Sridhar and Sathish, V. Chapter-7
		Ohta, Banana (*Musa* sp.)	–	–

1.5 GENOME EDITING TOOLS

Over the last 10 years, genome engineering technologies have rapidly advanced and are now commonly employed by research organizations. Their extensive use has a huge influence on fundamental biomedical research, health, and farming. In the following years, it is anticipated that even more diversified, efficient, and simple-to-use gene-editing technologies will be produced. ZFNs, TALENs, and CRISPR/Cas systems cause specific double-strand breaks (DSBs) in the genome, which may induce tiny insertions or deletions (indels) when mended by the NHEJ process (Carroll, 2014). If the targeting site is in the coding sequence, the inserted indels typically cause frameshifts, which result in gene disruption. The HEs, ZFNs, and TALENs use protein motifs to identify genomic target locations, and the molecular cassettes required for every target are extremely challenging to construct (Gaj et al., 2013). CRISPR, on the other hand, recognizes DNA by using base complementarity in between single-guide RNA (sgRNA) as well as the target DNA, which dramatically facilitates the cloning procedure (Jinek et al., 2012). For its ease, CRISPR is straightforward to extend to varied targets and is excellent for multiplex editing (Cong et al., 2013), in which multiple sites are edited concurrently by expressing multiple sgRNAs; as a consequence, it is the favored option for plant genome editing. Following the discovery of the first Cas9 protein utilized for GM (SpCas9) from *Streptococcus pyogenes*, several homologs with unique features have now been identified from other bacteria and being used for GM. Cas9 proteins from type II CRISPR systems identify G-rich PAM sequences and primarily generate DSBs with blunt ends (Cong et al., 2013), whereas Cas12a (Cpf1) and Cas12b (C2c1) proteins from type V CRISPR systems recognize T-rich PAM sequences and primarily generate DSBs with staggered ends (Cong et al., 2013; Zetsche et al., 2015). Furthermore, modified SpCas9 and Cas12a variants have been developed in order to broaden the target range and increase specificity (Hua et al., 2019).

1.5.1 ZFN'S TARGETED GENE SPECIFICITY AND APPLICATIONS

Although simplification of the ZFN-based gene disruption aids the use of ZFN-based gene disruption primarily for the treatment of cancer and HIV, there have been several experimental therapies that use ZFN-based gene disruption with the sole goal of discovering new treatments. Mouse has long been used as the animal model for chromosomal dissections of mammal's biology and infection due to the availability of existing development tools for targeted manipulation of its genome. The importance of studying human disease using mouse models is due to the fact that mice and humans have 95% of their genes in common. Mice are also an efficient and appropriate concept for human diseases. The research into the causes of human illnesses and the development of medical treatments, technologies, and procedures has benefited greatly from the work done using mouse models. Over the last decade, ZFNs have seen wide use in the manipulation of genomes in animals like as A variety of animal species ranging from Drosophila melanogaster, Caenorhabditis elegans, sea urchins, silkworms, zebrafish, frogs, rats, rabbits, pigs, cattle, hamsters, humans, and mice (Lakkireddy et al., 2014).

1.5.1.1 ZFN-BASED GENE KNOCKOUT BY NHEJ

The whole biosphere (from microbes to humans) has used several DSB repair strategies to make up for the cellular damage from DNA–DSBs. This break is addressed by an incomplete endogenous repair process leading to the loss of gene function and a knockout product. NHEJ accounts for a large percentage of replication-independent frame-shift changes in haploid and diploid NHEJ cells that are not regulated by NHEJ repression. The existence of NHEJ (non-homologous end joining) was critical for the development of all replication-independent frame-shift mutations that occurred in the wild type. By way of a 20% reduction in the gene being expressed, in cell lines, ZFN efficiency is achieved without using selection markers. HDR or NHEJ is used in DSB repair because it is error-prone and occurs due to the relatedness of the base pairs in the HUBS/STN region (Lakkireddy et al., 2014).

All eukaryotic organisms have acquired several more different methods to deal with damaged DNA double-strand breaks (DSBs) in chromosomes. Non-homologous end joining (NHEJ) accounts for a large percentage of replication-independent frame-shift mutations in haploid and diploid NHEJ cells that are not regulated by NHEJ repression. The NHEJ (non-homologous end joining) activity was absolutely required for the generation of all replication-independent frame-shift mutations that were detected in the wild type. A result from ZFN gene editing, known as ZFN-mediated knock-out, is equivalent to a 20% efficiency in cell lines without the requirement for any selection markers. HDR or NHEJ DNA repair is required to seal the DSBs that occur as a result of an error-prone or error-prone and homology-dependent repair. The HR-based ZFN-based gene editing IN cases where a researcher replaces one allele with just an engineered construct without affecting other loci in the genome, ZFNs are effective for editing the genome because they bind the homologous recombination (HR) machinery and replace damaged the DSB and use the provided DNA fragment as a framework. In addition to using the ZFN gene-targeting technique to manufacture highly effective HR vectors for varied intrinsic loci in cultured and basic mammalian cells, this method has also been used to synthesize HR vectors with varying internal loci in mammals and cultured cells. Mouse zygotes are capable of modifying the mammalian germ line in a single step. Recombinase target sites are widely employed in the embryonic stem cell industry. They are substrates for insertion of transgenic DNA or exchanges of DNA across supplier plasmids and genome loci, which is known as recognition mediated cassette exchange (RMCE). By analyzing the damaged chromosome, as well as the non-damaged fragment that is being added, the HR machinery evaluates the DNA segment and replaces the damaged portion with the new non-damaged fragment, resulting in gene addition, correction, or knockout. In regard to the genetic manipulation of human embryonic stem cells, the most critical tool is HR. Recently, researchers have begun using HR to insert reporter genes and functional cDNAs into sites of interest. There is a high degree of use of HR recombination for manipulating the mouse genome in the application of gene targeting in ES cells. Molecular gene modification, specifically using ZFNs, is made possible by using donor DNA to pre-deliver mRNA that codes for the ZFNs. Stable docking site mouse lines are generated by ES cell docking sites, which have been expanded to include enzymes suited for the mouse models. To avoid expensive maintenance, all homozygous zygotes need the docking line to be maintained. Although to yet, the most promising strategy for site-specific insertion in zygotes without change of the recipient genome is ZFNs-enhanced homologous recognition, this strategy has yet to provide any clinical success.

Using ZFNs to do gene therapy is a fast-emerging medical discipline in which genes are inserted into the body to provide patients with appropriate (or dysfunctional) genetic material to help them overcome illnesses. Genes are the determining factor for heredity and provide the fundamental biological code that controls the development of cells with unique activities. In order to properly execute the project, it will be critical to evaluate the risks involved with using ZFN technology in mice that most closely resemble human biology. Spontaneous genetic mutations in mice as well as gene targeting are tools used to learn about genes (Lakkireddy et al., 2014). When it comes to ZFN, the advantages and disadvantages may be mentioned.

1.5.1.2 PROS AND CONS

There are several advantages that accompany this gene-editing tool, among them the fact that it served as the first. This cleared the door for new areas of GE. Enabling us to learn quite a bit about genetics, GE, knockouts, and knock-ins, helped us to identify many things about the genetics, GM, knockouts, and knock-ins. The downside is that they have a very high proportion of mutations and cleavages that are off target. Due to the zinc finger's limiting the number of bases recognized, the zinc finger's specificity is reduced, resulting in a greater likelihood of errors. Studies have revealed that ZFN is foreign to the body, and also that it increases the immunogenicity of the cells in which it is present.

1.5.2 TALE NUCLEASES

There are bacterial effectors known as TALE proteins. A 2009 study led to the decoding of the DNA code used by TALE proteins, which led to the discovery of a TALE superfamily. Using this finding, researchers were able to rapidly design and produce their own bespoke TALENs, allowing them to manipulate practically any gene in only months. TALENs, which are modular and can be programmed to target many genomic locations, are constructed using an amino-terminal TALE DNA-binding domain joined to a carboxy-terminal FokI cleavage domain to widen the net and take in more (Schmidt and Siebert, 2001).

To function like ZFNs, dimerization of TALEN proteins is mediated by the FokI cleavage domain, which cuts inside a 12- to 19-bp spacer region that divides each TALE binding site. To be useful, the total number of bases in a TALE must be less than 12- to 20-bps, with additional bases causing the TALE to have greater genome-editing selectivity. The TALE-binding domain is made up of a succession of repeat domains, every 34 residues in length. To learn or build on the previously acquired information, DNA sequentially contacts each of the repeat variable di residues (RVDs), also known as the repeat variable domain (RVDs). Unlike zinc fingers, which have limited ability to identify DNA sequences due to overlapping target sites, each TALE repeat exclusively identifies a single base with little to no overlap between target sites and neighboring domains. For the most part, the RVDs that are most typically employed for the TALE array construction are NI for adenine, HD for cytosine, NG for thymine, and NN or HN for guanine or adenine. As simple as this is, there are many other techniques for creating TALE DNA-binding domains, with the simplest being Golden

Gate assembly. In more recent times, the advancement of TALEN assembly techniques has taken an interesting turn and has looked to use strategies to further improve their results, such as specificity profiling to unearth nonconventional RVDs that increase TALEN function (Schmidt and Siebert, 2001).

Compared to ZFNs, TALENs offer two distinct advantages for genome editing. Because there is no selection or directed evolution required to create TALE arrays, substantially lowering the amount of time and expertise required to assemble a functioning nuclease, TALE arrays may be rapidly and affordably constructed. Another advantage of TALENs is that they are known to display higher selectivity and less toxicity when compared to certain ZFNs. TALENs are much bigger than ZFNs, and have a highly repetitive structure, making it difficult to achieve effective transport of TALENs into cells by the use of lentivirus or a single AAV particle. Despite various other potential limitations, the utility of TALENs in both the coding and translational processes has increased greatly over the last few years due to their capacity to readily deliver both mRNA and protein, while alternative codon usage and amino acid degeneracy may also help in expressing RVD arrays that are less likely to be compromised by recombination (Schmidt and Siebert, 2001).

Developing 18,740 human protein-coding genes as a new target in the human genome is both a technical marvel and something that may support many new investigations and provide a stepping stone for future similarly ambitious undertakings. A subspecialty commercial manufacturer, is now offering TALE arrays (offered on an experimental basis) which are custom-designed to a researcher's specifications (Polizzi, 2013; Gaj et al., 2013).

1.5.2.1 TALENS PROS AND CONS

The positive aspects of it are that it is simpler than ZFN and it only attaches to a single nucleotide rather than three. ZFN is far more difficult approach compared to other contemporary techniques of genome editing. It has a drawback, since it is rather difficult to design, and it has issues in the distribution system. Furthermore, like ZFN does, it produces an immunological response as well. In addition, TALEN is time, effort, and consuming.

1.5.3 THE CRISPR/CAS

CRISPR was initially discovered in 1987 by microbiologists who discovered a string of repetitive segments interspersed by unique sequences in the genome of *E. coli* (Ishino et al., 1987). Similar CRISPR structures were later described for a wide range of bacteria and archaea, but still, it took almost two decades for the spacers to be recognized as being derived from invading DNAs. Spacer sequences resembled phage DNAs and so reflect a recollection of previous infections (Bolotin et al., 2005). RNAs are transcribed from CRISPR loci and then translated into short CRISPR-derived RNAs (crRNAs), which are required for the identification and start of directed slicing of foreign nucleic acids. This far, five major kinds of CRISPR-based bacterial immune systems have already been discovered. However, this list will be expanded in the future since presently unclassifiable variations must be defined (Makarova et al., 2015). CRISPR-associated (Cas) proteins that play a prominent role in the acquisition of new spacers, crRNA growth, and the elimination of contagious nucleic acid sequences, vary across the systems. CRISPR type II systems are composed of four Cas proteins, such as the multidomain Cas9 endonuclease, which has two nuclease domains, the RuvC-like and the HNH domain. By joining the crRNA to the complementary target sequence, a complex of crRNA and trans-activating crRNA (tracrRNA) combined with Cas9 recognizes the invading DNA. As a result, Cas9 causes a DSB, which results in the destruction of the foreign DNA molecule (Jinek et al., 2012). Only the Cas9 endonuclease, together with the crRNA and tracrRNA, are required to use a CRISPR system for the controlled insertion of DSBs for genome editing. Because the two RNAs were linked to a single guide RNA (sgRNA), editing any gene of interest requires just a simple cloning approach for the target sequence (Jinek et al., 2012). The sole constraint in the selection of a target gene is the requirement for a protospacer adjacent motif (PAM), which is dictated by the species origin of the Cas9 protein used. A "NGG" PAM at the 3 end of the targets is needed in conventional applications when Cas9 is obtained from *Streptococcus pyogenes*. This is not an issue since NGG sequences are not uncommon in any coding sequence. As a result, CRISPR/Cas9-based GM is now possible in practically all species.

1.5.3.1 CRISPR/CAS9 FOR PLANTS GENOME EDITING

Many transformation technologies have been successfully used in recent years to transport genetic material into plant cells. The most common technique for so many crops is *Agrobacterium tumefaciens*-mediated gene transfer. Virulence proteins produced on a helper plasmid are necessary to break the DNA between the left and right borders of a plasmid's transfer

DNA (T-DNA) and transport the T-DNA into afflicted plant cells, where it is subsequently transmitted to the nucleus through actin filaments (Yang et al., 2017). A transgenic may be stably integrated into the DNA and quickly translated into mRNA in the nucleus. CRISPR/Cas9 GM requires the sgRNA gene and the Cas9 endonuclease gene. A selectable marker can be used to choose stable changed plants producing Cas9 and the sgRNA, and the modification may happen at any phase of plant growth. Another strategy for transient expression or stable modification of plants is direct gene transfer into plant cells or tissues. This may be achieved by DNA microinjection, PEG absorption into protoplasts, electroporation, or particle bombardment (Altpeter et al., 2005). These non-bacterial techniques have been used on plants that could not be altered using *Agrobacterium*. As a consequence, just the sgRNA and Cas9 genes are needed for targeted editing, with no need for specific vectors. The first reports of CRISPR/Cas9 genome editing in plants were published in 2013.

Direct DNA transfer to protoplasts for temporary Cas9 and sgRNA synthesis was also used in Arabidopsis, tobacco, and rice for PDS genome editing. InDel mutations were detected in all cases, but with varied mutation frequency (Li et al., 2013; Nekrasov et al., 2013; Shan et al., 2013). The timely inefficient transformation of many field crops continues to be a barrier to effective genome editing in plant breeding. Approaches for ever more precise and efficient throughput procedures to allow genetic editing in crops should be developed in the future years (Altpeter et al., 2016). Recently, DNA-free gene targeting techniques based on *in vitro* pre-assembled Cas9-sgRNA ribonucleoproteins (RNPs) have been used (RNPs). RNPs were therefore targeted to immature embryos using particles, and crops with editing events were selected (Table 1.3) (Woo et al., 2015; Jung et al., 2018).

1.6 BIOETHICAL AND BIOSAFETY ISSUES RELATED TO TRANSGENIC CROPS

Genome-editing techniques have the ability not just to bring about a biotechnology revolution in agriculture and human disease, but to also lead to exploitation in a number of different ways, involving germline GM. Many specialists have expressed serious bioethical worries. While time will still be the ultimate judge of whether these innovations are a benefit or a curse, the approaches may affect the human species in a potentially catastrophic manner, and our upcoming human population may be persecuted in aspects we still do not comprehend.

TABLE 1.3 Genome Editing Tools in Crop Improvement

SL. No.	Technique	Crop	Trait	References
1.	ZFNs	*Arabidopsis thaliana*	Modification of genes	Osakabe et al. (2010)
		Tobacco (*Nicotiana tabacum*)	Gene editing	Zhang et al. (2010)
			Targeted integration into endogenous loci	Shukla et al. (2009)
		Maize (*Zea mays*)	Development of herbicide-tolerant and phytate reduction mai	Shukla et al. (2009)
		Rice (*Oryza sativa*)	Stacking of traits	Ainley et al. (2013)
			Stacking of traits	Cantos et al. (2014)
2.	TALEN	*Arabidopsis thaliana*	Targeted mutagenesis	Christian et al. (2013)
		Rice (*Oryza sativa*)	Suppression of transcriptional gene	Mahfouz et al. (2012)
		Wheat (*Triticum aestivum*)	Targeted mutagenesis and genome editing	Zhang et al. (2016)
		Maize (*Zea mays*)	Fragrant rice	Shan et al. (2015)
		Sugarcane (*Saccharum officinarum*)	Resistance to Powdery mildew disease	Wang (2014)
			Reduced epicuticular wax in leaves	Char (2015)
		Soybean	Compacting of cell wall	Kannan (2015)
		Potato	Improved saccharification efficiency and high oleic acid contents	Haun (2014)
		Tomato	High oleic, low linoleic contents	Demorest et al. (2016)
			Minimizing reducing sugars	Čermák (2015)
			Production of purple tomatoes rich in anthocyanin	

TABLE 1.3 *(Continued)*

SL. No.	Technique	Crop	Trait	References
3.	CRISPR	*Arabidopsis thaliana*	Multiplex genome editing	Mao et al. (2013)
		Nicotiana benthamiana	Gene insertion or replacement, knockout, resistance to biotic stress	Li et al. (2013)
		Nicotiana tabacum	Multiplex genome editing	Gao et al. (2015)
		Oryza sativa	Gene insertion or replacement, knockout, resistance to biotic stress	Mao et al. (2013)
		Glycine max	Gene knockout or editing, high yield and resistance to biotic stress	Endo (2016)
		Camelina sativa	Targeted mutagenesis	Li (2016)
		Solanum lycopersicum	Herbicide resistance	Sun (2017)
		Grapefruit	Increase in grain number, grain size and dense erect panicles	Wang (2016)
		Zea mays	High amylose content	Jacobs et al. (2015)
		Triticum aestivum	Resistance to blast diseases	Jiang et al. (2017)
			Targeted mutagenesis	Soyk et al. (2017)
			Improvement of nutritional value	Ortigosa (2018)
			High yield	Soyk et al. (2017)
			Resistance to Powdery mildew and bacterial speck	Jia (2016)
			Earlier harvest time	Jia (2017)
			Parthenocarpy	Svitashev et al. (2015)
			Resistance to citrus canker	Wang et al. (2014)
			Herbicide resistance	
			Resistance to biotic stress	

Besides unlawful germline alteration, major issues include ethics, eugenics assisting the strongest to exist, continuing clinical disputes regarding informed consent, theological controversy, and the potential development of clones, designer babies, and perhaps superhumans. Additionally, the present research considers gene editing as a potential future weapon of warfare. While many communities recognize the desire for a healthy newborn and the entitlement to the best possible care, the impending biotechnological revolution is inevitable and irrefutable. As a result, the urgent requirement necessitates a harmonic and controlled translation of required features of genome-editing-related innovations for medical science and other agricultural businesses. This will necessitate public consensus, expert discussions, the participation of biologists, the judgments of biomedical ethics experts, and ultimate instructions and supervision for the eventually permitted restricted use (Khan, 2019).

Questions have been voiced about the suitability of transgenic foods in terms of the possible hazard to public health and the environment (potential of transgene flow into the ecosystem and deleterious influence on biodiversity) (Mertens, 2008; Lovei and Bhn, 2010). The possible health risk of cytotoxicity and allergenicity connected with genetically engineered crops has long been a source of contention. Cry9c-expressing 'Starlink' maize, for example, was authorized for livestock feed and industrial utilization in the United States in 1998, but this is not accepted for human utilization due to the likelihood of being allergenic to people due to the protein's high stability (Bucchini and Goldman, 2002) and the possibilities to engage with the immune cells. Later, Aventis recalled this Lepidopteron insect-resistant maize, dubbed 'Starlink,' around the globe in 2000 due to the discovery of Cry protein residues in food items (EPA, 2017). However, a clear correlation between Cry9c and allergic responses in consumers could not be demonstrated. One of the most contentious and criticized studies (Seralini et al., 2012) found possible health risks in rats given transgenic NK-603 Roundup Ready maize, including high tumor incidences, chronic renal illness, elevated liver congestion and death in males, and higher female mortality. This work has been heavily criticized by the research world since its release, colloquially known as the 'Seralini scandal,' owing to defective research setup and inadequate data analysis, which resulted in the retraction of this chapter. The same group reprinted a virtually identical article in an extended version in 2014, emphasizing the need of performing long-term feeding studies to adequately investigate the sustainability of genetically engineered crops (Seralini et al., 2014). With a few exclusions, the majority of recent research on the health dangers of genetically engineered crops such as wheat, maize, and soybean on rats, birds; swine, amphibians, cattle, and monkeys have not shown harmful impacts of a genetically altered food diet on animal health (Domingo, 2016). Concerns have also been expressed concerning the likelihood of horizontal transmission of antibiotic-resistant marker genes from GM food to human and animal gut bacteria, which might lead to antibiotic resistance in the gut microflora (Netherwood et al., 2004; Heritage, 2004). However, the likelihood of this form of gene transfer is exceedingly remote. Furthermore, recent attempts have concentrated on generating marker-free transgenic agricultural plants to solve this problem (Tuteja et al., 2012). Pollen-mediated transgene transfer from genetically engineered crops to traditional cultivars and also wild species of numerous crops is a concern associated with detrimental environmental consequences (Yan et al., 2015). The prospective transgene flow from genetically engineered crops to their related species would lead to an increased risk of biodiversity loss, whereas gene flow to weeds could have negative consequences, such as the rise of new weed species and the presence of herbicide-resistant weeds known as 'superweeds.' This far, 16 glyphosate-resistant species of weeds have been recorded from GM cropping systems, out of 24 glyphosate-resistant weed species (Heap, 2014). Conyza canadensis is the most common weed plant, while *Amaranthus palmeri* and *Amaranthus tuberculatus* are the two most economically destructive glyphosate-resistant weeds in the world (Heap and Duke, 2018; Kumar et al., 2020).

1.7 COMMERCIALIZED GM CROPS

The commercialization of genetically modified crops started about two decades ago. Since then, the technique has expanded swiftly around the globe, both in developed and developing nations (Figure 1.1). GM crops were produced on 125 million acres in 25 countries in 2008. The United States (50%), Argentina (17%), Brazil (13%), India (6%), Canada (6%), and China (3%) had the highest proportion of GM crop area (James, 2008). Surprisingly, only Spain, a member of the European Union (EU), farms GM crops on a large scale. Despite the fact that a few other EU nations have authorized particular GM technology, the commercial market is currently small due to public acceptability issues and unattractive regulatory frameworks. Despite the increasing usage of GM crops on a global scale, the portfolio of accessible crop-trait combinations remains very restricted. Only just a few first-generation innovations have been marketed as of yet. HT in soybeans is the dominating technology, accounting for 53% of worldwide GM crop acreage in 2008. Currently, HT soybeans are farmed mostly in Brazil, Argentina, the United States, and other South American nations. This technique contributes for 70% of global soybean output. In 2008, GM maize was the second-most dominant crop, accounting for 30% of worldwide GM acreage and 24% of overall maize output (James, 2008). HT and insect resistance (IR) are involved in GM maize, both as distinct and layered technologies. Different genes from the soil bacteria *Bacillus thuringiensis* are responsible for IR (Bt). These Bt genes are responsible for controlling the European corn borer, the corn rootworm, and several stem borers (Romeis et al., 2008).

Cotton and canola are two more GM crops with large area shares. Bt cotton with bollworm and budworm resistance is very important in underdeveloped nations. India had the most Bt cotton acreage in 2008, with 7.6 million hectares, followed by China with 3.8 million ha. This method is also used in South Africa, Argentina, Mexico, and a few more nations. Bt and HT cotton are used in the United States, partly with stacked genes. Until recently, the majority of HT canola was cultivated in Canada and the United States. A few additional GM crops, such as HT alfalfa and sugar beet, as well as virus-resistant papaya and squash, have been permitted in specific nations, but have so far only covered very limited regions (Qaim, 2009).

1.8 APPLICATION OF GM CROPS

1.8.1 VITAMIN ENRICHMENT

Golden rice, produced by the International Rice Research Institute (IRRI), has more vitamin A and is intended to help those who are deficient in the vitamin. It has yet to be commercially cultivated in any nation as of January 2016.

1.8.2 TOXIN REDUCTION

Cassava that has been genetically engineered to have less cyanogen glucosides and more protein and other nutrients is currently being developed (called BioCassava). The USDA certified a potato in November 2014 that reduces bruising when frying and generates less acrylamide. They don't use genes from plants other than potatoes. The Russet was given this attribute. Burbank, Ranger Russet, and Atlantic cultivars are among the most popular.

1.8.3 RESISTANCE TO STRESS

Plants have been genetically modified to withstand non-biological stresses such as drought, cold, and excessive salt in the soil. DroughtGard maize, developed by Monsanto, was the first drought-resistant crop introduced in 2011.

The United States will approve the sale of a genetically modified crop. Drought resistance is achieved by altering the genes responsible for the crassulacean acid metabolism (CAM), a system that enables plants to live in low-water environments. Rice, wheat, soybeans, and poplar, all of which need a lot of water, might benefit from this to help them adjust to water scarcity faster. In salt-tolerant crops, a number of salinity tolerance mechanisms have been discovered. Rice, canola, and tomato crops, for example, have all been genetically engineered to boost their salt tolerance.

1.8.4 HERBICIDE RESISTANCE

Glyphosate is a pesticide that is used to kill plants by interfering with the shikimate. Glyphosate sensitivity was the most common GM trait in 1999 [needs to be updated]. The primary component in Roundup and other herbicides, glyphosate, destroys plants by causing interference in the shikimate system.

The aromatic amino acids tryptophan, tyrosine, and phenylalanine are synthesized by this route in plants. Animals, on the other hand, lack the shikimate pathway and instead rely on their food for aromatic amino acids. Glyphosate, in particular, inhibits the enzyme 5-enolpyruvylshikimate-3-phosphate synthase (EPSPS).

At the time, herbicides applied on grass and grain crops were extremely toxic and unsuccessful for narrow-leaved weeds, prompting the development of this feature. As a result, producing crops that can survive glyphosate spraying would lessen environmental and health hazards while also giving farmers a competitive advantage.

Some bacteria contain an EPSPS variant that is resistant to glyphosate suppression. One of them was identified from the glyphosate-resistant *Agrobacterium* strain CP4 (CP4 EPSPS). By connecting the 5' terminus of the CP4 EPSPS gene to a chloroplast transit peptide generated from the petunia

EPSPS, the gene was designed for plant expression. This transit peptide was chosen since it has already been shown to be effective.

Previously, it was possible to transfer bacterial EPSPS to other plants' chloroplasts. This CP4 EPSPS gene was cloned and successfully transfected into soybeans. PV-GMGTO4 was the plasmid that was utilized to transfer the gene to soybeans. It contains three bacterial genes, two CP4 EPSPS genes, and a marker gene for Escherichia coli's beta-glucuronidase (GUS). Using the particle acceleration approach, the DNA was delivered into the soybeans. The transformation was carried out using the soybean cultivar A54O3 (Qaim et al., 2018).

1.8.5 PREDICTED BOOM OVER COMING YEARS

When CRISPR/Cas is used to improve the features of various crops, it dramatically increases the probability of getting better results to conform to the breeding procedures that are commonly used. Crop enhancement of chosen crops by homozygous gene pyramiding may be produced in a single generation. Natural allelic differences are indistinguishable. CRISPR/Cas is an incredible technique that allows scientists to introduce new genes into their bodies. Changes are passed down across the generations for certain features.

It will be put into the genome at random, usually on a different locus or chromosome than the one it came from. The location of your choice and the transgenic plants obtained are mutant as well as hemizygous for the insert genes that are to be targeted. This suggests that selfing these plants might result in transgene-free offspring. It will minimize the possible risks and administrative regulations that may be imposed to transgenic plants plantations. The use of CRISPR/Cas9 to knock out negative regulator genes makes it much simpler to obtain desired results.

1.9 FUTURE DIRECTIONS

Undoubtedly the introduction of nucleases into the host cells, as well as the extremely specific characteristics of these nucleases poses a lot of challenges. However, the potential to effectively modify genes utilizing SSNs assures to vastly deepen our knowledge of genomic and the formation of agricultural crops with required traits. Among the most promising and innovative breeding approaches for agricultural enhancement is gene editing based on SSNs. Gene knockouts are useful for producing novel genetic variations, and genome editing may be used to create knockout events for important agricultural plants such as wheat and rice. SSN mutagenesis crops do not seem to have any foreign Genes in their genomes, and they are frequently unrecognizable from natural variations or those generated by traditional mutagenesis. As a result, they may escape the current restrictions governing GM crops. Furthermore, since SSNs could be utilized to insert single nucleotides or large segments of DNA at required genomic loci, the kinds of introduction produced by SSNs could prevent the position impacts associated with randomized insertion by classical transgenesis. Furthermore, if numerous transgenes are introduced at the same place, the genes will be transmitted as a single Mendelian locus, enabling the insertion of numerous transgenes into the desired genome. We anticipate that the advancement of genome modification in plants will open new promising paths for crop enhancement (Gao, 2015).

Future possible uses of GM technologies in agriculture include:

1. **Improvement in Nutrition:** Increased and wholesome profile of vitamins and fatty acids in crops.
2. **Stress Tolerance:** Heat, drought, and salt tolerance.
3. **Disease Resistance (DR):** Plants and especially trees more resistant to diseases such as fungal blight.
4. **Phytoremediation:** Plants that capture pollutants such as heavy metals from contaminated places.
5. **Biofuels:** Creation of plants having modified composition of cell wall for ensuring more effective conversion to biofuels (Byrne, 2014).

KEYWORDS

- **genetic engineering**
- **genetically modified organisms**
- **genome editing**
- **herbicide tolerance**
- **plant breeding**
- **plant transformation**
- **site-specific nucleases**

REFERENCES

Abdu, H., Robinson, D. A., Boettinger, J., & Jones, S. B., (2017). Engineering, electromagnetic induction mapping at varied soil moisture reveals field-scale soil textural patterns and gravel lenses. *Front. Agric. Sci. Eng., 4*(2), 135–145.

Adenle, A. A. J. B. R. N., (2011). Response to issues on GM agriculture in Africa: Are transgenic crops safe? *BMC Research Notes, 4*(1), 1–6.

Agrama, H. A., & Moussa, M., (1996). Mapping QTLs in breeding for drought tolerance in maize (*Zea mays* L.). *Euphytica, 91*(1), 89–97.

Agrama, H., Moussa, M., Naser, M., Tarek, M., & Ibrahim, A., (1999). Mapping of QTL for downy mildew resistance in maize. *Theor. App. Genet., 99*(3), 519–523.

Ainley, W. M., Sastry-Dent, L., Welter, M. E., Murray, M. G., Zeitler, B., Amora, R., Corbin, D. R., et al., (2013). Trait stacking via targeted genome editing. *Plant Biotechnol. J., 11*(9), 1126–1134.

Altpeter, F., Baisakh, N., Beachy, R., Bock, R., Capell, T., Christou, P., Daniell, H., et al., (2005). Particle bombardment and the genetic enhancement of crops: Myths and realities. *Mol. Breed., 15*(3), 305–327.

Altpeter, F., Springer, N. M., Bartley, L. E., Blechl, A. E., Brutnell, T. P., Citovsky, V., Conrad, L. J., et al., (2016). Advancing crop transformation in the era of genome editing. *The Plant Cell, 28*(7), 1510–1520.

Barone, E., Tramutola, A., Triani, F., Calcagnini, S., Di Domenico, F., Ripoli, C., Gaetani, S., et al., (2019). Biliverdin reductase-a mediates the beneficial effects of intranasal insulin in Alzheimer's disease. *Mol. Neurobiol., 56*(4), 2922–2943.

Bello, D., & Kadams, A., (2003). The role of biotechnology in crop improvement. *Niger. J. Biotechnol., 14*(1), 1–15.

Bolotin, A., Quinquis, B., Sorokin, A., & Ehrlich, S. D., (2005). Clustered regularly interspaced short palindrome repeats (CRISPRs) have spacers of extrachromosomal origin. *Microbiol., 151*(8), 2551–2561.

Borlaug, N. E., (1983). Contributions of conventional plant breeding to food production. *Sci., 219*(4585), 689–693.

Borlaug, N. E., (2002). Feeding a world of 10 billion people: the miracle ahead. *In Vitro Cellular & Developmental Biology. Plant, 38*(2), 221–228.

Briefs, I., (2017). Global status of commercialized biotech/GM crops in 2017: Biotech crop adoption surges as economic benefits accumulate in 22 years. *ISAAA Brief, 53*, 25, 26.

Byrne, P., (2014). *Genetically Modified (GM) Crops: Techniques and Applications*. Colorado State University. Libraries.

Cantos, C., Francisco, P., Trijatmiko, K. R., Slamet-Loedin, I., & Chadha-Mohanty, P. K., (2014). Identification of "safe harbor" loci in indica rice genome by harnessing the property of zinc-finger nucleases to induce DNA damage and repair. *Front. Plant Sci., 5*, 302.

Carroll, D., (2014). Genome engineering with targetable nucleases. *Ann. Rev. Biochem., 83*, 409–439.

Castro, A., Tacaliti, M., Giménez, D., Tocho, E., Dobrovolskaya, O., Vasicek, A., Collado, M., et al., (2008). Mapping quantitative trait loci for growth responses to exogenously applied stress induced hormones in wheat. *Euphytica, 164*(3), 719–727.

Čermák, T., Baltes, N. J., Čegan, R., Zhang, Y., & Voytas, D. F., (2015). High-frequency, precise modification of the tomato genome. *Genome Biol., 16*(1), 1–15.

Char, S. N., Unger-Wallace, E., Frame, B., Briggs, S. A., Main, M., Spalding, M. H., Vollbrecht, E., et al., (2015). Heritable site-specific mutagenesis using TALEN s in maize. *Plant Biotechnol. J., 13*(7), 1002–1010.

Chilton, M. D., Drummond, M. H., Merlo, D. J., Sciaky, D., Montoya, A. L., Gordon, M. P., & Nester, E. W., (1977). Stable incorporation of plasmid DNA into higher plant cells: The molecular basis of crown gall tumorigenesis. *Cell, 11*(2), 263–271.

Christian, M., Qi, Y., Zhang, Y., & Voytas, D. F., (2013). Targeted mutagenesis of *Arabidopsis thaliana* using engineered TAL effector nucleases. *G3: Genes. Genom, Geneti., 3*(10), 1697–1705.

Clasen, B. M., Stoddard, T. J., Luo, S., Demorest, Z. L., Li, J., Cedrone, F., Tibebu, R., et al., (2016). Improving cold storage and processing traits in potato through targeted gene knockout. *Plant Biotechnol. J., 14*(1), 169–176.

Cong, L., Ran, F. A., Cox, D., Lin, S., Barretto, R., Habib, N., Hsu, P. D., et al., (2013). Multiplex genome engineering using CRISPR/Cas systems. *Sci., 339*(6121), 819–823.

de Souza, L., & Oda, L. M. (2013). Long term toxicity of a Roundup herbicide and a Roundup-tolerant genetically modified maize. *Food and chemical toxicology: an international journal published for the British Industrial Biological Research Association, 53*, 440.

Demorest, Z. L., Coffman, A., Baltes, N. J., Stoddard, T. J., Clasen, B. M., Luo, S., Retterath, A., et al., (2016). Direct stacking of sequence-specific nuclease-induced mutations to produce high oleic and low linolenic soybean oil. *BMC Plant Bio., 16*(1), 1–8.

Domingo, J. L., (2016). Safety assessment of GM plants: An updated review of the scientific literature. *Food Chem. Toxicol., 95*, 12–18.

Duvick, D. N., (1986). Plant breeding: Past achievements and expectations for the future. *Econ. Bot., 40*(3), 289–297.

Endo, M., Mikami, M., & Toki, S., (2016). Biallelic gene targeting in rice. *Plant Physiol., 170*(2), 667–677.

Fagoaga, C., Rodrigo, I., Conejero, V., Hinarejos, C., Tuset, J. J., Arnau, J., Pina, J. A., et al., (2001). Increased tolerance to *Phytophthora citrophthora* in transgenic orange plants constitutively expressing a tomato pathogenesis related protein PR-5. *Mol. Breed., 7*(2), 175–185.

Filippucci, P., (2009). Heritage and methodology. *Heritage Studies: Methods and Approaches*, 319.

Fukuoka, S., & Okuno, K., (2001). QTL analysis and mapping of *pi21*, a recessive gene for field. Resistance to rice blast in Japanese upland rice. *Theor. App. Genet., 103* (2–3), 185–190.

Gadaleta, A., Giancaspro, A., Blechl, A. E., & Blanco, A., (2008). A transgenic durum wheat line that is free of marker genes and expresses 1Dy10. *J. Cereal Sci., 48*(2), 439–445.

Gaj, T., Gersbach, C. A., & Barbas, III. C. F., (2013). ZFN, TALEN, and CRISPR/Cas-based methods for genome engineering. *Trends in Biotechnol., 31*(7), 397–405.

Gao, C., (2015). Genome editing in crops: From bench to field. *Natl. Sci. Rev., 2*(1), 13–15.

Gao, J., Wang, G., Ma, S., Xie, X., Wu, X., Zhang, X., Wu, Y., et al., (2015). CRISPR/Cas9-mediated targeted mutagenesis in *Nicotiana tabacum*. *Plant Mol. Bio., 87*(1, 2), 99–110.

Gomez, S. M., Boopathi, N. M., Kumar, S. S., Ramasubramanian, T., Chengsong, Z., Jeyaprakash, P., Senthil, A., & Babu, R., (2010). Molecular mapping and location of QTLs for drought-resistance traits in indica rice (*Oryza sativa* L.) lines adapted to target environments. *Acta. Physio. Plant, 32*(2), 355–364.

Gómez-Pando, L., Eguiluz, A., Jimenez, J., Falconí, J., & Aguilar, E. H., (2009). Barley (*Hordeun vulgare*) and kiwicha (*Amaranthus caudatus*) improvement by mutation induction in Peru. *Induced Plant Mutations in the Genomics Era, Food and Agriculture Organization of the United Nations, Rome,* 371–374.

Griffiths, J. F., Griffiths, A. J., Wessler, S. R., Lewontin, R. C., Gelbart, W. M., Suzuki, D. T., & Miller, J. H., (2005). *An Introduction to Genetic Analysis*. Macmillan.

Han, K. M., Dharmawardhana, P., Arias, R. S., Ma, C., Busov, V., & Strauss, S. H., (2011). Gibberellin-associated cisgenes modify growth, stature and wood properties in Populus. *Plant Biotechnol. J., 9*(2), 162–178.

Haun, W., Coffman, A., Clasen, B. M., Demorest, Z. L., Lowy, A., Ray, E., Retterath, A., et al., (2014). Improved soybean oil quality by targeted mutagenesis of the fatty acid desaturase 2 gene family. *Plant Biotechnol. J., 12*(7), 934–940.

Heap, I., & Duke, S., (2018). Overview of glyphosate-resistant weeds worldwide. *Pest Manag. Sci., 74*(5), 1040–1049.

Heap, I., (2009). *The International Survey of Herbicide Resistant Weeds*. http://www.weedscience.com (accessed on 04 September 2022).

Herrera-Estrella, L., Depicker, A., Van, M. M., & Schell, J., (1983). Expression of chimaeric genes transferred into plant cells using a Ti-plasmid-derived vector. *Nature, 303*(5914), 209–213.

Holme, I. B., Dionisio, G., Brinch-Pedersen, H., Wendt, T., Madsen, C. K., Vincze, E., & Holm, P. B., (2012). Cisgenic barley with improved phytase activity. *Plant Biotechnol. J., 10*(2), 237–247.

Hua, K., Zhang, J., Botella, J. R., Ma, C., Kong, F., Liu, B., & Zhu, J. K., (2019). Perspectives on the application of genome-editing technologies in crop breeding. *Mol. Plant, 12*(8), 1047–1059.

Ishino, Y., Shinagawa, H., Makino, K., Amemura, M., & Nakata, A., (1987). Nucleotide sequence of the iap gene, responsible for alkaline phosphatase isozyme conversion in *Escherichia coli*, and identification of the gene product. *J. Bacteriol., 169*(12), 5429–5433.

Jacobs, T. B., LaFayette, P. R., Schmitz, R. J., & Parrott, W. A., (2015). Targeted genome modifications in soybean with CRISPR/Cas9. *BMC Biotechnol., 15*(1), 1–10.

James, C., (1997). Global status of transgenic crops in 1997. *ISAAA Briefs, 5*, 12.

Jauhar, P. P., (2006). Cytogenetic architecture of cereal crops and their manipulation to fit human needs: Opportunities and challenges. *Genetic Resources, Chromosome Engineering, and Crop Improvement, 2*, 1–25.

Jauhar, P. P., (2006). Modern biotechnology as an integral supplement to conventional plant breeding: The prospects and challenges. *Crop Sci., 46*, 1841–1859.

Jia, H., Orbovic, V., Jones, J. B., & Wang, N., (2016). Modification of the PthA4 effector binding elements in type I Cs LOB 1 promoter using Cas9/sg RNA to produce transgenic Duncan grapefruit alleviating XccΔpthA4: dCs LOB 1.3 infection. *Plant Biotechnol. J., 14*(5), 1291–1301.

Jia, H., Xu, J., Orbović, V., Zhang, Y., & Wang, N., (2017). Editing citrus genome via SaCas9/sgRNA system. *Front. Plant Sci., 8*, 2135.

Jiang, W. Z., Henry, I. M., Lynagh, P. G., Comai, L., Cahoon, E. B., & Weeks, D., (2017). Significant enhancement of fatty acid composition in seeds of the allohexaploid, *Camelina sativa*, using CRISPR/Cas9 gene editing. *Plant Biotechnol. J., 15*(5), 648–657.

Jinek, M., Chylinski, K., Fonfara, I., Hauer, M., Doudna, J. A., & Charpentier, E., (2012). A programmable dual-RNA–guided DNA endonuclease in adaptive bacterial immunity. *Sci., 337*(6096), 816–821.

Joint, F. A. O., (1977). *Manual on Mutation Breeding*. International Atomic Energy Agency (IAEA).

Jung, C., Capistrano-Gossmann, G., Braatz, J., Sashidhar, N., & Melzer, S., (2018). Recent developments in genome editing and applications in plant breeding. *Plant Breed., 137*(1), 1–9.

Kannan, B., Jung, J. H., Moxley, G. W., Lee, S. M., & Altpeter, F., (2018). TALEN-mediated targeted mutagenesis of more than 100 COMT copies/alleles in highly polyploid sugarcane improves saccharification efficiency without compromising biomass yield. *Plant Biotechnol. J., 16*(4), 856–866.

Kato-Nitta, N., Maeda, T., Inagaki, Y. et al. Expert and public perceptions of gene-edited crops: attitude changes in relation to scientific knowledge. Palgrave Commun 5, 137 (2019). https:doi.org/10.1057/s41599-019-0328-4

Kennedy, G. G., (2008). *Integration of Insect-Resistant Genetically MODIFIED Crops within IPM Programs*, 1–26.

Khan, S. H., (2019). Genome-editing technologies: Concept, pros, and cons of various genome-editing techniques and bioethical concerns for clinical application. *Mol. Ther. Nucleic Acids, 16*, 326–334.

Kharkwal, M., (1996). Accomplishments of mutation breeding in crop improvement in India. In: *Isotopes and Radiations in Agriculture and Environment Research.*

Khush, G. S., (2001). Green revolution: The way forward. *Nat. Rev. Genet., 2*(10), 815–822.

Klumper, W., & Qaim, M., (2014). A meta-analysis of the impacts of genetically modified crops. *PloS One, 9*(11), e111629.

Koltunow, A. M., & Tucker, M. R., (2003). Advances in apomixis research: Can we fix heterosis? In: *Plant Biotechnology 2002 and Beyond* (pp. 38–46). Springer.

Konzak, C., Wilson, M., & Franks, P., (1984). *Progress in the Evaluation, use in Breeding, and Genetic Analysis of Semi-Dwarf Mutants of Wheat, 15*(18), 39–50.

Kozgar, M. I., & Khan, S., (2009). Genetic improvement of chickpea through induced mutation. *J. Phytol., 1*(6).

Kulwal, P., Kumar, N., Gaur, A., Khurana, P., Khurana, J., Tyagi, A., Balyan, H., & Gupta, P., (2005). Mapping of a major QTL for pre-harvest sprouting tolerance on chromosome 3A in bread wheat. *Theor. App. Genet., 111*(6), 1052–1059.

Kumar, K., Aggarwal, C., Singh, I., & Yadava, P., (2018). Microbial genes in crop improvement. In *Crop Improvement Through Microbial Biotechnology* (pp. 39–56). Elsevier.

Kumar, K., Gambhir, G., Dass, A., Tripathi, A. K., Singh, A., Jha, A. K., Yadava, P., et al., (2020). Genetically modified crops: Current status and future prospects. *Planta, 251*(4), 1–27.

Lakkireddy, K., Podha, S., Krishna Satya, A., Kasturi, K. and KRS, S.R., (2014). Designer zinc finger nuclease: A versatile tool for genome editing in animals. *J. Biotechnol., 4*(1), 24–30.

Lanceras, J. C., Pantuwan, G., Jongdee, B., & Toojinda, T., (2004). Quantitative trait loci associated with drought tolerance at reproductive stage in rice. *Plant Physiol., 135*(1), 384–399.

Landrigan, P. J., Schechter, C. B., Lipton, J. M., Fahs, M. C., & Schwartz, J. J. E. H. P., (2002). Environmental pollutants and disease in American children: Estimates of morbidity, mortality, and costs for lead poisoning, asthma, cancer, and developmental disabilities. *Environ. Health Perspect., 110*(7), 721–728.

Lema, M., (2018). Marker assisted selection in comparison to conventional plant breeding. *Agric. Res. Technol., 14*, 555914.

Li, J. F., Norville, J. E., Aach, J., McCormack, M., Zhang, D., Bush, J., Church, G. M., & Sheen, J., (2013). Multiplex and homologous recombination-mediated genome editing in *Arabidopsis* and *Nicotiana benthamiana* using guide RNA and Cas9. *Nat. Biotechnol., 31*(8), 688–691.

Li, M., Li, X., Zhou, Z., Wu, P., Fang, M., Pan, X., Lin, Q., et al., (2016). Reassessment of the four yield-related genes Gn1a, DEP1, GS3, and IPA1 in rice using a CRISPR/Cas9 system. *Front. Plant Sci., 7*, 377.

Li, X., Wang, Z., Gao, S., Shi, H., Zhang, S., George, M., Li, M., & Xie, C., (2008). Analysis of QTL for resistance to head smut (*Sporisorium reiliana*) in maize. *Field Crops Res., 106*(2), 148–155.

Liu, L., Mu, P., Li, X., Qu, Y., Wang, Y., & Li, Z., (2008). Localization of QTL for basal root thickness in japonica rice and effect of marker-assisted selection for a major QTL. *Euphytica, 164*(3), 729–737.

Liu, Y., Subhash, C., Yan, J., Song, C., Zhao, J., & Li, J., (2011). Maize leaf temperature responses to drought: Thermal imaging and quantitative trait loci (QTL) mapping. *Env. Exp. Bot., 71*(2), 158–165.

Lou, J., Chen, L., Yue, G., Lou, Q., Mei, H., Xiong, L., & Luo, L., (2009). QTL mapping of grain quality traits in rice. *J. Cer. Sci., 50*(2), 145–151.

Lövei, G. L., Bøhn, T., & Hilbeck, A., (2007). Biodiversity, ecosystem services, and genetically modified organisms. *Biosafety First: Holistic Approaches to Risk and Uncertainty in Genetic Engineering and Genetically Modified Organisms* (pp. 161–180). Tapir Academic Press, Trondheim, Norway.

Lütken, H., Laura, M., Borghi, C., Ørgaard, M., Allavena, A., & Rasmussen, S. K., (2011). Expression of KxhKN4 and KxhKN5 genes in *Kalanchoe blossfeldiana* 'Molly' results in novel compact plant phenotypes: Towards a cisgenesis alternative to growth retardants. *Plant Cell Rep., 30*(12), 2267–2279.

Maggi, F., Tang, F. H., La Cecilia, D., & McBratney, A., (2019). Pest-chemgrids, global gridded maps of the top 20 crop-specific pesticide application rates from 2015 to 2025. *Sci. Data, 6*(1), 1–20.

Mahfouz, M. M., Li, L., Piatek, M., Fang, X., Mansour, H., Bangarusamy, D. K., & Zhu, J. K., (2012). Targeted transcriptional repression using a chimeric TALE-SRDX repressor protein. *Plant Mol. Biol., 78*(3), 311–321.

Makarova, K. S., Wolf, Y. I., Alkhnbashi, O. S., Costa, F., Shah, S. A., Saunders, S. J., Barrangou, R., et al., (2015). An updated evolutionary classification of CRISPR–Cas systems. *Nat. Rev. Microbiol., 13*(11), 722–736.

Mao, Y., Zhang, H., Xu, N., Zhang, B., Gou, F., & Zhu, J. K., (2013). Application of the CRISPR–Cas system for efficient genome engineering in plants. *Mol. Plant, 6*(6), 2008.

Melchiorre, M. G., Chiatti, C., Lamura, G., Torres-Gonzales, F., Stankunas, M., Lindert, J., Ioannidi-Kapolou, E., et al., (2013). Social support, socio-economic status, health and abuse among older people in seven European countries. *PloS One, 8*(1), e54856.

Mertens, M., (2008). Assessment of environmental impacts of genetically modified plants. *Implementation of the Biosafety Protocol Development of Assessment Bases,* 234.

Moreira, J. U. V., Bento, D. A. V., De Souza, A. P., De Souza, C. L., & Genetics, A., (2009). QTL mapping for reaction to *Phaeosphaeria* leaf spot in a tropical maize population. *Theor. App. Genet., 119*(8), 1361–1369.

Mujjassim, N. E., Mallik, M., Rathod, N. K. K., & Nitesh, S., (2019). Cisgenesis and intragenesis a new tool for conventional plant breeding: A review. *J. Pharmacogn. Phytochem., 8*, 2485–2489.

Murai, N., Kemp, J. D., Sutton, D. W., Murray, M. G., Slightom, J. L., Merlo, D. J., Reichert, N. A., et al., (1983). Phaseolin gene from bean is expressed after transfer to sunflower via tumor-inducing plasmid vectors. *Science, 222*(4623), 476–482.

Nandi, S., Subudhi, P., Senadhira, D., Manigbas, N., Sen-Mandi, S., & Huang, N., (1997). Mapping QTLs for submergence tolerance in rice by AFLP analysis and selective genotyping. *Mol. Gen. Genet., 255*(1), 1–8.

Navakode, S., Weidner, A., Lohwasser, U., Röder, M., & Börner, A., (2009). Molecular mapping of quantitative trait loci (QTLs) controlling aluminium tolerance in bread wheat. *Euphytica, 166*(2), 283–290.

Nekrasov, V., Staskawicz, B., Weigel, D., Jones, J. D., & Kamoun, S., (2013). Targeted mutagenesis in the model plant *Nicotiana benthamiana* using Cas9 RNA-guided endonuclease. *Nat. Biotechnol., 31*(8), 691–693.

Netherwood, T., Martín-Orúe, S. M., O'Donnell, A. G., Gockling, S., Graham, J., Mathers, J. C., & Gilbert, H. J., (2004). Assessing the survival of transgenic plant DNA in the human gastrointestinal tract. *Nat. Biotechnol., 22*(2), 204–209.

Ni, J., Wu, P., Senadhira, D., & Huang, N., (1998). Mapping QTLs for phosphorus deficiency tolerance in rice (*Oryza sativa* L.). *Theor. App. Genet., 97*(8), 1361–1369.

Ortigosa, U. A., Giménez-Ibáñez, S., Leonhardt, N., & Solano, R., (2019). Design of a bacterial speck resistant tomato by CRISPR/Cas9-mediated editing of SlJAZ2. *Plant Biotechnol. J.*

Osakabe, K., Osakabe, Y., & Toki, S., (2010). Site-directed mutagenesis in *Arabidopsis* using custom-designed zinc finger nucleases. *Proceedings of the National Academy of Sciences, 107*(26), 12034–12039.

Polizzi, K. M., (2013). What is synthetic biology? In: *Synthetic Biology* (pp. 3–6). Springer.

Presterl, T., Ouzunova, M., Schmidt, W., Möller, E. M., Röber, F. K., Knaak, C., Ernst, K., et al., (2007). Quantitative trait loci for early plant vigor of maize grown in chilly environments. *Theor. App. Genet., 114*(6), 1059–1070.

Przezbórska-Skobiej, L., & Siemiński, P., (2020). Gmos and global food security. *Roczniki (Annals), 2020*(2), 173–182.

Qaim, M., (2009). The economics of genetically modified crops. *Annu. Rev. Resour. Econ., 1*(1).

Qaim, M., (2019). Genetically modified crops. *Ann. Pol. Assoc. Agric. Aribus. Econ.*

Quarrie, S., Gulli, M., Calestani, C., Steed, A., & Marmiroli, N., (1994). Location of a gene regulating drought-induced abscisic acid production on the long arm of chromosome 5A of wheat. *Theor. App. Genet., 89*(6), 794–800.

Raina, A., Laskar, R. A., Khursheed, S., Amin, R., Tantray, Y. R., Parveen, K., & Khan, S., (2016). Role of mutation breeding in crop improvement-past, present and future. *Asian J. Agric. Res.,* 1–13.

Ray, J., Yu, L., McCouch, S., Champoux, M., Wang, G., & Nguyen, H., (1996). Mapping quantitative trait loci associated with root penetration ability in rice (*Oryza sativa* L.). *Theor. App. Genet., 92*(6), 627–636.

Reddy, B. V., Ramesh, S., & Reddy, P. S., (2006). Sorghum genetic resources, cytogenetics and improvement. *Genetic Resources, Chromosome Engineering, and Crop Improvement, 2*, 309–363.

Santra, D., Chen, X., Santra, M., Campbell, K., & Kidwell, K., (2008). Identification and mapping QTL for high-temperature adult-plant resistance to stripe rust in winter wheat (*Triticum aestivum* L.) cultivar 'Stephens.' *Theor. App. Genet., 117*(5), 793–802.

Schmidt, A., & Siebert, K. G., (2001). Albert–Software for scientific computations and applications. *Acta Math. Univ. Comenianae, 70*(1), 105–122.

Sears, E., (1981). *Transfer of Alien Genetic Material to Wheat*, 75–89.

Séralini, G. E., Clair, E., Mesnage, R., Gress, S., Defarge, N., Malatesta, M., Hennequin, D., & De Vendômois, J. S., (2014). Republished study: Long-term toxicity of a Roundup herbicide and a roundup-tolerant genetically modified maize. *Environ. Sci. Eur., 26*(1), 1–17.

Shan, Q., Wang, Y., Li, J., Zhang, Y., Chen, K., Liang, Z., Zhang, K., et al., (2013). Targeted genome modification of crop plants using a CRISPR-Cas system. *Nat. Biotechnol., 31*(8), 686–688.

Shan, Q., Zhang, Y., Chen, K., Zhang, K., & Gao, C., (2015). Creation of fragrant rice by targeted knockout of the Os BADH 2 gene using TALEN technology. *Plant Biotechnol. J., 13*(6), 791–800.

Shukla, V. K., Doyon, Y., Miller, J. C., DeKelver, R. C., Moehle, E. A., Worden, S. E., Mitchell, J. C., et al., (2009). Precise genome modification in the crop species Zea mays using zinc-finger nucleases. *Nat., 459*(7245), 437–441.

Singh, R. J., & Jauhar, P. P., (2006). *Genetic Resources, Chromosome Engineering, and Crop Improvement: Cereals* (Vol. 2). CRC press.

Soyk, S., Müller, N. A., Park, S. J., Schmalenbach, I., Jiang, K., Hayama, R., Zhang, L., et al., (2017). Variation in the flowering gene SELF PRUNING 5G promotes day-neutrality and early yield in tomato., *49*(1), 162–168.

Sridhar, D., & Sathish, V. (2018). *Chapter-7 Transgenic Assisted Fruit Crop Improvement*, 155.

Sun, Y., Jiao, G., Liu, Z., Zhang, X., Li, J., Guo, X., Du, W., et al., (2017). Generation of high-amylose rice through CRISPR/Cas9-mediated targeted mutagenesis of starch branching enzymes. *Front. Plant Sci., 8*, 298.

Svitashev, S., Young, J. K., Schwartz, C., Gao, H., Falco, S. C., & Cigan, A. M., (2015). Targeted mutagenesis, precise gene editing, and site-specific gene insertion in maize using Cas9 and guide RNA. *Plant Physiol., 169*(2), 931–945.

Tan, Y., Sun, M., Xing, Y., Hua, J., Sun, X., Zhang, Q., & Corke, H., (2001). Mapping quantitative trait loci for milling quality, protein content and color characteristics of rice using a recombinant inbred line population derived from an elite rice hybrid. *Theor. App. Genet., 103*(6, 7), 1037–1045.

Tian, Y., Zhang, H., Xu, P., Chen, X., Liao, Y., Han, B., Chen, Xet al., (2015). Genetic mapping of a QTL controlling leaf width and grain number in rice. *Euphytica, 202*(1), 1–11.

Upadhyay, S. K., & Sharma, S., (2014). SSFinder: High throughput CRISPR-Cas target sites prediction tool. *Biomed Res. Int., 2014*.

Vanblaere, T., Szankowski, I., Schaart, J., Schouten, H., Flachowsky, H., Broggini, G. A., & Gessler, C., (2011). The development of a cisgenic apple plant. *J. Biotech., 154*(4), 304–311.

Vasal, S. K., Riera-Lizarazu, O., & Jauhar, P. P., (2006). Genetic enhancement of maize by cytogenetic manipulation, and breeding for yield, stress tolerance, and high protein quality. *Genetic Resources, Chromosome Engineering, and Crop Improvement, 2*, 159–197.

Vijayalakshmi, K., Fritz, A. K., Paulsen, G. M., Bai, G., Pandravada, S., & Gill, B., (2010). Modeling and mapping QTL for senescence-related traits in winter wheat under high temperature. *Mol. Breed., 26*(2), 163–175.

Wang, F., Wang, C., Liu, P., Lei, C., Hao, W., Gao, Y., Liu, Y. G., & Zhao, K., (2016). Enhanced rice blast resistance by CRISPR/Cas9-targeted mutagenesis of the ERF transcription factor gene OsERF922. *PloS One, 11*(4), e0154027.

Wang, Y., Cheng, X., Shan, Q., Zhang, Y., Liu, J., Gao, C., & Qiu, J. L., (2014). Simultaneous editing of three homoeoalleles in hexaploid bread wheat confers heritable resistance to powdery mildew. *Nat. Biotechnol., 32*(9), 947–951.

Woo, J. W., Kim, J., Kwon, S. I., Corvalán, C., Cho, S. W., Kim, H., Kim, S. G., et al., (2015). DNA-free genome editing in plants with preassembled CRISPR-Cas9 ribonucleoproteins. *Nat. Biotechnol., 33*(11), 1162–1164.

Yan, S., Zhu, J., Zhu, W., Li, Z., Shelton, A. M., Luo, J., Cui, J., et al., (2015). Pollen-mediated gene flow from transgenic cotton under greenhouse conditions is dependent on different pollinators. *Sci. Rep., 5*(1), 1–9.

Yang, Q., Li, X., Tu, H., & Pan, S. Q., (2017). Agrobacterium-delivered virulence protein VirE2 is trafficked inside host cells via a myosin XI-K–powered ER/actin network. *Proceedings of the National Academy of Sciences, 114*(11), 2982–2987.

Zetsche, B., Gootenberg, J. S., Abudayyeh, O. O., Slaymaker, I. M., Makarova, K. S., Essletzbichler, P., Volz, S. E., et al., (2015). Cpf1 is a single RNA-guided endonuclease of a class 2 CRISPR-Cas system. *Cell, 163*(3), 759–771.

Zhang, F., Maeder, M. L., Unger-Wallace, E., Hoshaw, J. P., Reyon, D., Christian, M., Li, X., et al., (2010). High frequency targeted mutagenesis in *Arabidopsis thaliana* using zinc finger nucleases. *Proceedings of the National Academy of Sciences, 107*(26), 12028–12033.

Zhang, H., Gou, F., Zhang, J., Liu, W., Li, Q., Mao, Y., Botella, J. R., & Zhu, J. K., (2016). TALEN-mediated targeted mutagenesis produces a large variety of heritable mutations in rice. *Plant Biotechnol. J., 14*(1), 186–194.

CHAPTER 2

An Overview of Tools for Genome Editing: ZFNs, Mega Nucleases, and TALENs

LEKHA BHAGTANEY[1] and PRIYA SUNDARRAJAN[2]

[1]*Intern, Caius Research Laboratory, St. Xavier's College (Autonomous), Mumbai – 400001, Maharashtra, India*

[2]*Associate Professor, Department of Life Science and Biochemistry and Director, Caius Research Laboratory, St. Xavier's College (Autonomous), Mumbai – 400001, Maharashtra, India*

[*]*Corresponding author. E-mail: priya.s@xaviers.edu*

ABSTRACT

Targeted gene modification has facilitated the alteration of genomes by the introduction of new genes or correction of faulty genes and exploration of gene functions. Thereby rendering targeted genome editing a powerful tool in genetics. The phenomenon of random mutagenesis occurs naturally in organisms leading to genome modifications. The technique of genome editing has progressed from the use of mutagens (physical or chemical agents) that can alter sequences of DNA to well-engineered biological tools that can generate gene knock-in and knock-out. Zinc finger nucleases (ZFNs), transcription activator-like effector nucleases (TALENs), mega nucleases, and clustered regularly interspaced short palindromic repeats/Cas9 (CRISPR/Cas9) are some of these biological tools. Such multifaceted genome editing tools enable researchers to introduce modifications in an economic and rapid manner. These types of alterations of the genome that are sequence-specific can be

CRISPR/Cas-Mediated Genome Editing in Plants, Deepu Pandita, Anu Pandita (Eds.)

reproduced in a wide range of organisms and cell types. Since its application is diverse across many fields of study, important scientific revolutions have been observed in areas like agriculture, human gene therapy and synthetic biology. Specifically in plants, these tools have contributed in discovery of new traits and also in the development of plants with increased crop yields and resistance to biotic and abiotic stresses. The following chapter will discuss the functioning, delivery methods, and applications of the genome editing tools ZFNs, mega nucleases, and TALENs in depth. We will also discuss the possible harmful effects or non-target effects of these tools and the controls in place for prevention of these effects. The present-day applications of genome editing pose a risk to clinical therapies and the safety of the food supply worldwide. A brief discussion on the challenges of genome editing will also be covered in this chapter.

2.1 INTRODUCTION

When a modification occurs in the genome of an organism naturally, the process responsible for the change is random mutagenesis. These modifications often arise from single nucleotide polymorphisms (SNPs). Plant and animal breeders depend on SNPs for enhanced beneficial traits such as improved yield of grains in plants or improved production of milk in cows. Another such instance of random mutagenesis is seen in microorganisms like yeast, where there is the emergence of new strains that are used for fermentation. However, for such an event to occur, two parameters are of importance; the duration required for the modification to manifest and the process of selection of an improved strain of organism. Utilizing conventional methods such as transposons or mutagens (physical or chemical) can be time-consuming since thorough screening is needed to retrieve preferred mutations. In such a scenario, the field of genome editing is promising and holds a lot of potential in all disciplines of biology (Songstad et al., 2017).

The development of optimal and consistent genome editing tools which are able to make targeted modifications in cell genomes has been a major challenge. Nowadays, some of the preferred tools for genome editing include mega nucleases (MNs), ZFNs, transcription activator-like effector nucleases (TALENs), and clustered regularly interspaced short palindromic repeat/Cas9 (CRISPR/Cas9) (Guha et al., 2017). The diverse genetic variability achieved through the use of these tools is largely due to their capability to efficiently generate double-stranded breaks (DSBs) in the target sequence of

DNA. Cellular DNA repair pathways are activated on the induction of these DNA breaks, thereby leading to site-specific alterations in the genome. This pathway proceeds to non-homologous end joining (NHEJ), which produces knock-out through random insertions or deletions of bases. Otherwise, if a donor template that is homologous to the target site is present, base correction or gene integration through homology-directed repair (HDR) takes place. Additionally, these tools provide the foundation for synthetic transcription factors that have the potential to regulate the expression of any gene in the genome. This proves the versatility of these genome-editing tools (Gaj et al., 2016).

Genome editing tools offer a wide range of applications across many cell types and organisms. A significant amount of research is focused on medicine where there has been a generation of knock-out and knock-in animal model systems for in-depth study of many diseases. Nevertheless, the extent of knowledge generated to demonstrate genome editing in plants is impressive (Songstad et al., 2017). A key concern on the application of genome editing tools is potential non-target effects where sites other than the desired site get cleaved. Such an occurrence can lead to certain undesirable or detrimental mutations where thorough screening becomes necessary to detect cells with the preferred mutation or it can even be lethal. To reduce these effects, many techniques have been devised for temporary modulation of restriction enzymes (Guha et al., 2017).

In order to realize their full potential, genome editing tools need to overcome many persisting obstacles and challenges. Conception of effective delivery methods of genome editing tools, precision in gene editing, social acceptance, and policies of the government are some of the major challenges for these tools that need to be addressed (Mao et al., 2019).

2.2 GENOME EDITING TOOLS

Site-specific cleavage of DNA is achieved by targeted nucleases leading to activation of the DNA repair pathways – HR and NHEJ. Tools involved in genome editing enable the insertion, deletion, or replacement of a DNA sequence in an organism's genome (Figure 2.1). The objectives of genome editing technology include gene overexpression, interference with gene expression, replacement of a deficient gene or mutated gene correction.

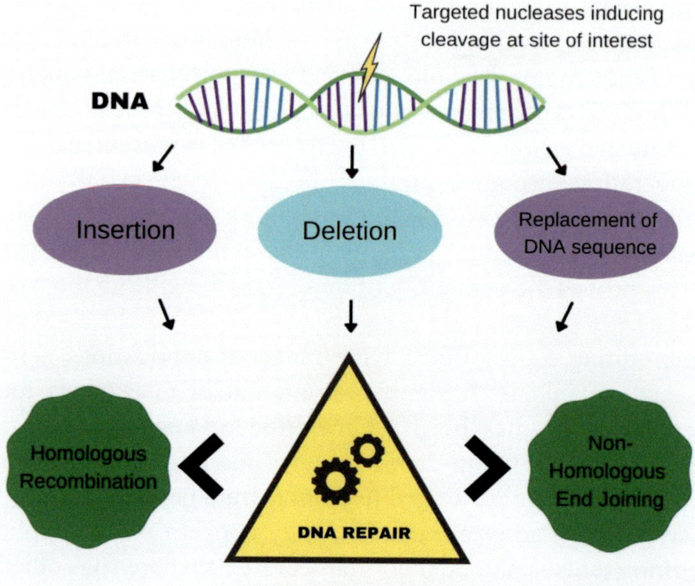

FIGURE 2.1 Mechanism of action of targeted nucleases.
Source: Original illustration. Designed by: Lekha Bhagtaney.

An important breakthrough in the field of genome editing was the demonstration of integration of a segment of DNA containing homology arms into the host genome via homologous recombination (HR). This procedure ensured desired changes to be introduced in a cell. However, there were several shortcomings in solely adopting HR, such as incorporations in unwanted sites in the genome and DNA segment getting integrated inefficiently. Therefore, scanning for modified cells was a tedious task as there was a low number of modified cells. It then becomes essential to enhance the existing procedure for better output. Later, it was discovered that targeted genome engineering can be more precise and efficient in eukaryotic cells with the introduction of a DSB at the desired site. Moreover, insertion of a synthetic DNA restriction enzyme into the cell can result in cuts in the double-stranded DNA (dsDNA) sequences at particular recognition sites. This way, both repair pathways of NHEJ and HR can be improved. For advancements in genome engineering, researchers have concentrated on developing procedures based on endonucleases to achieve the level of accuracy desired in the creation of DSBs.

At the moment, four major techniques of targeted genome editing are being utilized to expand applications in the field of medicine and agriculture:

MNs, ZFNs, TALENs, and CRISPR-associated protein 9 (CRISPR/Cas9) (Figure 2.2). In this chapter, we will be focusing on MNs, ZFNs, and TALENs (Khalil, 2020).

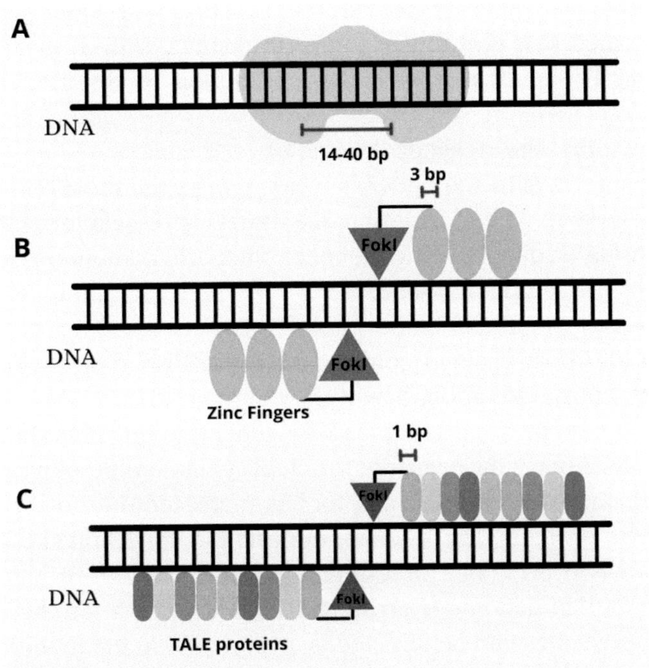

FIGURE 2.2 Genome editing tools: (A) meganucleases. They are endo-deoxyribonucleases that can cut 14–40 bp long recognition sites in dsDNA. (B) Zinc finger nucleases. They consist of two separate domains, DNA binding domains, i.e., zinc fingers where one finger comes in contact with 3 bp of DNA and cleavage domain, i.e., FokI, which is a natural restriction enzyme. (C) Transcription activator-like effector nucleases. They comprise of the cleavage domain of *Fok*I restriction enzyme and DNA binding domain of TALE proteins where 1 TALE protein binds to 1 bp of DNA.

Source: Original illustration. Designed by: Lekha Bhagtaney.

2.2.1 ZINC FINGER NUCLEASES (ZFNS)

One of the earliest chimerically developed genome editing tools were ZFNs that have proven to be versatile and efficient tools for genome editing. ZFNs consist of two separate domains: the DNA-binding domain and the cleavage domain. The observation that *Fok*I, a natural type IIS restriction enzyme, consists of detachable DNA-binding and cleavage domains led to the development of these synthetic nucleases. There is no sequence specificity observed in cleavage domains. Therefore, earlier experiments have shown that replacement of natural recognition domain with an engineered one could direct the cleavage. The most effective of these replacements was with Cys_2His_2 zinc fingers where each unit consists of 30 amino acids bound to one atom of zinc. Crystallographic studies of ZFNs with three fingers bound to DNA revealed that one finger comes in contact with 3 bp of DNA in a modular way. This indicated that novel assemblies of zinc fingers could be formed to target different sequences.

It is known that the cleavage domain of *Fok*I is activated only on dimerization. The interface of the dimer is weak, hence, the most optimal way to attain cleavage would be to assemble the fingers (two sets) oriented towards the neighboring sequences and connect them to a monomeric cleavage domain each. Utilization of a spacer (5 to 7 bp) between binding sites that are set in an inverted alignment and a small linker between the protein domains give rise to the most optimal configuration of ZFNs. Since a monomer is inactive without dimerization, it will not cut at single binding sites. This is a major advantage of the prerequisite for dimerization. If the fingers show satisfactory specificity the reagents for cleavage assemble only at the target site. Overall, the specificity of this tool is increased due to the criteria of binding of two proteins, for instance, in principle, an 18 bp site is specified by the use of two proteins (three-fingers) that is enough to recognize one target in a complex genome (Carroll, 2011). An observation that a series of consecutive 3 bp sites can be targeted is attributable to the modular recognition of corresponding zinc finger domains. This suggested that replacement of and intervention in the arrangement of the domains can increase specificity to sequences of interest thus allowing for unique sequences in the genome to be targeted. The construction of ZFNs relies on modular engineering, aggregation, and optimization of the zinc fingers so as to enhance specificity to the target sequence and further linking of individual zinc fingers to target longer sequences. Through the years, research and generation of zinc finger domains recognizing various triplet nucleotides has facilitated the selection and linking pattern of zinc fingers that can recognize the target sequence efficiently (Kamburova et al., 2017).

The key feature of ZFNs that make them advantageous is simplicity in design that can be assembled using a combination of established zinc fingers with known recognition patterns. However, with regular applications their drawbacks became apparent. Efficient cleavage of target sequence or "context-dependent specificity" is one of the biggest disadvantages. This means that their specificity is dependent not only on the target sequences but also on the adjoining sequences in the genome. Due to this several non-specified cleavages may take place and result in fragmentation of genome and instability. Hence, ZFNs are evidently toxic to cells due to their off-target cleavages. The improper interactions between domains or single nucleotide substitutions are responsible for most of the off-target cleavages caused by ZFNs (Khalil, 2020).

Designing ZFNs for successful targeted gene modifications is a challenging task. Assembling individual modules of zinc fingers in novel combinations for targets of interest can be effective. However, further enhancement in this procedure is required as effects of context can lead to its failure. Although time-consuming, selection of three-finger sets from libraries that have been partly randomized is another method being taken into consideration. Many groups have devised different strategies to simplify this method where some use two-finger module libraries for ZFN designing, others have come up with favorable neighbor combinations of fingers for certain DNA triplets, while some recommend independent fingers for unproblematic modular assembly that are a part of their collection.

Affinity and specificity of zinc finger sets are some of the concerns. A minimum of three fingers are needed in a ZFN to deliver sufficient affinity, however, the fingers have varying degrees of contributions. To overcome this issue, in many cases more fingers have been added (even up to six). DNA modification, the compactness of the chromatin structure, or other factors may limit accessibility to certain regions in the genome. Fortunately, the zinc finger model has been derived from natural transcription factors that identify their targets in the chromatin context. A few fingers can bind comfortably to triplets besides their preferred sequence and even the optimal fingers carry a certain affinity for related sequences. An improvement in affinity and specificity is observed on addition of fingers. However, a risk of off-target binding persists as a subset of fingers present in the polydactyl domain may mediate such a binding. A short linker facilitating separation of two-finger modules can enhance specificity (Carroll, 2011). Recently researchers have also described targeting techniques with broadened range of architectures of ZFNs. These newly developed architectures can expand configurational options available for designing ZFNs. The ZFNs designed based on these new architectures were successful in demonstrating specificity, efficiency, and precision in targeting three loci of therapeutic importance (Paschon et al., 2019).

2.2.2 MEGA NUCLEASES (MNS)

Mega nucleases (MNs) or homing endonucleases (HEs) are endodeoxyribonucleases that occur naturally in microorganisms and eukaryotic chloroplast and mitochondria (Khalil, 2020). On analysis of genome editing based on DSBs by several groups, it was found that an umpteen number of events were stimulated due to mega nuclease activity. Consequently, this technology has been used to understand the repair mechanism of DNA in a wide range of cell types and organisms. HEs consists of five families: LAGLIDADG, PD-(D/E)XK, His-Cys box, HNH, and GIY-YIG out of which LAGLIDADG is the most studied (Silva et al., 2011). Genes encoding MNs are usually integrated within the genes involved in self-splicing. Presence of both these elements with different molecular functions is mutually beneficial as the adjoining inteins and introns behave like invasive elements of DNA due to the endonuclease function and the endonuclease gene can be integrated within a particular coding sequence without the disruption of product due to the splicing function. These enzymes can cut particular recognition sites of dsDNA that are 14–40 bp in length which renders them highly accurate. The function of MNs is to facilitate lateral movement of segments of DNA within a cell as opposed to the function of restriction enzymes which protects bacterial cells against invading DNA. This mechanism is known as 'homing' therefore, MNs are also called homing nucleases.

MNs are excellent protein scaffolds for engineering of enzymes for genome modification attributing to their high specificity. In order to engineer enzymes based on MNs that can cleave the genome at the desired site with high specificity, an in-depth understanding of a mega nuclease's molecular recognition of DNA is mandatory. Mega nuclease families vary widely in their functional capability to recognize sequences of DNA. However, there are two requirements for this function to be successfully carried out (i) to minimize apparent toxicity to the host cell, recognition of target of sufficient length is necessary (ii) consideration of a small extent of sequence drift within the desired site. These requirements are orthogonal in nature. Through detection of structural characteristics of the DNA, the protein attains partial sequence specificity. This is the protein-DNA recognition mechanism known as indirect readout.

Many tools have been engineered based on templates of HEs, and these tools can cut sequences of DNA besides their former wild-type targets (Khalil, 2020). Partly rational approaches have been employed to engineer MNs where certain residues are mutated based on previous structural or functional understanding for generation of libraries with little variety. Several enzymes have been engineered by adopting the scaffold of I-CreI as a template and the DNA recognition properties of these enzymes have been modified on a small and a large scale successfully (Silva et al., 2011). A member of the MNs family, I-CreI has a homodimeric structure which targets and cleaves a 22 bp pseudo-palindromic site (5′-CAAAACGTCGTGAGACAGTTTG-3′). Engineering of enzymes with I-CreI as the scaffold revealed that indirect readout has an important role in the redesigning of protein-DNA interactions (Khalil, 2020). Another prototypical mega nuclease employed for genome editing is I-SceI. Recombination mediated by I-SceI is efficient and stable has been shown by a number of template-based experiments. Various alterations have been carried out effectively by I-SceI, such as gene targeting in plants, flies, and mosquitoes, recombination between repeated sequences or mutagenesis (Silva et al., 2011).

For application, this technology is highly advantageous as the target sequence length is long making MNs highly specific, a relatively smaller size makes the delivery smooth and generation of 3' overhang after cleavage leads to a more recombinant DNA. This way the possibility of cytotoxicity is less (Khalil, 2020). A major drawback of using natural MNs is the need to primarily introduce a known site for cleavage into the target region. Even though techniques involving pseudo-sites such as phiC31 integrase can be exploited, a higher degree of precision is required for cases involving therapy. In order to counter this disadvantage, engineered mega nucleases (EMNs) can be utilized, thereby allowing for a broader employment of homologous recombination (HR)-based technology (Silva et al., 2011).

2.2.3 TRANSCRIPTION ACTIVATOR-LIKE EFFECTOR NUCLEASES (TALENS)

Transcription activator-like effector nucleases (TALENs) are chimeric nucleases which function on the basis of distinct transcription activator-like effector (TALE) proteins that comprise of a set of tandem repeats leading to the activation of specific promoters in the genome of an organism. There is the presence of a central domain in TALE proteins that is in charge of – a domain that acts as a transcription activator of the target gene, nuclear localization signal (NLS) and DNA binding. DNA binding and specificity is imparted by the central repeat domain (CRD) that is a part of the DNA binding domain in TALE monomers. The CRD is composed of 34 amino acid long tandem repeats and a single repeat (34 amino acids) binds to one nucleotide in the target sequence. Highly variable amino acids at positions 12 and 13 of the repeat are known as the repeat variable diresidues (RVD) and recognize a particular nucleotide with degeneracy of binding multiple nucleotides with varying degrees of efficiency. The last tandem repeat is called a half-repeat as it is composed of 20 amino acids and binds to the nucleotide at the 3'-end of the target sequence. Engineering of TALE proteins to bind to a target sequence of interest is attainable yet research has shown that in order to retain the efficiency of TALENs, TALE recombinases (TALE-R) and TALE transcription factors (TALE-TF), the nucleotide at the furthest 5'-end of the target sequence bound by a TALE protein should always be Thymidine. Following the identification of prerequisites for DNA recognition mechanism of TALE proteins, the first quest was the development of chimeric TALENs. Therefore, the sequence coding for the DNA binding domain in TALE proteins was inserted in a plasmid vector which had initially been used to develop ZFN. The product was an artificial, chimeric site-specific nuclease genetic construct comprising of the cleavage domain of *Fok*I restriction enzyme and DNA binding domain of TALE proteins. With such a construct, it is possible to develop synthetic nucleases containing DNA binding domains that can target any unique sequence in the genome due to the presence of diverse RVDs (Kamburova et al., 2017). A few of the most frequently used RVDs are HN or NN for adenine or guanine, NG for thymine, HD for cytosine and NI for adenine (Gaj et al., 2016).

Constructing plasmids coding for long arrays of TALE repeats is idealistic due to the feature that TALE DNA binding domains are highly repetitive. Several methods have been introduced to surpass this limitation and assemble the TALE repeats in an economical and rapid way. Type IIS restriction enzymes are employed for many of the assembly methods so that both restriction and ligation can be performed in one reaction. Golden Gate assembly is a popular method to construct TALENs due to its rapid processing, simplicity, and practical approach. Type IIS restriction enzymes such as *Bbs*I, *BsmB*I, and *BSa*I are utilized to produce a 4bp 5'-overhang which is non-palindromic, by cleaving away from their recognition site. Redigestion through a restriction-ligation reaction will not occur as the ligated product would not contain the original restriction site of Type IIS restriction endonuclease. This enables for 10 TALE repeats to be ligated in a single reaction and for the restriction-ligation reaction to occur in the same reaction mixture in one step. PCR-based Golden Gate is a revised version of Golden Gate Assembly which involves PCR fragments rather than plasmid fragments. Through this approach one reaction can ligate 4–6 TALE repeat units. Until now, Golden Gate approach has been utilized to construct most of the TALENs targeting plant genes (Chaudhary et al., 2016).

Further, high throughput assembly methods for TALENs have also been developed, such as ligation-independent cloning, iterative capped assembly, and FLASH assembly. Improvements in TALEN assembly have concentrated on developing approaches that can enhance TALE specificity via TALEN activity-directed evolution by the discovery of unconventional RVDs through specificity profiling and generating chimeric nucleases with extended target specificity by fusing domains of TALEs to homing endonuclease variants.

The directed development of TALEN-like DNA binding proteins was utilized to enhance the specificity of TALENs by phage-assisted continuous evolution (PACE). This was further improved and used to produce genetically modified organisms (GMO). For the progression of DNA binding specificity and activity, employment of DNA binding PACE is a general approach. This technique can generate TALENs with high specificity in DNA cleavage and hence, can be used to optimize the accuracy of tools used for genome editing (Khalil, 2020).

The two important advantages of TALENs over ZFNs are as follows:

- The time and experience needed to construct the nucleases is greatly reduced as no directed evolution or selection is required to engineer TALE arrays; and
- A greater affinity towards target sequences or a substantial energetic penalty on mismatched pairing has been observed in TALENS thereby, exhibiting enhanced specificity and decreased toxicity when compared to ZFNs.

Being significantly larger than ZFNs in addition to their highly repetitive structure makes the viral delivery of TALENs quite difficult. To overcome this difficulty, TALENs have been delivered into cells as proteins or mRNA, even though amino acid degeneracy and alternative use of codons can be used to produce RVD arrays that are less prone to recombination. Adenoviral vectors have also demonstrated delivery of TALENs to cell types that are arduous to transfect (Gaj et al., 2016).

A drawback of using TALENs in plant genome editing is that they are unable to target methylated DNA. Specific gene silencing, cell differentiation, expression of genes, repair pathways, transcription, growth, and development of plants is dependent on DNA methylation in the genome. Plants adapt to stress and accomplish stable inheritance through hypermethylation and methylation of DNA. Therefore, achievement of these breeding outcomes is not feasible through the use of TALENs (Zhao and Wolt, 2017).

2.2.4 OFF-TARGET EFFECTS (OTES): OVERVIEW AND MINIMIZATION

After the event of genome editing, the changes observed can be categorized into untargeted mutations where modifications occur at regions that have no similarity to the target sequence and off-target edits that occur at regions that have sequence similarity to the target sequence (Graham et al., 2020).

Prediction of the level and possibility of the occurrence of off-target effects (OTEs) can be done by comparison of target sequence to the entire genome. Since only specific sequences are bound by genome editing tools, off-target edits potentially occur at sites with the most sequence similarity to the target site. Bioinformatics-based prediction algorithms have been developed to help in the identification of sites where potential off-target edits can occur. The protocols of genome editing tools are designed with the help of these programs (Cas-OFFinder, TALE-NT, PROGNOS) as they have been developed for most of the prominent genome editing tools. These programs help in predicting off-target edits in organisms with fully sequenced genomes and therefore have limited use in the prediction of off-target edits in organisms that do not have a comprehensive genome sequence data (Graham et al., 2020).

Due to ZFN's poor specificity and affinity, there is increased toxicity and a higher frequency of off-target edits when it is used for genome modifications. One of the members of a ZFN pair is responsible for the toxicity caused by off-target edits. The incorporation of substitutions in the dimer interface of the cleavage domain that inhibit homodimerization but enable heterodimers to form has vastly enhanced the effect. However, these alterations seldom lower the cleavage efficiency. Prevention of homodimerization while retaining activity have been observed in the new designs explored in recent studies. A low frequency of off-target mutation and cleavage is acceptable for use during genetic analysis of model organisms, because the preferred allele can most often be isolated through repetitive out-crossing. If ZFNs are to be used in human therapy, extreme caution must be exercised because the risk of undesirable genome modifications occurring is high. The ZFNs used in an ongoing clinical trial have been thoroughly tested and refined. Even at very low levels, it would be valuable to have a direct analysis of where off-target cleavages take place (Carroll, 2011).

TALENs are now widely recognized as a tool for site-directed mutagenesis. Direct recognition criteria between nucleotide sequences and protein repeats are used in the construction of TALENs, resulting in better editing efficiency and fewer OTEs. Construction of a TALEN protein specific to the target site as well as the nuclease protein is a challenging task as it is both expensive and laborious (Zhao and Wolt, 2017). However, unintended off target edits continue to be a significant area of concern. The bioinformatics prediction algorithms have also been used for TALENs where the user can configure the search criteria to work with any TAL effector nuclease or TAL effector design (Chaudhary et al., 2016). Previous efforts in the prevention of OTEs caused by ZFNs employed mutated *Fok*I domain leading to generation of heterodimeric variants thereby reducing ZFN homodimerization. This mutagenesis can also be applied to TALENs. Such heterodimeric variants, however, can lessen the activity at the target site probably through lowering the binding efficiency of the *Fok*I dimers. *Fok*I nickases have also been developed by *Fok*I mutagenesis. *Fok*I nickases have also been utilized effectively with TAL effectors (Lee et al., 2016).

2.3 DELIVERY METHODS OF GENOME EDITING TOOLS WITH FOCUS ON PLANTS

A significant advancement in the field of gene targeting has been observed through the development of newer, better techniques. The main hurdle of genome editing, however, remains the delivery of the tools to target cells or organisms.

2.3.1 AGROBACTERIUM-MEDIATED DELIVERY

One of the most popular delivery methods for genome editing in plants is the *Agrobacterium*-mediated delivery since it is economical and has a simple laboratory set up. Around 80% of DNA editing and mutations induced in experimental plants such as tobacco, *Nicotiana benthamiana*, *Arabidopsis*, *Brassica*, and several others have been brought about by *Agrobacterium*-mediated delivery. The constituents of a genome editing system, involving ZFNs or TALENs, are assembled in a single binary vector alongside a cassette containing selectable markers when targeting one gene locus. This binary vector is inserted in a strain of *Agrobacterium*. Mostly, ZFN and TALEN tools have been employed for single gene targeting in plant mutagenesis.

In an ideal situation, genome editing tools can be incorporated into binary constructs as cassettes for single gene or multiple gene targeting. Delivery of several binary vectors into an *Agrobacterium* strain is also feasible, leading to the transformation of plant cells. In some cases, two binary vectors containing editing tools for separate target sites can be integrated in two different strains of *Agrobacteria*, a mixture of which can be used to induce infection in the same plant cells. Whereas in other cases, a binary vector containing editing tools for multiple target sites integrated into a single *Agrobacterium* strain can be developed to induce infection. However, sizeable gene cassettes of genome editing reagents are a limitation for ZFNs and TALENs in multiplex genome editing (Ran et al., 2017).

During earlier studies, an assay was established in tobacco by researchers which was used for DNA insertion at the target site through HR leading to activation of selectable marker genes. The HR efficiency was boosted from 10^{-5} to 10^{-3} when the donor DNA was delivered along with a plasmid coding for I-SceI that recognized the target site. Further, in a study conducted on repair junctions found that *Agrobacterium*-delivered DNA coding for I-SceI could seldom integrate at the target site of I-SceI. This resulted in the implementation of MNs to introduce *Agrobacterium*-delivered DNA at the site of interest in the plant genome. MNs have successfully carried out targeted insertion of DNA in barley and maize in combination with selectable marker genes. However, this procedure is highly reliant on the specificity of the nuclease to a site in the genome which makes the implementation of MNs for targeted DNA insertion in plants limited. Therefore, emerging genome editing tools are being utilized for this purpose instead of MNs (Dong and Ronald, 2021).

A suspension of Agrobacteria containing donor DNA which infiltrates the intracellular spaces of plants is demonstrated in a gene transfer technique known as agroinfiltration. This is a quick and economical transformation technique. Researchers have suggested that vaccines can be developed through agroinfiltration technique by stamping antigens onto leaves. For instance, agroinfiltration of SEB genes into radish leaves led to the development of SEB vaccines. Co-expression of genes recognized for gene silencing and suppression of stress, heat shock and chemical additives can enhance transient gene expression based on agroinfiltration (Nandy et al., 2020). Scientists have also incorporated a heat shock promoter-driven ZFN gene and its target into the plant genome in order to evaluate ZFN activity with regards to the introduction of mutations at specified sites in *Arabidopsis*. The application of ZFNs for targeted mutagenesis in plants was demonstrated when, on induction of heat shock, ZFN was expressed leading to mutations at its recognition sites. TT4 genes, ADH1, ABA INSENSITIVE-4 (ABI4) and regulators of stress response are several such gene loci where ZFNs have induced mutagenesis in Arabidopsis, successfully (Nandy et al., 2020).

A study was conducted on the delivery of TALENs by the *Agrobacterium*-mediated method for transient expression in potato cultivars, *Solanum tuberosum* Russet Burbank and Shepody, both of which were vegetatively propagated. In this study, agroinfiltration through potato leaves was carried out to deliver two TALENS with differing molecular weights (22 and 27 aa repeat modules) and targeting two specific sites (genes for acid invertase and starch branching enzyme) in the genome. DNA sequencing and an assay for restriction site loss confirmed successful delivery of TALENs through agroinfiltration and also induction of mutations at both target sites (Ma et al., 2017).

It does, however, have a few disadvantages. Incorporation of the Ti plasmid backbone into the plant genome is difficult to prevent thereby leading to the generation of transgenic plants. Particularly in case of monocot plants, the success of this method is dependent on the host genotype. Additionally, the delivery of proteins, RNA or small segments of DNA are not feasible through this method. Also, in varied binary vectors, there is a limited potential to develop independent editing reagents (Ran et al., 2017).

2.3.2 VIRUS-MEDIATED DELIVERY

Virus-based vectors that replicate autonomously are another popular approach for the delivery of genome editing tools. Some of the most commonly used RNA viruses are tobacco rattle virus (TRV) and barley stripe mosaic virus (BSMV) for dicots and wheat streak mosaic virus (WSMV) for monocots. One of the most popular DNA viruses being used for this approach are geminiviruses. Geminiviruses are single-stranded DNA viruses and have been used as vectors for a variety of crops. Alteration of these viruses to contain heterologous coding sequences and attain protein expression has been observed in valuable crops like rye, oat, corn, barley, and wheat. Advancements in technology have allowed for researchers to integrate viral vectors and employ them to deliver genome editing tools and the reagents in plant cells (Zaidi and Mansoor, 2017). Otherwise, a binary vector in which the virus genome is integrated as a cDNA fragment can be inserted into the plant cell through agroinfection thereby delivering the genome editing tools (Figure 2.3) (Ran et al., 2017).

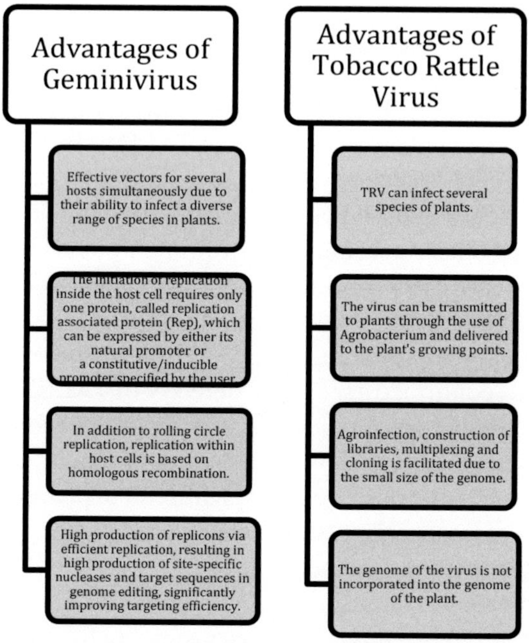

FIGURE 2.3 Advantages of geminivirus and tobacco rattle virus in the delivery of genome editing tools.

Source: Original diagram. Designed by: Lekha Bhagtaney.

Geminiviruses (family *Geminiviridae*) are prevalent worldwide and are able to infect a wide range of plants from important crops, fruits, vegetables to ornamental plants and common weeds. Geminiviruses have a ~2.8 Kb genome with 4 to 6 overlapping open reading frames in both sense and complimentary sense orientations.

In a study, ZFN and a repair template were delivered to tobacco cells using a deconstructed *Bean yellow dwarf virus* (BeYDV) which demonstrated gene targeting at integrated reporter gene. Targeted nucleases' transient expression confirmed by successful HDR was attributable to geminivirus-based replicons that replicated and accumulated efficiently. When tested further, it was found that BeYDV replicons could deliver TALENs due to their significant cargo capacity. In another study, gene targeting of a tomato gene and integration of strong promoter upstream to modulate the synthesis of anthocyanin was carried out using BeYDV-based replicons. When compared to gene delivery mediated by *Agrobacterium*, there was a ~12-fold increase in gene targeting frequencies (Zaidi and

Mansoor, 2017). Of late, TALENs and repair templates were delivered into cells using geminivirus-based replicons. The rate of in vivo targeting had improved considerably as the DNA of the geminivirus was used as the repair template, which can replicate in high numbers. Additionally, the geminiviral initiator proteins possess pleiotropic activity. TALENs are increasingly being delivered via viral vectors into plant cells because their advantages involve – TALEN genes not getting integrated with the plant genome, low toxicity, and increased levels of expression (Chaudhary et al., 2016).

TRV (genus Tobravirus, family Virgaviridae) is a pathogenic plant virus with a positive single stranded RNA (+ssRNA) that infects approximately 400 plant species across 50 families. The TRV genome has two constituents RNA1 or TRV1 and RNA2 or TRV2. Genes coding for movement protein (29 kDa), replicase proteins (134 kDa and 194 kDa) and viral movement are under TRV1 as well as a cysteine-rich protein with little information on its function. Whereas TRV2 is responsible for genes coding for the non-structural proteins and the coat protein (CP). However, TRV2 varies with different virus isolates.

For ZFN delivery, a non-transgenic method was employed where an expression system based on TRV was used for indirect transient delivery of ZFNs to plant cells leading to mutant plant production. A more effective use of TRV as a delivery vector is if the virus infects the germline cells of the plant. This would allow for harvesting of mutant seeds from such plants. A meganuclease was expressed in a TRV vector which was used to produce heritable and efficient mutations in dihydroflavonol-4-reductase (DFRa). This enzyme is dependent on NADPH and converts dihydroflavonols to their respective leucoanthocyanidins. The inactivation of this enzyme results in decreased purple pigmentation in organs where anthocyanin is usually accumulated, such as flower petals. The mutations were passed down through the M1 progeny, and some of them were passed down for at least two more generations (Zaidi and Mansoor, 2017).

Plant viruses have the potential to infect a wide range of plants both in natural and laboratory conditions. The introduction of viral vectors at different stages of growth in plants is a simple procedure and it can be followed by the transient expression of genome editing tools that do not incorporate into the plant genome. In addition to this, DNA repair templates can also be replicated using the replication system of the virus. Hence, viral-based vector delivery can be the ideal approach for genome editing (Ran et al., 2017).

2.3.3 PROTOPLAST TRANSFECTION

Digesting plant cell walls enzymatically gives rise to protoplasts. Through polyethylene glycol (PEG) treatment and electroporation, reagents, proteins, and DNA can be delivered to protoplasts. For regeneration of dicot plants, mesophyll protoplasts are utilized whereas for monocot plants protoplasts derived from embryogenic callus are utilized. Since it is possible to deliver several constituents through protoplast transfection, it is an ideal delivery method for genome editing tools. This method of delivery is advantageous as a high number of cells can be transfected and regenerated, thereby making them ideal for gene modification using repair templates. Additionally, the gene modifications relying on donor template repair (HR) can be recovered. However, difficulty in plant regeneration, particularly of monocot plants, is a major drawback of direct protoplast transfection.

Protoplast transfection can be used to carry out effective genome editing at the cell, tissue or whole plant level considering the regeneration potential of the host. In a study involving tobacco protoplasts with a faulty reporter gene of *GUS:NPTII*, the researchers incorporated a repair template along with ZFN. This resulted in repaired reporter gene through HR in the plant. Following this study, there has been research focusing on genome editing through protoplast transfection and regeneration in potato using TALENs. For instance, a study demonstrated the regeneration of potato plants with vacuolar invertase gene (*VInv*) knock-out from protoplasts transfected with TALENs. Refined chips produced from these potatoes exhibited lower levels of acrylamide and the tubers had negligible levels of reducing sugars. A standardized protoplast regeneration procedure is used for regeneration but it is functional in a limited number of species (Ran et al., 2017).

The transient assay in protoplast is reliable, quick, and accurate to evaluate the activity of nucleases in protoplasts and has been demonstrated in *Arabidopsis* and tobacco. A similar study was performed in rice and *Brachypodium* where a transient assay in protoplasts was developed. The expression vector pZHY051 was first used to clone TALEN repeat arrays to assess nuclease activity in protoplasts. PEG treatment was used to incorporate the constructs coding for TALENs into the protoplasts. Genomic DNA was isolated on two days of incubation. With further PCR and DNA sequencing, mutation frequencies induced by TALENs were analyzed (Sanagala et al., 2017).

2.3.4 *BIOLISTIC DELIVERY*

Bombarding cells with DNA-coated microprojectiles is a common method of delivery known as biological ballistics or biolistics. A gene gun is employed for the delivery of DNA, which has ample force to penetrate the plant cell wall leading to random integration of the DNA into the genome. Biolistic delivery is a popular method as it is not limited to certain species of plants, several constructs of DNA can be delivered at the same time and it is a fairly efficient method. The main drawback of this method is that with the existence of multiple copies of genes being introduced in transfected cells, there is the possibility of detrimental effects such as altered expression of genes and suppression of genes in the recovered plants.

There have been many studies on genome editing using biolistic delivery of DNA. Many researchers have reported successful insertion of DNA at the target site using biolistic delivery while failing to obtain the result using *Agrobacterium*-mediated delivery (Ran et al., 2017). Biolistic method has been used for studying important crops such as soybean. For instance, in a study, ZFN along with multiple DNA donors was delivered via biolistics to the targeted site – fatty acid desaturase 2-1a (FAD2-1a) in the soybean genome. Further, transgene integration was achieved through NHEJ. The targeted insertion was successful in both immature embryos and in tissue culture (Bonawitz et al., 2019).

2.4 APPLICATION OF GENOME EDITING TOOLS

By utilizing genome editing tools such as MNs, ZFNs, TALENs, and CRISPR/Cas9, a variety of different genome modifications can be obtained. These modifications involve replacement or correction of individual bases or genes, deletion or insertion of new genes and establishment of point mutations (Kamburova et al., 2017). Application of genome editing technology can lead to revolutionizing advancements in many fields including medicine, nutrition, and agriculture.

Genome editing tools can be used in drug developments, construction of models for diseases, functional analysis of genes or embryonic stem cell-based generation of transgenic animals. Research has been carried out on a variety of different organisms. Genetic modification (GM) of animals through genome editing tools can help scientists to comprehend the causes of diseases and exploration of strategies for treatment through the study of molecular mechanisms. TALENs and CRISPR technologies have been used at the level of cells in cattle, pigs, sheep, and chicken, whereas the use of ZFNs has been restricted to zebrafish. In earlier times, when restriction enzymes were used to modify DNA, the primary principles of gene therapy were introduced. Thereafter, much research has been focused on the treatment of genetic diseases. However, correction of multiple mutations is challenging. With the emergence of CRISPR technology, scientists are hopeful that the majority of human diseases may be eradicated as it is being applied in several fields like gene therapy and genetic improvement (Khalil, 2020).

2.4.1 APPLICATIONS OF GENOME EDITING TOOLS IN PLANTS

The introduction of genome editing has piqued the interest of agricultural researchers in particular. Generating superior crop varieties with higher yields is one of the main goals of using genome editing technologies in plants. Plants have been modified at the genomic level to obtain beneficial traits such as elimination of disadvantageous traits, pest, or disease resistance (DR), tolerance to stress, extended shelf life and enhanced nutritional value (Khalil, 2020). Additionally, implementing modifications in the plant genome also allows researchers to explore the significance of individual genes in the working of a cell or an organism.

Genome editing allows for the generation of novel crop varieties which has been one of the most challenging tasks in biotechnologies. These crops must have increased nutritional value, biotic, and abiotic stress resistance and have an enhanced yield. Therefore, modifications that are typically generated in plant breeding by the use of genome editing tools are:

- Generation of SNPs, corresponding to natural ones, through insertion of point mutations;
- Minor alterations to the function of a gene;
- Insertion of donor genes;
- Epigenetic editing;
- Generation of knock-out and gene pyramiding;
- Activation or repression of expression of certain genes.

An instance of the advantageous applications of genome editing technology in plants includes generation of dwarf trees of fruits which have a lot of promise for improving output by planting more densely and lowering labor costs. As a result, the utilization of fertilizers, pesticides, water, and land is reduced. Ethylene biosynthesis, which is mostly responsible for the fruit ripening process, can be inhibited via genome editing thereby generating new varieties with longer shelf life is another such example.

Creating cultivars resistant to various pests or pathogens is a significant application of genome editing technologies in plant breeding. This involves alteration of essential immunity stages in plants at levels that are dependent on the target crop. This resistance in crops is achieved by modification of – genes modulating the hormonal balance of plants, genes modulating the exchange between the target and effector, genes responsible for resistance and susceptibility genes (Kamburova et al., 2017).

A detailed description of the application of ZFNs, MNs, and TALENs in plant genome editing has been given in the following titles, and the summary has been provided in Table 2.1.

2.4.1.1 ZINC FINGER NUCLEASES (ZFNS)

ZFNs are widely used in plants for genome editing. Target mutagenesis through ZFNs is extensively used to produce gene knock-outs by impairing gene function (Razzaq et al., 2019). The model plants that were used to study the effectiveness of ZFNs as genome editing tools in plants were tobacco and *Arabidopsis*. Due to the extremely precise nature of the synthetic nucleases utilized in gene targeting, scientists believed that this technology could be extended to other crop species with improvements (Ahmar et al., 2020). Modification of target DNA sequence by utilizing ZFNs has been studied in various crops such as apple, rice, soybean, and maize.

A study was performed to attain the trait of herbicide tolerance (HT) in *Zea mays* (maize). Targeted insertion of the herbicide-tolerance gene PAT was carried out successfully using ZFN to target the locus IPK1. The resulting plants showed modified profiles of inositol phosphate in the developing seed along with HT. The functional roles of a gene coding for the soluble starch synthase isoform (*SSIVa*) in *Oryza sativa* (rice), which is important for starch biosynthesis, were understood using ZFNs. Mutations at the locus *SSIVa* induced by ZFNs in the plants resulted in reduced starch content and dwarfism (Kim et al., 2021). Another instance of the use of ZFNs in plant genome editing is in *Glycine max* (soybean). ZFN was used to target the DICER-like (DCL) genes in soybean. The findings revealed that due to ZFN-induced mutation, the mutation is remarkably efficient for the transmission of heredity. For a generation of new ZFN arrays, a quick and open method like assembly scaffold which is context-dependent can be used.

TABLE 2.1 Summary of the Use of Genome Editing Tools in Plants

Plant	Genome Editing Tool	Gene Modification/Resulting Trait	References
Zea mays	ZFN	Herbicide tolerance	Kim et al. (2021)
Oryza sativa	ZFN	Reduced starch content and dwarfism	Kim et al. (2021)
Glycine max	ZFN	Targeting DICER-like (DCL) genes to demonstrate the generation of heritable mutations in soybean.	Ahmar et al. (2020)
Brassica napus	ZFN	Activation of the transcriptional machinery of b-ketoacyl-ACP Synthase II	Ahmar et al. (2020)
Zea mays	Mega nuclease	Sterility in male plants	Razzaq et al. (2019)
Gossypium hirsutum	Mega nuclease	Herbicide tolerance	Razzaq et al. (2019)
Oryza sativa	TALEN	Bacterial blight resistance	Zhang et al. (2018)
Triticum aestivum	TALEN	Powdery mildew resistance	Kim et al. (2021)
Glycine max	TALEN	Enhanced oleic acid amounts	Kim et al. (2021)
Solanum lycopersicum	TALEN	Enhanced response of gibberellic acid	Ahmar et al. (2020)
Saccharum officinarum	TALEN	Improved efficiency of saccharification and composition of cell wall	Zhang et al. (2018)
Solanum tuberosum	TALEN	Immeasurable reducing sugar levels	Zhang et al. (2018)
Oryza sativa	TALEN	Enhancement of flavor	Zhang et al. (2018)
Zea mays	TALEN	Production of haploid plants to accelerate plant breeding processes	Zhang et al. (2018)
Brassica oleracea	TALEN	Early flowering	Zhang et al. (2018)

Additionally, previous research suggests that ZFNs, that have the ability to target gene families with single or multiple copies, can be employed in a technique involving induction of site-specific mutations in the paleopolyploid soybean genome.

According to another study, the transcriptional machinery of the gene b-ketoacyl-ACP Synthase II present in *Brassica napus* can be activated by the ZFN protein. A major factor that can assist in the manipulation of endogenous genes as well as traits in crops are synthetic ZFN-transcription factors (Ahmar et al., 2020). Some studies have also reported the use of ZFNs in modifying the gene coding for malate dehydrogenase in plants. This results in an increased yield in plants (Kamburova et al., 2017).

2.4.1.2 MEGA NUCLEASES (MNS)

Designing MNs specific to each target sequence separate from their natural targets is a difficult task. The DNA domain of MNs is intertwined with the catalytic domain. These domains cannot be separated from each other, thus, making MNs non-modular and difficult to redesign (Razzaq et al., 2019; Ahmar et al., 2020). Accordingly, their applications in plants are limited to the use of I-SceI and I-CreI which are naturally occurring MNs (Razzaq et al., 2019). Using modified MNs in maize, *Gossypium hirsutum* (cotton) and *Arabidopsis*, earlier studies revealed genome editing in plants (Ahmar et al., 2020; Khandagale and Nadaf, 2016). In maize, MNs were delivered through *Agrobacterium*-mediated method inducing gene knock-out to obtain the trait of sterility in male plants. The cotton plant genome was modified by trait-stacking through MNs which were delivered via particle bombardment to obtain HT in the plants (Razzaq et al., 2019). However, further investigation is required to improve this method, as manipulating MNs tends to be difficult. As a result, studies have concentrated on various genome editing technologies that are more easy, accurate, and efficient, such as ZFNs, TALENs, and CRISPR (Ahmar et al., 2020).

2.4.1.3 TRANSCRIPTION ACTIVATOR-LIKE EFFECTOR NUCLEASES (TALENS)

Genome editing mediated by TALENs has been reported in a variety of plant traits in several studies conducted. Potato browning, soybean oil quality, rice aroma, rice resistant to disease, and many other genes were targeted for functional research in model and other crop plants as pilot projects (Khandagale and Nadaf, 2016). TALEN was first used as a genome editing tool in rice where the gene for bacterial blight susceptibility (*OsSWEET14*) was modified. The mutant rice thus formed was bacterial blight resistant (Zhang et al., 2018). Targeted mutagenesis of *MLO* genes, which code for proteins that suppress defense against powdery mildew infections, present in *Triticum aestivum* (wheat) mediated by TALENs was carried out. The resulting mutant wheat crop demonstrated substantial resistance to powdery mildew. A study exhibited improvement in quality of seed oil of soybean by utilizing TALENs. The two genes targeted were FAD2-1A and FAD2-1B which code for the enzyme fatty acid desaturase-2. This enzyme helps to convert oleic acid to linoleic acid and hence, the two targeted genes were deleted. The mutant soybean could produce increased amounts of oleic acid (~4 times) than the original soybean plants (Kim et al., 2021). The signaling cascade regulation of gibberellic acid can be inhibited in *Solanum lycopersicum* (tomato) by the presence of the *DELLA* gene also known as *PROCERA* (*PRO*). However, editing of the *PRO* gene by TALEN under the regulation of an estrogen-inducible promoter yielded phenotypes with a continuously enhanced gibberellic acid response (Ahmar et al., 2020).

TALEN-mediated modifications in the *Saccharum officinarum* (sugarcane) genome resulted in improved efficiency of saccharification and composition of cell wall. TALENs were also used to knock-out the vacuolar invertase (VInv) gene in *Solanum tuberosum* (potato). The reducing sugars that accumulate during cold storage of potato tubers determine the quality of the product. The tubers thus produced by knocking out this gene had no measurable level of reducing sugars. Disruption of the gene betaine aldehyde dehydrogenase (BADH2) through TALENs gave rise to fragrant rice which helped to enhance the flavor. The process of plant breeding can be sped up if haploid plants that possess chromosomes of one parent are produced. A study was successful in producing haploid maize by generating frameshift mutations in MATRILINEAL (MTL) in the genome by the utilization of TALENs. In *Brassica oleracea* early flowering mutants can be obtained by TALEN-mediated targeting of FRIGIDA. These examples highlight the TALEN technology's enormous potential for enhancement of crop traits. Yet, constructing TALE repeats remains a hurdle, and the gene targeting efficiency of TALENs is inconsistent (Zhang et al., 2018).

2.5 CHALLENGES AND PROSPECTS

Genome editing is being used to modify several organisms including crops and animals. Although the resultant organisms are actually genetically modified, they are different from GMOs that were produced earlier in a number of ways. Generally, DNA from another species is not introduced and when required, it is inserted at a specific site in the genome. The modifications incorporated are most of the time, ones that could have occurred even without intervention. Additionally, an organism's whole genome sequencing can be performed to analyze OTEs. There is a lot of potential for generating healthier crops. Compared to breeding selection, genome editing allows a characteristic to be instituted in a single generation and avoiding disturbance of a favorable genetic history. Without causing monoculture, the same advantageous change can be introduced into several cultivars or breeds that are acclimated to diverse settings (Carroll, 2017).

The field of genome editing technology requires further enhanced prediction tools for OTEs. An understanding of the consequences of off-target edits is also required, which may affect the ecosystem where the mutant crop is grown or the consumers who receive the final product. For this, further development will be needed in genomic and other omics approaches, enhancement in models and tools of bioinformatics, as well as key data on the integration of knowledge at all sizes, from molecular to landscape. Along with this, improvement in the procedure of genome editing may relieve fears about the potential direct and indirect effects. Introducing the editing tools directly or by encoding the tools in DNA can be studied to understand the future outcome of editing reagents (Wolt, 2017).

A major concern is whether the modifications induced by genome editing which are highly efficient and mostly natural will be more widely accepted than the GMOs produced earlier. This is a significant barrier to overcome because the basis of the current concerns is related to dislike for power and commercial greed rather than evidence of detrimental consequences (Carroll, 2017). Providing assurance based on scientific evidence about the safety of genome editing tools in the production of valuable features in crops will be required to increase the trust of the public and knowledge of the technique, as well as to guide the proper regulatory approach for product commercialization (Wolt, 2017). Since the first GMO was developed in 1995, strict laws and fines have been adopted around the world to govern GM crops. Alternative strategies to improve crops include transgenic crop development and mutation breeding. Although traditional mutagenesis methodologies have helped in increasing yield and improving the quality of the crop, it is time-consuming and tedious. Hence, such strategies can no longer be employed as they will not be able to keep up with the increasing global food demands. Improvement of crops through transgenesis has been successful and provides many opportunities, however, the technology has raised questions regarding its possible consequences on human health and the environment. Therefore, there is presence of rigid regulatory procedures which are expensive and lengthy for marketing of GMO products.

Advancements in genome editing technologies have enhanced the quality and yield of crops by efficient production of targeted mutations. This technology does not include the incorporation of a transgene. Genome editing technology has the most similarity to traditional breeding as compared to other technologies and also requires the least amount of time. As a result, lots of money can be saved on producing and approving genome-edited crops, as opposed to costly and time-consuming field trials that ordinarily take years to regulate a GM crop. Genome editing tools such as ZFNs, TALENs, CRISPR/Cas9 and MNs have been used in various nations for the past few years and do not fall under GMO regulation legislation. To distinguish between GMOs and genome-edited plants, a solid and genuine regulatory policy is necessary. However, several countries have yet to define a strong regulatory policy for genome-edited plants. By implementation of such regulatory policies and clear scientific evidence of its safety, the applications of genome editing technology in the improvement of crops will get a boost (Razzaq et al., 2019).

2.6 CONCLUSION

Many tools, such as MNs, ZFNs, TALENs, and CRISPR/Cas9, are available in genome editing technology to assist us in making targeted changes to an organism's genome. Researchers can use such a valuable technology to analyze and modify the genetics of both plants and humans. While a number of challenges remain in the use of these tools, off-target edits, delivery of these tools, and finally applications, significant progress has been made in overcoming these challenges in recent years. To carry out successful site-specific modifications in the genome, the targeting mechanisms of MNs, ZFNs, and TALENs need to be fine-tuned further. In this regard, various mechanisms are being used, such as engineering MNs to target a broader range of sites by using I-CreI and I-SceI as templates, increasing the affinity and specificity of ZFNs by adding more zinc fingers, and delivering TALENs as mRNAs or proteins in cells to overcome delivery difficulties, among many others. Off-target edits can be significantly reduced with further advancements in bioinformatics prediction tools and enhancements to the genome editing procedure. People continue to express strong opposition to the acceptance of GMOs. As a result, in order for plant genome editing technology to gain wider acceptance, an appropriate regulatory framework based on scientific evidence demonstrating the safety of this technology must be established. The responsible use of this technology can help us in achieving a healthier future through the consumption of nutritious food and the eradication of life-threatening diseases.

KEYWORDS

- **homologous recombination**
- **homology-directed repair**
- **mega nucleases**
- **non-homologous end joining**
- **single nucleotide polymorphisms**
- **transcription activator-like effector nucleases**

REFERENCES

Ahmar, S., Saeed, S., Khan, M. H. U., Ullah, K. S., Mora-Poblete, F., Kamran, M., Faheem, A., et al., (2020). A revolution toward gene-editing technology and its application to crop improvement. *Int. J. Mol. Sci., 21*(16), 5665.

Bonawitz, N. D., Ainley, W. M., Itaya, A., Chennareddy, S. R., Cicak, T., Effinger, K., Jiang, K., et al., (2019). Zinc finger nuclease-mediated targeting of multiple transgenes to an endogenous soybean genomic locus via non-homologous end joining. *Plant Biotechnol. J., 17*(4), 750–761.

Carroll, D., (2011). Genome engineering with zinc-finger nucleases. *Genetics, 188*(4), 773–782.

Carroll, D., (2017). Genome editing: Past, present, and future. *Yale J. Biol. Med., 90*(4), 653–659.

Chaudhary, K., Pratap, D., & Sharma, P. K., (2016). Transcription activator-like effector nucleases (TALENs): An efficient tool for plant genome editing. *Eng. Life Sci., 16*(4), 330–337.

Dong, O. X., & Ronald, P. C., (2021). Targeted DNA insertion in plants. *Proc. Natl. Acad. Sci. U. S. A, 118*(22), e2004834117.

Gaj, T., Sirk, S. J., Shui, S. L., & Liu, J., (2016). Genome-editing technologies: Principles and applications. *Cold Spring Harb. Perspect. Biol., 8*(12), a023754.

Graham, N., Patil, G. B., Bubeck, D. M., Dobert, R. C., Glenn, K. C., Gutsche, A. T., Kumar, S., et al., (2020). Plant genome editing and the relevance of off-target changes. *Plant Physiol., 183*(4), 1453–1471.

Guha, T. K., Wai, A., & Hausner, G., (2017). Programmable genome editing tools and their regulation for efficient genome engineering. *Comput. Struct. Biotechnol. J., 15*, 146–160.

Kamburova, V. S., Nikitina, E. V., Shermatov, S. E., Buriev, Z. T., Kumpatla, S. P., Emani, C., & Abdurakhmonov, I. Y., (2017). Genome editing in plants: An overview of tools and applications. *Int. J. Agron., 2017*, 1–15.

Khalil, A. M., (2020). The genome editing revolution: Review. *J. Genet. Eng. Biotechnol., 18*(1), 68.

Khandagale, K., & Nadaf, A., (2016). Genome editing for targeted improvement of plants. *Plant Biotechnol. Rep., 10*(6), 327–343.

Kim, Y. C., Kang, Y., Yang, E. Y., Cho, M. C., Schafleitner, R., Lee, J. H., & Jang, S., (2021). Applications and major achievements of genome editing in vegetable crops: A review. *Front. Plant Sci., 12*, 688980.

Lee, C. M., Cradick, T. J., Fine, E. J., & Bao, G., (2016). Nuclease target site selection for maximizing on-target activity and minimizing off-target effects in genome editing. *Mol. Ther., 24*(3), 475–487.

Ma, J., Xiang, H., Donnelly, D. J., Meng, F. R., Xu, H., Durnford, D., & Li, X. Q., (2017). Genome editing in potato plants by agrobacterium-mediated transient expression of transcription activator-like effector nucleases. *Plant Biotechnol. Rep., 11*(5), 249–258.

Mao, Y., Botella, J. R., Liu, Y., & Zhu, J. K., (2019). Gene editing in plants: Progress and challenges. *Natl. Sci. Rev., 6*(3), 421–437.

Nandy, D., Maity, A., & Mitra, A. K., (2020). Target-specific gene delivery in plant systems and their expression: Insights into recent developments. *J. Biosci., 45*(1).

Paschon, D. E., Lussier, S., Wangzor, T., Xia, D. F., Li, P. W., Hinkley, S. J., Scarlott, N. A., et al., (2019). Diversifying the structure of zinc finger nucleases for high-precision genome editing. *Nat. Commun., 10*(1), 1133.

Ran, Y., Liang, Z., & Gao, C., (2017). Current and future editing reagent delivery systems for plant genome editing. *Sci. China Life Sci., 60*(5), 490–505.

Razzaq, A., Saleem, F., Kanwal, M., Mustafa, G., Yousaf, S., Imran, A. H. M., Hameed, M. K., et al., (2019). Modern trends in plant genome editing: An inclusive review of the CRISPR/Cas9 toolbox. *Int. J. Mol. Sci., 20*(16), 4045.

Sanagala, R., Moola, A. K., & Bollipo, D. R. K., (2017). A review on advanced methods in plant gene targeting. *J. Genet. Eng. Biotechnol., 15*(2), 317–321.

Silva, G., Poirot, L., Galetto, R., Smith, J., Montoya, G., Duchateau, P., & Pâques, F., (2011). Meganucleases and other tools for targeted genome engineering: Perspectives and challenges for gene therapy. *Curr. Gene Ther., 11*(1), 11–27.

Songstad, D. D., Petolino, J. F., Voytas, D. F., & Reichert, N. A., (2017). Genome editing of plants. *CRC Crit. Rev. Plant Sci., 36*(1), 1–23.

Wolt, J. D., (2017). Safety, security, and policy considerations for plant genome editing. *Prog. Mol. Biol. Transl. Sci., 149*, 215–241.

Zaidi, S. S. E. A., & Mansoor, S., (2017). Viral vectors for plant genome engineering. *Front. Plant Sci., 8*, 539.

Zhang, Y., Massel, K., Godwin, I. D., & Gao, C., (2018). Applications and potential of genome editing in crop improvement. *Genome Biol., 19*(1), 210.

Zhao, H., & Wolt, J. D., (2017). Risk associated with off-target plant genome editing and methods for its limitation. *Emerg. Top. Life Sci., 1*(2), 231–240.

CHAPTER 3

Clustered Regularly Interspaced Short Palindromic Repeats/CRISPR-Associated (CRISPR/Cas) Systems: Discovery, Structure, Classification, and General Mechanism

SUMAN GANGER,[1*] GEETANJALI HARALE,[2] and PARAMA MAJUMDAR[3]

[1]*Department of Microbiology, Vivekanand Education Society's College of Arts, Science and Commerce, Mumbai, Maharashtra, India*

[2]*Department of Biotechnology, Vivekanand Education Society's College of Arts, Science and Commerce, Mumbai, Maharashtra, India*

[3]*Department of Botany, Hislop College, Nagpur, Maharashtra, India*

[*]*Corresponding author. E-mail: sumanganger@gmail.com*

ABSTRACT

The discovery and implementation of the CRISPR-Cas systems have been a well-acknowledged milestone in the history of genetic engineering (GE). Exhilarating research has led to a systematic progress in the understanding of the system from the identification of CRISPR as short regularly spaced repeats (SRSRs) in a bacterial cell, to the recognition of CRISPR-Cas as a Prokaryotic RNAi-based Adaptive Immune System and finally to the discovery of its programmability which gives it enormous potential for applicability. The programmability of this gene editing tool is attributed to a well-characterized locus which can be divided into a leader sequence, followed by a CRISPR array and a termination sequence. The genes for

CRISPR/Cas-Mediated Genome Editing in Plants, Deepu Pandita, Anu Pandita (Eds.)

CRISPR-associated proteins (Cas proteins), which are positioned near the CRISPR array, code for the effector module. The architecture of effector modules is the basis of classification of the CRISPR-Cas systems; 13 families of core Cas proteins have been identified. The mechanism of the CRISPR-Cas system is fluid and adaptive. The first step is adaptation which occurs when the cell encounters a foreign DNA for the first time and identifies it as a potential target, leading to insertion of specific sequences in the CRISPR array as spacers. When the system encounters this foreign DNA once again, an effector/surveillance complex of Cas proteins and mature CRISPR RNA (crRNA) identifies it, forming a double stranded structure, which in turn recruits a Cas endonuclease that degrades the target DNA. This CRISPR-directed RNA interference (RNAi) forms the basis of the gene editing mechanism.

This chapter begins with a brief historical review of the discovery of the CRISPR-Cas system and its implementation as a genome editing tool, followed by a detailed and systematic explanation of its structure and general mechanism. The chapter also touches upon the relatively unexplored and enigmatic field of CRISPR Cas evolution. The objective is to give the readers a better perspective of this expeditiously evolving technique of genome manipulation.

3.1 INTRODUCTION

The ability to maneuver genes has been a watershed achievement not just for genetic engineering (GE), but also in the understanding of several biological processes as a whole. From agricultural advancements to human health, all sections have gained immensely from the advancement of gene editing tools.

Targeted Nuclease activity has been the primary player in the entire arena of genome manipulation. From the era of induced point mutations to the times of targeted change of virtually any genome, we have progressed our way, and this has opened doors to exciting new possibilities for genetic manipulations as a whole. Tools like ZFN, transcription activator-like effector nucleases (TALENs), and clustered regularly interspaced short palindromic repeats (CRISPR)-CRISPR-associated (Cas) have been the foundational technologies in this regard (Gaj et al., 2016).

ZFNs and TALENs are programmable technologies for producing double-stranded DNA (dsDNA) break that count on protein-based recognition of DNA sequences. Proteins designed for these tools have modular structures, i.e., multiple repeats of a few small domains. Different combinations of modules recognize specific DNA sequences. It works as a technology, but it necessitates a great deal of protein engineering. CRISPR/Cas systems, on the other hand, are RNA-programmed proteins, which means that a single protein (Cas) can be used to induce a break in any place of DNA by simply changing the sequence of the guide RNA (gRNA) associated with the Cas protein. CRISPR/Cas system here relies on RNA-based recognition rather than protein-based recognition of DNA (Carroll, 2011) or, in some cases, RNA (Zhu et al., 2018).

In prokaryotes, the CRISPR/Cas system is the only known adaptive and heritable immune system (Marraffini et al., 2010; Rath et al., 2015). Half of all known bacterial genomes and practically all archaeal genomes include this defense system against foreign mobile genetic elements (MGEs) like phages or conjugative plasmids (Al-Attar et al., 2011; Barrangou et al., 2007).

The exuberance of research in the field of CRISPR/Cas has revolutionized the applicability of genome editing. New understandings, applications, and modifications surface on scholarly search engines almost every week. In this chapter, we have tried to begin where it all started, and we attempt to take the readers through the history and origin, structure, classification, and mechanism of action. To sum up the concept, a note on the supplementary roles and a brief review of the current understanding of the evolution of CRISPR elements have been added.

3.2 HISTORY, ORIGIN, AND FURTHER DEVELOPMENTS

In the year 2012, some major progress in the avenue of gene editing methods made many heads turn. The development of two tools, i.e., – zinc-finger nucleases (ZFN) and TALENs proteins, were deemed among the 10 scientific breakthroughs of 2012 by the magazine Science. In an interesting selection of a phrase, the magazine called the two techniques as 'cruise missiles' that can be tamed to target a particular part of a genome and clip the dsDNA at that specific location only.

The disadvantage, however, was also reviewed almost simultaneously. Both ZFN and TALEN techniques required the synthesis of new gene sequences. Such genes will encode a particular protein for every novel site in the DNA that was to be changed, making it a cumbersome technology. In bacteria and archaea, a team of scientists from the University of California Berkeley's Department of Molecular and Cell Biology and the Laboratory for Molecular Infection Medicine in Sweden reported the discovery of a "programmable dual RNA-Guided-DNA Endonuclease system."

At this juncture, in the year 2012, a team of scientists from the University of California Berkeley's Department of Molecular and Cell Biology and the Laboratory for Molecular Infection Medicine in Sweden reported the discovery of a "programmable dual RNA-Guided-DNA Endonuclease system" in bacteria and archaea. These organisms have an adaptive defense system based on RNA that is known as CRISPR/CRISPR-associated (Cas) that protects organisms from invading viruses and plasmids. It was quite simple to use the CRISPR/Cas system since it only requires a short RNA molecule to program it for site-specific DNA recognition. Since this discovery in 2012, further evolution of this idea and advancements in the field have triggered breakthroughs in the sectors of health and healthy living, and improvement of crop plants and earned the researchers a Nobel Prize. And yet, it is believed to be just the dawn. Discussed below are some of the researchers and their work that has led to the current CRISPR/Cas knowledge. But the story begins with the identification of this element followed by the development of the idea of a genetic tool.

3.2.1 EVENTS IN THE IDENTIFICATION OF CRISPR/CAS

- **1987**

Identification of "Unusual" Repetitive Sequences

Yoshizumi Ishino et al.

In 1987, while studying three isozymes of alkaline phosphatase from *Escherichia coli,* a new type of repetitive sequence was discovered in prokaryotes by Yoshizumi Ishino and his co-workers. In the sequence, Ishino et al. reported the presence of an "unusual structure" at the 3'-end of the flanking region of iap gene. It was characterized by five highly homologous 29 nucleotide (nt) direct repeat (DR) sequences which were regularly spaced with 32 nucleotide sequences. The first sequence, which was found in the putative transcriptional end site, shared less similarity with the others (Ishino et al., 1987; Morange, 2015). It was predicted that iap mRNA was regulated by these regularly spaced direct repeats (RSDR). At the same time, independent studies in *E. coli* and *Salmonella typhimurium*, showed the presence of 20–40 nt repetitive sequences with dyad symmetry

(Palindromic Units: PUs) exhibiting species-specific consensus. It was projected that ~1% of the *E. coli* genome consisted of such PUs. A similar study by Nakata et al. further unearthed the presence of 29 nt RSDRs repeating 14 times with 32 or 33 nt patches separating them.

- **2000**

Characterization of 'Short Regularly Spaced Repeats'

Francisco J. M. Mojica et al.

In the year 2000, working in the Division of Microbiology, University of Alicante, Spain Francisco J. M. Mojica and his co-workers reported that these short, regularly spaced repeats were very homogeneous to the point of being identical within a genome. Heterogeneity, however, was seen in Archaea where *Pyrococcus abyssi, Archaeoglobus fulgidus, Aeropyrum pernix* and *Methanococcus jannaschii* were described by the authors as having less than 85% similarity.

These elements were referred to as a family of short, regularly spaced repeats (SRSRs). They were identified as clusters of repeated palindromic sequences of 20–40 base pairs (bp) separated by unique intervening 20–60 bp sequences (Mojica et al., 2016; Morange, 2015).

The sequence was reported to be conserved in closely related phylogenetic groups and high similarity percentage in domains revealed common origin. While the authors identified the palindromic nature of the tandem repeats, the location pattern was not identified with certainty. They could however recognize the positional coincidence of two clusters of SSR in the Archaeal species *Methanobacterium thermoautotrophicum* and *Pyrococcus horikoshii* (Mojica et al., 2000).

- **2002**

Reached the Term 'Clustered Regularly Interspaced Short Palindromic Repeats'

Ruud Jansen et al.

The year 2002 came with a breakthrough in the characterization of these direct repeats (DRs). Ruud Jansen and colleagues working in the

Department of Infectious Diseases and Immunology, Bacteriology Division, Netherlands, characterized these interspersed SSRs in bacteria. Their unit length was less than 200 bp and was classified as non-coding, intercistronic sequences that were routinely interspaced by non-repetitive sequences of equal size. Due to the identification of these characters with certainty, these workers coined the term clustered regularly interspaced short palindromic repeats (CRISPRs) to designate these sequences.

Further, a 300–500 bp leader sequence, flanking on one side, was also identified and a sequence similarity of about 80% was noted in phylogenetically related species. These leader sequences were several 100 base pairs long. A comparative study of the genes that flank the CRISPR loci in the genomes also revealed a clear homology. Four such genes were identified and called the CRISPR-associated genes or the Cas genes and defined as Cas1 to Cas4. The functions of the Cas proteins could not however be ascertained. The study of structural homology was used to predict the similarity of Cas3 proteins to the functional domains of the superfamily 2 of helicases and Cas4 proteins to the family of RecB exonucleases (Jansen et al., 2002).

- **2005**

Characterization of Spacers and the Way Ahead to Antisense RNA-based Bacterial Defense Against Bacteriophage

Francisco J. M. Mojica et al.; Alexander Bolotin et al.; C. Pourcel et al.

In spite of the development in CRISPR characterization, the identity of spacer sequences remained elusive for quite some time. In 2005, Francisco J.M. Mojica and his co-workers working in Spain, Alexander Bolotin and his co-workers at French National Research Institute for Agriculture and Christine Pourcel and her team in the laboratories of Paris Sud University simultaneously reported the characteristics of spacers while working in different sets of crenarchaeota, euryarchaeota, gram-positive, and gram-negative bacteria.

In a comparative study, all the groups hinted at the possible horizontal transfer that would be the reason for phylogenetic dissimilarities of the spacer sequences. Further, about 65% of the spacer homologs encountered, corresponded to the bacteriophages or conjugative plasmids (Mojica et al., 2005). The shared sequence homology with the phage suggested that CRISPRs may represent a memory of past 'genetic aggressions' (Pourcel et al., 2005b). Correlation between the number of spacers in a locus and the resistance observed in *Streptococcus thermophilus* to phage infection was thus extrapolated to derive the possible biological role of CRISPRs in defense against phage attack. Inhibition of phage gene expression by antisense RNA was suggested as the plausible mechanism. This was based on the observation that the spacers with homology to phage-coded sequences had either of the two orientations within a CRISPR locus, giving rise to antisense RNA (Bolotin et al., 2005b).

- **2006**

CRISPR/Cas may be a Prokaryotic RNAi based Immune System

Eugene V. Koonin and Kira S. Makarova et al.

The proximity of CRISPR and Cas genes was investigated, and on the basis of homology, it was predicted that this system was a defense mechanism, analogous to the eukaryotic RNAi systems. Several components of the Cas System were functionally compared to proteins, such as the endonuclease cleaving target mRNAs (slicer), the double-stranded RNA-specific helicase-nuclease (dicer) and the RNA-dependent RNA polymerase, all of which are pivotal in eukaryotic RNAi. However, no orthologous similarity was observed with the eukaryotic functional counterpart. It was hypothesized that these CRISPR inserts function as prokaryotic siRNAs (silencer RNA) and restrict infection by annealing with the target mRNAs leading to their degradation or translational silencing (Makarova et al., 2006).

- **2007**

Demonstration of CRISPR/Cas-based Acquired Immunity

Rodolphe Barrangou and Philippe Horvath

It was observed that the CRISPR Cas system was involved in a nucleic acid-based "immunity" system in prokaryotes where the spacer elements of CRISPR provide specificity and the enzymatic activity of Cas is associated with implementing resistance. It was further speculated by the authors that the Cas genes enforce resistance by insertion of additional CRISPR spacers and repeats, as part of an adaptive "immune" response. Two Cas genes in strains of *S. thermophilus* namely Cas5 and Cas7 were inactivated. Inactivation of Cas5 resulted in the loss of phage resistance. However, no CRISPR1 phage-resistant mutants could be isolated for Cas7 knockout. This could be due to Cas7's involvement in the synthesis and/or insertion of new spacers and repeats (Barrangou et al., 2007).

3.2.2 EVENTS IN THE IMPLEMENTATION OF CRISPR/CAS AS A GENOME EDITING TOOL

- **2012**

Programmability of CRISPR/Cas System

Jennifer Doudna and Emmanuelle Charpentier

Jennifer Doudna and Emmanuelle Charpentier were among the first to suggest that this microbial immunity mechanism may be harnessed to enable programmed genome editing. They suggested that the endonuclease was directed by two RNA structures, to cleave target DNA. They looked into the Cas9 protein, which was discovered to be an enzyme that cleaves target dsDNA by forming a base-paired structure between the activating tracrRNA (trans-activating CRISPR RNA) and the crRNA (CRISPR RNA). The authors examined the ability of Cas9 to cleave not only plasmid DNA but also any oligonucleotide duplex harboring a protospacer sequence corresponding to a mature crRNA while purifying the protein using an overexpression technique. This cleavage though was not directed by crRNA alone; addition of tracrRNA, which can base pair with the repeat sequence of crRNA, was found to be essential to trigger the cleavage of plasmid DNA by Cas9. Also, the tracrRNA:crRNA-guided Cas9 directed cleavage of short linear dsDNA was site-specific and produced blunt ends three base pairs upstream to the protospacer adjacent motif (PAM) sequence. Using customized tracrRNA:crRNA as a single transcript, Cas9 endonuclease could be programmed to target and cleave any dsDNA sequence of interest (sgRNA). It was based on RNA-programed Cas9, which exhibited greater versatility than Zinc-Finger Nucleases (ZFNs) and Transcription-Activator Like Effector Nucleases (TALENs) for gene targeting and genome editing applications (Wiedenheft et al., 2012; Jinek et al., 2012a). The comprehensive works of Jennifer Doudna and Emmanuelle Charpentier, earned them a Nobel Prize in the year 2020.

- **2013**

  Implementation of Programed CRISPR/Cas System

  Feng Zhang et al.

  Zhang and his co-workers used the type II prokaryotic CRISPR Cas system for RNA-guided site-specific DNA cleavage. They used two different type II systems to assess the short RNA-directed Cas9 nucleases, against the endogenous genomic loci in human and mouse cells resulting in their specific cleavage.

  Importantly, the authors discovered that several guide sequences encoded within a single CRISPR array can be used to edit different places within the eukaryotic genome at the same time, which is critical for broad applicability (Cong et al., 2013).

- **2013**

  Fine Tuning the CRISPR/Cas System for Genome Editing

  Stanley Qi et al.

  Stanley Qi and his team were pioneers in the development of understanding of CRISPR-based genome editing to magnify the possibilities in research and applications of this GE technology. They created the CRISPRi (interference) system, in which gRNA was co-expressed with a catalytically attenuated Cas9 lacking endonuclease activity. The DNA recognition complex was capable of interfering with transcriptional elongation, RNA polymerase binding, and transcription factor binding by combining the regulatory domains with the dCas9 (deactivated) protein, influencing various biological processes, and yielding various functional results. In their exhaustive work they also established that CRISPRi could be simultaneously used to regulate multiple genes using a dual-color fluorescence reporter system (Qi et al., 2013).

- **2016**

Programmable Editing of Target Base

David Liu et al.

Liu expanded the scope and efficiency of genome editing many folds, by identifying the programmable system for base editing in the target sequence, resulting in desired point mutations. This enabled a direct and permanent base substitution mutation, on the target DNA in a programmable manner. This could be done without cleaving the dsDNA backbone of a donor template. This was achieved by fusion of CRISPR/Cas9 with cytidine deaminase enzyme. CRISPR/Cas9 would then possess the ability to be programmed with a gRNA and cytidine deaminase would mediate the direct conversion of cytidine to uridine, thereby effecting a C→T transition, which in turn would lead to G→A substitution, by complementarity. The accuracy of this method of targeting was found to be within a window of approximately five nucleotides. This was a breakthrough since many of the human diseases are a result of such point mutations, which can now be efficiently rectified (Komor et al., 2016).

There has been a significant amount of work done since 2016, especially in the field of applications of CRISPR/Cas systems. Novel protocols and newer dimensions of research are being pursued and unearthed almost every day. In that parlance, the history of CRISPR may go up to yesterday. However, since most works of structure and classification that have been cited in the sections ahead are post-2016, we chose to culminate history at this point.

3.3 CRISPR/CAS LOCUS STRUCTURE

3.3.1 CRISPR/CAS LOCUS

It consists of CRISPR locus flanked by multiple Cas protein genes (Koonin et al., 2019b). CRISPR/Cas is a well-characterized locus, which can be distinctly divided into three parts; a leader sequence, followed by a cluster of short DRs interspersed with spacer sequences (also known as a CRISPR array) and a termination sequence (Figure 3.1). The repeats in an array are essentially identical in length and sequence. The length of the spacers is also consistent, while the sequence content is highly diverse (Godde et al., 2006; Grissa et al., 2007; Hille et al., 2016).

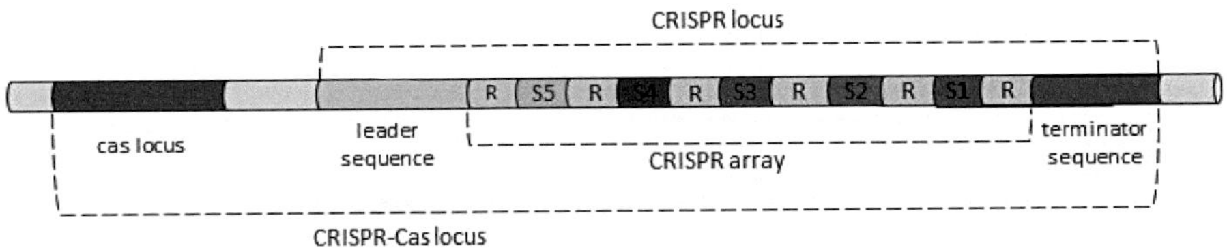

FIGURE 3.1 CRISPR/Cas locus structure.

Note: Cas locus is present mostly upstream to the CRISPR locus. It consists of multiple Cas genes oriented in a head-to-tail fashion. CRISPR locus comprises three parts: (i) leader sequence; (ii) alternate palindromic repeat (R) and spacer sequence (S) (together called CRISPR array); and (iii) terminator sequence. Here S5 denotes a recently added spacer in the array and S1 is the oldest spacer present in the CRISPR/Cas locus.

3.3.1.1 LEADER SEQUENCE

An AT-rich leader sequence is positioned upstream of the CRISPR array. The length of the leader sequence can range from 47 base pairs in bacteria to hundreds of base pairs in hyper-thermophilic archaea (Shah et al., 2011).

Leader region can be divided into two parts – core and extended leader. When studied across different species, the core leader was the conserved region, closer to the CRISPR locus. Extended leader, on the other hand, is conserved in numerous copies in the same genome or among related species. They were further upstream to the CRISPR locus. Also, extendable leader is longer than the core leader sequence (Alkhnbashi et al., 2016).

Two regulatory signals are present in the core leader region. Firstly, a promoter sequence of both archaea and bacteria was detected directly upstream to the CRISPR array (Heigwer et al., 2013; Lillestøl et al., 2009a). Secondly, a region essential for the acquisition of new spacers was identified. This region was discovered by deletion mutation studies in various organisms between 47 and 70. Post deletion, effects like infrequent addition of spacer and insertion of inverted spacers were observed during the acquisition stage. This leads to the conclusion that the leader is vital to the acquisition process (Díez-Villaseñor et al., 2013; Garrett et al., 2015; Erdmann et al., 2012).

The presence of leaderless CRISPR loci has been reported. These loci are transcriptionally defective, and also, they are inactive in adaptation. The process is not clear, but it is speculated that it can be either due to separation of the array from the loci or transposition (Alkhnbashi et al., 2016).

3.3.1.2 SPACER SEQUENCE

Originally it was observed that some spacer sequences match with viral genomes and plasmids (Kunin et al., 2007; Mojica et al., 2000a; Grissa et al., 2007; Lillestøl et al., 2009a). As a result, it was assumed that spacer regions from the extrachromosomal element were incorporated into the chromosome (Bolotin et al., 2005; Mojica et al., 2005; Pourcel et al., 2005a). Later this assumption was ascertained experimentally in *Streptococcus thermophilus*. Studies gave evidence that new spacer sequences are derived from phage genomes, acquired post-infection, imparting specific immunity against the phage (Jansen et al., 2002a; Terns et al., 2011).

A spacer sequence is similar to a repeat sequence in size, 20–72 base pairs (Grissa et al., 2007; Mojica et al., 2000a). New spacer-repeat units are included towards the leader sequence at the end of the first repeat sequence (Jansen et al., 2002a; Lillestøl et al., 2009a). Therefore, descendants of the same ancestor will have similar spacers at the distal end, and it will vary drastically towards the leader, a characteristic which can be used to understand phylogenetic relationships (Díez-Villaseñor et al., 2013).

3.3.1.3 REPEAT SEQUENCE

Structurally amid different species, on average, repeats vary from 21 to 47 base pairs, mostly around 32 base pairs (Godde et al., 2006; Grissa et al., 2007). Until recently, the nature of the repeat sequences was not considered of great significance as there were no comparative studies to evaluate their potential. Earlier studies described them as a highly variable sequence which exhibits diversities between organisms (Godde et al., 2006; Jansen et al., 2002a). However, Kunin et al. conducted a series of experiments to prove that these sequences can be grouped into clusters based on their similarities. They generated 33 clusters; some repeats distinctly related to bacteria and archaea (Kunin et al., 2007). This finding advocated the theory that CRISPR/Cas systems can be transferred horizontally amongst microorganisms (Haft et al., 2005; Makarova et al., 2006; Godde et al., 2006).

Mojica et al. discovered the presence of palindromic sequence in repeats leading them to hypothesize that repeats might have a functional RNA secondary structure (Makarova et al., 2011; Kunin et al., 2007). The presence of G-U pairs in the stem also suggests that the structures can exist in the locus' transcriptional products (Karginov et al., 2010a). When RNA folding software was used to predict intermolecular structure, it was found that these repeats resulted in a highly stable stem-loop structure. Even amongst the clusters, with different sequences, compensatory changes were observed to preserve the structure of the base pairing. It was also shown that Cas subtypes and repeat clusters had a clear association (Kunin et al., 2007).

3.3.2 CAS PROTEIN

CRISPR-associated (Cas) genes are positioned near the CRISPR array, about 530 base pairs away (Makarova et al., 2015). The genes associated with CRISPR are commonly oriented in a head-to-tail manner as if they co-transcribe (Haft et al., 2005). Out of 1,949 Cas loci analyzed, 1,484 (75%) were next to a CRISPR array, 383 loci (22%) were present in genomes which has CRISPR located far from any array and 82 were present in the genome which lack CRISPR array. Currently, 13 families of core Cas proteins have been identified (Makarova et al., 2015). Table 3.1 describes a few properties of the core Cas proteins.

TABLE 3.1 Core Cas Protein and Their Properties

Core Protein	Properties	References
Cas1	• Metal-dependent deoxyribonuclease • Also functions as integrase	Haft et al. (2005); Karginov et al. (2010a); Makarova et al. (2015)
Cas2	• ssRNA endonuclease specific to U-rich region • dsDNAse	Karginov et al. (2010a); Makarova et al. (2015)
Cas3	• Metal-dependent deoxyribonuclease (specificity for ds oligonucleotide) • ss DNA nuclease • ATP-dependent helicase activity • Nucleases • crRNA-guided degradation of invading nucleic acids	Haft et al. (2005); Karginov et al. (2010a); Makarova et al. (2015)
Cas4	• RecB-family exonuclease • Cleaves ssDNA (5' to 3' direction or both direction)	Haft et al. (2005); Makarova et al. (2015)

TABLE 3.1 *(Continued)*

Core Protein	Properties	References
Cas5	• Part of effector complex • Binds with Cas7 and crRNA to function as ribonucleases. • Contributes in phage resistance using existing spacers	Karginov et al. (2010a); Makarova et al. (2015)
Cas6	• Metal-independent endoribonuclease • Possible repair-associated mysterious protein (RAMP)	Haft et al. (2005); Makarova et al. (2015)
Cas7	• Part of effector complex • Involved in crRNA binding	Lundgren et al. (2015); Makarova et al. (2015)
Cas8	• Part of effector complex • Involved in PAM recognition	Lundgren et al. (2015); Makarova et al. (2015)
Cas9	• It has two domains: RuvC nuclease and HNH domain • RuvC nucleases: cleaves ssDNA • HNH Domain: an endonuclease • Generate crRNA • Cleaves target DNA • Involved in PAM recognition	Lundgren et al. (2015); Makarova et al. (2015)
Cas10	• Subunit of cascade/CRISPR RAMP* (Cmr) complex • Single stranded specific deoxyribonuclease	Wei et al. (2019); Lundgren et al. (2015)
Cas11	• Belongs to CRISPR-associated endonuclease Cas1 family • Metal-ion binding • DNA Binding • Endodeoxyribonuclease activity	Shmakov et al. (2018)
Cas12	• Nuclease with RuvC domain	Yan et al. (2019)
Cas13	• It has two higher eukaryotes and prokaryotes nucleotide-binding (HEPN) domains: nuclease activity • Non-specific degradation of surrounding transcript post recognition	Yan et al. (2019)

*RAMP: Repeat-associated mysterious proteins (Wang et al., 2012).

3.4 CLASSIFICATION OF CRISPR/CAS

Donning the role of adaptive immune response (Makarova et al., 2006; Mojica et al., 2000b), which in itself is a dynamic and complex process, CRISPR/Cas systems adapt to evolve into frequent rearrangements of their locus architecture resulting in heterogeneity. The horizontal gene transfer (HGTs) mechanisms seen in prokaryotes further complicate their locus structure. The CRISPR repeat sequences have also been shown to favor homologous recombination (HR) and cause genomic rearrangements (DeBoy et al., 2006). To have a comprehensive classification for such a dynamic and evolving system is tedious.

By means of a 'polythetic' methodology built on shared evidence from comparative genomic, phylogenetic, and structural analysis, Makarova et al. in 2011 gave an exhaustive classification encompassing three main types of CRISPR/Cas systems (type I, type II, and type III). These types were further sub-classified into 11 subtypes (I-A to I-F, II-A to II-C, and III-A to III-B) (Chylinski et al., 2014; Makarova et al., 2011). The three types are discrete with the presence of unique marker proteins: Cas3, Cas9, and Cas10 for the type I, type II, and type III CRISPR/Cas systems, respectively. Types I and III are found in both bacteria and archaea, but Type II is only found in bacteria (Makarova et al., 2011, 2015). There is no complete set of Cas proteins that can play the flawless role of phylogenetic markers. Among Cas proteins, Cas1 is considered as the most evolutionary protein (Wiedenheft et al., 2009). Still, Cas1 fails to accurately describe the relationship among the CRISPR/Cas system owing to their partly independent evolutions for different protein modules (Koonin et al., 2019a). It is a never-ending race between the MGEs like phages and plasmids to transform and overcome the defense put in by the cell and CRISPR/Cas immune systems to adapt to such breaches by modifying their genetic code to develop smatter effector proteins to compete or attack (Takeuchi et al., 2012).

Based on the available genomic data and sequence similarity searches, extensive structural research on Cas proteins and associated RNAs, experimental data obtained for molecular mechanism and biochemistry, organization of Cas operon and ancillary genes the most recent classification of CRISPR-Cas loci splits these systems into two classes – Class 1 and Class 2 (Makarova et al., 2020) (Figure 3.2). The Hierarchy currently followed in CRISPR/Cas classification begins with Class, which is divided into types, types divided into subtypes and subtypes into variants (Makarova et al., 2020). Class 1 systems (including type I, III, and IV systems) recruit multi-subunit effector complexes and class 2 systems (comprising type II, V, and VI systems) make use of single protein effector complexes (Mohanraju et al., 2016) (Table 3.2). Class 1 systems are more complex and account for the majority (about 90%) of CRISPR/Cas systems (Pickar-Oliver et al., 2019). As it appears from numerous studies, Class 2 systems, due to their single effector complex, have far exceeding applications of gene editing and genetic screening as compared to Class 1 (Tang et al., 2018). Cas9, the frequently used genome engineering endonuclease belongs to the type II system. Other than Class I and Class II, there are unusual and derived CRISPR/Cas systems, which populate the CRISPR world, which will further require a thorough analysis and understanding (Shmakov et al., 2018; Makarova et al., 2020).

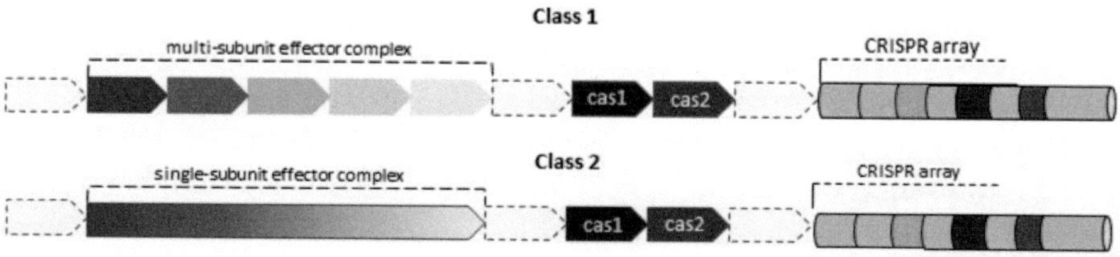

FIGURE 3.2 General structure of CRISPR/Cas Class 1 and 2 system based on effector protein complex.

Note: The location and orientation of Cas genes in the operon varies among the types and subtypes. Most Cas operons are situated close to the CRISPR array; however, they can be present elsewhere in the genome. (■▶) Cas genes are responsible for the effector complex. (⬚) Cas genes have a role in different aspects of CRISPR/Cas mechanism. Cas1 and Cas2 are essential for spacer acquisition in Types I and II systems.

Each type, which is based on the architecture of the effector complex, is further divided into many subtypes, which have variation among the position of the PAMs and the numbers, order, and uniqueness of the Cas genes. Discovery of derived CRISPR/Cas variants generally associated with MGEs supplement the classification of subtypes into variants (Makarova et al., 2020). Type III and type VI are RNA-targeting CRISPR/Cas types (Zhu et al., 2018).

Updated classification of the CRISPR/Cas system is as follows (Koonin et al., 2017; Makarova et al., 2018):

- Class I CRISPR/Cas system – three types:
    - Type I – seven subtypes:

        - Subtypes I-A to I-G;
        - Variant I-F1, I-F2 and I-F3.
    - Type III – six subtypes:
        - III-A to III-F.
    - Type IV – 3 subtypes:
        - IV-A to IV-C.
- Class II CRISPR/Cas system – three types:
    - Type II – three subtypes:
        - II-A to II-C;
        - variant II-C1, IIC2.
    - Type V – 10 subtypes:
        - V-A to V-I and V-K;
        - variant V-B1, V-B2, V-F1, V-F2, V-F3;
        - uncharacterized variant V-U1.
    - Type VI – four subtypes:
        - VI-A to VI-D;
        - variant VI-B1 and VI-B2.

Newer variants have been discovered that are currently unclassifiable or uncharacterized which suggests incorporation of additional types and subtypes in the near future (Makarova et al., 2015, 2020). The proximal location of conserved and well-characterized Cas genes to the CRISPR array makes it convenient to discover repeat sequences, which are spread across the whole genome. Nevertheless, what if the Cas information is unavailable? The availability of metagenomics data has led the hunt for repeat sequences in a Cas-independent approach for the CRISPR classification loci. A machine learning approach has enabled documentation of unclassified loci missed by traditional Cas-based methodologies. This repeat-based approach though in its infancy provides a resourceful tactic for classification of CRISPR loci, such as metagenomes and fragmented genome assemblies (Nethery et al., 2021). Classification of the CRISPR/Cas system is a mammoth task to accomplish and further updates on the classification are bound to happen. Makarova and her colleagues have worked greatly on the 'Atlas' of CRISPR classification and their reviews give in-depth information on the CRISPR/

Cas loci types and subtypes (Koonin et al., 2019a; Makarova et al., 2015, 2018).

TABLE 3.2 Classification of CRISPR/Cas Systems based on Effector Module and Distribution of the Core Cas Genes Families as Reported in the Different Stages of CRISPR/Cas Mechanism

		Class 1			Class 2		
Effector module		Multi-Subunit Complexes with Several Cas Proteins			Single Large Multidomain Protein		
		Types based on Effector Modules Architecture (Signature Proteins)					
Role in Mechanism	**Gene Type**	**I**	**III**	**IV**	**II**	**V**	**VI**
Spacer acquisition	Cas1	Y	Y	D	Y	D	D
	Cas2	Y	Y	D	Y	D	D
	Cas4	D	–	–	D	D	–
Pre-crRNA processing*	Cas6	Y	D	D	–	–	–
		–	–	–	–	Cas12	Cas13
Target binding with crRNA*	Cas5	Y	Y	Y	–	–	–
	Cas7	Y	Y	Y	–	–	–
	Cas8	Y	–	–	–	–	–
	Cas9	–	–	–	Y	–	–
	Cas12	–	–	–	–	Y	–
	Cas13	–	–	–	–	–	Y
Target cleaving*	Cas3	Y	–	–	–	–	–
	Cas10	–	Y	–	–	–	–
	–	–	–	–	Cas9	Cas12	Cas13

Key: Y: Gene required; D: Dispensable gene and/or missing.

**More proteins involved.*

Source: Koonin et al. (2017, 2019a); Makarova et al. (2020).

3.5 FLUID MECHANISM OF CRISPR/CAS IMMUNITY

The mechanism of defense by CRISPR/Cas systems can be allocated into three different but intertwined stages: (i) adaptation (spacer acquisition), (ii) expression-processing/maturation (crRNA biogenesis), and (iii) interference (degradation of invading nucleic acid) (Amitai et al., 2016; Marraffini, 2015; Zhu et al., 2018) (Figure 3.3).

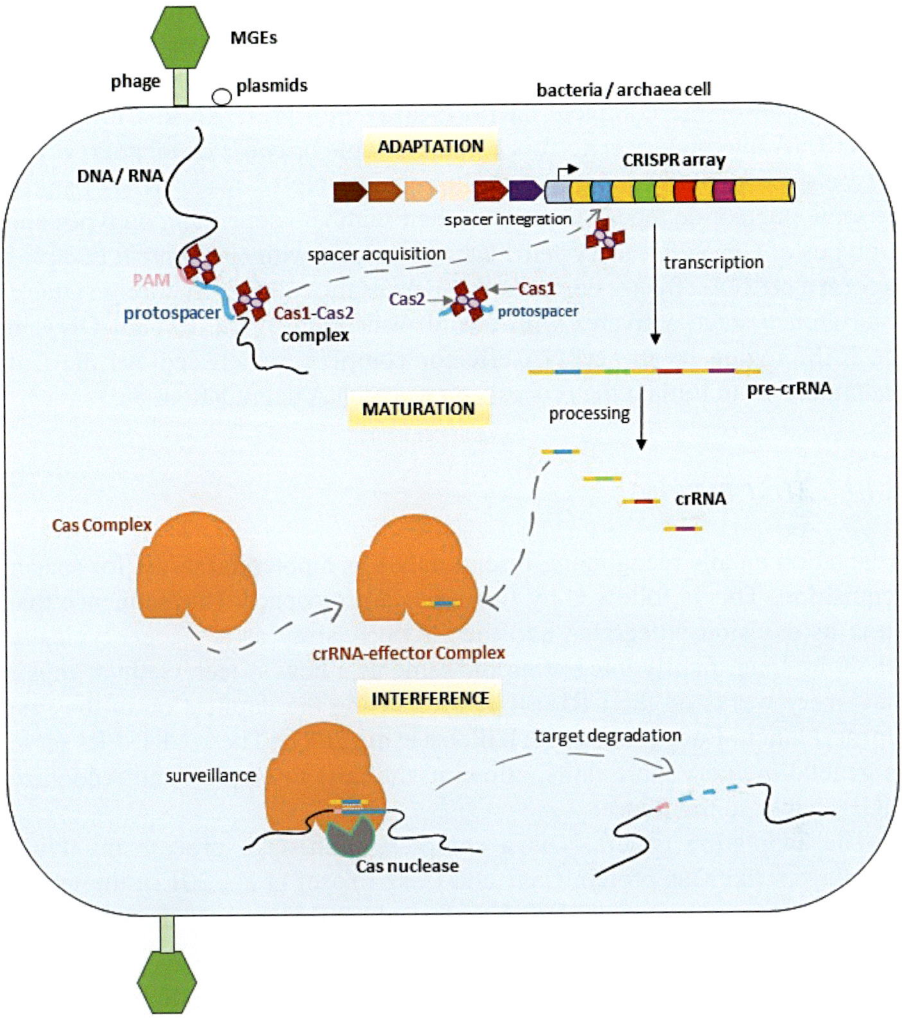

FIGURE 3.3 Overview of CRISPR/Cas immunity.

Upon entry of mobile genetic elements (MGEs) like phage or plasmid, adaptation stage commences. Two Cas1 dimers and a single Cas2 dimer form a Cas1–Cas2 heterohexameric complex that scans the incoming DNA for a specific 2–6 nucleotide sequence called Protospacer-Adjacent Motifs (PAM). Cas1 endonuclease cleaves the adjacent region of PAM called Protospacer. Depending upon the types of CRISPR/Cas system, other proteins like Cas9 or Cas4 also play a significant role in spacer acquisition (not shown in the figure). The adaptation stage ends with the integration of the protospacer sequence at the first repeat from the leader sequence end in the CRISPR array resulting in duplication of the repeat. Maturation stage begins with transcription of the complete CRISPR-array into pre-crRNA. Curation of pre-crRNA into mature crRNA occurs in a single or two-step manner, which is type dependent. The mature crRNA (comprising of spacer RNA flanked by some nucleotide repeat sequences, their numbers depending on types and subtypes of CRISPR/Cas system) later associates with Cas protein complex to form a crRNA-effector complex that patrols the cell for any foreign attack. Interference stage activates with the identification of the foreign DNA at the PAM region by the crRNA-effector complex, which deploys the Cas endonuclease to initiate the process of target DNA degradation.

3.5.1 ADAPTATION

Adaptation entails recognizing foreign DNA as a potential target for spacer acquisition. This is followed by fishing of a protospacer (the sequence that upon its excision, integrates into the CRISPR array as a spacer) typically 30–40 bp long, finally integrating the same as a new spacer. Cellular repair machinery works on the CRISPR array, causing the duplication of the first repeat (Ivančić-Bace et al., 2015; Killelea et al., 2017). These CRISPRs serve as genetic memory store units, allowing the host to employ and recognize MGE attacks in the future.

The adaptation generally is a complex, multistage process involving key players like Cas protein Cas1 and Cas2 (Yosef et al., 2012), the leader sequence (McGinn et al., 2016; Yosef et al., 2012) and the first CRISPR repeat (Wei et al., 2015; Yosef et al., 2012). The other Cas proteins (Heler et al., 2015; Wei et al., 2015) and host factor (Nuñez et al., 2016; Yoganand et al., 2017) involved in the spacer acquisition further differ between organisms and subtypes of CRISPR/Cas system (Amitai et al., 2016).

Cas1 is an endonuclease (Kim et al., 2013; Wiedenheft et al., 2009) which is crucial for spacer DNA acquisition (Nuñez et al., 2014). Cas2 is also an endonuclease (Nam et al., 2012) but studies prove that the nuclease activity is not essential for spacer acquisition (Nuñez et al., 2015; Wang et al., 2015) but have a role as structural subunit (Amitai et al., 2016). Cas1 or Cas2 deletions, on the other hand, prevent spacer acquisition, implying that both Cas1 and Cas2 are required (He et al., 2018). In most cases, the same operon encodes Cas1 and Cas2 proteins (Koonin et al., 2017). In the adaptation stage, they play a dual role of acquiring protospacer DNA and integrating it into the CRISPR array.

Two Cas1 dimers and a single Cas2 dimer combine to generate a hetero-hexameric Cas1–Cas2 protein complex (Nuñez et al., 2014). This complex recognizes only protospacers that are closest to the PAM. PAM is a 2–6 nt type-specific sequence that acts as an antigenic signature and during the interference stage, is required for cleaving off the target DNA (Bolotin et al., 2005a; Mojica et al., 2009; Shah et al., 2013). In type I, the Cas1–Cas2 complex alone promotes PAM-dependent spacer acquisition (Yosef et al., 2012) though the recent studies demonstrate the need of Cas4 in subtype I-A where the mechanisms are still unclear (Zhang et al., 2019). Type II CRISPR/Cas systems additionally require Cas9 (Heler et al., 2015; Yunzhou et al., 2015). The PAM sequence does not tag along with the protospacer for integration in the bacterial CRISPR array ensuring that the nuclease does not cleave self-DNA (Jinek et al., 2012b).

Cas1–Cas2 protein complex further inserts the protospacer into the CRISPR array at the first repeat (Leenay et al., 2017; Mojica et al., 2009; Nuñez et al., 2015; Wang et al., 2015) creating a new spacer and duplication of the repeat, polarizing this spacer near the leader end (Pourcel et al., 2005b). This arrangement makes a chronologically ordered array suggesting their sequestration (Amitai et al., 2016). The integrase property of Cas1–Cas2 complex then performs the nicking activity in the CRISPR array, targeting the two ends of the first repeat in a two-step mechanism similar to retroviral integration and DNA transposition, according to studies (Nuñez et al., 2015). In Type I and Type II CRISPR-Cas systems, the Cas1 and Cas2 proteins are vital for spacer acquisition, but they do not appear to play any role in the maturation and interference stages (Koonin et al., 2017; Makarova et al., 2015). Recent studies also indicate the presence of Cas1 and Cas2 genes in Type II, IV, V, and VI; however, their role is dispensable, suggesting redundancy of the role by other players in the process (Koonin et al., 2019a).

It is important to note that, there could be an unintentional procurement of spacers from bacterial own DNA instead of from invading foreign DNA. This would be suicidal, as it would result in the CRISPR-Cas interference mechanism cutting the self-DNA called CRISPR/Cas autoimmunity.

Some of the well-established strategies identified in the cell, preventing this autoimmunity are the mutations in the Cas gene rendering them inactive, mutations of the repeats adjacent to self-derived spacers and escape mutations in the PAM (Stern et al., 2010). Gil Amitai and Rotem Sorek in their review "CRISPR/Cas adaptation: insights into the mechanism of action" describe the adaptation process exhaustively (Amitai et al., 2016).

3.5.2 MATURATION

There are three steps in the biosynthesis of mature crRNAs. Leader sequence harbors the promoter from which transcription begins into long precursor-CRISPR transcripts (pre-crRNA). Repeats within the pre-crRNA are sites of cleavage, which produce CRISPR RNAs (crRNAs), encompassing the complete spacer region bordered by partial repeat sequences. Secondary cleavage may be required in some subtypes to produce active mature or processed crRNAs (Charpentier et al., 2015; Lillestøl et al., 2009b) also called prokaryotic silencing RNAs (psiRNAs) (Hale et al., 2008) or gRNAs (Karvelis et al., 2013).

crRNA biogenesis comprises highly tuned processing mechanisms which have been evolved in various CRISPR Cas systems. A specific endoribonuclease derived from Cas6 family, either alone or in combination with other Cas proteins cleaves the pre-crRNA within the repeat areas in type I and type III (Carte et al., 2008; Haurwitz et al., 2010; Nam et al., 2012). Whereas in type II, the trans-acting small RNA (tracrRNA) produces a dual-RNA which is complementary to the repeats. The formation of this dsRNA recruits the endogenous RNase III leading to its cleavage with the help of protein Cas9 (Charpentier et al., 2015).

In Type I, a protein of the Cas6 family or Cas5d processes the pre-crRNA to mature crRNAs. The palindromic repeats in pre-crRNA are either unstructured or form hairpin structures that the nucleases Cas6 and Cas5 (Charpentier et al., 2015) recognize and cleaves an 8 nucleotides conserved sequence, upstream to the repeat-spacer boundary (Brouns et al., 2008; Carte et al., 2008). Types I-C, I-E, and I-F do not go through a second maturation stage, but Types I-A, I-B, and I-D, as well as Types II and III do (Charpentier et al., 2015).

A trans-activating CRISPR RNA (tracrRNA) gene, upstream of the Cas operon transcribes a unique noncoding RNA that is analogous to the repeat sequences in the CRISPR array in the type II CRISPR/Cas system. The tracrRNA then anneals to the pre-crRNA repeats forming double stranded RNA regions, inviting the housekeeping endoribonuclease III (RNase III) to cleave in the presence of Cas9 (Deltcheva et al., 2011; Jiang et al., 2017). This coordination results in many short crRNAs corresponding to the interval between repeats (spacers) (Brouns et al., 2008; Carte et al., 2008; Karginov et al., 2010b). A nuclease prunes the 5' end of the crRNA and reduces the length of the guide sequence, creating active and mature crRNAs, in a second maturation process whose components and mechanisms are still unknown (Charpentier et al., 2015; Jiang et al., 2017).

In Type III, unstructured pre-crRNA is bound to the single Cas6 endonuclease, which cleaves within each repetition to form intermediate crRNAs with 5' and 3' repeat-derived termini. crRNAs are then loaded to the Csm (Type III-A/D) or Cmr (Type III-B/C) complexes, where they are further matured by cutting the 3' repeat-derived sequence with yet-to-be-identified nucleases (Charpentier et al., 2015).

The synthesis of mature crRNAs has evolved with modest differences in the CRISPR/Cas types. Pre-crRNAs may now be distinguished from other cytosolic RNAs thanks to unique RNA recognition mechanisms. CrRNAs consist of a spacer and a piece of the neighboring DR act as a guide to direct the CRISPR ribonucleoprotein (crRNP) effectors to the target for degradation. Emmanuelle Charpentier and her team lucidly narrates the crRNA maturation steps in their review article "Biogenesis pathways of RNA guides in archaeal and bacterial CRISPR/Cas adaptive immunity" (Charpentier et al., 2015).

3.5.3 INTERFERENCE

To interrogate the foreign nucleic acid, CRISPR/Cas systems rely on the assembly of mature crRNAs with Cas proteins to produce crRNA-effector complexes or Surveillance complexes. This interference mechanism wipes out the matching DNA or RNA sequences thereby halting the spread and proliferation of foreign MGEs (Jiang et al., 2015; Oost et al., 2014; Wiedenheft et al., 2012).

Different CRISPR/Cas types have different organizational structures for Cas proteins forming effector ribonucleoprotein (RNP) complexes that are involved in crRNA-mediated silencing (Wiedenheft et al., 2012). The mature crRNA, in cooperation with several Cas proteins, forms the CRISPR-associated complex for antiviral defense (Cascade) in Type I systems. Cascade then binds to complementary DNA target sequences and recruits Cas3 (a trans-activating nuclease-helicase), which unwinds and cleaves the DNA (Sinkunas et al., 2013; Westra et al., 2012). The CRISPR locus of the Type I-E CRISPR/Cas system from *E. coli* K12, for example, contains eight

Cas genes (Brouns et al., 2008; Jore et al., 2011). Cas1 and Cas2 are necessary for spacer acquisition by creating a stable heterocomplex (Nuñez et al., 2014; Yosef et al., 2012), whereas duplex unwinding and DNA destruction is executed by Cas3 (Westra et al., 2012). Around 11 Cas proteins [Cse1_1 (CasA):Cse2_2(CasB):Cas7_6(CasC):Cas5e_1(CasD):Cas6e_1(CasE)] combine to produce a 405-kDa Cascade complex that includes a 61-nt crRNA (Jore et al., 2011). This assembly then targets the raiding nucleic acids with a sequence similar to the crRNA. At sites proximal to the PAM, base pairing between the seed region of the crRNA spacer and the complementary target protospacer is required for effective interference. The seed region of the crRNA in Cascade/I-E includes nucleotides 1–5 and 7–8 at the spacer's 5′ border. The Cascade target DNA binding affinity is reduced by mutations in the PAM sequence and mismatches in the seed region (Semenova et al., 2011). The nuclease-helicase Cas3 is then recruited by Cascade to catalyze DNA target destruction (Jiang et al., 2015).

In comparison to Type I systems, Type II CRISPR loci have fewer Cas genes — Cas1, Cas2, Csn2, and the Type II signature gene, Cas9. It was experimentally proved that Cas9-mediated DNA interference depends on the presence of tracrRNA. Cas9 endonuclease is recruited by the mature tracrRNA:crRNA hybrid structure, which then signals it to cut foreign DNA harboring a 20-nt crRNA complementary sequence preceding the PAM sequence (Karvelis et al., 2013). When produced as a single-guide RNA (sgRNA) chimera, this dual tracrRNA:crRNA guide structure cleared the path for CRISPR/Cas9 to become a dominant tool for genome engineering (Jinek et al., 2012b). The recognition of PAM is required for target DNA binding. Base pairing proceeds towards the distal end of the target region across a 12-nucleotide seed sequence (Sternberg et al., 2014), followed by DNA strand displacement and R-loop creation for Cas9 enzymatic activity (Jiang et al., 2015).

Type III CRISPR/Cas RNP complexes are homologous to class 1 CRISPR/Cas systems, which are built up of four to six different Cas proteins (Zhu et al., 2018). The backbone of their overall structure, which is made up of two twisted helical protein filaments (Csm2 or Cmr5), consists of numerous copies of the Cas7 protein and many copies of the small component Cas11 (Kazlauskiene et al., 2016; Staals et al., 2014). The hallmark protein Cas10, together with Cas5, caps one end of the RNP, whereas distinct Cas7 variants cap the other end of the complex. In addition, the RNP contains a mature 39-nt crRNA consisting of 31-nt guide segments and a repeat-derived 8-nt sequence called the 5'-handle (5' tag). The crRNAs spacer-derived sequence spans the RNP's backbone, whilst the 5'-handle is firmly bound within the Cas10/Cas5 cap (Osawa et al., 2015; Zhu et al., 2018). The mechanism of DNA cleavage by the type III effectors still remains unclear (Nishimasu et al., 2017). The structural elucidation of numerous Cas proteins and crRNP complexes, as well as biochemical studies, indicated that type I and type III CRISPR/Cas surveillance complexes have a common architectural core. Despite their extraordinary sequence variety, individual subunits generally perform comparable activities (Plagens et al., 2015). The processes by which CRISPR/Cas surveillance complexes identify and destroy target DNA are still open to decode.

This surveillance period also witnesses the discrimination done between the self and non-self-DNA by the crRNA-effector complexes in CRISPR/Cas systems (Oost et al., 2014). Type I and type II systems employ a "non-self-activation" method that entails protein-mediated recognition of a PAM, which then "turns on" the interference. In type I, a conformational shift causes a nuclease to be recruited to the crRNP complex, while type II causes intrinsic crRNP nuclease activity to be induced (Sashital et al., 2012; Sternberg et al., 2014). Type III systems do not appear to use a PAM-based mechanism; instead, they employ a "self-inactivation" technique that includes base pairing between the crRNAs 5' tag and the repeat sequence in the CRISPR locus on the host chromosome. Base pairing appears to stop the interference process in this site, probably by preventing the nuclease from being recruited (Marraffini et al., 2010). Phages do evolve rapidly, and studies show them encoding many anti-CRISPR proteins (Acrs), which mainly targets components of effector complexes, thereby escaping surveillance (Bondy-Denomy et al., 2013).

3.6 SUPPLEMENTARY ROLES OF CRISPR/CAS SYSTEM

The CRISPR locus appears to have additional activities in various systems, in addition to its protective roles. The repeats in their natural form boons a prospect for homology-driven genome rearrangements (DeBoy et al., 2006). *In vitro*, the Cas1 protein from *E. coli* demonstrated nuclease activity against single-stranded and branched DNAs from Holliday Junctions and replication forks. Cas1 interaction with recB, recC, and ruvB suggested its role in the DNA repair mechanism of the cell (Babu et al., 2011). *Francisella novicida* promotes its virulence and attenuates its host pro-inflammatory innate immune response to bacterial lipoprotein (BLP) by CRISPR/Cas mediated repression of the BLP transcript (Sampson et al., 2013). In one study, CRISPR region and associated Cas genes involvement terminates the lysogenized cell from exhibiting the group behaviors like biofilm formation and swarming motility (Zegans et al., 2009), suggesting its role in endogenous cellular processes.

3.7 EVOLUTION OF CRISPR/CAS SYSTEMS

The CRISPR/Cas systems as seen from the above present complex molecular mechanisms, and as such, unquestionably seem as the products of a complicated progression of evolutionary events. Dwelling into their origins or synthesis is very intriguing. Genetic analysis of Cas genes showed that Cas1 may not be necessarily coded within CRISPR/Cas loci (Makarova et al., 2013) rather in some they were found on region of DNA flanked by terminal inverted repeats (TIRs), having DNA polymerase as additional genes within them suggesting transposable elements or large MGEs (Koonin et al., 2015). Cas1 catalyzing spacer integration into CRISPR arrays overlaps the integrase mechanism of transposons. This led to the new class called casposons wherein Cas1 were dubbed as caspases (Krupovic et al., 2014, 2016). The known casposons also encode nucleases homologous to Cas4 (Krupovic et al., 2016). The current architecture of CRISPR/Cas could be the result of an unintentional addition of a casposons into the region of an ancestral innate immunity locus, its immobility, and further deletion of some of its genes (Koonin et al., 2019a).

The repeats and leader sequence could have arisen from the primordial casposons' TIRs or a duplicated target site (Krupovic et al., 2017). A close examination of class 2 effector protein sequences reveals strong similarities to nucleases encoded by other MGE classes (Chylinski et al., 2014). Common nuclease protein families, showing RuvC-like endonuclease domain, which belongs to the RNase H fold, are also part of the type II and type V effectors of Cas9 and Cas12 (Majorek et al., 2014). They have a sequence similarity to the IS605 and other comparable transposons' for RuvC-domain-containing TnpB proteins (Shmakov et al., 2015). Studies show multiple recruitment of nucleases of Class II effectors from MGEs (Koonin et al., 2019a).

Type III CRISPR/Cas systems possess an additional feature of acquiring spacers from RNA genomes that are converted into DNA before integration by a reverse transcriptase (RT) enzyme associated with CRISPR (Silas et al., 2016). These RTs are linked to the RTs of group II introns, according to a phylogenetic study of the RT superfamily (Toro et al., 2017). Class 2 systems emerged from class 1 systems with the introduction of transposable elements encoding diverse nucleases, making them the unsung heroes of the genome editing toolbox (Mohanraju et al., 2016).

A review by Eugene V. Koonin and Makarova titled 'Origins and evolution of CRISPR/Cas systems' present an enigmatic piece of literature. They concluded that at least four distinct MGE types played a role in CRISPR/Cas evolution. Adaptation modules arose from Casposons while the RT required for a wide range of type III came from the group II introns. Type II and type V effectors originated from Non-autonomous IS605-like transposons and a toxin-antitoxin (TA) module that likely supplied Cas2 (Koonin et al., 2019a; Makarova et al., 2020).

With humongous information uncovered every week in CRISPR/Cas world, a better understanding of its evolution will aid in exploring this proven gene-editing tool with more precision and in a multifaceted approach to its applications.

KEYWORDS

- **acquisition**
- **adaptation**
- **Cas protein**
- **CRISPR history**
- **interference**
- **maturation**

REFERENCES

Al-Attar, S., Westra, E. R., Van, D. O. J., & Brouns, S. J. J., (2011). Clustered regularly interspaced short palindromic repeats (CRISPRs): The hallmark of an ingenious antiviral defense mechanism in prokaryotes. *Biol. Chem., 392*(4), 277–289.

Alkhnbashi, O. S., Shah, S. A., Garrett, R. A., Saunders, S. J., Costa, F., & Backofen, R., (2016). Characterizing leader sequences of CRISPR loci. *Bioinformatics, 32*(17), i576–i585.

Amitai, G., & Sorek, R., (2016). CRISPR-Cas adaptation: Insights into the mechanism of action. *Nat. Rev. Microbiol., 14*, 67–76.

Babu, M., Beloglazova, N., Flick, R., Graham, C., Skarina, T., Nocek, B., Gagarinova, A., et al., (2011). A dual function of the CRISPR-Cas system in bacterial antivirus immunity and DNA repair. *Mol. Microbiol., 79*(2), 484–502.

Barrangou, R., Fremaux, C., Deveau, H., Richards, M., Boyaval, P., Moineau, S., Romero, D. A., & Horvath, P., (2007). CRISPR provides acquired resistance against viruses in prokaryotes. *Science, 315*(5819), 1709–1712.

Bolotin, A., Quinquis, B., Sorokin, A., & Dusko, E. S., (2005). Clustered regularly interspaced short palindrome repeats (CRISPRs) have spacers of extrachromosomal origin. *Microbiology, 151*(8), 2551–2561.

Bondy-Denomy, J., Pawluk, A., Maxwell, K. L., & Davidson, A. R., (2013). Bacteriophage genes that inactivate the CRISPR/Cas bacterial immune system. *Nature, 493*(7432), 429–432.

Brouns, S. J. J., Jore, M. M., Lundgren, M., Westra, E. R., Slijkhuis, R. J. H., Snijders, A. P. L., Dickman, M. J., et al., (2008). Small CRISPR RNAs guide antiviral defense in prokaryotes. *Science, 321*(5891), 960–964.

Carroll, D., (2011). Genome engineering with zinc-finger nucleases. *Genetics, 188*(4), 773–782.

Carte, J., Wang, R., Li, H., Terns, R. M., & Terns, M. P., (2008). Cas6 is an endoribonuclease that generates guide RNAs for invader defense in prokaryotes. *Genes Dev., 22*(24), 3489–3496.

Charpentier, E., Richter, H., Van, D. O. J., & White, M. F., (2015). Biogenesis pathways of RNA guides in archaeal and bacterial CRISPR-Cas adaptive immunity. *FEMS Microbiol. Rev., 39*(3), 428–441.

Chylinski, K., Makarova, K. S., Charpentier, E., & Koonin, E. V., (2014). Classification and evolution of type II CRISPR-Cas systems. *Nucleic Acids Res., 42*(10), 6091–6105.

Cong, L., Ran, F. A., Cox, D., Lin, S., Barretto, R., Habib, N., Hsu, P. D., et al., (2013). Multiplex genome engineering using CRISPR/Cas systems. *Science, 339*(6121), 819–823.

DeBoy, R. T., Mongodin, E. F., Emerson, J. B., & Nelson, K. E., (2006). Chromosome evolution in the thermotogales: Large-scale inversions and strain diversification of CRISPR sequences. *J. Bacteriol., 188*(7), 2364–2374.

Deltcheva, E., Chylinski, K., Sharma, C. M., Gonzales, K., Chao, Y., Pirzada, Z. A., Eckert, M. R., et al., (2011). CRISPR RNA maturation by trans-encoded small RNA and host factor RNase III. *Nature, 471*(7340), 602–607.

Díez-Villaseñor, C., Guzmán, N. M., Almendros, C., García-Martínez, J., & Mojica, F. J. M., (2013). CRISPR-spacer integration reporter plasmids reveal distinct genuine acquisition specificities among CRISPR-Cas I-E variants of *Escherichia coli*. *RNA Biol., 10*(5), 792–802.

Erdmann, S., & Garrett, R. A., (2012). Corrigendum: Selective and hyperactive uptake of foreign DNA by adaptive immune systems of an archaeon via two distinct mechanisms. *Mol. Microbiol., 86*(3), 757.

Gaj, T., Sirk, S. J., Shui, S., & Liu, J., (2016). Genome-editing technologies: Principles and applications. *Cold Spring Harb. Perspect. Biol., 8*(a203754), 1–20.

Garrett, R., Shiraz, S., Erdmann, S., Liu, G., Mousaei, M., León-Sobrino, C., Wenfnfang, P., et al., (2015). CRISPR-Cas adaptive immune systems of the sulfolobales: Unravelling their complexity and diversity. *Life, 5*, 783–817.

Godde, J. S., & Bickerton, A., (2006). The repetitive DNA elements called CRISPRs and their associated genes: Evidence of horizontal transfer among prokaryotes. *J. Mol. Evol., 62*(6), 718–729.

Grissa, I., Vergnaud, G., & Pourcel, C., (2007). The CRISPRdb database and tools to display crisprs and to generate dictionaries of spacers and repeats. *BMC Bioinformatics, 8*, 1–10.

Haft, D. H., Selengut, J., Mongodin, E. F., & Nelson, K. E., (2005). A guild of 45 CRISPR-associated (Cas) protein families and multiple CRISPR/Cas subtypes exist in prokaryotic genomes. *PLoS Comput. Biol., 1*(6), 0474–0483.

Hale, C., Kleppe, K., Terns, R. M., & Terns, M. P., (2008). Prokaryotic silencing (Psi)RNAs in *Pyrococcus furiosus*. *RNA, 14*(12), 2572–2579.

Haurwitz, R. E., Jinek, M., Wiedenheft, B., Zhou, K., & Doudna, J. A., (2010). Sequence- and structure-specific RNA processing by a CRISPR endonuclease. *Science, 329*(5997), 1355–1358.

He, Y., Wang, M., Liu, M., Huang, L., Liu, C., Zhang, X., Yi, H., et al., (2018). Cas1 and Cas2 from the type II-C CRISPR-Cas system of *Riemerella anatipestifer* are required for spacer acquisition. *Front. Cell. Infect. Microbiol., 8*(195), 1–13.

Heigwer, F., Kerr, G., Walther, N., Glaeser, K., Pelz, O., Breinig, M., & Boutros, M., (2013). E-TALEN: A web tool to design TALENs for genome engineering. *Nucleic Acids Res., 41*(20).

Heler, R., Samai, P., Modell, J. W., Weiner, C., Goldberg, G. W., Bikard, D., & Marraffini, L. A., (2015). Cas9 specifies functional viral targets during CRISPR-Cas adaptation. *Nature, 519*(7542), 199–202.

Hille, F., & Charpentier, E., (2016). CRISPR-Cas: Biology, mechanisms and relevance. *Philos. Trans. R. Soc. B Biol. Sci., 371*(1707).

Ishino, Y., Shinagawa, H., Makino, K., Amemura, M., & Nakata, A., (1987). Nucleotide sequence of the Iap gene, responsible for alkaline phosphatase isozyme conversion in *Escherichia Coli*, and identification of the gene product. *J. Bacteriol., 169*(12), 5429–5433.

Ivančić-Bace, I., Cass, S. D., Wearne, S. J., & Bolt, E. L., (2015). Different genome stability proteins underpin primed and naïve adaptation in *E. coli* CRISPR-Cas immunity. *Nucleic Acids Res., 43*(22), 10821–10830.

Jansen, R., Embden, Jan. D. A. V., Gaastra, W., & Schouls, L. M., (2002). Identification of genes that are associated with DNA Repeats in prokaryotes. *Mol. Microbiol., 43*(6), 1565–1575.

Jansen, R., Van, E. J. D. A., Gaastra, W., & Schouls, L. M., (2002). Identification of a novel family of sequence repeats among prokaryotes. *Omics J. Integr. Biol., 6*(1), 23–33.

Jiang, F., & Doudna, J. A., (2015). The structural biology of CRISPR-Cas systems. *Curr. Opin. Struct. Biol., 30*, 100–111.

Jiang, F., & Doudna, J. A., (2017). CRISPR – Cas9 structures and mechanisms. *Annu. Rev. Biophys., 46*, 505–529.

Jinek, M., Chylinski, K., Fonfara, I., Hauer, M., Doudna, J. A., & Charpentier, E., (2012). A Programmable dual-RNA-guided DNA endonuclease in adaptive bacterial immunity. *Science, 337*(6096), 816–821.

Jore, M. M., Lundgren, M., Van, D. E., Bultema, J. B., Westra, E. R., Waghmare, S. P., Wiedenheft, B., et al., (2011). Structural basis for CRISPR RNA-guided DNA recognition by cascade. *Nat. Struct. Mol. Biol., 18*(5), 529–536.

Karginov, F. V., & Hannon, G. J., (2010). The CRISPR system: Small RNA-guided defense in bacteria and archaea. *Mol. Cell, 37*(1), 7–19.

Karvelis, T., Gasiunas, G., Miksys, A., Barrangou, R., Horvath, P., & Siksnys, V., (2013). CrRNA and TracrRNA guide Cas9-mediated DNA interference in *Streptococcus thermophilus*. *RNA Biol., 10*(5), 841–851.

Kazlauskiene, M., Tamulaitis, G., Kostiuk, G., Venclovas, Č., & Siksnys, V., (2016). Spatiotemporal control of type III-A CRISPR-Cas immunity: Coupling DNA degradation with the target RNA recognition. *Mol. Cell, 62*(2), 295–306.

Killelea, T., & Bolt, E. L., (2017). CRISPR-Cas adaptive immunity and the three Rs. *Biosci. Rep., 37*(4), 1–9.

Kim, T. Y., Shin, M., Huynh, T. Y. L., & Kim, J. S., (2013). Crystal structure of cas1 from *Archaeoglobus fulgidus* and characterization of its nucleolytic activity. *Biochem. Biophys. Res. Commun., 441*(4), 720–725.

Komor, A. C., Kim, Y. B., Packer, M. S., Zuris, J. A., & Liu, D. R., (2016). Programmable editing of a target base in genomic DNA without double-stranded DNA Cleavage. *Nature, 533*(7603), 420–424.

Koonin, E. V., & Krupovic, M., (2015). Evolution of adaptive immunity from transposable elements combined with innate immune systems. *Nat. Rev. Genet., 16*(3), 184–192.

Koonin, E. V., & Makarova, K. S., (2019). Origins and evolution of CRISPR-Cas systems. *Phil. Trans. R. Soc. B, 374*(20180087), 1–16.

Koonin, E. V., Makarova, K. S., & Zhang, F., (2017). Diversity, classification and evolution of CRISPR-Cas systems. *Curr. Opin. Microbiol., 37*, 67–78.

Krupovic, M., & Koonin, E. V., (2016). Self-synthesizing transposons: Unexpected key players in the evolution of viruses and defense systems. *Curr Opin Microbiol., 31*, 25–33.

Krupovic, M., Beguin, P., & Koonin, E. V., (2017). Casposons: Mobile genetic elements that gave rise to the CRISPR-Cas adaptation machinery. *Curr Opin Microbiol., 38*, 36–43.

Krupovic, M., Makarova, K. S., Forterre, P., Prangishvili, D., & Koonin, E. V., (2014). Casposons: A new superfamily of self-synthesizing DNA transposons at the origin of prokaryotic CRISPR-Cas immunity. *BMC Biol., 12*(1), 36.

Kunin, V., Sorek, R., & Hugenholtz, P., (2007). Evolutionary conservation of sequence and secondary structures in CRISPR repeats. *Genome Biol., 8*(4).

Leenay, R., & Beisel, C., (2017). Deciphering, communicating, and engineering the CRISPR PAM. *J. Mol. Biol., 429*(2), 177–191.

Lillestøl, R. K., Shah, S. A., Brügger, K., Redder, P., Phan, H., Christiansen, J., & Garrett, R. A., (2009). CRISPR families of the crenarchaeal genus sulfolobus: Bidirectional transcription and dynamic properties. *Mol. Microbiol., 72*(1), 259–272.

Lundgren, M., Charpentier, E., & Fineran, P. C., (2015). *CRISPR: Methods and Protocols*. Humana Press.

Majorek, K. A., Dunin-Horkawicz, S., Steczkiewicz, K., Muszewska, A., Nowotny, M., Ginalski, K., & Bujnicki, J. M., (2014). The RNase H-like superfamily: New members, comparative structural analysis and evolutionary classification. *Nucleic Acids Res., 42*(7), 4160–4179.

Makarova, K. S., Grishin, N. V., Shabalina, S. A., Wolf, Y. I., & Koonin, E. V., (2006). A putative RNA-interference-based immune system in prokaryotes: Computational analysis of the predicted enzymatic machinery, functional analogies with eukaryotic RNAi, and hypothetical mechanisms of action. *Biol. Direct, 1*(7), 1–26.

Makarova, K. S., Haft, D. H., Barrangou, R., Brouns, S. J. J., Charpentier, E., Horvath, P., Moineau, S., et al., (2011). Evolution and classification of the CRISPR-Cas systems. *Nat. Rev. Microbiol., 9*(6), 467–477.

Makarova, K. S., Wolf, Y. I., & Koonin, E. V., (2013). Comparative genomics of defense systems in archaea and bacteria. *Nucleic Acids Res., 41*(8), 4360–4377.

Makarova, K. S., Wolf, Y. I., & Koonin, E. V., (2018). Classification and nomenclature of CRISPR-Cas systems: Where from here? *CRISPR J., 1*(5), 325–336.

Makarova, K. S., Wolf, Y. I., Alkhnbashi, O. S., Costa, F., Shah, S. A., Saunders, S. J., Barrangou, R., et al., (2015). An updated evolutionary classification of CRISPR-Cas systems. *Nat. Rev. Microbiol., 13*(11), 722–736.

Makarova, K. S., Wolf, Y. I., Iranzo, J., Shmakov, S. A., Alkhnbashi, O. S., Brouns, S. J. J., Charpentier, E., et al., (2020). Evolutionary classification of CRISPR–Cas systems: A burst of class 2 and derived variants. *Nat. Rev. Microbiol., 18*(2), 67–83.

Makarova, K., & Koonin, E. V., (2015). Annotation and classification of CRISPR-Cas systems Kira. *Methods Mol. Biol., 1311*, 47–75.

Marraffini, L. A., & Sontheimer, E. J., (2010). Self versus non-self discrimination during CRISPR RNA-directed immunity. *Nature, 463*(7280), 568–571.

Marraffini, L. A., (2015). CRISPR-Cas immunity in prokaryotes. *Nature, 526*(7571), 55–61.

McGinn, J., & Luciano, M. A., (2016). CRISPR-Cas systems optimize their immune response by specifying the site of spacer integration. *Mol. Cell, 64*(3), 616–623.

Mohanraju, P., Makarova, K. S., Zetsche, B., Zhang, F., Koonin, E. V., & Van, D. O. J., (2016). Diverse evolutionary roots and mechanistic variations of the CRISPR-Cas systems. *Science, 353*(6299), 556–568.

Mojica, F. J. M., & Rodriguez-Valera, F., (2016). The discovery of CRISPR in archaea and bacteria. *FEBS J., 283*(17), 3162–3169.

Mojica, F. J. M., Díez-Villaseñor, C., García-Martínez, J., & Almendros, C., (2009). Short motif sequences determine the targets of the prokaryotic CRISPR defense system. *Microbiology, 155*(3), 733–740.

Mojica, F. J. M., Díez-Villaseñor, C., García-Martínez, J., & Soria, E., (2005). Intervening sequences of regularly spaced prokaryotic repeats derive from foreign genetic elements. *J. Mol. Evol., 60*(2), 174–182.

Mojica, F. J. M., Díez-Villaseñor, C., Soria, E., & Juez, G., (2000). Biological significance of a family of regularly spaced repeats in the genomes of archaea, bacteria and mitochondria. *Mol. Microbiol., 36*(1), 244–246.

Morange, M., (2015). What history tells us XXXVII. CRISPR-Cas: The discovery of an immune system in prokaryotes. *J. Biosci., 40*(2), 221–223.

Nam, K. H., Ding, F., Haitjema, C., Huang, Q., DeLisa, M. P., & Ke, A., (2012). Double-Stranded endonuclease activity in *Bacillus halodurans* clustered regularly interspaced short palindromic repeats (CRISPR)-associated Cas2 protein. *J. Biol. Chem., 287*(43), 35943–35952.

Nam, K. H., Haitjema, C., Liu, X., Ding, F., Wang, H., DeLisa, M. P., & Ke, A., (2012). Cas5d protein processes pre-CrRNA and assembles into a cascade-like interference complex in subtype I-C/Dvulg CRISPR-Cas system. *Structure, 20*(9), 1574–1584.

Nethery, M. A., Korvink, M., Makarova, K. S., Wolf, Y. I., Koonin, E. V., & Barrangou, R., (2021). CRISPRclassify: Repeat-based classification of CRISPR loci. *CRISPR J., 4*(4), 558–574.

Newsom, S., Parameshwaran, H. P., Martin, L., & Rajan, R., (2021). The CRISPR-Cas mechanism for adaptive immunity and alternate bacterial functions fuels diverse biotechnologies. *Front. Cell. Infect. Microbiol., 10*, 1–10.

Nishimasu, H., & Nureki, O., (2017). Structures and mechanisms of CRISPR RNA-guided effector nucleases. *Curr. Opin. Struct. Biol., 43*, 68–78.

Nuñez, J. K., Bai, L., Harrington, L. B., Hinder, T. L., & Doudna, J. A., (2016). CRISPR immunological memory requires a host factor for specificity. *Mol. Cell, 62*(6), 824–833.

Nuñez, J. K., Harrington, L. B., Kranzusch, P. J., Engelman, A. N., & Doudna, J. A., (2015). Foreign DNA capture during CRISPR-Cas adaptive immunity. *Nature, 527*(7579), 535–538.

Nuñez, J. K., Kranzusch, P. J., Noeske, J., Wright, A. V., Davies, C. W., & Doudna, J. A., (2014). Cas1-Cas2 complex formation mediates spacer acquisition during CRISPR-Cas adaptive immunity. *Nat. Struct. Mol. Biol., 21*(6), 528–534.

Nuñez, J. K., Lee, A. S. Y., Engelman, A., & Doudna, J. A., (2015). Integrase-mediated spacer acquisition during CRISPR-Cas adaptive immunity. *Nature, 519*(7542), 193–198.

Osawa, T., Inanaga, H., Sato, C., & Numata, T., (2015). Crystal structure of the Crispr-Cas RNA silencing Cmr complex bound to a target analog. *Mol. Cell, 58*(3), 418–430.

Pickar-Oliver, A., Black, J. B., Lewis, M. M., Mutchnick, K. J., Klann, T. S., Gilcrest, K. A., Sitton, M. J., et al., (2019). Targeted transcriptional modulation with type I CRISPR–Cas systems in human cells. *Nat. Biotechnol., 37*, 1493–1501.

Plagens, A., Richter, H., Charpentier, E., & Randau, L., (2015). DNA and RNA interference mechanisms by CRISPR-Cas surveillance complexes. *FEMS Microbiol. Rev., 39*(3), 442–463.

Pourcel, C., Salvignol, G., & Vergnaud, G., (2005). CRISPR elements in yersinia pestis acquire new repeats by preferential uptake of bacteriophage DNA, and provide additional tools for evolutionary studies. *Microbiology, 151*(3), 653–663.

Qi, L. S., Larson, M. H., Gilbert, L. A., Doudna, J. A., Weissman, J. S., Arkin, A. P., & Lim, W. A., (2013). Repurposing CRISPR as an RNA-guided platform for sequence-specific control of gene expression. *Cell, 152*(5), 1173–1183.

Rath, D., Amlinger, L., Rath, A., & Lundgren, M., (2015). The CRISPR-Cas immune system: Biology, mechanisms and applications. *Biochimie., 117*, 119–128.

Sampson, T. R., Saroj, S. D., Llewellyn, A. C., Tzeng, Y. L., & Weiss, D. S., (2013). A CRISPR-Cas system mediates bacterial innate immune evasion and virulence. *Nature, 497*(7448), 254–267.

Sashital, D. G., Wiedenheft, B., & Doudna, J. A., (2012). Mechanism of foreign DNA Selection in a bacterial adaptive immune system. *Mol. Cell, 46*(5), 606–615.

Semenova, E., Jore, M. M., Datsenko, K. A., Semenova, A., Westra, E. R., Wanner, B., Van, D. O. J., et al., (2011). Interference by clustered regularly interspaced short palindromic repeat (CRISPR) RNA is governed by a seed sequence. *Proc. Natl. Acad. Sci., 108*(25), 10098–10103.

Shah, S. A., & Garrett, R. A., (2011). CRISPR/Cas and Cmr modules, mobility and evolution of adaptive immune systems. *Res. Microbiol., 162*(1), 27–38.

Shah, S. A., Erdmann, S., Mojica, F. J. M., & Garrett, R. A., (2013). Protospacer recognition motifs. *RNA Biol., 10*(5), 891–899.

Shmakov, S. A., Makarova, K. S., Wolf, Y. I., Severinov, K. V., & Koonin, E. V., (2018). Systematic prediction of genes functionally linked to CRISPR-Cas systems by gene neighborhood analysis. *Proc. Natl. Acad. Sci. U. S. A, 115*(23), E5307–E5316.

Shmakov, S., Abudayyeh, O. O., Makarova, K. S., Wolf, Y. I., Jonathan, S., Semenova, E., Minakhin, L., et al., (2015). Discovery and functional characterization of diverse class 2 CRISPR-Cas systems. *Mol Cell, 60*(3), 385–397.

Silas, S., Mohr, G., Sidote, D. J., Markham, L. M., Sanchez-Amat, A., Bhaya, D., Lambowitx, A. M., & Fire, A. Z., (2016). Direct CRISPR spacer acquisition from RNA by a natural reverse-transcriptase-Cas1 fusion protein. *Science, 351*(6276), 1–31.

Sinkunas, T., Gasiunas, G., Waghmare, S. P., Dickman, M. J., Barrangou, R., Horvath, P., & Siksnys, V., (2013). *In vitro* reconstitution of cascade-mediated CRISPR immunity in *streptococcus thermophilus*. *EMBO J., 32*(3), 385–394.

Staals, R. H. J., Zhu, Y., Taylor, D. W., Kornfeld, J. E., Sharma, K., Barendregt, A., Koehorst, J. J., et al., (2014). RNA targeting by the type III-A CRISPR-Cas Csm complex of thermus thermophilus. *Mol. Cell, 56*(4), 518–530.

Stern, A., Keren, L., Wurtzel, O., Amitai, G., & Sorek, R., (2010). Self-targeting by CRISPR: Gene regulation or autoimmunity? *Trends Genet., 26*(8), 335–340.

Sternberg, S. H., Redding, S., Jinek, M., Greene, E. C., & Doudna, J. A., (2014). DNA interrogation by the CRISPR RNA-guided endonuclease Cas9. *Nature, 507*(7490), 62–67.

Takeuchi, N., Wolf, Y. I., Makarova, K. S., & Koonin, E. V., (2012). Nature and intensity of selection pressure on crispr-associated genes. *J. Bacteriol., 194*(5), 1216–1225.

Tang, Y., & Fu, Y., (2018). Class 2 CRISPR/Cas: An expanding biotechnology toolbox for and beyond genome editing 06 biological sciences 0604 genetics. *Cell Biosci., 8*(59), 1–13.

Terns, M. P., & Terns, R. M., (2011). CRISPR-based adaptive immune systems. *Curr. Opin. Microbiol., 14*(3), 321–327.

Toro, N., Martínez-Abarca, F., & González-Delgado, A., (2017). The reverse transcriptases associated with CRISPR-Cas systems. *Sci. Rep., 7*(7089), 1–7.

Van, D. O. J., Westra, E. R., Jackson, R. N., & Wiedenheft, B., (2014). Unravelling the structural and mechanistic basis of CRISPR–Cas systems. *Nat. Rev. Microbiol., 12*(7), 479–492.

Wang, J., Li, J., Zhao, H., Sheng, G., Wang, M., Yin, M., & Wang, Y., (2015). Structural and mechanistic basis of PAM-dependent spacer acquisition in CRISPR-Cas systems. *Cell, 163*(4), 840–853.

Wang, R., & Li, H., (2012). The mysterious RAMP proteins and their roles in small RNA-based immunity. *Protein Sci., 21*(4), 463–470.

Wei, W., Zhang, S., Fleming, J., Chen, Y., Li, Z., Fan, S., Liu, Y., et al., (2019). Mycobacterium tuberculosis type III-A CRISPR/Cas system CrRNA and its maturation have atypical features. *FASEB J., 31*, 1496–1509.

Wei, Y., Chesne, M. T., Terns, R. M., & Terns, M. P., (2015). Sequences spanning the leader-repeat junction mediate CRISPR adaptation to phage in *Streptococcus thermophilus. Nucleic Acids Res., 43*(3), 1749–1758.

Westra, E. R., Van, E. P. B. G., Künne, T., Wong, S. P., Staals, R. H. J., Seegers, C. L. C., Bollen, S., et al., (2012). CRISPR Immunity relies on the consecutive binding and degradation of negatively supercoiled invader DNA by cascade and Cas3. *Mol. Cell, 46*(5), 595–605.

Wiedenheft, B., Sternberg, S. H., & Doudna, J. A., (2012). RNA-guided genetic silencing systems in bacteria and archaea. *Nature, 482*(7385), 331–338.

Wiedenheft, B., Zhou, K., Jinek, M., Coyle, S. M., Ma, W., & Doudna, J. A., (2009). Structural basis for DNase activity of a conserved protein implicated in CRISPR-mediated genome defense. *Structure, 17*(6), 904–912.

Yan, F., Wang, W., & Zhang, J., (2019). CRISPR-Cas12 and Cas13: The lesser known siblings of CRISPR-Cas9. *Cell Biol. Toxicol., 35*(6), 489–492.

Yoganand, K. N. R., Sivathanu, R., Nimkar, S., & Anand, B., (2017). Asymmetric positioning of Cas1-2 complex and integration host factor-induced DNA bending guide the unidirectional homing of protospacer in CRISPR-Cas type I-E system. *Nucleic Acids Res., 45*(1), 367–381.

Yosef, I., Goren, M. G., & Qimron, U., (2012). Proteins and DNA elements essential for the CRISPR adaptation process in *Escherichia coli. Nucleic Acids Res., 40*(12), 5569–5576.

Zegans, M. E., Wagner, J. C., Cady, K. C., Murphy, D. M., Hammond, J. H., & O'Toole, G. A., (2009). Interaction between bacteriophage DMS3 and host CRISPR region inhibits group behaviors of *Pseudomonas aeruginosa. J. Bacteriol., 191*(1), 210–219.

Zhang, Z., Pan, S., Liu, T., Li, Y., & Peng, N., (2019). Cas4 nucleases can effect specific integration of CRISPR spacers zhufeng. *J. Bacteriol., 201*(12), 1–16.

Zhu, Y., Klompe, S. E., Vlot, M., Van, D. O. J., & Staals, R. H. J., (2018). Shooting the messenger: RNA-targeting CRISPR-Cas systems. *Biosci. Rep., 38*(BSR20170788), 1–11.

CHAPTER 4

Computational Tools and Approaches for CRISPR/Cas Technology

HUMAIRA SHAH, AUQIB MANZOOR, and ASHRAF DAR*

Department of Biochemistry, University of Kashmir, Srinagar – 190006, Jammu and Kashmir, India

**Corresponding author. E-mail: ashrafdar@kashmiruniversity.ac.in*

ABSTRACT

The CRISPR/Cas endonuclease system, generally used as a defense mechanism in bacteria and archaea against invading phages, has emerged as a potential editing tool in genomics and epigenomics. Because of its accuracy, rapidity, and effectiveness, the system has gained a wide spectrum of applications in the fields of genetic engineering (GE), genome research, biotechnology, therapeutics, and agricultural sciences. The outcome of genome editing experiments with the CRISPR/Cas system largely depends on the nature of the target site and the design of sgRNA. However, due to poor cutting efficacy at desired target sites in general and off-target cleavage activity, the development and applicability of CRISPR/Cas systems get compromised. Therefore, a large number of computational tools have been developed that determine the off-target effects (OTEs) beforehand in a proposed experiment. A large number of computational tools have been developed for designing sgRNAs and prediction of cutting efficacy and binding specificity. A comparatively novel class of tools utilizes differences in repair results to forecast the outcome of CRISPR/Cas experiment. Furthermore, a different group of tools has been designed to help in the analysis of the editing results from sequencing data. This chapter brings together thorough

CRISPR/Cas-Mediated Genome Editing in Plants, Deepu Pandita, Anu Pandita (Eds.)

information about the optimized computational tools and approaches that facilitate the designing of CRISPR/Cas gene editing experiments at various stages from sgRNA synthesis to analysis of editing outcomes.

4.1 INTRODUCTION

The CRISPR/Cas, clustered regularly interspaced short palindromic repeats and CRISPR-associated system has been discovered recently and is the most widely used genomic perturbation platforms in molecular biology and biotechnology. Originally regarded as the prokaryotic immune system, CRISPR system works as a defense system in bacteria against invading phages (Mojica and Montoliu, 2016). During the last decade, the simplified versions of bacterial CRISPR system have been developed and vigorously used in genome editing functions. Since its initial discovery as a genome editing program, the functions of CRISPR toolbox have been diversified by discoveries of new CRISPR-associated endonucleases (Cas proteins and their variants) that have different target motif specifications, fidelities, and editing mechanisms (Clement et al., 2020). The toolbox has now been engineered in various ways to facilitate genomic, epigenomic, and RNA editing with unprecedented rapidity, ease, and accuracy in different organisms, especially in eukaryotic cells and organisms (Figure 4.1) (Jinek et al., 2013).

The most common genomic modifications carried out by using CRISPR tool kits include activation of gene expression (CRISPR activation [CRISPRa]), repression of gene expression (CRISPR interference [CRISPRi]), epigenome editing by fusion of Cas protein(s) to epigenetic modifiers (Thakore et al., 2016), and changing nucleotide sequence without leaving a double-strand break (DSB) (base editing and prime editing) (Anzalone et al., 2019; Rees and Liu, 2018). The genome editing and target specificity of CRISPR systems are programmed with the help of a short guide RNA (sgRNA) transcribed from the CRISPR locus and an RNA dictated Cas9 endonuclease. The working mechanism is based on the fact that Cas9 is recruited to the target sequence by a 20-nt sgRNA that is located upstream of PAM. PAM sequence is a short DNA motif of 2–6 nucleotides long (Zhou et al., 2016). The recruited Cas9 protein then produces a double-stranded break (DSB) in the target genomic sequence. The nick generated is error-prone and is further repaired by DNA repair mechanisms in the host cell (Mali et al., 2018; Jasin and Haber, 2016). The CRISPR system can be used to cause frameshift mutations or random indels (insertions/deletions) at desired genomic locus. These mutations arise by repairing of DSBs by host nonhomologous end joining (NHEJ) DNA repair machinery (Clement et al., 2020). Exogenous DNA templates are used to generate desired genetic insertions by exploiting host of homology-directed repair (HDR) pathway (Riesenberg and Maricic, 2018; Vartak and Raghavan, 2015).

New versions of CRISPR-based gene modifiers that edit the DNA sequence without DSBs have also been developed. Catalytically non-functional Cas9 (dCas9) is fused to effector domains of transcriptional activators (e.g., p65, VPR) and repressors (e.g., KRAB) to promote activation or repression (CRISPRa/i) respectively of a particular genetic locus (Qi et al., 2013). Fusing dCas9 with other epigenetic modifiers (e.g., DNMT3A) helps in blockade of transcriptional initiation or elongation through steric hindrance.

The application of CRISPR/Cas system demands a complicate workflow which involves selection of specific CRISPR/Cas system, designing of gRNA, transfection, establishment of single cell clone, screening of clone, and organized editing strategies. The different variants of Cas endonucleases from bacterial and archaea have been identified and functionally characterized. The information is available in various database and web portals (Tang et al., 2019). From the beginning, various bioinformatics and computational tools have played a significant role in the discovery and application aspects of CRISPR/Cas systems. The CRISPR/Cas experimental design and the investigation of its results also necessarily involve different programs, algorithms, and databases. These tools help to design gRNA for a given Cas protein that increases its hybridization to the specific genomic location beyond the target site. Designing a gRNA sequence that will direct the Cas to the target position while minimizing the binding of these sgRNAs to nonspecific positions with the same sequences in the genetic material of the interested organism is the most crucial step of the experiment (Torres-Perez, 2019). Another important factor that is taken into account while designing the sgRNA is the efficiency of sgRNA for the target site, which has been proven by various studies to be dependent on features of sgRNA sequence, genetics, and epigenetics of the target genome. Therefore, a large number of computational platforms has been put forward for developing and evaluating highly active sgRNAs (Liu et al., 2019). The CRISPR-associated computational tools are also used in the evaluation of the results of the CRISPR/Cas experiment. For example, in order to estimate the genomic effects of the introduced changes, next-generation sequencing is used, which is further analyzed by computational programs (Torres-Perez et al., 2019). A schematic outline of the CRISPR/Cas gene editing experiment and representative groups of involved computational tools and databases is given in Figure 4.1.

In the last few years, many attempts have been made to extend the use of the CRISPR/Cas system to different applications like disease treatment, drug development and biotechnology. To facilitate the application of CRISPR technology, different web-accessible computational and bioinformatics tools have been developed from time to time for all stages of CRISPR/Cas experiments. These web-based computational approaches have helped in remarkable technological advancements in CRISPR field (Tables 4.1 and 4.2).

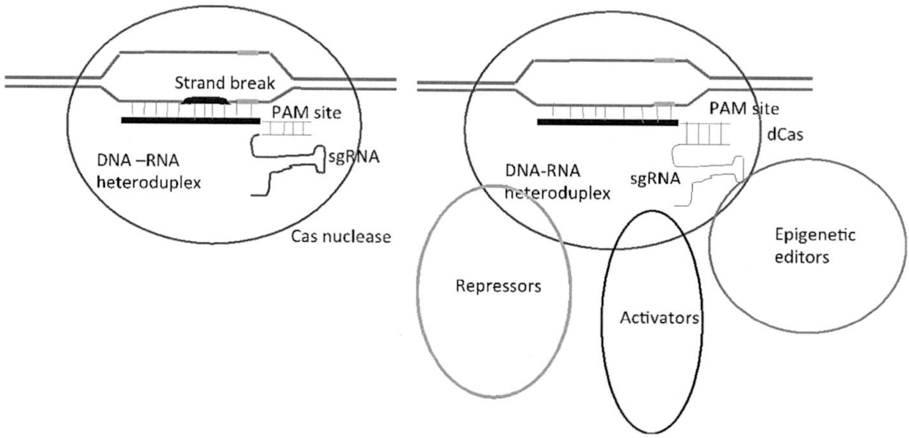

FIGURE 4.1 Working mechanism and function of CRISPR/Cas.

TABLE 4.1 List of CRISPR Computational Tools and Their Weblinks

Computational Tool	Web Link
CRISPRFinder	https://crispr.i2bc.paris-saclay.fr/Server/
PILER-CR	http://www.drive5.com/pilercr/
CRISPRDetect	http://crispr.otago.ac.nz/CRISPRDetect/
CRISPRTarget	http://bioanalysis.otago.ac.nz/CRISPRTarget/
MinCED	https://github.com/ctSkennerton/minced
MetaCRAST	https://github.com/molleraj/MetaCRAST
Crass	https://ctskennerton.github.io/crass/
metaCRT	https://omics.informatics.indiana.edu/CRISPR/
CRF	http://bioinfolab.miamioh.edu/crf/home.php
PatScan	https://patscan.seconDarymetabolites.org/
CRT	http://www.room220.com/crt/
CRISPRmap	http://rna.informatik.unifreiburg.de/CRISPRmap
HMMER	http://hmmer.org/
CRISPRdisco	https://github.com/CRISPRlab/CRISPRdisco
CRISPRCasFinder	https://crisprcas.i2bc.parissaclay.fr

TABLE 4.1 *(Continued)*

Computational Tool	Web Link
MacSyFinder	https://github.com/gem-pasteur/macsyfinder
CRISPRBank	http://bioanalysis.otago.ac.nz/CRISPRBank/
CRISPRone	https://omics.informatics.indiana.edu/CRISPRone/
CRISPRCasdb	https://crisprcas.i2bc.parissaclay.fr/
CHOPCHOP	https://chopchop.cbu.uib.no/
E-CRISP	http://www.e-crisp.org/
Cas-OFFinder	http://www.rgenome.net/cas-offinder/
CCTop	https://crispr.cos.uni-heidelberg.de/
SgRNAScorer1.0	https://sgrnascorer.cancer.gov/sgRNAScorerV1/
sgRNAdesigner	https://portals.broadinstitute.org/gpp/public/analysis-tools/sgrna-design
CRISPOR	http://crispor.org/
WU-CRISPR	https://bio.tools/wu-crispr
CRISPRScan	https://www.crisprscan.org/
CROP-IT	http://cheetah.bioch.virginia.edu/AdliLab/CROP-IT/homepage.html
GuideScan	http://www.guidescan.com/
CASPER	https://github.com/TrinhLab/CASPER
DeepCpf1	http://deepcrispr.info/
CRISPRpred	https://github.com/khaled-buet/CRISPRpred
CRISPRater	http://crispr.cos.uni-heidelberg.de/
Azimuth	https://github.com/MicrosoftResearch/Azimuth
PROTOSPACER	http://www.protospacer.com/
SSC	http://cistrome.org/SSC/
CRISPR-ERA	http://crispr-era.stanford.edu/
Crisflash	https://github.com/crisflash
Flashfry	https://aaronmck.github.io/FlashFry/
DeepCRISPR	http://www.deepcrispr.net/
SgRNACas9	http://www.biootools.com/
CRISTA	http://crista.tau.ac.il/
CasOT	http://eendb.zfgenetics.org/casot/
FORECasT	https://partslab.sanger.ac.uk/FORECasT
SPROUT	https://zou-group.github.io/SPROUT
InDelphi	https://www.crisprindelphi.design/
Lindel	https://shendurelab.github.io/Lindel/
CRISPResso2	https://crispresso.pinellolab.partners.org/
Cas-Analyzer	http://www.rgenome.net/cas-analyzer/
CRISPR-GA	http://crispr-ga.net

TABLE 4.1 *(Continued)*

Computational Tool	Web Link
TIDE and TIDER	https://tide.nki.nl/
CRISPR-GE	http://skl.scau.edu.cn
CRISPR-PLANT	https://www.genome.arizona.edu/crispr2
CRISPR-P	http://crispr.hzau.edu.cn/CRISPR2/
EuPaGDT	http://grna.ctegd.uga.edu
CRISPRscan	https://www.crisprscan.org
CRISPETa	http://crispeta.crg.eu
Azimuth	https://github.com/MicrosoftResearch/Azimuth

TABLE 4.2 Chronology of the Evolution of Web-based CRISPR/Cas System Tools

2012	CRISPR Cas9 was first put into use for genome editing
2013	Crispr.mit.edu
	Zinc Finger Targeter
	Cas9 guide RNA design
2014	CRISPR-P
	CHOPCHOP v1
	E-CRISPR
	FlyCRISPR
	Microhomology
	CRISPR-PLANT
2015	Cas-Designer
	CCTop
	EuPaGDT
	CRISPR-ERA
2016	CRISPOR
	CHOPCHOP v2
2017	CRISPETa
	CRISPR-P 2.0
	CRISPR-GE
	CRISPR-FOCUS
2018	CRISPR-PLANT v2
	BE-Designer
2019	CHOPCHOP v3
	FORECasT
	Lindel
2020	PnB designer

4.2 CRISPR/CAS SYSTEM IDENTIFICATION

CRISPR RNAs (crRNAs) and Cas effectors, completely regulate the specificity and genome editing efficacy of CRISPR/Cas systems, thus recognition and categorization of CRISPR/Cas system which includes various classes of crRNAs and Cas endonucleases is very important requirement for ensuring desired functions.

4.2.1 TOOLS FOR IDENTIFICATION OF CRISPR ARRAYS

The main part of the CRISPR/Cas-engineered system is crRNA which is synthesized from CRISPR arrays. Thus, identification of potent CRISPR arrays necessarily dictates the engineering precision in the system. Various computational tools and algorithms are used to identify CRISPR arrays with the help of sequence information. PatScan, a computational tool designed much before the discovery of CRISPR/Cas system, identifies the regions similar to the given sequence (Dsouza et al., 1997). Its application is limited by the fact that it recognizes common repeats not specific for CRISPRs, and thus is unable to differentiate the spacers and repeats in the complete CRISPR array. Various CRISPR recognizers were designed later on such as, CRISPRFinder; PILER-CR. CRISPRFinder identifies the maximal repeats that are holded by the non-repeating sequences with the same length (Grissa et al., 2007; Edgar, 2007; Dsouza et al., 1997). PILER-CR is a command-line program which recognizes presumed CRISPR arrays by screening common hits of the given genome to itself and filters them using sequence resemblance, length distribution, and conservation. CRISPRDetect is a web-based command line program designed on k-mer and extension strategy (Biswas et al., 2016). It is more precise in recognition of short and degenerated repeats by scanning the alternative repeats under a low identity threshold in long spacers. However, there is the probability of false segmentation of the long elemental CRISPRs. The anticipated spacers are directly used by CRISPRTarget to predict target regions. CRISPR forecasting has been implied to metagenomic data and thus tools like MinCED, MetaCRAST, Crass, and metaCRT were designed (Skennerton, 2016; Moller and Liang, 2017; Skennerton, 2013; Rho et al., 2012). MinCED, Crass, and MetaCRT apply the *de novo* identification. MinCED identifies spacers in reads without assembly and Crass assembles the reads into arrays. In comparison, metaCRT combines the *de novo and* reference-based identification (Rho et al., 2012). Another reference-based method, MetaCRAST investigates the pair with given sequences that are recognized by different tools like CRF, CRISPRFinder, PILER-CR (Moller and Liang, 2017). Table 4.3 contains the comparison of different representational tools for identifying CRISPR arrays.

4.2.2 TOOLS FOR IDENTIFICATION OF CAS PROTEIN

New tools have been designed to include Cas effector nuclease identifier to increase the classification capability and help in the efficient detection of CRISPR/Cas system. These programs correlate the given Cas nuclease with the sequences in the given protein database and identify the supposed Cas with the help of homologous sequence searcher such as HMMER and BLAST (Camacho et al., 2009; Eddy, 2011). CRISPRmap uses CRISPRFinder and CRT for CRISPR array recognition and HMMER for annotating Cas enzyme (Lange et al., 2013). Another tool, CRISPRdisco incorporates MinCED and BLAST for same functions (Crawley et al., 2018). CRISPRCasFinder algorithm uses CRISPRFinder for CRISPR array recognition and tool MacSyFinder for Cas identification (Abbey et al., 2014). There are other tools that collect the anticipated Cas nucleases and CRISPRs such as CRISPRCasdb, CRISPRBank, and CRISPRone (Naito et al., 2015; Zhang and Ye, 2017; Pourcel et al., 2019).

4.3 GUIDE RNA (GRNA) DESIGN AND DEVELOPMENT

Theoretically, the Cas enzyme screens the PAM sequence, and single guide RNA (sgRNA) identifies the desired loci and turns the endonuclease function of Cas on to cut at these positions. But the cutting potency differs widely in various target DNA sites and cell lines, indicating that various factors affect the binding and cleaving ability of the Cas-sgRNA complex. Various researches have shown that these include the features of gRNA sequence (composition of sequence, position of nucleotides, GC content), genomic, and epigenomic features (accessibility to chromatin, expression of gene) and energetics properties (secondary structure of RNA, free energy, melting temperature,). Thus, large numbers of computational programs have been designed for developing greatly efficient and precise guide RNAs (gRNAs) (Tables 4.4 and 4.5).

TABLE 4.3 Main Tools for Identifying CRISPR Arrays

Tool	Input	Output	Platform	Disadvantage	Advantage
PatScan	DNA/ protein sequences	Repeat sequences	Web server	• Cannot differentiate CRISPRS from other types of repeats	• Provide web server • Predicts genomic patterns
CRISPRFinder	DNA sequences	Repeat and spacer sequences	Web server	• Do not take repeat mutation into account • Detection of small and degenerate repeats is poor	• Provides reliable CRISPRS
PILER-CR	DNA sequences	Repeat sequences Cluster by similarity and position	Standalone program	• Cannot filter out tandem repeat sequences • Not user friendly	• Provide Classification of CRISPRS • Handle indels in repeats • Execute rapidly
CRT	DNA sequences	Repeat and spacer sequences	Standalone program	• Detection of small and degenerate repeats is poor • Not user friendly	• Use simple data structure • High reliability
CRISPRDetect	DNA sequences or species name	Repeat and spacer sequences Mutations Potential Cas genes	Web server and Standalone program	• Not fast when dealing with large query set mis-split large query set into short arrays	• Sensitive to short and degenerate arrays • Provide additional information such as array direction and variation

TABLE 4.4 Representative Computational Tools for sgRNA Design

Tool	Data Source	Nuclease Supported	Input Requirement	On Target Indication	Off-Target Specificity	Genome Specificity
CHOPCHOP	Any species	Custom PAMs	Genomic region Gene ID Gene symbol	ML	Yes	Yes
E-CRISP	55 kinds of vertebrates, invertebrates, plants	Custom PAMs	RefSeq Gene ID Genomic region	ML	Yes	Yes
Cas-OFFinder	Any species	16PAMs	CrRNA sequences	PR	Yes	Yes
CCTop	Any species	11 PAMs	DNA sequences	ML	Yes	Yes
sgRNA Scorer 1.0	Any species	SpCas9 StCas9	DNA sequences	Rule set 2.0	No	No
sgRNA designer	Any species	SoCas9 SaCas9	DNA sequence Transcript ID Gene ID Gene symbol	Rule set 2.0	CFD	No
CRISPOR	Any species	9 PAMs	DNA sequence Gene region	ML	Yes	Yes
WU-CRISPR	Any species	SpCas9	Gene symbol Gene ID DNA sequence	ML	Yes	Yes
CRISPR-ERA	9 kinds of vertebrates and invertebrates	SpCas9	–	FR	Yes	–

TABLE 4.4 *(Continued)*

Tool	Data Source	Nuclease Supported	Input Requirement	On Target Indication	Off-Target Specificity	Genome Specificity
CRISPR scan	24 kinds of vertebrates, invertebrates, and plants	SpCas9 LbCpf1 AsCpf1	DNA sequence Gene ID Gene symbol	ML	Yes	Yes
CROP-IT	Human and mouse	SpCas9	Gene region	FR	Yes	Yes
Guide scan	Six kinds of vertebrates and invertebrates	LbCpf1 SpCas9 AsCf1	Gene region Gene symbol	Rule set 2	CFD	Yes
CASPER	Any species	Custom PAM	DNA sequence	CASPER	CASPER	No

TABLE 4.5 Computational Tools for Evaluation of Guide RNA Efficacy

Tool	Nuclease Supported	Input Source	PAMs	Basic Features
E-CRISP	Cas9	55 kinds of vertebrates, invertebrates, plants	User-defined	SC, EF
CASPER	Cas9, cpf1	Any species	TTTN, NGG	SC
DeepCpf1	Cpf1	Human	User-defined	SC, EGF
CRISPRpred	Cas9	Mouse, human	User-defined	SC, EP
CRISPRater	Cas9	Human	NGG, NRG, NGA	SC, GF
WU-CRISPR	Cas9	Human, mouse	NGG	SC, EP
CRISPRscan	Cas9, cpf1	Zebrafish	NGG	SC
Azimuth	Cas9	Human, mouse, rat	NGG, NNGRR	SC, GF, EP
Evaluate CRISPR	Cas9	Drosophila	–	SC
SSC	Cas9	Mouse, human	NGG	SC

4.3.1 OVERVIEW OF SGRNA DESIGN TOOLS

Because of significant differences in the efficiency of sgRNAs, it is important to understand the relation between the sgRNA sequence features and the efficacy of Cas enzyme cutting. sgRNA designers can be divided into three major categories, machine learning-based, pattern recognition-based, and feature recognition-based.

4.3.1.1 MACHINE LEARNING-BASED TOOLS FOR SGRNA

These tools use mathematical approach for designing optimal sgRNA. These tools mostly have differences in algorithms and input information. For example, to predict the on-target efficiency of sgRNA, Rule Set 1 model was designed with the help of data obtained from 1,841 sgRNA for three human and six mouse genes (Doench et al., 2014). In this model a SVM mode is used to predict the value, and a supervised learning method analyzes data in a general linear manner. Rule Set 1 is specifically helpful to understand guide characteristics that affect the cutting ability of Cas enzyme. The score from this model is in the range of 0–1, where a higher value indicates the effectiveness of sgRNA. This model has been used in various tools like CHOPCHOP, CRISPETa, E-CRISP, CRISPOR, CLD, and PROTOSPACER (Montague et al., 2014; Heigwer et al., 2014,

2016; Labun et al., 2016; MacPherson and Scherf, 2015; Haeussler et al., 2016; Pulido-Quetglas et al., 2017). Later on, Rule set 2 was proposed by including more features like counts of position independent nucleotides and target site position in the gene to improve the model (Doench et al., 2015). This modified model is used in several tools, including Guide Picker, Guide Scan, CRISPOR, and CHOPCHOP (Perez et al., 2015; Montague et al., 2015; Labun et al., 2016; Haeussler et al., 2016; Hough et al., 2017). The designing of sgRNAs for CRISPR knockout and CRISPRa/i experiments make use of these optimized models.

Important differences in nucleotide preference of sgRNA target positions in CRISPRa/i and CRISPR KO were revealed by screening these libraries in mammals for various CRISPR-based editing experiments (Xu et al., 2015). Elastic Net algorithm has found selective bias for cytosine for sgRNA and has been put in use to develop models for CRISPRa/i and CRISPR KO. This model has also been applied in various platforms like SSC, CRISPR-FOCUS, CHOPCHOP, CRISPOR, CRISPR-DO (Heigwer et al., 2014; Cao et al., 2017; Labun et al., 2016; Montague et al., 2015; Ma et al., 2016). The sgRNA activity and stability *in vivo* was found to be dependent on its different characteristics in the zebrafish genome where it was found that activity of sgRNA increased with increase of guanine content and decrease of adenine content (Haeussler et al., 2016). The randomized logistic regression model was used to choose important features that decide the efficacy of sgRNA and a linear regression model is used to forecast the activity of a known sgRNA and integrate it into the online tool CRISPRscan (Haeussler et al., 2016).

To evaluate sgRNA on-target activity, online tool called sgRNA scorer has been developed that is based on SVM model). CHOPCHOP and CRISPOR used this model for prediction of on-target efficiency. A modified model of the sgRNA scorer 2.0 tool was established that increased the specificity, prediction ability and additional prediction for multiple Cas9 orthologs, for example, AsCpf and SaCas9 (Chari et al., 2017).

4.3.1.2 PATTERN RECOGNITION-BASED TOOLS FOR SGRNA

These tools depend on base-pairing principle. These databases look for a stretch of nucleotides composed of a precise PAM sequence and approximately 20-bp candidate sgRNA that is complementary to the given sequence in a particular genome. CRISPR seek, flyCRISPR, SSFinder, and GT-Scan take pattern recognition into account (Zhu et al., 2014;

Gratz et al., 2013; Upadhyay and Sharma, 2014; O'brien and Bailey, 2014). Another tool, Crisflash has increased the accuracy of sgRNAs by including given somatic mutation information into pattern matching (Jacquin et al., 2019).

4.3.1.3 FEATURE RECOGNITION-BASED TOOLS FOR SGRNA

Binding of sgRNA to the target site and Cas endonuclease's cutting depends on multiple features, such as genomic context, chromatin accessibility, gene position and expression, RNA-DNA thermodynamics, and the presence of nucleotide bulges (Sledzinski et al., 2020). The presence of nucleosomes inhibits Cas9 target site cutting and DNase I hypersensitivity sites and epigenome markers change sgRNA binding efficiency. Various predictive models have been designed to take these features into account, e.g., DeepCRISPR, CRISTA, uCRISPR or CROP-IT, (Abadi et al., 2017; Singh et al., 2015; Chuai et al., 2018; Zhang et al., 2019). An R package termed as PredictSGRNA was designed to evaluate about efficacy of sgRNA by acquiring information from oligonucleotide pattern and nucleosome occupancy models. This model methodically evaluates various features obtained from different nucleic acids, thermodynamics, and secondary structure models on CRISPR datasets (Kuan et al., 2017). CRISTA (CRISPR Target Assessment) algorithm which is based on machine learning pattern and determines the tendency of a specific gene site to be bound by specified sgRNA. It depends on a Random Forest regression algorithm, and allows the evaluation of the characteristics that define the change in cleavage efficiency. The model accounts for the possibility of buldges within the DNA site or sgRNA and includes a large number of characteristics that are specific to the genomic content, attributes that determine thermodynamics of the sgRNA, and features concerning the combined similarity between the sgRNA and the target genome (Zhang et al., 2019). *DeepCRISPR*, an extensive computational tool that uses *deep unsupervised representation learning* technique, unifies sgRNA efficiency and specificity prediction into one platform. It automates the identification of underlying sequence and epigenetic characteristics that may influence sgRNA knockout efficiency in a data-driven way. The model learns which features are important in standardized sgRNA construction using few training samples. The identified features can be used for efficient sgRNA design (Chuai et al., 2018).

CRISPR editing efficacy is to a large extent dependent on energetics correlated with the forming of complex between the target DNA, sgRNA, and Cas enzyme (Szczelkun et al., 2014; Wu et al., 2014; Xu et al., 2017). The differences among energetic models can be eliminated by observing energy parameters associated with RNA–RNA, DNA–DNA, and DNA–RNA duplexes. CRISPRoff and CRISPRspec, help in correct off-targeting analysis of gRNA selections and help to use the CRISPR/Cas9 system with higher efficacy. CRISPRoff assigns values to predicted off-targets and CRISPRspec measures the on-target specificity of the gRNA. These tools provide evidences that support the fact that binding energies play a very important role in an inverse relationship between the on-target cleavage efficiency and specificity of the CRISPR system. CRISPRpred considers the locations of nucleotides and secondary structures of sgRNAs to analyze the cleavage efficiency, of sgRNAs (Doench et al., 2014; Rahman and Rahman, 2017). A free energy bioinformatic model called uCRISPR has been designed for analyzing the editing efficiency of Cas9 as well as off-target effects (OTEs). This tool takes into account three factors to analyze the Cas9 editing efficacy: the free-energy change of formation for the R-loop structure under Cas9 binding, the native-structure selection from the conformational ensemble for the complete gRNA, and the comparative activities of different PAMs (Zhang et al., 2019).

4.3.2 TOOLS FOR PREDICTION OF CRISPR SPECIFICITY

Cas9 nucleases may cut at undesired genomic locations and give rise to deleterious mutations because of sgRNA recognizing DNA motifs with few mismatches (Zhang et al., 2015). Experimental methods which include whole-genome sequencing, e.g., DISCOVER-seq, HTGTS, GUIDE-seq, Digenome-seq, BLESS, CIRCLE-seq, and the modified VIVO strategy can be used for off-target identification (Tsai et al., 2015; Wienert et al., 2015; Kim et al., 2015; Ran et al., 2015; Frock et al., 2015; Akcakaya et al., 2018). These methods are limited by cost and labor intensiveness. To overcome those limitations, new *in silico* tools have been designed over the years (Table 4.6).

TABLE 4.6 Computational Tools for Evaluating Guide RNA Specificity

Tool	Nuclease Supported	Important Features	Approach
CasOT	Cas9	Infinite mismatch number, paired-gRNA mode	Alignment based
CRISTA	Cas9	MLA, SC, EF	Scoring based
CCTOP	Cas9, cpf1	Score based on mismatch numbers	Scoring based
Cas-OFFinder	Custom	Infinite mismatch number, GPU acceleration, web support	Alignment based
Flashfry	Custom	infinite mismatch number, various on target and off-target values, annotation	Alignment based
CFD	Cas9	Enables noncanonical PAM	Scoring based
DeepCRISPR	Cas9	Deep learning, sequence composition and epigenetic features	Scoring based
SgRNA Cas9	Cas9	Risk evaluation, max 5 mismatches, paired Grna mode	Alignment based
Crisflash	Cas9	Variant data support, unlimited mismatch number	Alignment based
CRISPRoff	Cas9	Energetic property and sequence composition	Scoring based

Alignment-based methods have been used to foresee the specificity of sgRNAs which are alienated to a particular genomic site and off-target regions and positions are rejected, based on conventional or specially formulated algorithms. This method is mostly used in identifying all possible nonspecific positions *in silico.* The possible nonspecific positions are recognized by alienating sgRNAs to the given gene depending on the similarity of sequences. Two conventional databases that are used for alignment of sequences and are capable of nonspecific detection are Bowtie and BWA (Langmead et al., 2009; Li and Durbin, 2009; Haeussler et al., 2016; Wilson et al., 2018). Some models make use of these tools, and adjust the parameters according to the requirements for prediction of non-specific (off-target) sites. Bowtie is utilized by CCTop to find nonspecific sites by first recognizing PAM sites, then with adjusted parameters of Bowtie, similarity, and dissimilarity in protospacer are searched (Stemmer et al., 2015). CCTop permits only five or fewer mismatches in protospacer because DSB induction is stopped by more than five mismatches. CHOPCHOP also uses Bowtie with -v and -L parameters for probing gRNA core regions to recognize off-target sites (Montague et al., 2014). CRISPOR utilizes BWA to recognize all possible nonspecific sites and can detect all verified off-targets (Haeussler et al., 2016). These conventional tools suffer few limitations which include the inability to recognize small PAMs. These tools were established for use in next-generation sequencing. Also, these tools are inappropriate for recognition of all possible off-targets sites as they allow restricted mismatches in their seed region. Similarly, Bowtie is not able to detect all potential off-targets but is able to detect off-targets up to one mismatch only (Doench et al., 2016; Langmead and Salzberg, 2012).

Besides these conventional tools, various improved models and programs have been designed for detecting non-specific sites. Cas-OFFinder is a prominent tool for the detection of potential off-targets and its benefit is that it can detect infinite mismatches, length of gRNAs or type of PAM. It is able to foresee off-target sites with a single base pair insertion or deletion (Bae et al., 2014). It has been found as a better option for recognizing all possible off-target sites with any Cas endonuclease. CasOT, another tool not only helps in finding possible off-targets with up to six mismatches in a given sequence, but also helps in finding Cas9 on-target sites from a given genomic sequence. It also helps in finding out if the off-targets are present within a coding region (Xiao et al., 2014). sgRNAcas9 makes use of high-speed mapping tool SeqMap to recognize off-targets and categorize all sites in three groups to create a data output of leading contender gRNAs (Jiang and Wong, 2008; Xie et al., 2014). Crisflash and Flashfry are two novel alignment based computational algorithm. Crisflash quickly outline gRNAs and improve gRNA precision by using input variable data (Jacquin et al., 2019). Crisflash makes use of tree-based algorithms. Flashfry briskly explores off-target sites and comes up with data like marking of off-target sites, on- and off-target scores, and GC content for contender gRNAs (McKenna and Shendure, 2018). It is a very useful tool because of its rapidity and comprehensive results (Doench et al., 2014, 2016; Hsu et al., 2013).

MIT score was put forward for the analysis of off-targets of gene editing in which more than 700 gRNA forms and insertion/deletion mutations brought about by SpCas9 (Streptococcus pyogenes Cas9) at more than 100 estimated genomic off-target loci was analyzed (Hsu et al., 2013). A weight matrix is calculated for the off-target determining ability for every gRNA, by assessing the contribution of individual mismatch sites and numbers in a desired 20 base pairs. MIT score accounts for greater than 50% of the deviation in the cleaving rate. Many CRISPR sgRNA design algorithms like CRISPOR and CHOPCHOP have integrated with MIT score (Haeussler et al., 2016; Labun et al., 2016). CFD, cutting frequency determination, second recognized score for off-target assessment includes PAM features in their scoring criteria to counter the sgRNAs binding to genomic loci with non-conventional PAMs like NAG and NCG (Doench et al., 2016). The interconnection between sgRNAs and off-targets is also examined with both sgRNAs with mismatches and insertion/deletion in target sequences is incorporated. CFD score corroborated with GUIDE-seq performs better than MIT score and is approved by CRISPscan, GuideScan, and CRISPOR (Moreno-Mateos et al., 2015; Perez et al., 2017; Haeussler et al., 2016).

Applied models are not able to analyze the off-targets successfully, as they only examine trivial aspects, while on the contrary, learning-based methods use an amalgam of diverse characteristics to assemble composite models and predict off-targets better. Based on machine learning, different models have been developed for off-target detection. CRISTA (CRISPR Target Assessment) predicts off-targets using BWA tool and applies amalgam of features like chromatin structure, GC content, PAM type, DNA methylation, RNA secondary structure for better cleavage prediction (Abadi et al., 2017). CRISTA performs better than MIT, CFD, and CCTop. Linear regression model known as Elevation developed by Doench lab predicts off-target function of sgRNAs. This model takes both sequence features and accessibility to chromosome into observation (Listgarten et al., 2018). Scores for each off-target site and an accumulative score for gRNAs are predicted by Elevation. This method works better than CCTop, CFD, MIT. But the problem with this tool is that it can determine the off-targets for human exome only and cannot be used to determine off-targets in genomes of other organisms. DeepCRISPR another computational tool that predicts both sgRNA on-target and off-target sites by unifying them in a single framework using deep learning (Chuai et al., 2018). It recognizes all practicable sequences and epigenomic features that can influence the gRNA knock-out accuracy by taking information from large data. This tool outperforms all other off-target prediction tools (Doench et al., 2016; Hsu et al., 2013; Singh et al., 2015; Stemmer et al., 2015).

4.4 PREDICTIONS TOOLS

A new class of algorithms have been developed that use the knowledge that DNA repair is a deciding factor in CRISPR/Cas experiment success. These platforms aid in estimating Cas edit outcomes by analyzing microhomologies and by understanding the differences in the repair results (Table 4.7). The repair pathways particularly involve non-homologous DNA end joining pathway (NHEJ) (Ghezraoui et al., 2014), which is prone to error and may result in random insertion and deletion mutations (INDELs), homology-directed repair pathway (HDR) (Ye et al., 2015), which depends on a donor DNA template that is homologous to the nucleotide sequence surrounding the DSB site to achieve the correct edit or precise correction; and micro-homology-mediated end joining (MMEJ) (Sakuma et al., 2016), where the single-stranded ends generated by the nuclease are base-paired at the micro-homologies that are present both before and after the DSB. These findings were considered by the designers of bioinformatic algorithms concerned with the thorough characterization of repair outputs and upgradation of the experimental setup. Two main models were used to analyze the result of the mutation. First, some MLA-based algorithms, such as InDelphi, FORECasT, and Lindel attained a large prediction on the distribution of mutations by using the features of sequence (Shen et al., 2018; Allen et al., 2018; Chen et al., 2019). InDelphi, a machine learning model comprises of library of more than 2000 Cas9 gRNAs paired with DNA target locus that anticipates genotypes and rate 1- to 60-base-pair deletions and 1-base-pair insertions with high precision in five human and mouse cell lines. It predicts that 5–11% of Cas9 gRNAs targeting the human genome yield a single genotype comprising greater than or equal to 50% of all major editing products and accurately predicts the spectrum of cut-site proximal genotypic products resulting from double-stranded break repair at a target DNA site of interest. The ability to predict Cas9-mediated products enables new precision genome editing research applications and facilitates existing applications, such as performing efficient bi-allelic gene knockout and predicting end-joining by-products of HDR. FORECasT, another online database for Cas9 edit outcome analysis uses artificial constructs with a gRNA that is expressed downstream the human U6 promoter and desired DNA sequence that is complementary to the sgRNA bordered by varying nucleotide sequences. A large number of these vectors were transfected into cell lines continuously expressing Cas9, and leads to alterations of the given target loci. An online tool named SPROUT was designed using an endogenous model. In this tool, 1,656 genomic sites were aimed in primary T lymphocyte cells separated from 18 donors. This tool used an RNP complex that is composed of Cas9 protein, crRNA, and the tracrRNA (Leenay et al., 2018). In these experiments, cut-site-proximal repair outcomes were identified following PCR amplification of the desired loci with their surrounding areas and deep-sequencing, which shows a sequence-dependent mutation profile. Finally, sequencing data information was used to improve tools designed by the researchers to accurately predict Double Stranded Break repair results, and lastly, the precision of prediction

TABLE 4.7 Comparison of Prediction Tools

Tool	Indel Frequency	Average Indel Length	Sequence of Prediction	Cell Type	Input Requirement	Alignment of Predicted Sequences	Inframe Frequency
FORECasT	Yes	No	Yes	No	gRNA, PAM,	Yes	Yes
SPROUT	Yes	Yes	No	No	Target DNA sequences,	No	No
In Delphi	Yes	No	Yes	HEK293, HCT116, K562, mESC, U2OS	PAM sequence	Yes	Yes

was proved. Each tool displayed large accuracy of repair profile prediction for a given cell type.

4.5 OUTCOMES OF ANALYSIS TOOLS

There are different laboratory approaches and accessory computational tools available for the assessment of the effectiveness of target alteration, which is an important step in the usage of CRISPR technology. A large number of algorithms and databases have been designed to aid the posterior (post-experiment) CRISPR/Cas9 edit result analysis, in which special importance is given to off-target assessment (Table 4.8). The post-experiment assessment programs allow the evaluation and imaging of CRISPR/Cas9 results in table form with appropriate graphs. These tools depict the rate of insertions/deletions and repair pathways either non-homologous end joining (NHEJ) or homology-directed repair, facilitating the data reading (Sledzinski et al., 2020).

The tracking of indels by decomposition (TIDE) algorithm analyzes gene edits by quantifying Sanger sequencing results of the target site giving a complete figure of all indels that took place in a sample (Brinkman et al., 2014). The input data to the algorithm are sequencing traces generated from PCR products of genomic DNA, one from Cas-treated cells and other from untreated cells, which are easily uploaded to the website. The result of editing experiments is observed by analysis of the peaks, which indicates what proportion of the cells has integrated the desired change. For knockout experiments, random pattern of overlapping peaks indicates the presence of a large number of repair products which gives an estimation of the ratio of frameshifted to in-frame repair products, in addition to other results (Hanna and Doench, 2020). TIDER—Tracking of Insertions, Deletions, and Recombination events is an advanced version of TIDE and helps to evaluate the efficiency of desired target genome engineering and determines the rate of homology-directed repair incidents (Brinkman et al., 2018). These programs are simple, require readily available laboratory instrument and are comparatively economical. Another online algorithm developed to measure base editing efficiency from fluorescence-based Sanger sequencing, is termed EditR which requires single Sanger sequencing file and gRNA sequence as input. This tool also helps to understand the position and type of base editing (Kluesner et al., 2018).

Cas-Analyzer, a faster JavaScript-based platform available online for analysis works completely as web browser and profoundly reduces time by

TABLE 4.8 Post Prediction Tools

Tool	Type of Analysis	Output	File Type	Base Editing Experiment	Template Mediated Editing	Supported Nuclease
CRISPResso2	NGS	Mutation length and locations NHEJ and HDR rate Sequence alignment with reference Allele-specific analysis	FASTQ	Yes	Yes	Cas9, Cpf1
Cas-analyzer	NGS	Mutation length and locations NHEJ and HDR rate-sequence alignment with reference	FASTQ	No	Yes	NmCas9 StCas9 SaCas9 SpCas9 LbCpf1 AsCpf1 CjCas9
CRISPR-GA	NGS	Mutation length and locations NHEJ and HDR rate	FASTQ	No	Yes	Cas9
TIDE	Sanger sequencing	Mutation length and locations NHEJ and HDR rate	ABI	No	Yes	SpCas9 SaCas9 St1Cas9 NmCas9 AsCpf1 FnCpf1 LbCpf1

TABLE 4.8 *(Continued)*

Tool	Type of Analysis	Output	File Type	Base Editing Experiment	Template Mediated Editing	Supported Nuclease
TIDER	Sanger sequencing	Mutation length and locations NHEJ and HDR rate	ABI	No	yes	NmCas9 SpCas9 FnCpf1 SaCas9 St1Cas9 AsCpf1 LbCpf1

avoiding step of data upload to server. It also supports different nucleases used by researchers, which include the paired nickases. It provides web interfaces for users to input sequencing data, as well as customizing the important interpretation parameters. The tool is highly useful for new users because of its graphical user interface that guides them through the appropriate study parameters (Park et al., 2017). Another tool that helps in the interpretation of sequencing data is CRISPResso2, a modified form of the CRISPResso program (Clement et al., 2019; Pinello et al., 2016). CRISPResso2, performs editing rate analysis and helps in allele-specific analysis of data by alienating reads to different control sequences. It also computes statistics individually for different alleles. CRISPResso2 is widely used in base editing experiments because it thoroughly analyzes base substitutions. The main shortcoming of this database is the limitation of input information that can be uploaded for interpretation (Sledzinski et al., 2020). CRISPR-genome analyzer (CRISPR-GA) is a platform that uses next generation sequencing data to analyze and identify insertion/deletions and homologous recombination (HR) incidents at the desired locus (Güell et al., 2014). CRISPR-GA analysis comprises five distinct stages, which include quality control, mapping, indel calling, HR, and NHEJ estimation and graphical representation, which is used to calculate the efficacy of editing, the editing rate of homology direct repair (HDR) and NHEJ and to generate charts with insertion/deletion positions and lengths.

Databases and algorithms are also used in the interpretation of large-scale, pooled CRISPR screening data. MAGeCK and BAGEL are the platforms that have been designed for these functions. They use a command line interface, and thus cannot be used by less experienced users (Sledzinski et al., 2020). MAGeCK, was one of the first and most demanded tools for CRISPR screen interpretation and requires raw sgRNA input reads to perform its analysis (Li et al., 2014). BAGEL was developed to use preceding information describing essential and nonessential genes as a gold standard reference set for the identification of new important genes. This tool results in high sensitivity and specificity for identifying new essential genes. sgRNA raw read counts or LFC of guides is utilized as input data in this computational tool (Hart and Moffat, 2014). Recently, designed tools that provide internet-based and convenient interfaces for the deconvolution of data of a large number of perturbations collectively, such as CRISPR AnalyzeR, PinAPL-Py, and CRISPRcloud2, have been designed for users with no programming experience (Jeong et al., 2018; Spahn et al., 2017; Winter et al., 2017). These tools require uploading FASTQ files with Next Generation Sequencing data and a file of gRNA library to start the program. These tools statically analyze the input and give the output results which include thorough analysis and imaging of data besides quality control.

4.6 CONCLUSION

The implementation of CRISPR/Cas as a gene editing tool requires the Cas-sgRNA complex activity, the single-guide RNA (sgRNA), formation of complex between gRNA and the trans-activating crRNA (tracrRNA). The complex guides the Cas protein towards particular gene region to be either deleted or modified that depends on the DNA repairing pathway that follows. One important concern for the application of the CRISPR/Cas system is the off-target activities due to the incomplete match between the sgRNA and random sequences resembling the assumed one, which can lead to deleterious gene changes. Bioinformatics is indispensable in understanding the complicated working mechanism and expansion of CRISPR systems. Every stage of CRISPR/Cas systems (CRISPR/Cas system recognition and gRNA design) and post-experimental evaluation (post-experiment prediction, and analysis) is assisted by different computational programs (Figure 4.2). But, none of the tools can claim to provide a complete solution for their respective assignment, and there is no complete approach that can predict, classify, and completely characterize CRISPR/Cas systems. Thus, users have to combine a few programs to achieve their individual objectives, which require some information about previously available tools, their working mechanism, and drawbacks. Thus, there is demand for more optimized and efficient bioinformatic tools at each step of CRISPR/Cas design, which requires information about genetic factors and thermodynamics of DNA and RNA along with the activity of Cas-sgRNA complex which will take CRISPR/Cas system to an advanced level.

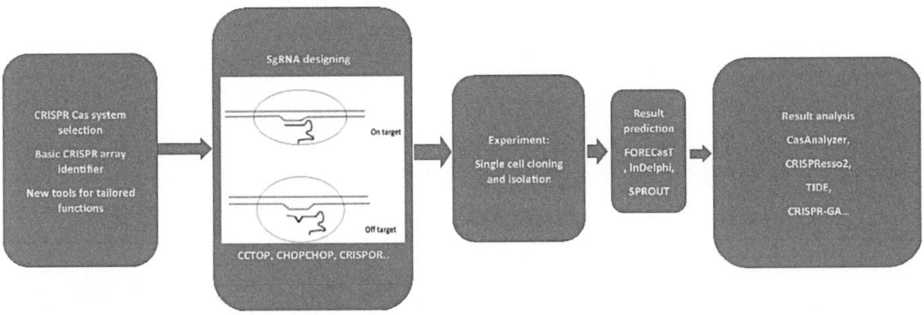

FIGURE 4.2 Outline of CRISPR/Cas experiment.

KEYWORDS

- **computational tools**
- **CRISPR-Cas gene editing**
- **direct repeat**
- **mobile genetic elements**
- **regularly spaced direct repeats**
- **sgRNAs**

REFERENCES

Abadi, S., Yan, W. X., Amar, D., & Mayrose, I., (2017). A machine learning approach for predicting CRISPR/Cas9 cleavage efficiencies and patterns underlying its mechanism of action. *PLoS Comput Biol., 13*(10), e1005807.

Abby, S. S., Neron, B., Menager, H., Touchon, M., & Rocha, E. P., (2014). MacSyFinder: A program to mine genomes for molecular systems with an application to CRISPR/Cas systems. *PLoS One, 9*, e110726.

Akcakaya, P., Bobbin, M. L., Guo, J. A., Malagon-Lopez, J., Clement, K., & Garcia, S. P., (2018). *In vivo* CRISPR editing with no detectable genome-wide off-target mutations. *Nature, 561*, 416–419.

Allen, F., Crepaldi, L., Alsinet, C., Strong, A. J., Kleshchevnikov, V., & De Angeli, P., (2018). Predicting the mutations generated by repair of Cas9-induced double-strand breaks. *Nat. Biotechnol., 37*, 64–72.

Anzalone, A. V., Randolph, P. B., Davis, J. R., Sousa, A. A., Koblan, L. W., Levy, J. M., Chen, P. J., et al., (2019). Search-and-replace genome editing without double-strand breaks or donor DNA. *Nature, 576*, 149–157.

Bae, S., Park, J., & Kim, J. S., (2014). Cas-OFFinder: A fast and versatile algorithm that searches for potential off-target sites of Cas9 RNA-guided endonucleases. *Bioinformatics, 30*(10), 1473–1475.

Biswas, A., Staals, R. H., Morales, S. E., Fineran, P. C., & Brown, C. M., (2016). CRISPRDetect: A flexible algorithm to define CRISPR arrays. *BMC Genomics, 17*, 356.

Brinkman, E. K., Chen, T., Amendola, M., & Van, S. B., (2014). Easy quantitative assessment of genome editing by sequence trace decomposition. *Nucleic Acids Res., 42*(22), e168.

Brinkman, E. K., Kousholt, A. N., Harmsen, T., Leemans, C., Chen, T., Jonkers, J., & Van, S. B., (2018). Easy quantification of template-directed CRISPR/Cas9 editing. *Nucleic Acids Res., 1, 46*(10), e58.

Camacho, C., Coulouris, G., Avagyan, V., Ma, N., Papadopoulos, J., Bealer, K., & Madden, T. L., (2009). BLAST+: Architecture and applications. *BMC Bioinformatics, 10*, 421.

Cao, Q., Ma, J., Chen, C. H., Xu, H., Chen, Z., Li, W., & Liu, X. S., (2017). CRISPR-FOCUS: A web server for designing focused CRISPR screening experiments. *PLoS One, 12*(9), e0184281.

Chen, W., McKenna, A., Schreiber, J., Haeussler, M., Yin, Y., Agarwal, V., Noble, W. S., & Shendure, J., (2019). Massively parallel profiling and predictive modeling of the outcomes of CRISPR/Cas9-mediated double-strand break repair. *Nucleic Acids Res., 47*, 7989–8003.

Chuai, G., Ma, H., Yan, J., Chen, M., Hong, N., Xue, D., Zhou, C., et al., (2018). DeepCRISPR: Optimized CRISPR guide RNA design by deep learning. *Genome Biol., 19*(1), 80.

Clement, K., Hsu, J. Y., Canver, M. C., Joung, J. K., & Pinello, L., (2020). Technologies and computational analysis strategies for CRISPR applications. *Mol Cell, 79*(1), 11–29.

Clement, K., Rees, H., Canver, M. C., Gehrke, J. M., Farouni, R., Hsu, J. Y., Cole, M. A., et al., (2019). CRISPResso2 provides accurate and rapid genome editing sequence analysis. *Nat Biotechnol., 37*(3), 224–226.

Crawley, A. B., Henriksen, J. R., & Barrangou, R., (2018). CRISPRdisco: An automated pipeline for the discovery and analysis of CRISPR/Cas systems. *CRISPR J., 1*, 171–181.

Doench, J. G., Fusi, N., Sullender, M., Hegde, M., Vaimberg, E. W., Donovan, K. F., Smith, I., et al., (2016). Optimized sgRNA design to maximize activity and minimize off-target effects of CRISPR/Cas9. *Nat Biotechnol., 34*(2), 184–191.

Doench, J. G., Hartenian, E., Graham, D. B., Tothova, Z., Hegde, M., Smith, I., Sullender, M., et al., (2014). Rational design of highly active sgRNAs for CRISPR/Cas9-mediated gene inactivation. *Nat Biotechnol., 32*(12), 1262–1267.

Dsouza, M., Larsen, N., & Overbeek, R., (1997). Searching for patterns in genomic data. *Trends Genet., 13*, 497, 498.

Eddy, S. R., (2011). Accelerated profile HMM searches. *PLoS Comput. Biol., 7*, e1002195.

Edgar, R. C., (2007). PILER-CR: Fast and accurate identification of CRISPR repeats. *BMC Bioinformatics, 8*, 18.

Frock, R. L., Hu, J., Meyers, R. M., Ho, Y. J., Kii, E., & Alt, F. W., (2015). Genome-wide detection of DNA double-stranded breaks induced by engineered nucleases. *Nat. Biotechnol., 33*, 179–186.

Ghezraoui, H., Piganeau, M., Renouf, B., Renaud, J. B., Sallmyr, A., Ruis, B., Oh, Set al., (2014). Chromosomal translocations in human cells are generated by canonical nonhomologous end-joining. *Mol. Cell, 55*, 829–842.

Gratz, S. J., Cummings, A. M., Nguyen, J. N., Hamm, D. C., Donohue, L. K., Harrison, M. M., Wildonger, J., & O'Connor-Giles, K. M., (2013). Genome engineering of *drosophila* with the CRISPR RNA-guided Cas9 nuclease. *Genetics, 194*(4), 1029–1035.

Grissa, I., Vergnaud, G., & Pourcel, C., (2007). CRISPRFinder: A web tool to identify clustered regularly interspaced short palindromic repeats. *Nucleic Acids Res., 35*, W52–57.

Güell, M., Yang, L., & Church, G. M., (2014). Genome editing assessment using CRISPR genome analyzer (CRISPR-GA). *Bioinformatics, 30*(20), 2968–2970.

Haeussler, M., Schönig, K., Eckert, H., Eschstruth, A., Mianné, J., Renaud, J. B., Schneider-Maunoury, S., et al., (2016). Evaluation of off-target and on-target scoring algorithms and integration into the guide RNA selection tool CRISPOR. *Genome Biol., 17*(1), 148.

Hanna, R. E., & Doench, J. G., (2020). Design and analysis of CRISPR–Cas experiments. *Nat. Biotechnol., 38*, 813–823.

Hart, T., & Moffat, J., (2016). BAGEL: A computational framework for identifying essential genes from pooled library screens. *BMC Bioinformatics, 17*, 164.

Heigwer, F., Kerr, G., & Boutros, M., (2014). E-CRISP: Fast CRISPR target site identification. *Nat. Methods, 11*(2), 122, 123.

Heigwer, F., Zhan, T., Breinig, M., Winter, J., Brügemann, D., Leible, S., & Boutros, M., (2016). CRISPR library designer (CLD): Software for multispecies design of single guide RNA libraries. *Genome Biol., 17*, 55.

Hough, S. H., Kancleris, K., Brody, L., Humphryes-Kirilov, N., Wolanski, J., Dunaway, K., Ajetunmobi, A., & Dillard, V., (2017). Guide picker is a comprehensive design tool for visualizing and selecting guides for CRISPR experiments. *BMC Bioinformatics, 18*(1), 167.

Hsu, P. D., Scott, D. A., Weinstein, J. A., Ran, F. A., Konermann, S., Agarwala, V., Li, Y., et al., (2013). DNA targeting specificity of RNA-guided Cas9 nucleases. *Nat. Biotechnol., 31*(9), 827–832.

Jacquin, A. L. S., Odom, D. T., & Lukk, M., (2019). Crisflash: Open-source software to generate CRISPR guide RNAs against genomes annotated with individual variation. *Bioinformatics, 35*, 3146–3147.

Jasin, M., & Haber, J. E., (2016). The democratization of gene editing: Insights from site-specific cleavage and double-strand break repair. *DNA Repair (Amst), 44*, 6–16.

Jeong, H. H., Kim, S. Y., Rousseaux, M. W. C., Zoghbi, H. Y., & Liu, Z., (2018). *CRISPRCloud2: A Cloud-Based Platform for Deconvolving CRISPR Screen Data*. bioRxiv.

Jiang, H., & Wong, W. H., (2008). SeqMap: Mapping massive amount of oligonucleotides to the genome. *Bioinformatics, 24*(20), 2395, 2396.

Jinek, M., East, A., Cheng, A., Lin, S., Ma, E., & Doudna, J., (2013). RNA-programmed genome editing in human cells. *Elife., 2*, e00471.

Kim, D., Bae, S., Park, J., Kim, E., Kim, S., Yu, H. R., Hwang, J., Kim, J. I., & Kim, J. S., (2015). Digenome-seq: Genome-wide profiling of CRISPR/Cas9 off-target effects in human cells. *Nat. Methods, 12*, 237–243.

Kluesner, M. G., Nedveck, D. A., Lahr, W. S., Garbe, J. R., Abrahante, J. E., Webber, B. R., & Moriarity, B. S., (2018). EditR: A method to quantify base editing from sanger sequencing. *CRISPR J., 1*(3), 239–250.

Labun, K., Montague, T. G., Gagnon, J. A., Thyme, S. B., & Valen, E., (2016). CHOPCHOP v2: A web tool for the next generation of CRISPR genome engineering. *Nucleic Acids Res., 44*, W1:W272–276.

Lange, S. J., Alkhnbashi, O. S., Rose, D., Will, S., & Backofen, R., (2013). CRISPRmap: An automated classification of repeat conservation in prokaryotic adaptive immune systems. *Nucleic Acids Res., 41*, 8034–8044.

Langmead, B., & Salzberg, S. L., (2012). Fast gapped-read alignment with Bowtie 2. *Nature methods*, *9*(4), 357–359.

Langmead, B., Trapnell, C., Pop, M., & Salzberg, S. L., (2009). Ultrafast and memory-efficient alignment of short DNA sequences to the human genome. *Genome Biol., 10*(3), R25.

Leenay, R. T., Aghazadeh, A., Hiatt, J., Tse, D., Hultquist, J. F., Krogan, N., Wu, Z., Marson, A., et al., (2018). *Systematic Characterization of Genome Editing in Primary T Cells Reveals Proximal Genomic Insertions and Enables Machine Learning Prediction of CRISPR/Cas9 DNA Repair Outcomes.* bioRxiv. doi: https://doi.org/10.1101/404947.

Li, H., & Durbin, R., (2009). Fast and accurate short read alignment with burrows-wheeler transform. *Bioinformatics, 25*(14), 1754–1760.

Li, W., Xu, H., Xiao, T., Cong, L., Love, M. I., Zhang, F., Irizarry, R. A., et al., (2014). MAGeCK enables robust identification of essential genes from genome-scale CRISPR/Cas9 knockout screens. *Genome Biol., 15*(12), 554.

Listgarten, J., Weinstein, M., Kleinstiver, B. P., Sousa, A. A., Joung, J. K., Crawford, J., Gao, K., et al., (2018). Prediction of off-target activities for the end-to-end design of CRISPR guide RNAs. *Nat. Biomed. Eng., 2*(1), 38–47.

Liu, G., Zhang, Y., & Zhang, T., (2019). Computational approaches for effective CRISPR guide RNA design and evaluation. *Comput. Struct. Biotechnol. J., 18*, 35–44.

Ma, J., Köster, J., Qin, Q., Hu, S., Li, W., Chen, C., Cao, Q., et al., (2016). CRISPR-DO for genome-wide CRISPR design and optimization. *Bioinformatics., 32*(21), 3336–3338.

MacPherson, C. R., & Scherf, A., (2015). Flexible guide-RNA design for CRISPR applications using protospacer workbench. *Nat. Biotechnol., 33*(8), 805, 806.

Mali, P., Yang, L., Esvelt, K. M., Aach, J., Guell, M., DiCarlo, J. E., Norville, J. E., & Church, G. M., (2013). RNA-guided human genome engineering via Cas9. *Science, 339*(6121), 823–826.

McKenna, A., & Shendure, J., (2018). FlashFry: A fast and flexible tool for large-scale CRISPR target design. *BMC Biol., 16*(1), 74.

Mojica, F. J. M., & Montoliu, L., (2016). On the origin of CRISPR/Cas technology: From prokaryotes to mammals. *Trends Microbiol., 24*, 811–820.

Moller, A. G., & Liang, C., (2017). MetaCRAST: Reference-guided extraction of CRISPR spacers from unassembled metagenomes. *PeerJ., 5*, e3788.

Montague, T. G., Cruz, J. M., Gagnon, J. A., Church, G. M., & Valen, E., (2014). CHOPCHOP: A CRISPR/Cas9 and TALEN web tool for genome editing. *Nucleic Acids Res., 42*, W401–407.

Moreno-Mateos, M. A., Vejnar, C. E., Beaudoin, J. D., Fernandez, J. P., Mis, E. K., Khokha, M. K., & Giraldez, A. J., (2015). CRISPRscan: Designing highly efficient sgRNAs for CRISPR/Cas9 targeting *in vivo*. *Nat Methods, 12*(10), 982–988.

Naito, Y., Hino, K., Bono, H., & Ui-Tei, K., (2015). CRISPRdirect: Software for designing CRISPR/Cas guide RNA with reduced off-target sites. *Bioinformatics, 31*, 1120–1123.

O'brien, A., & Bailey, T. L., (2014). GT-scan: Identifying unique genomic targets. *Bioinformatics, 30*, 2673–2675.

Park, J., Lim, K., Kim, J. S., & Bae, S., (2017). Cas-analyzer: An online tool for assessing genome editing results using NGS data. *Bioinformatics, 33*(2), 286–288.

Perez, A. R., Pritykin, Y., Vidigal, J. A., Chhangawala, S., Zamparo, L., Leslie, C. S., & Ventura, A., (2017). GuideScan software for improved single and paired CRISPR guide RNA design. *Nat. Biotechnol., 35*(4), 347–349.

Pinello, L., Canver, M. C., Hoban, M. D., Orkin, S. H., Kohn, D. B., Bauer, D. E., & Yuan, G. C., (2016). Analyzing CRISPR genome-editing experiments with CRISPResso. *Nat. Biotechnol., 34*(7), 695–697.

Pourcel, C., Touchon, M., Villeriot, N., Vernadet, J. P., Couvin, D., Toffano-Nioche, C., & Vergnaud, G., (2020). CRISPRCasdb a successor of CRISPRdb containing CRISPR arrays and Cas genes from complete genome sequences, and tools to download and query lists of repeats and spacers. *Nucleic Acids Res., 8*(48), D1:D535–D544.

Pulido-Quetglas, C., Aparicio-Prat, E., Arnan, C., Polidori, T., Hermoso, T., Palumbo, E., Ponomarenko, J., et al., (2017). Scalable design of paired CRISPR Guide RNAs for genomic deletion. *PLoS Comput. Biol., 13*(3), e1005341.

Qi, L. S., Larson, M. H., Gilbert, L. A., Doudna, J. A., Weissman, J. S., Arkin, A. P., & Lim, W. A., (2013). Repurposing CRISPR as an RNA-guided platform for sequence-specific control of gene expression. *Cell, 152*, 1173–1183.

Rahman, M. K., & Rahman, M. S., (2017). CRISPRpred: A flexible and efficient tool for sgRNAs on-target activity prediction in CRISPR/Cas9 systems. *PloS One, 12*(8), e0181943.

Ran, F. A., Cong, L., Yan, W. X., Scott, D. A., Gootenberg, J. S., Kriz, A. J., Zetsche, B., et al., (2015). *In vivo* genome editing using *Staphylococcus aureus* Cas9. *Nature, 520*(7546), 186–191.

Rees, H. A., & Liu, D. R., (2018). Base editing: Precision chemistry on the genome and transcriptome of living cells. *Nat. Rev. Genet., 19*, 770–788.

Rho, M., Wu, Y. W., Tang, H., Doak, T. G., & Ye, Y., (2012). Diverse CRISPRs evolving in human microbiomes. *PLoS Genet., 8*, e1002441.

Riesenberg, S., & Maricic, T., (2018). Targeting repair pathways with small molecules increases precise genome editing in pluripotent stem cells. *Nat Commun., 9*, 2164.

Sakuma, T., Nakade, S., Sakane, Y., Suzuki, K. T., & Yamamoto, T., (2016). MMEJ-assisted gene knock-in using TALENs and CRISPR/Cas9 with the PITCh systems. *Nat Protoc., 11*, 118–133.

Shen, M. W., Arbab, M., Hsu, J. Y., Worstell, D., Culbertson, S. J., Krabbe, O., Cassa, C. A., et al., (2018). Predictable and precise template-free CRISPR editing of pathogenic variants. *Nature, 563*(7733), 646–651.

Singh, R., Kuscu, C., Quinlan, A., Qi, Y., & Adli, M., (2015). Cas9-chromatin binding information enables more accurate CRISPR off-target prediction. *Nucleic Acids Research, 43*(18), e118.

Skennerton, C. T., (2016). *MinCED: Mining CRISPRs in Environmental Datasets*. https://github.com/ctskennerton/Crass, http://www.room220.com/crt/.

Skennerton, C. T., Imelfort, M., & Tyson, G. W., (2013). Crass: Identification and reconstruction of CRISPR from unassembled metagenomic data. *Nucleic Acids Res., 41*, e105.

Sledzinski, P., Nowaczyk, M., & Olejniczak, M., (2020). Computational tools and resources supporting CRISPR/Cas experiments. *Cells, 9*(5), 1288.

Spahn, P. N., Bath, T., Weiss, R. J., Kim, J., Esko, J. D., Lewis, N. E., & Harismendy, O., (2017). PinAPL-Py: A comprehensive web application for the analysis of CRISPR/Cas9 screens. *Sci. Rep., 7*, 15854.

Stemmer, M., Thumberger, T., Del, S. K. M., Wittbrodt, J., & Mateo, J. L., (2015). CCTop: An intuitive, flexible and reliable CRISPR/Cas9 target prediction tool. *PLoS One, 10*(4), e0124633.

Szczelkun, M. D., Tikhomirova, M. S., Sinkunas, T., Gasiunas, G., Karvelis, T., Pschera, P., Siksnys, V., & Seidel, R., (2014). Direct observation of R-loop formation by single RNA-guided Cas9 and Cascade effector complexes. *Proc Natl Acad Sci U S A, 111*(27), 9798–9803.

Tang, Z., Chen, S., Chen, A., He, B., Zhou, Y., Chai, G., Guo, F., & Huang, J., (2019). CasPDB: An integrated and annotated database for Cas proteins from bacteria and archaea. *Database: The Journal of Biological Databases and Curation*, baz093.

Thakore, P. I., Black, J. B., Hilton, I. B., & Gersbach, C. A., (2016). Editing the epigenome: Technologies for programmable transcription and epigenetic modulation. *Nat. Methods, 13*, 127–137.

Torres-Perez, R., Garcia-Martin, J. A., Montoliu, L., Oliveros, J. C., & Pazos, F., (2019). WeReview: CRISPR tools—Live repository of computational tools for assisting CRISPR/Cas experiments. *Bioengineering, 6*, 63.

Tsai, S. Q., Nguyen, N. T., Malagon-Lopez, J., Topkar, V. V., Aryee, M. J., & Joung, J. K., (2017). CIRCLE-seq: A highly sensitive *in vitro* screen for genome-wide CRISPR/Cas9 nuclease off-targets. *Nat Methods, 14*, 607–614.

Tsai, S. Q., Zheng, Z., Nguyen, N. T., Liebers, M., Topkar, V. V., & Thapar, V., (2015). GUIDE-seq enables genome-wide profiling of off-target cleavage by CRISPR/Cas nucleases. *Nat. Biotechnol., 33*, 187–197.

Upadhyay, S. K., & Sharma, S., (2014). SSFinder: High throughput CRISPR/Cas target sites prediction tool. *Biomed Res. Int.,* 742482.

Vartak, S. V., & Raghavan, S. C., (2015). Inhibition of nonhomologous end joining to increase the specificity of CRISPR/Cas9 genome editing. *FEBS J., 282*(22), 4289–4294.

Wienert, B., Wyman, S. K., Richardson, C. D., Yeh, C. D., Akcakaya, P., & Porritt, M. J., (2019). Unbiased detection of CRISPR off-targets *in vivo* using DISCOVER-Seq. *Science, 364*, 286–289.

Wilson, L., O'Brien, A. R., & Bauer, D. C., (2018). The current state and future of CRISPR/Cas9 gRNA design tools. *Frontiers in Pharmacology, 9*, 749.

Winter, J., Schwering, M., Pelz, O., Rauscher, B., Zhan, T., Heigwer, F., & Boutros, M., (2017). *CRISPR Analyzer: Interactive Analysis, Annotation and Documentation of Pooled CRISPR Screens*. bioRxiv.

Wu, X., Scott, D. A., Kriz, A. J., Chiu, A. C., Hsu, P. D., Dadon, D. B., Cheng, A. W., et al., (2014). Genome-wide binding of the CRISPR endonuclease Cas9 in mammalian cells. *Nature Biotechnology, 32*(7), 670–676.

Xiao, A., Cheng, Z., Kong, L., Zhu, Z., Lin, S., Gao, G., & Zhang, B., (2014). CasOT: A genome-wide Cas9/gRNA off-target searching tool. *Bioinformatics, 30*(8), 1180–1182.

Xie, S., Shen, B., Zhang, C., Huang, X., & Zhang, Y., (2014). sgRNAcas9: A software package for designing CRISPR sgRNA and evaluating potential off-target cleavage sites. *PLoS One, 9*(6), e100448.

Xu, H., Xiao, T., Chen, C. H., Li, W., Meyer, C. A., Wu, Q., Wu, D., et al., (2015). Sequence determinants of improved CRISPR sgRNA design. *Genome Res., 25*(8), 1147–1157.

Xu, X., Duan, D., & Chen, S. J., (2017). CRISPR/Cas9 cleavage efficiency correlates strongly with target-sgRNA folding stability: From physical mechanism to off-target assessment. *Sci Rep., 7*, 143.

Ye, L., Wang, C., Hong, L., Sun, N., Chen, D., & Chen, S., (2018). Programmable DNA repair with CRISPRa/i enhanced homology-directed repair efficiency with a single Cas9. *Cell Discov., 4*, 46.

Zhang, D., Hurst, T., Duan, D., & Chen, S. J., (2019). Unified energetics analysis unravels SpCas9 cleavage activity for optimal gRNA design. *Proc. Natl. Acad. Sci. U S A, 116*(18), 8693–8698.

Zhang, Q., & Ye, Y., (2017). Not all predicted CRISPR/Cas systems are equal: Isolated Cas genes and classes of CRISPR like elements. *BMC Bioinformatics, 18*, 92.

Zhang, X. H., Tee, L. Y., Wang, X. G., Huang, Q. S., & Yang, S. H., (2015). Off-target effects in CRISPR/Cas9-mediated genome engineering. *Mol. Ther. Nucl. Acids,* 4.

Zhou, M., Li, D., Huan, X., Manthey, J., Lioutikova, E., & Zhou, H., (2016). Mathematical and computational analysis of CRISPR Cas9 sgRNA off-target homologies. *IEEE International Conference on Bioinformatics and Biomedicine (BIBM),* 449–454.

Zhu, L. J., Holmes, B. R., Aronin, N., & Brodsky, M. H., (2014). CRISPRseek: A Bioconductor package to identify target-specific guide RNAs for CRISPR/Cas9 genome-editing systems. *PLoS One, 9*, e108424.

Part II

Genome Editing of Crop Plants with CRISPR/Cas Systems: Past, Present, and Future

CHAPTER 5

Genome Editing Using CRISPR/Cas9 System for Crop Improvement, Nutrient Enrichment, and Abiotic Stress Tolerance in Rice

ADITYA BANERJEE and ARYADEEP ROYCHOUDHURY*

Department of Biotechnology, St. Xavier's College (Autonomous), 30, Mother Teresa Sarani, Kolkata – 700016, West Bengal, India

Corresponding author. E-mail: aryadeep.rc@gmail.com

ABSTRACT

Genome editing is a revolutionary technology for improving rice cultivation as a whole. Traditional breeding approaches are extremely cumbersome and time-consuming, whereas overexpression of beneficial transgenes involves ecological and biosafety concerns. Hence, precision tools like clustered regularly interspaced short palindromic repeats (CRISPR) along with CRISPR-associated protein Cas9 (CRISPR/Cas9) are being regularly used to enable *in vivo*, subtle manipulations in the genetic architecture of rice genotypes for improvement in yield, nutrient uptake and storage, and environmental stress tolerance. The present chapter discusses the advancements of CRISPR/Cas9 applications in the above-mentioned areas and also illustrates the success stories of genome editing in safeguarding the principles of global food security and sustainable development.

5.1 INTRODUCTION

Oryza sativa (rice) is the principal staple food crop of a huge population residing in South East Asia, Africa, and America (Clarke and Zhang, 2013).

CRISPR/Cas-Mediated Genome Editing in Plants, Deepu Pandita, Anu Pandita (Eds.)

In spite of its varied roles in providing beneficial nutrition to the global population, this crop is extremely susceptible to environmentally challenging conditions like salinity, drought, heat, chilling, fluoride, heavy metal pollution, harmful irradiation (UV-B, gamma rays), etc. (Banerjee and Roychoudhury, 2021a, b; Banerjee et al., 2020a). Such sub-optimal conditions severely deteriorate the nutritional quality and productivity of rice (Hussain et al., 2017; Razzaq et al., 2020). Tolerance against environmental stresses along with improvements in rice quality has been marginally accomplished through breeding approaches which are extremely time-consuming and labor-intensive (Mann, 1999; Ansari et al., 2017). The use of transgenic technology via overexpression or knock-down of a single or multiple genes has been rigorously used to engineer stress tolerance in rice and also biofortify its grains. However, the transgenic crops have not yet been cultivated exhaustively in rice fields due to political interventions, public refusals, and ecosystem impacts (Prakash et al., 2011). Hence, the technology for *in vivo* genome editing and engineering is currently being popularly used to accelerate stress resistance and biofortification in rice (Endo and Toki, 2020). Genome editing in rice has been generally mediated by clustered regularly interspaced short palindromic repeats (CRISPR) along with CRISPR-associated protein Cas9 (CRISPR/Cas9), which has been widely regarded as a novel and versatile tool for easy and diverse genetic engineering (GE) of crop plants (Terada et al., 2002; Cong et al., 2013; Miao et al., 2013; Shan et al., 2013; Ma et al., 2015; Shimatani et al., 2017; Zong et al., 2017). Apart from having diverse uses as a GE tool, CRISPR/Cas9 also exhibits high specificity and accuracy during multiplexing (Wang and Wang, 2017). This has been regarded to be essential for intricately altering the sequences of multiple genes at multiple genomic loci (Endo et al., 2019a). These genetic manipulations are induced by double stranded breaks (DSBs) triggered by CRISPR/Cas9 at desired locations, which are then mended by homology directed repair (HDR) or non-homologous end joining (NHEJ) repair machineries present within cells (Li et al., 2018a). Due to the non-requirement of donor template and higher frequency in plants, NHEJ-mediated repair is more rapidly and inherently undertaken by the cells, compared with HDR (Puchta, 2004).

Execution of CRISPR/Cas9 applications in rice seedlings has gained appreciable impetus due to efficient functional characterization of the underlying genomic mechanism in this model crop plant (Li et al., 2012; Feng et al., 2013). This was obviously possible due to the small genomic size of rice and evolutionary and syntenic association with related cereals (Zafar et al., 2020). The present chapter concisely focuses on the reported roles and success stories of CRISPR/Cas9 technology in improving rice quality, enriching the nutrient reserves, and generating abiotic stress adaptation.

5.2 AN OVERVIEW OF THE CRISPR/CAS9 SYSTEM

CRISPR/Cas9 utilizes Cas9-mediated cleavage of the genomic template, forming DSBs which is ushered by the guide RNA (gRNA) and a nucleotide triplet nomenclature as protospacer adjacent motif (PAM) NGG, where N is any nucleotide base (Romero and Gatica-Arias, 2019). Genes involved in pigment synthesis in rice have been targeted using CRISPR/Cas9 to engineer lines with albino phenotype (Feng et al., 2013).

For ensuring multiplex genome editing, several gRNAs from a gene were generated using the tRNA processing system, where tandemly arrayed tRNA and gRNA are combined and association of every gRNA with a target spacer sequence is mediated (McCarty et al., 2020). This is achieved due to the cleavage of polycistronic tRNA-gRNA (PTG) by cellular RNase P and RNase Z to form a large population of gRNAs attached to specific signature sequences. The gRNAs thus guide Cas9 proteins within the cells to specific target sequences where DSB formation is initiated upon cleavage (Xie et al., 2015). Kim et al. (2019) observed CRISPR/Cas9 technology to exhibit very high efficacy and perfection in rice plants. During HDR activation, the tRNA processing machinery provided the donor repair template (DRT) (Li et al., 2018b). Butt et al. (2017) reported the synthesis of a chimeric single guide RNA (sgRNA) as gRNA—gRNA Scaffold—pre-tRNA—DRT—Terminator which was itself able to promote DSBs at specific sites and also deliver DRTs for activating HDR.

5.3 CRISPR/CAS9 MEDIATED INCREASE IN RICE PRODUCTIVITY

Productivity and development of grains are controlled by multiple genes referred to as quantitative trait loci (QTLs), which might be manipulated using CRISPR/Cas9 technology to generate transgenic rice lines with significantly improved yield (Sedeek et al., 2019). Genes associated with grain development and seed setting like *grain number 1a* (*GN1a*), *dense and erect panicle 1* (*DEP1*), *grain size 3* (*GS3*), and *ideal plant architecture 1* (*IPA1*) have been successfully edited via CRISPR/Cas9. These genes respectively regulated the number of grain formation, development of panicles, size of seeds and overall seedling architecture, respectively (Wang et al., 2015;

Wang and Wang, 2017; Nan et al., 2018; Zhao et al., 2019). Manipulation of the yield-related genes using genome editing resulted in the formation of variable phenotypes, thus illustrating the efficacy of the technology in editing even complex traits associated with QTL functioning (Li et al., 2016).

Shen et al. (2018) showed that CRISPR/Cas9 could also disrupt the QTLs like *GS3* and *GN1a* in popular rice cultivars, resulting in variable effects on grain yield due to background genomic heterogeneity. Multiplex genome editing was also performed to manipulate the sequences of agronomic QTLs like *GS3*, *GN1a* and *grain width 2* (*GW2*). Significantly higher grain formation (30–68% more) per panicle was observed in the rice triple mutants generated via the above multiplex genome editing experiments (Zhou et al., 2019). Such approaches are also being presently propagated to generate high yielding elite cultivars of rice. Likewise, Huang et al. (2018) used genome editing tools to design yield-associated alleles that were genetically superior to their wild type counterparts. The CRISPR/Cas9-mediated improvement in allele functioning included the genes *GN1a* and *DEP1*. It was observed that the rice lines containing improved versions of both alleles produced much higher yields than wild type plants, thus illustrating the ability of genome editing to manipulate the genetic sequence and improve the overall functioning of complex genes (Huang et al., 2018). Cross-talks between the stress phytohormone abscisic acid (ABA) and grain production in rice were also revealed through CRISPR/Cas9-dependent genome editing. It was observed that specific manipulations in ABA receptor-encoding genes like *pyrabactin resistance like 1* (*PYL1*), *PYL4* and *PYL6* increased the yield by 31%, compared with the wild type variety when both sets were cultivated in the field (Miao et al., 2018). This also indicated at the requirement to genetically modulate phytohormone-associated pathways to alter the metabolome and improve the productivity of elite rice cultivars.

5.4 BIOFORTIFICATION OF RICE USING CRISPR/CAS9

Rice cultivars are the most cultivated cereal crops due to their acceptance as the staple food grain for more than 50% population of the world (Arvanitoyannis and Tserkezou, 2008; Mohammed et al., 2020). This is an enormous challenge, since the majority of the poor and low-income families within this percentage of population rely entirely on rice grain consumption for fulfilling their overall nutritional requirement (Zhu, 2018). Hence, enriching the grains with nutrients using GE would definitely fulfill the nutrient requirement of the vast population which will consume the biofortified rice grains (Bandumula, 2018; Rathna et al., 2019).

5.4.1 INCREASING ESSENTIAL OIL CONTENT USING CRISPR/CAS9

Genome editing using CRISPR/Cas9 has been utilized to promote the biosynthesis and accumulation of essential components like oleic acid within the rice bran oil. It has been previously reported that elevated levels of the fatty acid, i.e., oleic acid are associated with reduced autoimmune, coronary, and inflammatory diseases (Sales-Campos et al., 2013). Furthermore, consumption of rice grains enriched with oleic acid has also been found to maintain good health and physical activity in children and adults (Latha and Nasirullah, 2014).

The enzyme fatty acid desaturase 2 (FAD2) catalyzes the formation of linoleic acid from oleic acid in rice (Zhiguo et al., 2019). Maximum transcript level of *FAD2-1* in the grains revealed direct correlation of this isoform of FAD2 with linoleic acid production in rice seeds (Dar et al., 2017; Miao et al., 2019; Salimonti, 2020). Lowering of linoleic acid formation would lead to higher oleic acid accumulation which is more beneficial to consumers than linoleic acid (Jones et al., 2008; Carillo et al., 2011). Abe et al. (2018) inferred that functional disruption of *OsFAD2-1* via genome editing could be performed to achieve high oleic acid accumulation with the rice bran oil. CRISPR/Cas9 mediated knockout of *FAD2-1* doubled the oleic acid level within the grains derived from the mutant rice lines. Furthermore, the content of linoleic acid was also found to be undetectable, thus illustrating the efficacy and specificity of CRISPR/Cas9 technology (Abe et al., 2018). Overall, a mutant rice line was engineered which produced oleic acid-rich rice bran oil.

5.4.2 INCREASING Β-CAROTENE AND AMYLOSE CONTENT USING CRISPR/CAS9

Enriching rice grains with β-carotene has been already attempted via gene pyramiding for the GE of golden rice (Tang et al., 2009; Moghissi et al., 2016; Kettenburg et al., 2018; Zhu et al., 2018; Stokstad, 2019). Carotenoids are important antioxidants and source of vitamin A, which is associated with physiological development and proper vision (Chapman, 2012; Khalil et al., 2017; Dowling, 2020). Endo et al. (2019b) used CRISPR/Cas9 technology to manipulate the *orange* (*OR*) gene in rice. This particular gene has been associated with the production of β-carotene in the curd of cauliflower (Osorio, 2019; Liu et al., 2020). Genetic editing of the *OR* gene in rice resulted in an increased accumulation of β-carotene in the callus tissues of cultured mutant rice lines (Endo et al., 2019b).

Increased amylose level in a food source indicates that it is also enriched with nutritious starch, which is involved in maintaining the good health of the consumers (Regina et al., 2006; Jiang et al., 2010). Sun et al. (2017) genetically edited the genes regulating amylose production, like *starch branching enzyme I* (*SBEI*) and *SBEIIb* in rice using CRISPR/Cas9. Mutation in the *SBEI* gene significantly triggered amylose level by 25%, whereas accumulation of resistant starch was promoted by 9.8%, compared with wild-type seedlings (Sun et al., 2017). This report characteristically revealed the potential of CRISPR/Cas9 technology to generate high amylose-containing biofortified rice.

5.4.3 INCREASING ANTIOXIDANT CONTENT USING CRISPR/CAS9

Proanthocyanidins are precursors to important non-enzymatic antioxidants named anthocyanins which are responsible for the coloration of rice seeds (Kim et al., 2007). The genes responsible for reddish color in rice grains are *Rc* and *Rd*. While *Rc* is translated into a basic helix-loop-helix (bHLH) protein which regulates the reddening of the pericarp, *Rd* encodes dihydroflavonol-4-reductase involved in the formation of leucoanthocyanidins from dihydroflavonols (Sweeney et al., 2006; Furukawa et al., 2007; Li et al., 2017). Presence of intact *Rc* and *Rd* genes leads to the formation of reddish grains in *Oryza rufipogon*, whereas the absence of a 14 bp sequence within *Rc* at the seventh exon results in loss-of-function of the gene, leading to the formation of normal white grains in domesticated rice cultivars (Roy and Shil, 2020). Zhu et al. (2019) re-inserted the deleted 14 bp sequence within *Rc* gene using CRISPR/Cas9 by bringing them in-frame within the open reading frame. As a result, grains produced by the mutant rice lines accumulated enhanced levels of proanthocyanidins and anthocyanins and exhibited red coloration as well. Furthermore, all agronomic parameters in the mutant rice lines were similar to wild-type seedlings (Zhu et al., 2019). Thus, it was established that CRISPR/Cas9-mediated reversion of the *Rc* allele biofortified the rice grains without suppressing agronomic and reproductive development.

5.4.4 *INCREASE IN AROMA LEVEL USING CRISPR/CAS9*

Rice grains with high aroma content are economically lucrative commodities due to their rich fragrance, superior texture, and palatability (Banerjee et al., 2019). The aromatic cultivars exhibit suppressed expression of *betaine aldehyde dehydrogenase 2* (*BADH2*), resulting in high accumulation of 2-acetyl-1-pyrroline (2AP), which is the major chemical volatile responsible for aroma generation (Banerjee et al., 2020b). Shao et al. (2017) introduced an extra thymine nucleotide within the first exon sequence of *BADH2* in the fragrant rice cultivar Zhonghua 11. This stimulated 2AP production and improved the aroma profile of the genetically edited rice lines (Shao et al., 2017).

5.5 ABIOTIC STRESS TOLERANCE (AT) IN RICE USING CRISPR/CAS9

5.5.1 *SALINITY*

Salinization is one of the most predominant types of environmental challenge for sustainable agriculture of rice (Tisarum et al., 2020). Usually, a large number of salt-tolerant cultivars like Pokkali, Nonabokra, Matla, Jarava, etc., have been cultivated across the salt-affected regions with the hope of some grain production (Prusty et al., 2018). However, these rice varieties are neither high yielding nor have elite phenotypes. As a result, *in vivo* crop improvement using genome editing has been used to manifest salt tolerance in elite and susceptible genotypes. It was reported that specific targeting of *RR22* via CRISPR/Cas9 technique significantly triggered salt-tolerant responses and phenotype in the engineered mutant lines (Farhat et al., 2019). With respect to wild-type sets, the genetically manipulated lines exhibited higher and efficient adaptation to salt stress (Zhang et al., 2019).

5.5.2 *DROUGHT AND LOW TEMPERATURE*

Desiccation or drought invariably leads to the accumulation of ABA which regulates guard cell signaling to close the stomatal aperture and prevent water loss. The *SNF1-related protein kinase 2* gene, *SAPK2* is an important player in the ABA signal transduction pathway (Roychoudhury and Banerjee, 2017). Lou et al. (2017) revealed the importance of *SAPK2* in dictating drought tolerance in rice seedlings. Loss-of-function mutation in *SAPK2* using CRISPR/Cas9 promoted severe oxidative stress and drought sensitivity in the mutant lines, thus illustrating the crucial regulatory roles of *SAPK2* during ABA-dependent drought resistance in rice (Lou et al., 2017).

In another study, Zeng et al. (2020) engineered a knock-out mutant of the MYELOBLASTOSIS 30 (MYB30)-encoding transcription factor gene in rice using CRISPR/Cas9. It was shown that the genetically manipulated lines were more resistant to cold, compared with wild-type plants, and the mutant seedlings also exhibited superior productivity and agronomic traits with respect to the wild-type counterparts during cold stress (Zeng et al., 2020).

5.5.3 HEAVY METAL TOXICITY

Rice seedlings are highly susceptible to toxic concentrations of heavy metals like cadmium (Cd), arsenic (As), lead (Pb), nickel (Ni), etc. (Clemens et al., 2013; Banerjee et al., 2020a). The transporter encoded by *natural resistance-associated macrophage protein 5* (*Nramp5*) is responsible for Cd translocation within the rice biomass (Ullah et al., 2018). Tang et al. (2017) utilized CRISPR/Cas9 to eliminate the *Nramp5* gene, due to which the engineered mutant rice lines bioaccumulated significantly lower levels of Cd, compared with the wild-type plants even under field trials. Similarly, Yang et al. (2019) also showed reduced Cd uptake in the mutants of japonica cultivars where the *Nramp5* gene was knocked out via CRISPR/Cas9. However, reduced import of manganese (Mn) ions compromised the agronomic and yield-related traits in these mutant rice lines (Yang et al., 2019). Thus, it was inferred that CRISPR/Cas9 targeting should be performed by considering the associated physicochemical changes in the metabolome to assure optimum rice yield.

5.6 CONCLUSION

CRISPR/Cas9 is a technology with high potential in crop improvement, since they allow *in vivo* manipulation of the genetic architecture with exact precision. This has eliminated the requirement of elaborate breeding programs or the need to insert beneficial transgenes from other genotypes or species. Genome editing has appreciably helped to improve rice quality, nutrition, and tolerance to a variety of environmental stresses. This has been achieved by small alterations performed in specific genes associated with the phenotypes. Improved rice varieties through using such ecologically safe procedures will definitely help the large proportion of consumers to get sufficient nutrition through the staple food crop. This is obviously a huge success for ensuring food security and sustainable development, since rice is the most important staple food crop among all other cereals.

5.7 FUTURE PERSPECTIVES

Genome editing has immense prospects in revolutionizing rice agriculture and productivity across the world. Future perspectives should include more elaborate uses of CRISPR/Cas9 across several landraces and genotypes of rice so that these could be efficiently cultivated across the globe as and when required. More exhaustive research featuring the use of genome editing tools should be applied in generating abiotic stress-tolerant rice lines. Future investigations should also focus on the use of CRISPR/Cas variants on generating crop genotypes that would bioaccumulate low levels of harmful xenobiotics within the edible grains. This would definitely help the farmers to cultivate seedlings in polluted fields and produce rice grains that can be safely consumed with minimal risk.

ACKNOWLEDGMENTS

Financial assistance from the Science and Engineering Research Board, Government of India, through the grant [EMR/2016/004799] and the Department of Higher Education, Science, and Technology and Biotechnology, Government of West Bengal, through the grant [264(Sanc.)/ST/P/S&T/1G-80/2017] to Dr. Aryadeep Roychoudhury is gratefully acknowledged. Mr. Aditya Banerjee is thankful to the University Grants Commission, Government of India, for providing a Senior Research Fellowship in the course of this work.

KEYWORDS

- **abiotic stress**
- **biofortification**
- **CRISPR/Cas9**
- **double-stranded breaks**
- **genome editing**
- **homology-directed repair**
- **tolerance**

REFERENCES

Abe, K., Araki, E., Suzuki, Y., Toki, S., & Saika, H., (2018). Production of high oleic/low linoleic rice by genome editing. *Plant Physiol. Biochem., 131*, 58–62.

Ansari, A., Wang, C., Wang, J., Wang, F., Liu, P., Gao, Y., et al., (2017). Engineered dwarf male-sterile rice: A promising genetic tool for facilitating recurrent selection in rice. *Front Plant Sci., 8*, 2132.

Arvanitoyannis, I. S., & Tserkezou, P., (2008). Cereal waste management: Treatment methods and potential uses of treated waste. In: Arvanitoyannis, I. S., (ed.), *Waste Management for the Food Industries: A Volume in Food Science and Technology* (pp. 629–702). Academic Press, Elsevier, United Kingdom.

Bandumula, N., (2018). Rice production in Asia: Key to global food security. *Proc. Natl. Acad. Sci. India Sect. B Biol. Sci., 88*, 1323–1328.

Banerjee, A., & Roychoudhury, A., (2021a). Maghemite nano-fertilization promotes fluoride tolerance in rice by restoring grain yield and modulating the ionome and physiome. *Ecotoxicol. Environ. Saf., 215*, 112055.

Banerjee, A., & Roychoudhury, A., (2021b). Differential lead-fluoride and nickel-fluoride uptake in co-polluted soil variably affects the overall physiome in an aromatic rice cultivar. *Environ. Pollut., 268*, 115504.

Banerjee, A., Ghosh, P., & Roychoudhury, A., (2019). Salt acclimation differentially regulates the metabolites commonly involved in stress tolerance and aroma synthesis in indica rice cultivars. *Plant Growth Regul., 88*, 87–97.

Banerjee, A., Ghosh, P., & Roychoudhury, A., (2020b). Differential regulation of genes co-involved in aroma production and stress amelioration during salt acclimation in indica rice cultivars. *Biologia, 75*, 495–506.

Banerjee, A., Samanta, S., Singh, A., & Roychoudhury, A., (2020a). Deciphering the molecular mechanism behind stimulated co-uptake of arsenic and fluoride from soil, associated toxicity, defence and glyoxalase machineries in arsenic-tolerant rice. *J. Hazard. Mater., 390*, 121978.

Butt, H., Eid, A., Ali, Z., Atia, M. A., Mokhtar, M. M., Hassan, N., et al., (2017). Efficient CRISPR/Cas9-mediated genome editing using a chimeric single-guide RNA molecule. *Front. Plant Sci., 8*, 1441.

Carrillo, C., Del, M. C. M., & Alonso-Torre, S. R., (2011). Oleic acid versus linoleic and α-linolenic acid. Different effects on Ca^{2+} signaling in rat thymocytes. *Cell Physiol. Biochem., 27*, 373–380.

Chapman, M. S., (2012). Vitamin a: History, current uses, and controversies. *Semin. Cutan. Med. Surg., 31*, 11–16.

Clarke, J. L., & Zhang, P., (2013). Plant biotechnology for food security and bioeconomy. *Plant Mol. Biol., 83*, 1–3.

Clemens, S., Aarts, M. G., Thomine, S., & Verbruggen, N., (2013). Plant science: The key to preventing slow cadmium poisoning. *Trends Plant Sci., 18*, 92–99.

Cong, L., Ran, F. A., Cox, D., Lin, S., Barretto, R., Habib, N., et al., (2013). Multiplex genome engineering using CRISPR/Cas systems. *Science, 339*, 819–823.

Dar, A. A., Choudhury, A. R., Kancharla, P. K., & Arumugam, N., (2017). The *FAD2* gene in plants: Occurrence, regulation, and role. *Front. Plant Sci., 8*, 1789.

Dowling, J. E., (2020). Vitamin A: Its many roles-from vision and synaptic plasticity to infant mortality. *J. Comp. Physiol. A Neuroethol. Sens. Neural. Behav. Physiol., 206*, 389–399.

Endo, A., Saika, H., Takemura, M., Misawa, N., & Toki, S., (2019b). A novel approach to carotenoid accumulation in rice callus by mimicking the cauliflower orange mutation via genome editing. *Rice, 12*, 1–5.

Endo, M., & Toki, S., (2020). Genome editing in rice. *Rice (N.Y.), 13*, 27.

Endo, M., Mikami, M., Endo, A., Kaya, H., Itoh, T., Nishimasu, H., Nureki, O., & Toki, S., (2019a). Genome editing in plants by engineered CRISPR-Cas9 recognizing NG PAM. *Nat. Plants*, 5, 14–17.

Farhat, S., Jain, N., Singh, N., Sreevathsa, R., Dash, P. K., Rai, R., et al., (2019). CRISPR-Cas9 directed genome engineering for enhancing salt stress tolerance in rice. *Semin. Cell Dev. Biol., 96*, 91–99.

Feng, Z., Zhang, B., Ding, W., Liu, X., Yang, D. L., Wei, P., et al., (2013). Efficient genome editing in plants using a CRISPR/Cas system. *Cell Res., 23*, 1229–1232.

Furukawa, T., Maekawa, M., Oki, T., Suda, I., Iida, S., Shimada, H., Takamure, I., & Kadowaki, K., (2007). The *Rc* and *Rd* genes are involved in proanthocyanidin synthesis in rice pericarp. *Plant J., 49*, 91–102.

Huang, L., Zhang, R., Huang, G., Li, Y., Melaku, G., Zhang, S., et al., (2018). Developing superior alleles of yield genes in rice by artificial mutagenesis using the CRISPR/Cas9 system. *Crop J., 6*, 475–481.

Hussain, S., Zhang, J. H., Zhong, C., Zhu, L. F., Cao, X. C., Yu, S. M., Bohr, J. A., et al., (2017). Effects of salt stress on rice growth, development characteristics, and the regulating ways: A review. *J. Integr. Agric., 16*, 2357–2374.

Jiang, H., Lio, J., Blanco, M., Campbell, M., & Jane, J. L., (2010). Resistant-starch formation in high-amylose maize starch during kernel development. *J. Agric. Food Chem., 58*, 8043–8047.

Jones, P. J., Jew, S., & AbuMweis, S., (2008). The effect of dietary oleic, linoleic, and linolenic acids on fat oxidation and energy expenditure in healthy men. *Metabolism, 57*, 1198–1203.

Kettenburg, A. J., Hanspach, J., Abson, D. J., & Fischer, J., (2018). From disagreements to dialogue: Unpacking the golden rice debate. *Sustain. Sci., 13*, 1469–1482.

Khalil, S., Bardawil, T., Stephan, C., Darwiche, N., Abbas, O., Kibbi, A. G., Nemer, G., & Kurban, M., (2017). Retinoids: A journey from the molecular structures and mechanisms of action to clinical uses in dermatology and adverse effects. *J. Dermatolog. Treat., 28*, 684–696.

Kim, B. G., Kim, J. H., Min, S. Y., Shin, K. H., Kim, J. H., Kim, H. Y., Ryu, S. N., & Ahn, J. H., (2007). Anthocyanin content in rice is related to expression levels of anthocyanin biosynthetic genes. *J. Plant Biol., 50*, 156–160.

Kim, Y. A., Moon, H., & Park, C. J., (2019). CRISPR/Cas9-targeted mutagenesis of *Os8N3* in rice to confer resistance to *Xanthomonas oryzae* pv. *oryzae*. *Rice, 12*, 67.

Latha, R. B., & Nasirullah, D. R., (2014). Physico-chemical changes in rice bran oil during heating at frying temperature. *J. Food Sci. Technol., 51*, 335–340.

Li, C., Zong, Y., Wang, Y., Jin, S., Zhang, D., Song, Q., Zhang, R., & Gao, C., (2018a). Expanded base editing in rice and wheat using a Cas9-adenosine deaminase fusion. *Genome Biol.,* 19, 59.

Li, J., Zhang, X., Sun, Y., Zhang, J., Du, W., Guo, X., Li, S., et al., (2018b). Efficient allelic replacement in rice by gene editing: A case study of the *NRT1.1B* gene. *J. Integr. Plant Biol., 60*, 536–540.

Li, M., Li, X., Zhou, Z., Wu, P., Fang, M., Pan, X., et al., (2016). Reassessment of the four yield-related genes *Gn1a*, *DEP1*, *GS3*, and *IPA1* in rice using a CRISPR/Cas9 system. *Front. Plant Sci., 7*, 377.

Li, T., Liu, B., Spalding, M. H., Weeks, D. P., & Yang, B., (2012). High-efficiency TALEN-based gene editing produces disease-resistant rice. *Nat. Biotechnol., 30*, 390–392.

Li, Y., Liu, X., Cai, X., Shan, X., Gao, R., Yang, S., Han, T., et al., (2017). Dihydroflavonol 4-reductase genes from *Freesia hybrida* play important and partially overlapping roles in the biosynthesis of flavonoids. *Front. Plant Sci., 8*, 428.

Liu, Y., Ye, S., Yuan, G., Ma, X., Heng, S., Yi, B., Ma, C., et al., (2020). Gene silencing of *BnaA09.ZEP* and *BnaC09.ZEP* confers orange color in *Brassica napus* flowers. *Plant J., 104*, 932–949.

Lou, D., Wang, H., Liang, G., & Yu, D., (2017). OsSAPK2 confers abscisic acid sensitivity and tolerance to drought stress in rice. *Front. Plant Sci., 8*, 993.

Ma, L., Zhu, F., Li, Z., Zhang, J., Li, X., Dong, J., et al., (2015). TALEN based mutagenesis of *lipoxygenase LOX3* enhances the storage tolerance of rice (*Oryza sativa*) seeds. *PLoS One, 10*, e0143877.

Mann, C. C., (1999). *Crop Scientists Seek a New Revolution*. American Association for the Advancement of Science.

McCarty, N. S., Graham, A. E., Studená, L., & Ledesma-Amaro, R., (2020). Multiplexed CRISPR technologies for gene editing and transcriptional regulation. *Nat. Commun., 11*, 1281.

Miao, C., Xiao, L., Hua, K., Zou, C., Zhao, Y., Bressan, R. A., et al., (2018). Mutations in a subfamily of abscisic acid receptor genes promote rice growth and productivity. *Proc. Natl. Acad. Sci. U.S.A, 115*, 6058–6063.

Miao, J., Guo, D., Zhang, J., Huang, Q., Qin, G., Zhang, X., et al., (2013). Targeted mutagenesis in rice using CRISPR-Cas system. *Cell Res., 23*, 1233–1236.

Miao, X., Zhang, L., Hu, X., et al., (2019). Cloning and functional analysis of the *FAD2* gene family from desert shrub *Artemisia sphaerocephala*. *BMC Plant Biol., 19*, 481.

Moghissi, A. A., Pei, S., & Liu, Y., (2016). Golden rice: Scientific, regulatory and public information processes of a genetically modified organism. *Crit. Rev. Biotechnol., 36*, 535–541.

Mohammed, S. G., Bossa, S., & Qoronfleh, M. W., (2020). Grains. *Adv. Neurobiol., 24*, 377–394.

Nan, J., Feng, X., Wang, C., Zhang, X., Wang, R., Liu, J., Yuan, Q., et al., (2018). Improving rice grain length through updating the *GS3* locus of an elite variety Kongyu 131. *Rice, 11*, 21.

Osorio, C. E., (2019). The role of orange gene in carotenoid accumulation: Manipulating chromoplasts toward a colored future. *Front. Plant Sci., 10*, 1235.

Prakash, D., Verma, S., Bhatia, R., & Tiwary, B. N., (2011). Risks and precautions of genetically modified organisms. *Int. Sch. Res. Notices,* 369573.

Prusty, M. R., Kim, S. R., Vinarao, R., Entila, F., Egdane, J., Diaz, M. G. Q., & Jena, K. K., (2018). Newly identified wild rice accessions conferring high salt tolerance might use a tissue tolerance mechanism in leaf. *Front. Plant Sci., 9*, 417.

Puchta, H., (2004). The repair of double-strand breaks in plants: Mechanisms and consequences for genome evolution. *J. Exp. Bot., 56*, 1–14.

Rathna, P. T., Eliazer, N. A. R. L., Ravichandran, K., et al., (2019). Nutritional and functional properties of colored rice varieties of South India: A review. *J. Ethn. Food, 6*, 11.

Razzaq, A., Ali, A., Safdar, L. B., Zafar, M. M., Rui, Y., Shakeel, A., Shaukat, A., et al., (2020). Salt stress induces physiochemical alterations in rice grain composition and quality. *J. Food Sci., 85*, 14–20.

Regina, A., Bird, A., Topping, D., Bowden, S., Freeman, J., Barsby, T., et al., (2006). High-amylose wheat generated by RNA interference improves indices of large-bowel health in rats. *Proc. Natl. Acad. Sci. U.S.A, 103*, 3546–3551.

Romero, F. M., & Gatica-Arias, A., (2019). CRISPR/Cas9: Development and application in rice breeding. *Rice Sci., 26*, 265–281.

Roy, S. C., & Shil, P., (2020). *Black Rice Developed Through Interspecific Hybridization (O. Sativa x O. rufipogon): Origin of Black Rice Gene from Indian Wild Rice*. bioRxiv. doi.org/10.1101/2020.12.25.423663.

Roychoudhury, A., & Banerjee, A., (2017). Abscisic acid signaling and involvement of mitogen activated protein kinases and calcium-dependent protein kinases during plant abiotic stress. In: Pandey, G., (ed.), *Mechanism of Plant Hormone Signaling Under Stress* (Vol. 1, pp. 197–241). John Wiley & Sons, Inc., USA.

Sales-Campos, H., Souza, P. R., Peghini, B. C., Da Silva, J. S., & Cardoso, C. R., (2013). An overview of the modulatory effects of oleic acid in health and disease. *Mini Rev. Med. Chem., 13*, 201–210.

Salimonti, A., Carbone, F., Romano, E., Pellegrino, M., Benincasa, C., et al., (2020). Association study of the 5'UTR intron of the *FAD2-2* gene with oleic and linoleic acid content in *Olea europaea* L. *Front. Plant Sci., 11*, 66.

Samanta, S., Banerjee, A., & Roychoudhury, A., (2021). Exogenous melatonin regulates endogenous phytohormone homeostasis and thiol-mediated detoxification in two indica rice cultivars under arsenic stress. *Plant Cell Rep., 40*, 1585–1602..

Sedeek, K. E. M., Mahas, A., & Mahfouz, M., (2019). Plant genome engineering for targeted improvement of crop traits. *Front. Plant Sci., 10*, 114.

Shan, Q., Wang, Y., Li, J., Zhang, Y., Chen, K., et al., (2013). Targeted genome modification of crop plants using a CRISPR-Cas system. *Nat. Biotechnol.,* 31, 686–688.

Shao, G., Xie, L., Jiao, G., Wei, X., Sheng, Z., Tang, S., et al., (2017). CAS9-mediated editing of the fragrant gene *Badh2* in rice. *Chin. J. Rice Sci., 31*, 216–222.

Shen, L., Wang, C., Fu, Y., Wang, J., Liu, Q., Zhang, X., et al., (2018). QTL editing confers opposing yield performance in different rice varieties. *J. Integr. Plant Biol., 60*, 89–93.

Shimatani, Z., Kashojiya, S., Takayama, M., Terada, R., et al., (2017). Targeted base editing in rice and tomato using a CRISPR-Cas9 cytidine deaminase fusion. *Nat. Biotechnol.,* 35, 441–443.

Stokstad, E., (2019). After 20 years, golden rice nears approval. *Science, 366*, 934.

Sun, Y., Jiao, G., Liu, Z., Zhang, X., Li, J., Guo, X., et al., (2017). Generation of high amylose rice through CRISPR/Cas9-mediated targeted mutagenesis of starch branching enzymes. *Front. Plant Sci., 8*, 298.

Sweeney, M. T., Thomson, M. J., Pfeil, B. E., & McCouch, S., (2006). Caught red-handed: *Rc* encodes a basic helix-loop-helix protein conditioning red pericarp in rice. *Plant Cell, 18*, 283–294.

Tang, G., Qin, J., Dolnikowski, G. G., Russell, R. M., & Grusak, M. A., (2009). Golden rice is an effective source of vitamin A. *Am. J. Clin. Nutr., 89*, 1776–1783.

Tang, L., Mao, B., Li, Y., Lv, Q., Zhang, L., Chen, C., et al., (2017). Knockout of *OsNramp5* using the CRISPR/Cas9 system produces low Cd accumulating indica rice without compromising yield. *Sci. Rep., 7*, 14438.

Terada, R., Urawa, H., Inagaki, Y., Tsugane, K., & Iida, S., (2002). Efficient gene targeting by homologous recombination in rice. *Nat. Biotechnol., 20*, 1030–1034.

Tisarum, R., Theerawitaya, C., Samphumphuang, T., Polispitak, K., Thongpoem, P., Singh, H. P., & Cha-um, S., (2020). Alleviation of salt stress in upland rice (*Oryza sativa* L. ssp. *indica* cv. Leum Pua) using arbuscular mycorrhizal fungi inoculation. *Front. Plant Sci., 11*, 348.

Ullah, I., Wang, Y., Eide, D. J., et al., (2018). Evolution, and functional analysis of natural resistance-associated macrophage proteins (NRAMPs) from *Theobroma cacao* and their role in cadmium accumulation. *Sci. Rep., 8*, 14412.

Wang, B., & Wang, H., (2017). IPA1: A new "green revolution" gene? *Mol. Plant, 10*, 779–781.

Wang, J., Xu, H., Li, N., Fan, F., Wang, L., Zhu, Y., & Li, S., (2015). Artificial selection of *Gn1a* plays an important role in improving rice yields across different ecological regions. *Rice, 8*, 37.

Xie, K., Minkenberg, B., & Yang, Y., (2015). Boosting CRISPR/Cas9 multiplex editing capability with the endogenous tRNA-processing system. *Proc. Natl. Acad. Sci. U.S.A, 112*, 3570–3575.

Yang, C. H., Zhang, Y., & Huang, C. F., (2019). Reduction in cadmium accumulation in Japonica rice grains by CRISPR/Cas9-mediated editing of *OsNRAMP5*. *J. Integr. Agric., 18*, 688–697.

Zafar, K., Sedeek, K. E. M., Rao, G. S., Khan, M. Z., Amin, I., Kamel, R., et al., (2020). Genome editing technologies for rice improvement: Progress, prospects, and safety concerns. *Front. Genome Ed., 2*, 5.

Zeng, Y., Wen, J., Zhao, W., Wang, Q., & Huang, W., (2020). Rational improvement of rice yield and cold tolerance by editing the three genes *OsPIN5b*, *GS3*, and *OsMYB30* with the CRISPR-Cas9 system. *Front. Plant Sci., 10*, 1663.

Zhang, A., Liu, Y., Wang, F., Li, T., Chen, Z., Kong, D., et al., (2019). Enhanced rice salinity tolerance via CRISPR/Cas9-targeted mutagenesis of the *OsRR22* gene. *Mol. Breed., 39*, 47.

Zhao, M., Zhao, M., Gu, S., Sun, J., Ma, Z., Wang, L., et al., (2019). *DEP1* is involved in regulating the carbon-nitrogen metabolic balance to affect grain yield and quality in rice (*Oryza sativa* L.). *PLoS One, 14*, e0213504.

Zhiguo, E., Chen, C., Yang, J., Tong, H., Li, T., Wang, L., & Chen, H., (2019). Genome-wide analysis of fatty acid desaturase genes in rice (*Oryza sativa* L.). *Sci. Rep., 9*, 19445.

Zhou, J., Xin, X., He, Y., Chen, H., Li, Q., Tang, X., et al., (2019). Multiplex QTL editing of grain-related genes improves yield in elite rice varieties. *Plant Cell Rep., 38*, 475–485.

Zhu, F., (2018). Anthocyanins in cereals: Composition and health effects. *Food Res. Int., 109*, 232–249.

Zhu, Q., Zeng, D., Yu, S., Cui, C., Li, J., Li, H., Chen, J., et al., (2018). From golden rice to aSTARice: Bioengineering astaxanthin biosynthesis in rice endosperm. *Mol. Plant, 11*, 1440–1448.

Zhu, Y., Lin, Y., Chen, S., Liu, H., Chen, Z., Fan, M., et al., (2019). CRISPR/Cas9-mediated functional recovery of the recessive rc allele to develop red rice. *Plant Biotechnol. J., 17*, 2096–2105.

Zong, Y., Wang, Y., Li, C., Zhang, R., Chen, K., Ran, Y., Qiu, J. L., Wang, D., & Gao, C., (2017). Precise base editing in rice, wheat and maize with a Cas9-cytidine deaminase fusion. *Nat. Biotechnol., 35*, 438–440.

CHAPTER 6

Genome Editing of Wheat with CRISPR/Cas Systems: Past, Present, and Future

SUMAN BAKSHI,[1*] UTTAM KUMAR,[2] PRADEEP BHATI,[2] and SANJAY J. JAMBHULKAR[1]

[1]Nuclear Agriculture and Biotechnology Division, Bhabha Atomic Research Center, Mumbai – 400085, Maharashtra, India

[2]Borlaug Institute for South Asia, Ladhowal, Ludhiana, Punjab – 141008, India

[*]Corresponding author. E-mail: sumansud@barc.gov.in

ABSTRACT

Targeted genome editing is a fast-paced technology for inducing mutations in the genes of interest, especially in the complex genome of bread wheat. The foremost requirement to understand gene function is to either modify/edit it or knock out the gene completely. The major bottleneck to understand gene function in wheat is the existence of homologous loci at three sub-genomes which are difficult to edit simultaneously. CRISPR/Cas9-based genome editing opened up new prospects to tailor crop traits and to generate desirable variability, which could be exploited in crop improvement. Targeted genome editing has been advanced in preciseness, minimizing off-targets, transgene-free, multiplexing, higher editing efficiency, and better transformation protocols since its first application in wheat. Significant developments in the field of precision editing and brief mention of altered traits and achievements are discussed.

6.1 INTRODUCTION

Staple food crops feed million mouths a day is one of the major sources of sustenance of human life on earth. Wheat is one of the staple cereals grown

CRISPR/Cas-Mediated Genome Editing in Plants, Deepu Pandita, Anu Pandita (Eds.)

over 216.0 million hectares with a production of 765.7 million tons (FAO Stat). The production of wheat crop is under threat due to the rapid pace of population increase, depreciation of cultivable land and continually changing environment. The rate at which wheat improvement occurred in past harmonized with utilization of natural genetic variation in the crop. However, further improvement is conceivable with the availability of functional genetic variability and this require massive efforts to induce locus-specific mutations which became reality with advances in field of mutagenesis especially the discovery of precise editing tool-the clustered regularly interspaced short palindromic repeats (CRISPR) and the CRISPR-associated tool (CRISPR-Cas9).

Bread wheat is an allohexaploid (AABBDD) with three homoeologous genomes from three diploid ancestors (Petersen et al., 2006). The difficulties posed by the very large size of the genome with highly repetitive DNA content of 80 to 90% resulted in obscuring of mutations induced. The genes located on the three genomes are similar in their coding regions; hence have redundancy in their function. The possibility of inducing mutations simultaneously in the genes located on all three genomes is very less. Therefore, genome editing technologies offered a perpetual solution to get rid of complications posed by complex wheat genome. Genomes editing technologies refers to the creation of precise genome modifications such as insertion, deletion, substitution, or knockouts at a predetermined specific locus in the genome. Site-specific nucleases such as zinc finger nucleases (ZFNs), transcription activator-like effector nucleases (TALENs) and CRISPR/associated (Cas) proteins are capable of inducing modifications at specific loci in the genome (CRISPR-Cas9). However, CRISPR-Cas9 method has been immensely used in crop plants due to its ease, versatility, and precision over other site-specific nucleases (SSNs) (Chen et al., 2019). Prokaryotes harbor CRISPR-CRISPR-associated protein 9 as a component of their immune system, and it captures a portion of DNA of invading organisms in their CRISPR locus and retains as a memory (Jinek et al., 2012). The CRISPR-Cas9 system is modified for their application as a genome editing tool. The modified CRISPR-Cas9 system has two components which are comprised of crRNA-tracrRNA duplex (sgRNA) and the Cas9 endonuclease. The function of sgRNA is to identify the target DNA and to guide Cas9 proteins to create double-strand breaks (DSBs) at the preselected site of the DNA. SpCas9 from *Streptococcus pyogenes*is is functionally characterized and contains two domains namely HNH and RuvC-like. These domains coordinate to cleave DSBs in the DNA and subsequently repaired by cells repair mechanism either through homologous directed recombination (HR) or non-homologous end joining (NHEJ) and might involve both repair mechanisms

(Danner et al., 2017). The prime requirement for CRISPR/Cas9 editing is the existence of a protospacer-adjacent motif (PAM) adjacent to the sequence of interest and the PAM for SpCas9 is NGG. Several modifications have been introduced to the CRISPR/Cas editing, which has improved the efficiency and preciseness of inducing alterations without off-target mutations. This chapter summarizes the advances and developments in the CRISPR/Cas-based genome editing system and traits modified in wheat crop.

6.2 ADVANCES IN GENOME EDITING TECHNOLOGY

Genome editing technology has undergone several changes to increase its effectiveness and efficiency. These technological advances included excision of specific bases, elimination of off-targets and transgene integration at random in the genome, improved delivery systems, better endonucleases, and efficient transformation efficiency. The CRISPR/Cas editing system for its successful application in wheat has been improved in several ways, and year-wise chronological developments are depicted in Figure 6.1.

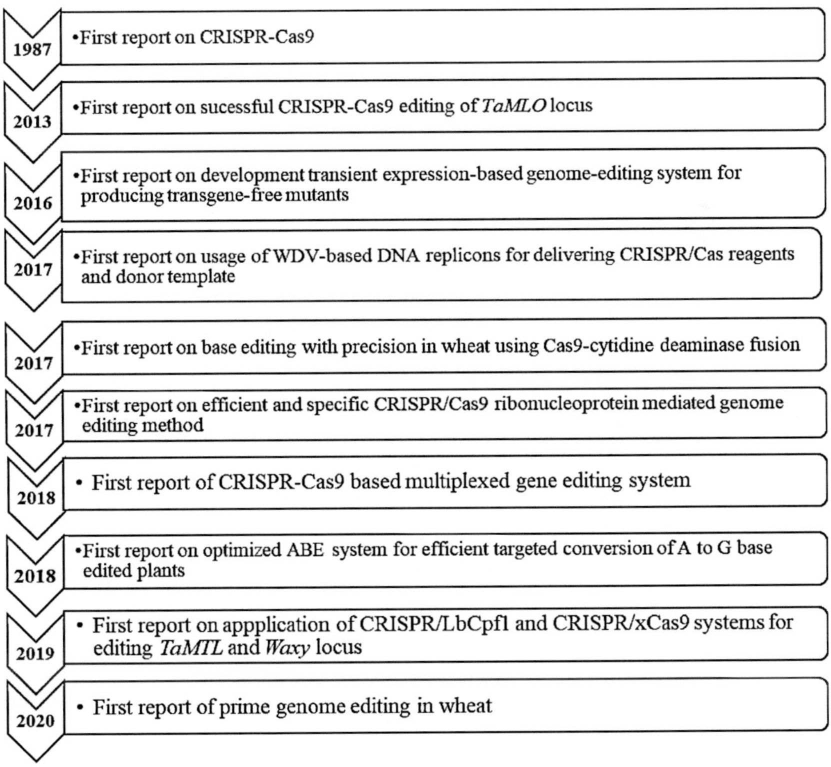

FIGURE 6.1 Advances in Targeted Genome Editing in Wheat

6.2.1 BASE EDITING

CRISPR-Cas-based system is a proven tool to induce knock-out mutants with high efficiency. However, the mutations induced by CRISPR-Cas editing are random insertions and deletions in the gene. The tool lacks the ability to produce single nucleotide changes in the genome. Homology-directed (HD) repair could be a substitute to deliver a donor template with desired changes; however, HD repair has very low efficiency. Therefore, a new technique known as base editing is evolved, which has the capability to cause direct, irreparable conversion of a single base of target DNA into another base in a programmable manner without imposing double-strand DNA backbone cleavage or involving a donor template. Base editing resulted in transitions which involve four types of point mutations with the specific base editors (Gaudelli et al., 2017; Komor et al., 2016; Nishida et al., 2016). The enzyme used for base editing is modified Cas9 in which acidic aspartate amino acid substituted with alanine (D10A) in the RuvC active domain to produce nickase Cas9 and other one is dead Cas9 (dCas9) where both domains RuvC and HNH are inactivated to produce a nuclease deficient Cas9. The dCas9 has the ability to bind the target locus precisely as directed by single guide RNA. However, it could not be possible for dCas9 to cleave DNA strands. The two types of editing involve usage of cytosine base (CB) editor in which nickase Cas9 is fused with enzyme cytidine deaminase and in other nickase Cas9 is fused with adenosine deaminase and known as adenosine base (AB) editor. In CB editing, the enzyme deaminates a target cytosine into uracil and it becomes Uracil-Guanine pair which is a mismatch. The nickase Cas9 generates a nick in guanine containing DNA strand to imagine repair machinery so that it anticipates guanine as mismatch. Therefore, the repair system excise guanine and an adenine are coordinated against uracil. In this CB editing, a CG pair is converted to TA pair. In AB editing, the enzyme deaminates the adenine (A) base into inosine (I) and IT base pair is produced. The nickase Cas9 creates a nick in thymine encompassing DNA strand and the cell's mismatch repair system eliminates thymine as incompatible base. This was followed by inosine to be anticipated as guanine and therefore cytosine is replaced against inosine. The Inosine-Cytosine pair is converted to Guanine-Cytosine pair.

6.2.2 PRIME EDITING

The prime editing is an excellent method to create desired insertions, deletions, specific transitions and transversions to edit targeted loci. This technique involves prime editing guide RNA (pegRNA) and prime editor.

Prime editing gRNA contained a spacer, a primer binding site, and a template sequence to be reverse transcribed. The prime editor comprises of Cas9 nickase and reverse transcriptase (RT) (Anzalone et al., 2019). The prime edited construct which harbors prime editor and prime editing gRNA as a complex reach to the target DNA site. The Cas9 nickase creates incisions in one of DNA strands with protospacer-adjacent motif sequence and produces a flap a 3'end, which is then hybridized to primer binding site of prime editing gRNA. The RT extends the nicked DNA strand by copying the sequence from prime editing gRNA and inserts the chosen edit in a strand of DNA. The generation of modified strand with edited sequence and other original strand with no edit forms a heteroduplex. This is followed by second nick in unedited strand using sgRNA and ultimately repaired by using sequence information from modified strand with edited sequence and causing insertion of chosen edit in both strands of target DNA. The first report of editing using plant prime editors in wheat protoplast resulted into transversions namely A to T and C to G in *Ubiqtin*, G to C in *TaGW2* and *TaGASR7*, C to G in *TaLOX2*, T to G in *TaMLO* and C to A in *TaDME1* gene (Lin et al., 2020).

6.2.3 TRANSGENE AND OFF-TARGET FREE WHEAT MUTANTS

Many of plant species are recalcitrant to transformation techniques and involved integration of foreign DNA into the transformed cells resulting into off-target mutations. Zhang et al. (2016) reported transient expression based CRISPR/Cas9 editing for developing transgene-free homozygous mutants in wheat. Mutant plants are regenerated after transiently expressing CRISPR/Cas9 DNA or *in vitro* transcripts of Cas9-coding sequence and gRNA into wheat callus cells. The study further established that genome editing methods and tissue culture protocols require no herbicide selection and are efficient in both bread wheat and durum wheat. Transgene free mutants were developed for six genes namely *TaGASR7*, *TaGW2*, *TaDEP1*, *TaNAC2*, *TaPIN1*, and *TaLOX2* in wheat.

The conventional CRISPR/Cas9 methodology involved delivery of CRISPR/Cas9 DNA vector constructs into plant cells through *Agrobacterium* mediated transfer or particle bombardment, which then cleave target sequences and induce mutations. There is the likelihood of CRISPR/Cas9 cassettes upon breakdown get integrated into non-targeted sites in the plant genome. These non-targeted insertions have been found to cause transgene integrations and off-target mutations. This poses a great risk regarding the safety of genome edited crops for human and animal health. Liang et al. (2017) developed CRISPR/Cas9 ribonucleoproteins (RNPs) and delivered into immature embryo cells of bread wheat by particle bombardment. The RNP facilitated editing has been found to be efficient with substantial reduction in off-target mutations in the edited cell and recovered plants are free from any unwanted mutations.

6.2.4 *GENE EDITING VIA REPLICON-BASED SYSTEM*

The most common delivery system to edit plant genomes are *Agrobacterium* transfer-DNA, particle bombardment or stable integration of nuclease-encompassing cassettes and DNA repair templates into the genome. Gil-Humanes et al. (2017) devised a wheat dwarf virus (WDV) based DNA replicon for gene targeting in wheat and other cereal crops. Application of WDV replicon for delivering CRISPR/Cas9 cassette and DNA repair templates resulted in gene targeting of *Ubiquitin* locus with 12 times higher frequency over other delivery methods. The replicon-based delivery system facilitated multiplexed gene targeting through homologous recombination (HR) at all six homeoalleles of bread wheat.

6.2.5 *CAS VARIANTS*

The Cpf1 system from *Prevotella* and *Francisella* 1 is type V CRISPR/Cas12a systems found more accurate and efficient for genetic editing system. Cpf1 has smaller size endonuclease and require short CRISPR-RNA with better working efficiency as compared Cas9. The Cpf1 system circumvents the trans-activating crRNA (tracrRNA) obligation for the maturation of crRNAs. Cpf1 operates through T-rich protospacer-adjacent motif site (5′-TTTN-3′) at the 5′ end of the protospacer region. Cpf1 proteins generate DSBs with 5’ overhangs as compared to the blunt ends formed by type II CRISPR/Cas9 systems. Cpf1 protein has RNase III activity which carries pre-crRNA processing and utilized for multiplex genome engineering via tandemly arranged pre-crRNA-coding construct. Further widening the genome editing tools, Cas9 nucleases with modified PAM sequences and improved cleavage specificity have been engineered. One of these engineered xCas9, which has a better ability to identify range of PAM sequences and hence, reduce random unintended mutations in the genome. Liu et al. (2020a) utilized these modified Cas nucleases (LbCpf1 and xCas9) for targeted editing of *TaMTL* and *TaWaxy* in wheat.

6.2.6 *SIMULTANEOUS TARGETING OF MORE GENES*

Wheat is an allohexaploid with three sub-genomes (A, B, D) and requires a scheme to induce mutations at all three homeologs of locus simultaneously. There are three multiplexing methods which include tandem repeats of separate U3 or U6 promoters (TSRP), tRNA processing mechanism, and ribozyme processing system. In the TSRP system, strong Pol III promotors namely TaU3, TaU6.1, and TaU6.3 have been used for independent expression of single-guided RNA. Using this method in wheat, Zhang et al. (2019a) edited four regulatory genes which are *TaCKX2-1*, *TaGLW7*, *TaGW2,* and *TaGW8* for grain traits in wheat. The other method used for delivering and expression of multiple genes under control of TaU3 promoter in wheat is tRNA-sgRNA construct (Xie et al., 2015). CRISPR/Cas9 facilitated targeted editing by the ribozyme-gRNA-ribozyme (RGR) system which is driven by the single promoter and successfully used in plant (Tang et al., 2016; Cermák et al., 2017). Li et al. (2020b) compared all three systems of multiplex editing in wheat and established that the tRNA and ribozyme are more effective for multiplex genome editing. However, the ribozyme-based system has shown more effectiveness and edited several homoeologous genes at multiplex chromosomes and resulted in loss of function mutants.

6.3 EFFORTS TO IMPROVE TRAITS THROUGH TARGETED EDITING

CRISPR-Cas editing developed as a favored reverse genetic approach to improve wheat crop due to its flexibility, versatility, efficiency, and cost-effective approach for targeted sequence-specific modifications. The CRISPR/Cas system has undergone several adaptations and the most noteworthy are based editing (Komor et al., 2016; Gaudelli et al., 2017) and prime editing (Anzalone et al., 2019) which facilitated precise single nucleotide substitutions in every possible combination and widened the scope of technique. The widespread applications of CRISPR-CRISPR-associated protein (CRISPR/Cas) editing in wheat is evident from targeting simple to complex traits and further broadening the germplasm of wheat crop with stable mutations at all homeologs of a gene (Table 6.1).

6.3.1 HERBICIDE-TOLERANT CROP VARIETIES

Herbicide-tolerant crop varieties are prerequisite for economical control of weeds and sustaining quality and quantity of crop produce. A reasonable amount of success is attained in developing wheat tolerant against imidazolinone (IMI) herbicides by inducing point mutations using chemical mutagens (Newhouse et al., 1992). The newly emerged base editing tool has the advantage of introducing point mutations in the loci conferring tolerance to herbicides. Zhang et al. (2019b) used this tool to edit acetolactate synthase (ALS) and acetyl-coenzyme-A carboxylase genes controlling tolerance to sulfonylurea, IMI, and aryloxyphenoxy propionate-type herbicides in wheat has been developed. In this system, base editing at the wheat ALS Pro-174 codon (TaALS-P174) made wheat resistant to nicosulfuron herbicide which has been used in MS growth medium for selection. This TaALS-P174 editor coupled with editors for other targets sequences resulted in coediting in the nicosulfuron-resistant plants and selection for nicosulfuron resistance in growth medium also enhanced the frequency of other edited targets.

6.3.2 FACILITATING GENES TRANSFER FROM WILD RELATIVES

Wild relatives of wheat are one of the important sources of contributing important genes for several traits to cultivated wheat. The limitation to the successful transfer of traits is the low frequency of homoeologous crossover between chromosomes of cultivated wheat and wild relatives. Deletion mutants for *ph1b* gene have been exploited for enhancing homologous crossover frequency in wheat. It has been further deciphered that meiotic crossover gene ZIP4 is located in the 70Mb deletion region of chromosome 5B and responsible for limiting homologous crossover. Hence, *ph1b* phenotype is the outcome of *Tazip4-B2* deletion mutant. Rey et al. (2018) reported editing of *ZIP4* gene of wheat using CRISPR-Cas9 system and the *Tazip4 ph1b* mutant upon crossing with *Aegilops variablis* showed homoeologous pairing. Using these deletion mutants, it has been established that homoeologous cross over frequency is increased in the presence of Mg^{2+} ion and magnesium supplementation to improve genetic introgressions from wild relative to cultivated wheat.

6.3.3 ACHIEVING COMMERCIALLY EXPLOITABLE MALE STERILITY

Stable male sterile system is the prime requirement for hybrid seed production. Taigu and Dwarf sterility system known to exist in wheat is under control of a dominant gene *Ms2* located on short arm of 4-D chromosome. When the genotype with *Ms2* gene hybridized with fertile parent, then only 50% of plant are male fertile in F_1 generation. Tang et al. (2021) edited *Ms2* gene in male sterile lines and recovered male fertile plants with improved plant traits and better quality. This strategy could be exploited for fertility restoration of male sterile durum wheat and triticale lines endowed with *Ms2* gene. Okada et al. (2019) applied CRISPR/Cas9 editing to cause biallelic frameshift mutation in a male fertility locus *Ms1* and generated complete male sterility in cultivars Fielder and Gladius which opened an avenue to rapidly generate male sterility in any cultivars for exploitation of heterosis.

TABLE 6.1 Application of Targeted Genome Editing in Wheat

Target Gene	Trait Altered	Delivery Mode	Explant Used	Mutations Induced	References
TaMLO	Powdery mildew resistance	PEG-mediated transfection	Protoplast	Insertions–deletions	Shan et al. (2013)
TaMLO-A1	Powdery mildew resistance	Biolistic transformation	Immature embryos	Insertions–deletions	Wang et al. (2014)
TaLOX2	Lipoxygenase 2, chloroplastic precursor	PEG-mediated	Protoplast	Insertions–deletions	Shan et al. (2014)
TaLOX2	Increased storability	Particle bombardment	Immature embryos	C-to-T conversion	Zong et al. (2017)
TaEDR1	Resistance to powdery mildew	Biolistic transformation	Immature embryos	Insertions–deletions	Zhang et al. (2017)
Ubiquitin	Regulatory protein	WDV-derived replicon	Protoplast	HDR	Gil-Humanes et al. (2017)
TaPDS	Desaturation of phytoene to zeta-carotene and inositol oxygenase	*Agrobacterium*-mediated	Immature embryos	Insertions–deletions	Howells et al. (2018)
α-gliadin	Low gluten	Biolistic transformation	Immature scutella	Insertions–deletions	Sanchez-Leon et al. (2018)
TaMs45	Male sterility	*Agrobacterium*-mediated	Immature embryos	Insertions–deletions	Singh et al. (2018)
TaZIP4	Increased homoeologous CO frequency	*Agrobacterium*-mediated	Immature embryos	Insertions–deletions	Rey et al. (2018)
TaGW2	Grain size and thousand grain weight	Particle bombardment	Immature embryos	Insertions–deletions	Wang et al. (2018c)
α-gliadin, γ-gliadin	Low immunogenic wheat	*Agrobacterium*-mediated	Immature embryos	Insertions–deletions	Jouanin et al. (2019)
Ta*GW7*	Grain shape and weight	Particle bombardment	Immature embryos	Insertions–deletions	Wang et al. (2019)
Qsd1	Seed dormancy	*Agrobacterium*-mediated	Immature embryos	Insertions–deletions	Abe et al. (2019)
TaMs1	Male sterility	*Agrobacterium*/ bombardment	Immature embryos	Insertions–deletions-Knock out	Okada et al. (2019)

TABLE 6.1 *(Continued)*

Target Gene	Trait Altered	Delivery Mode	Explant Used	Mutations Induced	References
TaHRC	*Fusarium* head blast resistance	Particle bombardment	Immature embryos	Knock out	Su et al. (2019)
TaMs2	Male sterility	*Agrobacterium*	Immature embryos	Insertions–deletions	Tang et al. (2021)
TaCENH3	High haploid induction rate	Particle bombardment	Immature embryos	Indels, restored frameshift	Lv et al. (2020)
TaNP1	Male sterility	Particle bombardment	Immature embryos	Insertions–deletions	Li et al. (2020a)
Q-locus	Spike morphogenesis and grain threshability	*Agrobacterium*-mediated	Immature embryos	Insertions–deletions	Liu et al. (2020b)
TaSBEIIa	High amylose and resistant starch	Bombardment	Immature embryos	Insertions–deletions	Li et al. (2021)
TaASN2	Low asparagine	Particle bombardment	Immature embryos	Insertions–deletions	Raffan et al. (2021)
TaARE1	Nitrogen use efficiency	Particle bombardment	Immature embryos	Knock out	Zhang et al. (2021)
TaPDS, *TaINOX*	*Phytoene desaturase and inositol oxygenase*	*Agrobacterium*-mediated	Immature embryos	Insertions–deletions	Upadhyay et al. (2013)
TaDREB2, *TaERF3*	DREB: drought tolerance ERF: root elongation and root hair development	PEG mediated	Protoplast	Knock out	Kim et al. (2017)
TaLox2, *TaUbiL1*	Lox2: lipoxygenase 2 enzymes for hydrolysis of polyunsaturated fatty acids	Neon transfection system	Microspore transfection	Indels, frameshift	Bhowmik et al. (2018)
TaALS, *TaMTL*	ALS: Herbicide resistance, *MTL: sperm-specific phospholipase 1*	Particle bombardment	Immature embryos	C-to-T conversion Base editing	Zong et al. (2018)
TaDEP1, *TaGW2*	Panicle growth, grain weight	Particle bombardment	Immature embryos	A-to-G conversion	Li et al. (2018)

TABLE 6.1 *(Continued)*

Target Gene	Trait Altered	Delivery Mode	Explant Used	Mutations Induced	References
*Ta*ALS, *Ta*ACCase	Herbicide resistance	Particle bombardment	Immature embryos	C to T and C to G conversion, Base editing	Zhang et al. (2019b)
TaMTL, *TaWaxy*	Haploid induction, amylose content	*Agrobacterium*-mediated	Immature embryos	Insertions–deletions	Liu et al. (2020a)
TaGW2, *TaLpx-1*, *TaMLO*	GW2: grain traits, Lpx-1: 9-lipoxygenase affect resistant to *Fusarium graminearum; TaMLO:* resistance to powdery mildew	Particle bombardment	Immature embryos	Insertions–deletions	Wang et al. (2018b)
TaDA1/ TaPDS/ TaNCED1	DA1-grain size PDS: *Phytoene desaturase NCED-9-cis-epoxy-carotenoid dioxygenase*	*Agrobacterium* transformation	Immature embryos	Insertions–deletions	Li et al. (2020b)
TaABCC6, *TaNFXL1*, *TansLTP*, *TansLTP94*	ABCC6:ABC transporter, NFXL: Nuclear transcription Xbox binding like1, nsLTP, nsLTP9.4 – a non-specific lipid transfer protein associated with FHB resistance	PE mediated	Protoplast	Knock out	Cui et al. (2017)
TaCKX2-1, *TaGLW7*, *TaGW2*, *TaGW8*	Grain traits	*Agrobacterium* transformation	Immature embryos	Insertions–deletions	Zhang et al. (2019a)

TABLE 6.1 *(Continued)*

Target Gene	Trait Altered	Delivery Mode	Explant Used	Mutations Induced	References
TaPinb, *TaDA1*, *TaDA2*, *TaNCED*, *TaLPR2*	*Pinb*: Puroindoline b, *TaDA1* and *TaDA2*-grain development, *TaNCED1*: 9-cis epoxycarotenoid dioxygenase, *TaLPR2*-low phosphate response	*Agrobacterium* transformation	Immature embryos	Insertions–deletions	Zhang et al. (2019c)
TaGASR7, *TaGW2*, *TaDEP1*, *TaNAC2*, *TaPIN1*, *TaLOX2*	*GASR7*: grain length and width, GW2: grain weight, DEP1: Panicle growth and grain yield, NAC2: shoot branching, PIN1: auxin-dependent adventitious root emergence and tillering, LOX2: storability of wheat grains	Particle bombardment	Immature embryos	Insertions–deletions	Zhang et al. (2016)
Ubi10, *TaGW2*, *TaGASR7*, *TaDME1*, *TaLOX2*, *TaMLO*	DME: endosperm gene imprinting	PEG-mediated	Protoplast	Point mutations, through prime editing	Lin et al. (2020)
TaQsd1, *TaARE1*, *TaNPT1*, *TaSBEIIa*, *TaSPDT*, *TAIPA1*	Qsd1-Seed dormancy ARE1-NUE NPT1, *IPA1*-plant type SBEIIa-high amylose SPDT-phosphorus transport	Particle bombardment	Immature embryos	Insertions–deletions	Luo et al. (2021)

Similarly, *TaMs45* gene responsible for strictosamide synthase-like enzyme is needed for male fertility was exploited to generate variety of mutations in the homoeologous of *Ms45* gene individually and also simultaneously in all three homeologs in wheat (Singh et al., 2018). Further analysis of these genetic mutations showed all three *Ms45* homeologs caused male fertility except triple homozygous mutants which affect the development of pollens and lead to male sterility. In another study by Li et al. (2020a) reported usage of optimized CRISPR/Cas9 system for inducing triple homozygous mutations for male sterility which can be commercially used for hybrid seed production. Targeted editing of *TaNP1* in wheat which is an ortholog of *OsNP1* in rice and *ZmIPE1* in maize and known to translate into a putative glucose-methanol-choline oxidoreductase in tapetum of the pollen required for male fertility. CRISPR/Cas9 editing system was optimized to edit three homoeoalleles of *TaNP1* and introduced complete sterility in wheat (Li et al., 2020a). However, male fertility could be maintained with one copy of wild type *TaNP1* allele.

6.3.4 HAPLOID INDUCTION

Studies in maize and rice reported increased frequency of haploids by inducing mutations in the gene *MATRILINEAL* (*MTL*), which synthesizes specific phospholipases in the pollen (Kelliher et al., 2017; Liu et al., 2017; Yao et al., 2018). Orthologs of *MATRILINEAL* gene in wheat, *TaMTL* were found to be located on 4A, 4B, and 4D chromosomes. CRISPR/Cas9 editing of *TaMTL* causes maternal haploid formation in double and triple knockouts at a frequency of 10% and 11.8–31.6%, respectively (Liu et al., 2020a). The paternal haploid induction scheme has an underlying mechanism of modified *Centromeric Histone* (*CENH3*). Genome-edited *TaCENH3α*-heteroalleles in B and D sub-genomes followed by identification of paternal haploid inducer lines with haploid induction rate ≈7% has been reported in wheat by Lv et al. (2020).

6.3.5 NITROGEN USE EFFICIENT WHEAT

Developing nitrogen use efficient staple food crops is farmer and environment friendly approaches especially in the era of depleting soil fertility. Modern dwarf wheat cultivars utilize 40% of applied nitrogen. Wang et al. (2018a) characterized *are1* locus in rice mutant defective in assimilation of nitrogen. The cytokinin response1 repressor1 (*ARE1*) is a negative regulator of nitrogen assimilation and null mutation at this locus contribute to delay in senescence, increased nitrogen uses efficiency (NUE) and higher yield under low nitrogen conditions. Zhang et al. (2021) developed a range of partial to triple null allele transgene-free mutant wheat lines using CRISPR-Cas9 editing of *TaARE1* locus. The lines which have null-mutations at only D-loci keeping other two loci intact (AABBdd), and null alleles at both A and B-locus keeping intact alleles at D-locus (aabbDD) displayed higher NUE, delayed leaf senescence and higher seed yield without any negative effects on growth compared to control plants. The results of the study established that it is feasible to modify homeoalleles of *TaARE1* to engineer nitrogen use efficient wheat with improved grain yield to grow under low nitrogen fertilizer conditions.

6.3.6 DISEASE RESISTANCE (DR)

Powdery mildew caused by *Blumeria graminis f. sp. Tritici* results in substantial yield loss in wheat. It is controlled through fungicide sprays and resistant genes in the varieties. Nevertheless, resistant genes are effective against specific isolates of fungal pathogen, and mutations in pathogen resulted into development of new virulent strains. Thus, mutation of a loci contributing susceptibility to the powdery mildew pathogen has been used for developing broad-spectrum and durable resistance in barley (Jørgensen, 1992; Kusch and Panstruga, 2017) and wheat (Acevedo-Garcia et al., 2017). The first fruitful attempt in wheat to knock out *TaMLO* locus conferring resistance to powdery mildew was carried out by Shan et al. (2013); and Wang et al. (2014). Later, Zhang et al. (2017) targeted *EDR1* (enhanced disease resistance) locus, a negative regulator of resistance against powdery mildew and has conserved structure across species. Genome editing with CRISPR/Cas9 system for simultaneous alteration in all three homeologs of wheat *EDR1* has been developed. The regenerated resistant CRISPR-edited plants obtained are free of any off-target mutations and no pleiotropic effect on any other trait. Other disease against which CRISPR editing is successfully applied is *Fusarium* head blight (FHB) caused by *Fusarium graminearum*. FHB toxins, namely deoxynivalenol found in infected grains is the cause of concern for food safety and reduce the market acceptability. Therefore, inducing alterations at loci contributing susceptibility is one of the strategies to develop resistance against FHB pathogen. CRISPR-Cas9 editing of

TaHRC locus induced a frameshift mutation and a stop codon at 264th place from initiation of the start codon. Editing of the locus resulted into loss of function alteration and improved FHB resistance (Su et al., 2019).

6.3.7 GRAIN WEIGHT TRAIT

Grain size and weight are major factors influencing seed yield. Homeologs of *TaGW2* negatively influence grain size and gain weight in wheat. Mutations caused by CRISPR-Cas9-based genome editing contribute dosage-dependent effect on thousand grain weight (TGW). Wang et al. (2018c) found that CRISPR-Cas9 edited mutations in single genome showed 5.5%, double homologue mutants in any combination resulted into 12.1% and triple mutants shown 16.3% average increase in TGW over wild type cultivar. Trait targeted variation for TGW will help crop scientists to combine mutant alleles in desired combination for refining the TGW traits in their selections. Other locus which has been targeted for causing mutations for grain size and weight is *TaGW7*. CRISPR-Cas9 editing of *TaGW7* homoeologs showed that mutations in either B and D genome copies or on both genomes has positive effect resulting in increase of grain width and grain weight, however, negative effect on grain length (Wang et al., 2019). The editing of genes for grain size and weight is dosage dependent with double mutants in homeologs showed larger effect than single copy mutants.

6.3.8 QUALITY IMPROVEMENT

Wheat products rich in amylose and resistant starch content are desirable for human health and largely reduce the risk of metabolic disorders. Efforts have been made in the past using target induced local lesions in genome (TILLING) (Hazard et al., 2012) and RNAi (Regina et al., 2006) to develop wheat rich in amylose and resistant starch content. The new gene editing technique particularly CRISPR/Cas9 has become the method of choice in wheat due to its capability to alter all three homeoloci located on subgenome A, B, and D. Li et al. (2021) altered *TaSBEII* locus using CRISPR-Cas9 and developed series of transgene-free mutant lines in winter and spring wheat genotype. The obtained mutant lines have homeologs edited at single subgenomes (AAbbDD, AABBdd), at two subgenomes (aabbDD, AAbbdd, aaBBdd) and at all three subgenomes (aabbdd) in winter wheat cultivar ZM whereas in spring wheat only two types of mutants with *TaSBEII* locus edited in two subgenome (AAbbdd) and modified in all sub genomes increased of amylose content and resistant starch compared to non-edited control. Other flour quality determining trait is amylose synthesis in the presence of grain bound starch synthase-1 which is encoded by the Waxy gene. CRISPR/SpCas9 has been used effectively to edit *TaWaxy* homeologs located on 4A, 4B, and 4D chromosome (Liu et al., 2020a).

6.3.9 COELIAC SAFE WHEAT

Wheat grain is used for making flour which is used to prepare dough for end-use purposes. The dough viscoelastic properties responsible for various end-use purposes are determined by gluten proteins. However, these gluten proteins are known to cause coeliac disease in 1–2% of genetically susceptible population in the world. The only solution for people with this auto-immune disorder is a gluten-free diet. Gluten-free foods lack viscoelasticity and many additives were added to develop rheological properties as that of wheat. The development of wheat genotypes which have reduced or no immunogenic epitomes and retain their processing quality is the requirement for individual with this disorder. Immunoreactive response is due peptide sequences present on α, γ and ω gliadins and some of low molecular weight glutenins whereas high molecular weight glutenins are mostly safe for gluten sensitive individuals. CRISPR/Cas9 editing has been used to mutate conserved regions of α-gliadin genes to produce low immunogenic wheat genotypes in the durum and bread wheat (Sanchez-Leon et al., 2018). Jouanin et al. (2019) edited bread wheat cultivar 'Fielder' with constructs comprising Cas9 and single guide RNAs combinations causing precise mutations at the region upstream or within the epitomes in the α-gliadin and γ-gliadin genes.

6.3.10 REDUCED FREE ASPARAGINE IN WHEAT GRAIN

Presence of free asparagine in food products is a matter of health concern. Free asparagine in food upon cooking and processing gets converted into acrylamide which is a potent carcinogen for all age groups. Wheat has five asparagine synthetase genes and among those TaASN2 has highest expression in the embryo (Curtis et al., 2019). Raffan et al. (2021) used CRISPR/Cas9 editing to induce mutation in *TaASN2* gene to lower down the expression of asparagine synthetase gene (*TaASN2*) in wheat grain. The amount of free asparagine is significantly reduced in the grains of plants with edited all the six *TaASN2* homeoalleles compared to non-edited control plants and one of the plants has shown 90% reduction in free asparagine in second generation of transformed plants.

6.3.11 ALTERED SEED DORMANCY

Sprouting of grains in spike prior to harvest owing to untimely rains is one of the threats experienced in the field due to climate change. Seed dormancy for a long period in barley is observed due to recessively controlled quantitative trait locus (QTL) of seed dormancy 1, *Qsd1* (Takeda and Hori, 2007). Barley has similarity in genome structure with wheat and many orthologous genes shares comparable functions in both species. However, *Qsd1* orthologs have not been studied in bread wheat through induced mutagenesis because difficulty to introduce mutation at all three homeoalleles in an individual plant. Abe et al. (2019) attained mutations in all three homeoloci of *TaQsd1* locus through CRISPR/Cas9 mediated targeted editing using *Agrobacterium* facilitated gene transfer. The mutant with all three recessive homeoalles has shown longer dormancy period compared to wild type parent. Loss of function mutations in altered phenotype has been induced due to frameshift of nucleotide bases at 14th exon *TaQsd1* sequence.

6.3.12 ABIOTIC STRESS TOLERANCE (AT)

Abiotic stress conditions such as heat and drought affect wheat crop yield and quality. Transcriptional signaling of various stress response mechanisms got activated or deactivated on the occurrence of drought or heat stress in plants. Two transcription factors viz. dehydration responsive element binding protein 2 (Morran et al., 2011) and ethylene-responsive factor 3 (Cheng et al., 2016) are known to contribute to drought tolerance. In a genome editing study based on CRISPR/Cas9 system for transcriptional factors, *TaDREB2* and *TaERF3* was performed in wheat protoplast (Kim et al., 2018). Targeted modification of both drought-induced transcription factors was achieved with transient expression sgRNA and Cas9 protein in wheat protoplast however, off-target mutations were also observed.

6.4 MULTI TRAIT EDITING

Targeted editing of more than one gene requires improvement in existing CRISPR/Cas9 system to enhance its specificity and efficiency to cause mutations in desired genes. Wang et al. (2018b) has designed genome editing construct by uniting tandemly arranged tRNA-gRNA units under control of single promotor to introduce genetically transmissible mutations in the *TaGW2*, *TaLpx-1*, and *TaMLO* genes of bread wheat. The loss of function mutations induced in all three copies of the *TAGW2* gene-directed to a significant increase of the TGW, grain area, grain width, and grain length. Further, CRISPR/Cas9 editing activity could be seen in following generations suggesting and gives opportunity to recover additional mutations. Li et al. (2020b) developed multiplex genome editing construct involving Pol III promoter TaU3 to drive the expression of tRNA-gRNAs units to introduce three genes namely *TaDA1*, *TaPDS*, and *TaNCED1* through *Agrobacterium*-mediated transformation in wheat. Zhang et al. (2019a) emphasized on use of modified *Agrobacterium*-mediated delivery of CRISPR/Cas9 system in which codon-optimized Cas9 under control of promotor of maize ubiquitin gene and a gRNA cassette under control of TAU6 promotor in a single binary vector. Targeted genome editing of four-grain regulatory genes *TaCKX2-1*, *TaGLW7*, *TaGW2*, and *TaGW8* using modified transformation method found to be highly effective for perceiving essential mutations in T_0, T_1, and T_2 generations without identifying any off-target mutation. This study demonstrated that optimized CRISPR/Cas9 editing remain effective for causing mutations in future generation of transformed plants. Zhang et al. (2019c) applied CRISPR/Cas9 system to achieve efficient mutagenesis in five gene's namely: *TaPinb*, *TaDA1*, *TaDA2*, *TaNCED1* and *TaLPR2* through *Agrobacterium*-mediated transformation of a sgRNA targeting single, double, and all three genomes' homologues in wheat. The Cas9/sgRNA existence in the transformed lines has been found to cause additional mutation in following generations; however, transformed lines without Cas9/sgRNA have been found to retain the desired mutations. The induced mutations displayed stable inheritance in the following generations and editing efficiency of transformed mutant lines increased across generations. Luo et al. (2021) applied a universal multi-gene polycistronic tRNA editing technology in which six target genes viz. *TaQsd1, TaARE1, TaNPT1, TaSBEIIa, TaSPDT*, and *TAIPA1* were used to construct synchronously targeting multigene editing vector of two, three, four, and five gene combinations. Transformed plants obtained were transgene-free with targeted mutagenesis of up to 15 genomic loci and stacked the favorable alleles in an elite variety of winter wheat.

6.5 CONCLUSIONS

Wheat being a staple food crop and grown over diverse environments has always been a priority for improvement using innovative methods such as haploid induction, alien gene transfer, male sterility system, induced mutagenesis targeted induced local lesions in genome and transgenic technology. However, each of these techniques has limited success due to complexity of wheat genome and existence of three homologous genes in sub-genome A, B, and D. Other major bottleneck is limited genetic variability in wheat germplasm due to single hybridization event in the history of wheat origin and domestication resulted into cultivated wheat. CRISPR/Cas system has shown potential to overcome both these limitations and therefore wheat becomes a favored crop for precision-based genome editing. The CRISPR/Cas-based genome editing turns into an exemplary cutting-edge technology for understanding gene functions and trait improvement in wheat crop. The genetic heritable deviations induced through precise genome editing has wide spectrum ranged from insertions, deletions, single base pair conversions (transitions or transversions), frameshift mutations, gene/allele replacement to large deletions resulting into gain or loss of function mutants. Therefore, this editing technology is utilized in wheat to modify most of commercial traits in a span of 7–8 years. The locus-specific editing for most of economically important traits in wheat included disease resistance, abiotic stress resistance, resistance to pre-harvest sprouting, NUE, herbicide resistance, gluten-free wheat, reduced amount of free asparagine, male sterility, grain shape, TGW, and grain yield. The paramount technological advancement in perspective of crop improvement occurred in CRISPR/Cas approach is its adaptation for inducting mutations at single, two, three, and many genes simultaneously. Pyramiding of two or more genes in a popular cultivar through any conventional approach is exceedingly difficult due dominant/recessive nature, masking effect, gene silencing and then multi-generation hybridization and selfing required to achieve the trait expression. The precise genome editing techniques with its modification which involved using efficient promotors and gene multiplexing system made it possible to modify two, three, four, five, and up to six genes in wheat. CRISPR/Cas approach with various modifications has the potential to improve desired traits in wheat and strengthen wheat germplasm to deal with anticipated problems to sustain production in the period of climate change for sustainable food security.

KEYWORDS

- **CRISPR/Cas9**
- **multiplexing**
- **non-homologous end joining**
- **prime editing**
- **transgene-free genome editing**
- **zinc finger nucleases**

REFERENCES

Abe, F., Haque, E., Hisano, H., Tanaka, T., Kamiya, Y., Mikami, M., Kawaura, K., et al., (2019). Genome-edited triple-recessive mutation alters seed dormancy in wheat. *Cell Rep., 28,* 1362–1369. doi: 10.1016/j.celrep.2019.06.090.

Acevedo-Garcia, J., Spencer, D., Thieron, H., Reinstädler, A., Hammond-Kosack, K., Phillips, A. L., & Panstruga, R., (2017). *mlo*-based powdery mildew resistance in hexaploid bread wheat generated by a non-transgenic TILLING approach. *Plant Biotechnol. J., 15,* 367–378. doi: 10.1111/pbi.12631.

Anzalone, A. V., Randolph, P. B., Davis, J. R., Sousa, A. A., Koblan, L. W., Levy, J. M., Chen, P. J., et al., (2019). Search-and-replace genome editing without double-strand breaks or donor DNA. *Nature, 576*, 149–157. doi: 10.1038/s41586-019-1711-4.

Bhowmik, P., Ellison, E., Polley, B., Bollina, V., Kulkarni, M., Ghanbarnia, K., Song, H., et al., (2018). Targeted mutagenesis in wheat microspores using CRISPR/Cas9. *Sci. Rep., 8*, 6502. doi: 10.1038/s41598-018-24690-8.

Cermák, T., Curtin, S. J., Gil-Humanes, J., Cegan, R., Kono, T. J. Y., Konecná, E., Belanto, J. J., et al., (2017). A multipurpose toolkit to enable advanced genome engineering in plants. *Plant Cell, 29*, 1196–1217.

Chen, K., Wang, Y., Zhang, R., Zhang, H., & Gao, C., (2019). CRISPR/Cas genome editing and precision plant breeding in agriculture. *Annu. Rev. Plant Biol., 70*, 667–697. doi: 10.1146/annurev-arplant-050718-100049.

Cheng, S., Zhou, D. X., & Zhao, Y., (2016). WUSCHEL-related homeobox gene WOX11 increases rice drought resistance by controlling root hair formation and root system development. *Plant Signal. Behav., 11,* e1130198. doi: 10.1080/15592324.2015.1130198.

Cui, X., (2017). *Targeted Gene Editing Using CRISPR/Cas9 in a Wheat Protoplast System.* PhD Thesis, University of Ottawa; Xiucheng Cui.

Curtis, T. Y., Raffan, S., Wan, Y., King, R., Gonzalez-Uriarte, A., & Halford, N. G., (2019). Contrasting gene expression patterns in grain of high and low asparagine wheat genotypes in response to Sulphur supply. *B.M.C. Genomics, 20,* 628. doi: 10.1186/s12864-019-5991-8.

Danner, E., Bashir, S., Yumlu, S., Wurst, W., Wefers, B., & Kühn, R., (2017). Control of gene editing by manipulation of DNA repair mechanisms. *Mamm. Genome, 28*, 262–274. doi: 10.1007/s00335-017-9688-5.

Gaudelli, N. M., Komor, A. C., Rees, H. A., Packer, M. S., Badran, A. H., Bryson, D. I., & Liu, D. R., (2017). Programmable base editing of Al T to Gl C in genomic DNA without DNA cleavage. *Nature, 551*(7681), 464–471. doi: 10.1038/nature24644.

Gil-Humanes, J., Wang, Y., Liang, Z., Shan, Q., Ozuna, C. V., Sánchez-León, S., Baltes, N. J., et al., (2017). High-efficiency gene targeting in hexaploid wheat using DNA replicons and CRISPR/Cas9. *Plant J., 89,* 1251–1262. doi: 10.1111/tpj.13446.

Hazard, B., Zhang, X., Colasuonno, P., Uauy, C., Beckles, D. M., & Dubcovsky, J., (2012). Induced mutations in the *starch branching enzyme II* (*SBEII*) genes increase amylose and resistant starch content in durum wheat. *Crop Sci., 52*(4), 1754–1766. doi: 10.2135/cropsci2012.02.0126.

Howells, R. M., Craze, M., Bowden, S., & Wallington, E. J., (2018). Efficient generation of stable, heritable gene edits in wheat using CRISPR/Cas9. *B.M.C. Plant Biol., 18,* 215. doi: 10.1186/s12870-018-1433-z.

Jinek, M., Chylinski, K., Fonfara, I., Hauer, M., Doudna, J. A., & Charpentier, E., (2012). A Programmable dual-RNA-guided DNA endonuclease in adaptive bacterial immunity. *Science, 337*(6096), 816–821. doi: 10.1126/science.1225829.

Jørgensen, I. H., (1992). Discovery, characterization and exploitation of Mlo powdery mildew resistance in barley. *Euphytica, 63,* 141–152. doi: 10.1007/BF00023919.

Jouanin, A., Schaart, J. G., Boyd, L. A., Cockram, J., Leigh, F. J., Bates, R., Wallington, E. J., et al., (2019). Outlook for coeliac disease patients: Towards bread wheat with hypoimmunogenic gluten by gene editing of α- and γ-gliadin gene families. *B.M.C. Plant Biol., 19,* 333. doi: 10.1186/s12870-019-1889-5.

Kelliher, T., Starr, D., Richbourg, L., Chintamanani, S., Delzer, B., Nuccio, M. L., Green, J., et al., (2017). MATRILINEAL, a sperm-specific phosphor-lipase, triggers maize haploid induction. *Nature, 542*(7639), 105–109. doi: 10.1038/nature20827.

Kim, D., Alptekin, B., & Budak, H., (2018). CRISPR/Cas9 genome editing in wheat. *Funct. Integr. Genomics, 18,* 31–41. doi: 10.1007/s10142-017-0572-x.

Komor, A. C., Kim, Y. B., Packer, M. S., Zuris, J. A., & Liu, D. R., (2016). Programmable editing of a target base in genomic DNA without double stranded DNA cleavage. *Nature, 533*(7603), 420–424. doi: 10.1038/nature17946.

Kusch, S., & Panstruga, R., (2017). mlo-based resistance: An apparently universal "weapon" to defeat powdery mildew disease. *Mol. Plant Microbe Interact., 30,* 179–189. doi: 10.1094/MPMI-12-16-0255-CR.

Li, C., Zong, Y., Wang, Y., Jin, S., Zhang, D., Song, Q., Zhang, R., & Gao, C., (2018). Expanded base editing in rice and wheat using a Cas9adenosine deaminase fusion. *Genome Biol., 19,* 59. https://doi. doi: 10.1186/s13059-018-1443-z.

Li, J. Y., Jiao, G., Sun, Y., Chen, J., Zhong, Y., Yan, L., Jiang, D., et al., (2021). Modification of starch composition, structure and properties through editing of *TaSBEIIa* in both winter and spring wheat varieties by CRISPR/Cas9. *Plant Biotechnol. J., 19,* 937–951. doi: 10.1111/pbi.13519.

Li, J., Wang, Z., He, G., Ma, L., & Deng, X. W., (2020a). CRISPR/Cas9-mediated disruption of TaNP1 genes results in complete male sterility in bread wheat. *J. Genet. Genomics, 47,* 263–272. doi: 10.1016/j.jgg.2020.05.004.

Li, J., Zhang, S., Zhang, R., Gao, J., Qi, Y., Song, G., Li, W., et al., (2020b). Efficient multiplex genome editing by CRISPR/Cas9 in common wheat. *Plant Biotechnol. J., 19,* 427–429. doi: 10.1111/pbi.13508.

Liang, Z., Chen, K., Li, T., Zhang, Y., Wang, Y., Zhao, Q., Liu, J., et al., (2017). Efficient DNA-free genome editing of bread wheat using CRISPR/Cas9 ribonucleoprotein complexes. *Nat. Commun., 8,* 14261. doi: 10.1038/ncomms14261.

Lin, Q., Zong, Y., Xue, C., Wang, S., Jin, S., Zhu, Z., Wang, Y., et al., (2020). Prime genome editing in rice and wheat. *Nat. Biotechnol., 38*(5), 582–585. doi: 10.1038/s41587-020-0455-x.

Liu, C., Li, X., Meng, D., Zhong, Y., Chen, C., Dong, X., Xu, X., et al., (2017). A 4-bp insertion at *ZmPLA1* encoding a putative phosphor-lipase a generates haploid induction in maize. *Mol. Plant, 10*(3), 520–522. doi: 10.1016/j.molp.2017.01.011.

Liu, H., Wang, K., Jia, Z., Gong, Q., Lin, Z., Du, L., Pei, X., & Ye, X., (2020a). Efficient induction of haploid plants in wheat by editing of *TaMTL* using an optimized agrobacterium-mediated CRISPR system. *J. Exp. Bot., 71*(4), 1337–1349. doi: 10.1093/jxb/erz529.

Liu, H., Wang, K., Tang, H., Gong, Q., Du, L., Pei, X., & Ye, X., (2020b). CRISPR/Cas9 editing of wheat TaQ genes alters spike morphogenesis and grain thresh ability. *J. Genet. Genomics, 47*(9), 563–575. doi: 10.1016/j.jgg.2020.08.004.

Luo, J., Li, S., Xu, J., Yan, L., Ma, Y., & Xia, L., (2021). Pyramiding favorable alleles in an elite wheat variety in one generation by CRISPR/Cas9-mediated multiplex gene editing. *Mol. Plant, 14*(6), 847–850. doi: 10.1016/j.molp.2021.03.024.

Lv, J., Yu, K., Wei, J., Gui, H., Liu, C., Liang, D., Wang, Y., et al., (2020). Generation of paternal haploids in wheat by genome editing of the centromeric histone CENH3. *Nat. Biotechnol., 38*(12), 1397–1401. doi: 10.1038/s41587-020-0728-4.

Morran, S., Eini, O., Pyvovarenko, T., Parent, B., Singh, R., Ismagul, A., Eliby, S., et al., (2011). Improvement of stress tolerance of wheat and barley by modulation of expression of DREB/CBF factors. *Plant Biotechnol. J., 9,* 230–249. doi: 10.1111/j.1467-7652.2010.00547.x.

Newhouse, K. E., Smith, W. A., Starrett, M. A., Schaefer, T. J., & Singh, B. K., (1992). Tolerance to imidazolinone herbicides in wheat. *Plant Physiol., 100,* 882–886. doi: 10.1104/pp.100.2.882.

Nishida, K., Arazoe, T., Yachie, N., Banno, S., Kakimoto, M., Tabata, M., Mochizuki, M., et al., (2016). Targeted nucleotide editing using hybrid prokaryotic and vertebrate adaptive immune systems. *Science, 353*(6305), doi: 10.1126/science.aaf8729.

Okada, A., Arndell, T., Borisjuk, N., Sharma, N., Watson-Haigh, N. S., Tucker, E. J., Baumann, U., et al., (2019). CRISPR/Cas9-mediated knockout of Ms1 enables the rapid generation of male-sterile hexaploid wheat lines for use in hybrid seed production. *Plant Biotechnol. J., 17,* 1905–1913. doi: 10.1111/pbi.13106.

Petersen, G., Seberg, O., Yde, M., & Berthelsen, K., (2006). Phylogenetic relationships of *Triticum* and *Aegilops* and evidence for the origin of the A, B, and D genomes of common wheat (*Triticum aestivum*). *Mol. Phylogenet. Evol., 39,* 70–82. doi: 10.1016/j.ympev.2006.01.023.

Raffan, S., Sparks, C., Huttly, A., Hyde, L., Martignago, D., Mead, A., Hanley, S. J., et al., (2021). Wheat with greatly reduced accumulation of free asparagine in the grain, produced by CRISPR/Cas9 editing of asparagine synthetase gene *TaASN2*. *Plant Biotechnol. J., 19,* 1602–1613. doi: 10.1111/pbi.13573.

Regina, A., Bird, A., Topping, D., Bowden, S., Freeman, J., Barsby, T., Kosar-Hashemi, B., et al., (2006). High-amylose wheat generated by RNA interference improves indices of large-bowel health in rats. *Proc. Natl. Acad. Sci. U. S. A, 103,* 3546–3551. doi: 10.1073/pnas.0510737103.

Rey, M. D., Martín, A. C., Smedley, M., Hayta, S., Harwood, W., Shaw, P., & Moore, G., (2018). Magnesium increases homoeologous crossover frequency during meiosis in *ZIP4* (*Ph1* Gene) mutant wheat-wild relative hybrids. *Front. Plant Sci., 9,* 509. doi: 10.3389/fpls.2018.00509.

Sánchez-León, S., Gil-Humanes, J., Ozuna, C. V., Giménez, M. J., Sousa, C., Voytas, D. F., Barro, F., & Low-Gluten, N. T., (2018). Low-gluten, nontransgenic wheat engineered with CRISPR/Cas9. *Plant Biotechnol. J., 16,* 902–910. doi: 10.1111/pbi.12837.

Shan, Q., Wang, Y., Li, J., & Gao, C., (2014). Genome editing in rice and wheat using the CRISPR/Cas system. *Nat. Protoc., 9*, 2395–2410. doi: 10.1038/nprot.2014.157.

Shan, Q., Wang, Y., Li, J., Zhang, Y., Chen, K., Liang, Z., Zhang, K., et al., (2013). Targeted genome modification of crop plants using a CRISPR-Cas system. *Nat. Biotechnol., 31,* 686–688. doi: 10.1038/nbt.2650.

Singh, M., Kumar, M., Albertsen, M. C., Young, J. K., & Cigan, A. M., (2018). Concurrent modifications in the three homeologs of *Ms45* gene with CRISPR-Cas9 lead to rapid generation of male sterile bread wheat (*Triticum aestivum* L.). *Plant Mol. Biol., 97,* 371–383. doi: 10.1007/s11103-018-0749-2.

Su, Z., Bernardo, A., Tian, B., Chen, H., Wang, S., Ma, H., Cai, S., et al., (2019). A deletion mutation in *TaHRC* confers *Fhb1* resistance to fusarium head blight in wheat. *Nat. Genet., 51*, 1099–1105. doi: 10.1038/s41588-019-0425-8.

Takeda, K., & Hori, K., (2007). Geographical differentiation and diallel analysis of seed dormancy in barley. *Euphytica, 153,* 249–256. doi: 10.1007/s10681-006-9260-y.

Tang, H., Liu, H., Zhou, Y., Liu, H., Du, L., Wang, K., & Ye, X., (2021). Fertility recovery of wheat male sterility controlled by Ms2 using CRISPR/Cas9. *Plant Biotechnol. J., 19,* 224–226. doi: 10.1111/pbi.13482.

Tang, X., Zheng, X., Qi, Y., Zhang, D., Cheng, Y., Tang, A., Voytas, D. F., et al., (2016). A single transcript CRISPR-Cas9 system for efficient genome editing in plants. *Mol. Plant, 9*, 1088–1091.

Upadhyay, S. K., Kumar, J., Alok, A., & Tuli, R., (2013). RNA-guided genome editing for target gene mutations in wheat. *G3* (*Bethesda*), *9*(3), 2233–2238. doi: 10.1534/g3.113.008847.

Wang, Q., Nian, J., Xie, X., Yu, H., Zhang, J., Bai, J., Dong, G., et al., (2018a). Genetic variations in *ARE1* mediate grain yield by modulating nitrogen utilization in rice. *Nat. Commun., 9,* 735. doi: 10.1038/s41467-017-02781-w.

Wang, W., Pan, Q., He, F., Akhunova, A., Chao, S., Trick, H., & Akhunov, E., (2018b). Transgenerational CRISPR-Cas9 activity facilitates multiplex gene editing in allopolyploid wheat. *C.R.I.S.P.R. J., 1,* 65–74. doi: 10.1089/crispr.2017.0010.

Wang, W., Pan, Q., Tian, B., He, F., Chen, Y., Bai, G., Akhunova, A., Trick, H. N., & Akhunov, E., (2019). Gene editing of the wheat homologs of *TONNEAU1*-recruiting motif encoding gene affects grain shape and weight in wheat. *Plant J., 100,* 251–264. doi: 10.1111/tpj.14440.

Wang, W., Simmonds, J., Pan, Q., Davidson, D., He, F., Battal, A., Akhunova, A., et al., (2018). Gene editing and mutagenesis reveal inter-cultivar differences and additivity in the contribution of *TaGW2* homoeologues to grain size and weight in wheat. *Theor. Appl. Genet., 131,* 2463–2475. doi: 10.1007/s00122-018-3166-7.

Wang, Y., Cheng, X., Shan, Q., Zhang, Y., Liu, J., Gao, C., & Qiu, J. L., (2014). Simultaneous editing of three homoeoalleles in hexaploid bread wheat confers heritable resistance to powdery mildew. *Nat. Biotechnol., 32,* 947–951. doi: 10.1038/nbt.2969.

Xie, K., Minkenberg, B., & Yang, Y., (2015). Boosting CRISPR/Cas9 multiplex editing capability with the endogenous tRNA-processing system. *Proc. Natl. Acad. Sci. U. S. A, 112*, 3570–3575. doi: 10.1073/pnas.1420294112.

Yao, L., Zhang, Y., Liu, C., Liu, Y., Wang, Y., Liang, D., Liu, J., et al., (2018). OsMATL mutation induces haploid seed formation in indica rice. *Nat. Plants, 4*(8), 530–533. doi: 10.1038/s41477-018-0193-y.

Zhang, J., Zhang, H., Li, S., Li, J., Yan, L., & Xia, L., (2021). Increasing yield potential through manipulating of an ARE1 ortholog related to nitrogen use efficiency in wheat by CRISPR/Cas9. *J. Integr. Plant Biol.* doi: 10.1111/jipb.13151.

Zhang, R., Liu, J., Chai, Z., Chen, S., Bai, Y., Zong, Y., Chen, K., et al., (2019b). Generation of herbicide tolerance traits and a new selectable marker in wheat using base editing. *Nat. Plants, 5,* 480–485. doi: 10.1038/s41477-019-0405-0.

Zhang, S., Zhang, R., Gao, J., Gu, T., Song, G., Li, W., Li, D., et al., (2019c). Highly efficient and heritable targeted mutagenesis in wheat via the *Agrobacterium tumefaciens*-Mediated CRISPR/Cas9 system. *Int. J. Mol. Sci., 20*(17), 4257. doi: 10.3390/ijms20174257.

Zhang, Y., Bai, Y., Wu, G., Zou, S., Chen, Y., Gao, C., & Tang, D., (2017). Simultaneous modification of three homoeologs of *TaEDR1* by genome editing enhances powdery mildew resistance in wheat. *Plant J., 91*, 714–724. doi: 10.1111/tpj.13599.

Zhang, Y., Liang, Z., Zong, Y., Wang, Y., Liu, J., Chen, K., Qiu, J. L., & Gao, C., (2016). Efficient and transgene-free genome editing in wheat through transient expression of CRISPR/Cas9 DNA or RNA. *Nat. Commun., 7,* 12617. doi: 10.1038/ncomms12617.

Zhang, Z., Hua, L., Gupta, A., Tricoli, D., Edwards, K. J., Yang, B., & Li, W., (2019a). Development of an agrobacterium-delivered CRISPR/Cas9 system for wheat genome editing. *Plant Biotechnol. J., 17,* 1623–1635. doi: 10.1111/pbi.13088.

Zong, Y., Song, Q., Li, C., Jin, S., Zhang, D., Wang, Y., Qiu, J. L., & Gao, C., (2018). Efficient C-to-T base editing in plants using a fusion of nCas9 and human APOBEC3A. *Nat. Biotechnol., 36,* 950–953. doi: 10.1038/nbt.4261.

Zong, Y., Wang, Y., Li, C., Zhang, R., Chen, K., Ran, Y., Qiu, J. L., et al., (2017). Precise base editing in rice, wheat and maize with a Cas9-cytidine deaminase fusion. *Nat. Biotechnol., 35,* 438–440. doi: 10.1038/nbt.3811.

CHAPTER 7

Genome Editing of Maize with CRISPR/Cas Systems: Past, Present, and Future

BUSHRA HAFEEZ KIANI

Department of Biological Sciences (Female Campus), International Islamic University, Islamabad – 44000, Pakistan, E-mail: bushrahafeez.kiani@gmail.com

ABSTRACT

Zea mays is one of the chief cereals that can grow under changing environments. Some of the most important uses of maize are feed, food, and unrefined material in industries. It not only provides economic benefit at a global level but also the main source of income for many farmers. Recent advancements in genotyping techniques, data sciences, new breeding approaches, and modifications such as genome editing, doubled haploid technology, sequencing, and genotyping technology contribute towards *Zea mays* cultivation. There has been an increase in crop production since the last century, thanks to genetic innovations like transgenesis, breeding, and hybridization. Genome editing is the latest progress in the field of genetics through which enhancement of crops at the genetic level can be done and can be utilized for targeted additions, deletions, and corrections in the genes. The most recent, simple, accurate, and revolutionized genome editing technique is CRISPR (clustered regularly interspaced short palindromic repeats) linked with protein 9 (Cas9). In this chapter, we will focus on the past, present, and future of genome editing techniques applied to maize and the use of the CRISPR/Cas9 system.

CRISPR/Cas-Mediated Genome Editing in Plants, Deepu Pandita, Anu Pandita (Eds.)

7.1 INTRODUCTION

Zea mays L. is the most cultivated crop all over the world. It became the most advantageous crop around the globe because of its diverse applications and as it is easy to cultivate both in changing environmental and soil conditions. Feed for animals, organic matter for chemical and food industries, renewable energy sources are some of the other uses of maize other than human consumption (Pegoraro et al., 2011). Scientists are specifically working to change the genome in order to increase physical and physiological traits by genetic techniques. Irradiation like gamma or fast-moving neutrons and chemical mutagens like ethyl methane sulfonate (EMS) are used normally for modifying and editing genes in Zea mays. In order to initiate mutations in plant genetic makeup, these techniques are used during the exposure and repair process of DNA. Mutation breeding is normally not as accurate. It can result in both positive and negative consequences and no control on areas of the genome that are modified. One more technique that is used normally in the case of *Zea mays* genetics is transposon tagging, in which specified transposons are applied to introduce mutations and thus allow furthermore discovery (Walbot, 2000). Time taking and uneconomical is the main reasons that are drawbacks of this technique (Brutnell, 2002). It can also result in random mutations and difficulty in performing large screens (Feng et al., 2016). The restriction of random mutagenesis promotes research on targeted genome modification techniques. With the evolution in such techniques in the last 10 years, gene editing fidelity has been increased sufficiently up to 1,000 times (Puchta et al., 1993; Svitashev et al., 2016).

The manufacturing of nucleases to propagate double standard breaks (DSB) in DNA at a required site is the key idea behind targeted genome editing. Commonly, sites, where DNA is broken, can develop into mutation by endogenous repair mechanisms in the non-homologous end by joining (NHEJ) or by externally added homologous DNA repair template in homologous directed repair (HDR). The first-generation targeted gene-editing approaches are zinc finger nucleases (ZFNs) and transcription activators like effector nucleases (Joung and Sander, 2013). Even though these approaches are successful to a certain extent, they have some pitfalls. Altering ZFNs and transcription activators like effector nucleases is a difficult task as both of them are required in pairs, as the two upstream and downstream regions of a particular region on the chromosome have to be focused (Beumer et al., 2013). In order to edit many targets by multiplexing, ZFNs, and transcription activators like effector nucleases are needed. To produce double-stranded breaks on a specific position, genetically engineered ZFNs and transcription activators like effector nucleases are used (Kim and Kim, 2014).

One of the advanced approaches is the CRISPR/Cas9 system (Jinek et al., 2012; Gilbert et al., 2013; O'Connell et al., 2014). This technique has transformed the area of genome editing as it has a simple design, adjustability of operation, and is economical (Doudna and Charpentier, 2014; Murugan et al., 2017). Basic requirements are Cas9 nuclease protein, single guide RNA having a target site of 20 base pairs, and hairpin structure. There are multiple potential ways to use the CRISPR/Cas9 technique for genome improvement of maize (Figure 7.1).

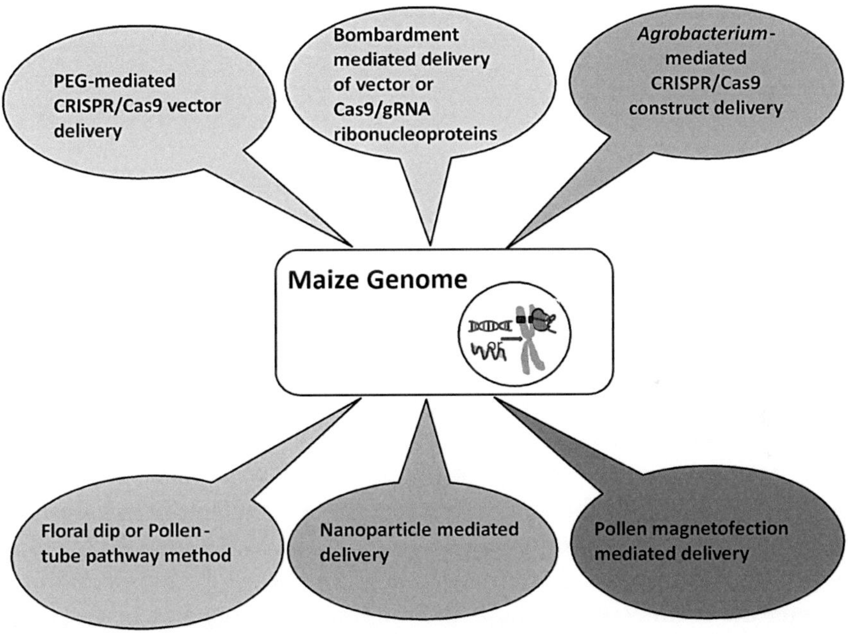

FIGURE 7.1 Existing and potential future CRISPR/Cas9 delivery methods in maize.

Source: Adapted with permission Sandhya et al. (2020). https://creativecommons.org/licenses/by/4.0/

A double-strand DNA break is introduced next to the PAM sequence in the 20-base pair target-specific genomic region indicated by the sgRNA. Surprisingly, in single copy sections of the corn genome, approximately 20 base pairs are required to guarantee the specificity of the allele. Because of its compact design, the construction of sgRNA is simple (Newman and Ausubel, 2016; Shan et al., 2014). The CRISPR/Cas9 technology can be delivered in maize cells temporarily or stably. This approach also allows for multiplexing (Shan et al., 2014). CRISPR/Cas9 can potentially be used for targeting methylated DNA, making it an adaptable technology for editing the plant's genome (Bortesi and Fischer, 2015). In comparison to transcription activator-like effector nuclease or ZFN, the targeting effectiveness of the CRISPR/Cas9 technology is also significantly higher (Reis et al., 2014).

There are no restrictions on species with the CRISPR/Cas9 technology. This technology has been used in a variety of plant families and genera, including *Arabidopsis thaliana* (Jiang et al., 2014), orange (Jia and Wang, 2014), tomato (Brooks et al., 2014), the sorghum genus (Sun et al., 2015), potato (Wang et al., 2015), tobacco (Gao et al., 2015), wheat (Zhang et al., 2016), and rice (Mazumdar et al., 2016). This chapter focuses on the present, past, and future state of gene editing in maize through CRISPR/Cas9 technology.

7.2 THE PRESENT, PAST, AND FUTURE OF MAIZE GENE EDITING THROUGH CRISPR/CAS9

7.2.1 THE PAST STATUS OF GENOME EDITING IN MAIZE THROUGH CRISPR/CAS9

Luckily, CRISPR/Cas technology can be applied to alter the genetic makeup of maize plants (Svitashev et al., 2015, 2016; Shi et al., 2017; Char et al., 2017). Biosynthesis of Lignin, RNA metabolism, Tolerance capacity for drought, secondary metabolism, tolerance capability against herbicides, male sterility, and grain composition are some of those characteristics that can be changed by the application of this approach. By knowing the order of region adjoining the double-strand break, the required sequence can be inserted in the large double-strand template by using single or double-stranded DNA (dsDNA) to alter targeted small number of codons, application of two cuts across the intervening region so that it is dropped out and hence sequence can be replaced, by enabling double-stranded template of the gene of interest having sequence same as genomic regions distal to both cuts and template-free gene knockout are some of those editing types that could be used. Besides these characteristics, CRISPR can use homologous recombination (HR) to help in assembling multiple transgenic traits in order to construct a great number of site-specific integration landing sites. The simplest edit is template-free gene knockouts, after which targeted insertion of transgene having gene marker is inserted. Even in limited resources experiments containing DNA templates can be successfully administered by use of large double-strand DNA molecules or single-stranded DNA oligos (Nelson and Rines, 1962).

7.2.1.1 GENE KNOCKOUT

WX1 gene of maize expresses starch synthase that is used in the formation of kernels. Maize contains homozygous alleles for loss-of-function due to the WX1 gene that is responsible for changes in the starch composition of grain. In normal conditions, wild-type maize grain consists of 75% and 25% of amylopectin and amylose, respectively whereas Wx1/Wx1 lines have about 100% of amylopectin as grain starch (Sprague and Hixon, 1943). Due to these changes in starch consistency in kernels, they have a candle wax-like appearance and that is why they are called "Waxy corn." This cornstarch that is made up of Waxy corn has a lot of functions like stability, increasing uniformity, and giving texture to different food products. It can also be used in both textile and adhesive industries and also in the paper industry as a binder in manufacturing processes. According to the first-ever research on Waxy corn in China in 1909 (Collins, 1909), Wx1 alleles that occur naturally are characterized on the basis of molecular structure, many are small deletions that affect the sequence responsible for coding (Wessler et al., 1990).

The wax products that are in use today have two major drawbacks. The use of ancient genetic information is their main flaw. By the last stage of this process, Wx1 alleles introgress in elite genetics. As compared to conventionally used products, the use of waxy products is far behind because introgression takes a great time to complete. Moreover, in comparison with the waxy identically same hybrids, waxy hybrids have fewer yields. Such a drop in the yield might be due to incomplete trait introgression, in which non-elite genes surrounding the Wx1 trait are accidentally transported during the process of introgression. CRISPR/Cas has the capability to solve both of these drawbacks in gene deletion. Fundamentally, altering genetic makeup in elite germplasm has the capability of being economical and fast as compared to trait introgression and hence could be used at a commercial level for modern genetics. Furthermore, problems created due to reduced yield could be avoided if Wx1 deletions induced as a result of elite germplasm do not have defects related to trait introgression.

Using a no-template two Cas9–gRNA method, the complete removal of the WX1 gene was possible by CRISPR/Cas. A Cas9–gRNA targeted the cleavage downstream of the stop codon, while the other targeted the sequence upstream of the transcriptional start site. The area among both targets was cut, and the residual distant chromosomal endpoints were united via the non-homologous end joining (NHEJ) mechanism for DNA repair, yielding a Wx1 null allele with the Waxy phenotype.

Genome modification is carried out by bombarding gold-coated immature embryos that are having six types of plasmids, like two gRNAs, NPT2 selectable marker, the *Streptococcus pyogenes* Cas9 gene, maize WUS gene, and maize ODP2 gene. Phosphotransferase protein is translated by the NPT2 gene that is responsible for kanamycin and G418 resistance (Fraley, 1983). From the bombardment of 100 embryos approximately 2–29 plants are regenerated which gives the regeneration frequency of 2%–29% across genotypes. By PCR amplification regenerated plants can be inspected for the existence of particular traits of a focused part that follows Illumina-based DNA sequencing. In general, out of a total of 100 plants, 80 plants that are regenerated will be having a mutation in individual Cas9–gRNA target sites, and 10 plants will go through deletions between the two Cas9–gRNA targets. Approximately, plants that were biallelic and have deletions in WX1 alleles are 10 in number. Plants that were having required deletions of Cas9–gRNA sites were crossed with the plant that was used in transformation and the descendants were then self-pollinated or crossed. In the next generation, plants that were having desired characteristics and no other bombarded plasmid DNA were selected. In the start, quantitative PCR, which was designed for elements of six plasmids that were administered during the bombardment, was used to detect the absence of plasmid DNA, followed by SBS, which has the capability to detect even small DNA fragments. About 80% of Wx1 mutations causing deletions in previously mentioned plants could be segregated by using gene markers and other types of plasmid DNA.

As anticipated, the self-pollination of hemizygous plants gives 1:3 for the waxy phenotype (recessive) according to the law of segregation. We can get direct production of CRISPR/Cas-derived Waxy hybrids as Wx1 deletions were produced in more than 10 different elite inbreds. The average efficiency was about 0.5% which includes molecular attrition, greenhouse attrition, and transformation frequency, that is one line that is having required mutations and no helper DNA was produced for each 200 immature corn embryos that were bombarded. Yet, for various experiments efficiency was changing. Prior analysis of CRISPR/Cas-derived lines showed exact phenotype apart from particularly waxy traits. Quantitative determination of the starch composition of these lines can be governed by established aspects and found to be identical with formerly distinguished Wx1 null alleles.

7.2.1.2 GENE EDIT

Herbicides containing sulfonylurea block acetolactate synthase (ALS), which inhibits the production of branched amino acid in crops (Haughn et al., 1988; Lee et al., 1988; Yu and Powles, 2014). Mutations in a singular amino acid of ALS gene from proline to serine can induce sulfonylurea herbicide chlorsulfuron resistance (Haughn et al., 1988; Walter, 2014). ALS1 and ALS2 genes are relatively similar and found in maize (Burr, 1991). However, discrepancies in nucleotide sequences enable the construction of Cas9–gRNA targets that are restricted to one or the other. The ALS2 gene was the precise target of the CAs9-gRNA that guided the cleavage of 26 nucleotides upstream of the targeted modification. This swaps a codon that code for amino acid, serine rather than the amino acid, proline.

Immature embryos coated with gold particles having 127-nucleotide single-strand DNA oligo repair template and five plasmids that contain the *S. pyogenes* Cas9 gene, maize ODP2 gene, gene encoding the gRNA, the maize WUS gene, and MoPAT-DsRED selectable gene marker are bombarded to get done modification in ALS gene. Gene marker MoPAT-DsRED is responsible for herbicide bialaphos resistance and red color of cells (Ananiev et al., 2009). For this experiment, red color was observed instead of bialaphos selection to administer the successful delivery of DNA.

Almost 800 immature embryos were bombarded and then moved to chlorsulfuron media for the purpose of direct selection of desired traits. Resistant Calli became a source for regeneration of three independent plants that all have considered ALS2 gene edits. ALS1 remains the same, as decided after sequencing of PCR amplicons spanning the site of editing. Self-pollination was done in hemizygous plants that were having required mutations. These seeds were then sown in flats and chlorsulfuron spray was done after every 28 days. As supposed, chlorsulfuron resistance was shown by the modified ALS2 allele.

A single-stranded DNA oligonucleotide which was 127 base pairs long was chosen as an editing repair template in this research. A 794 base pair long dsDNA oligonucleotide yielded identical results. Plants that show resistance to chlorsulfuron was produced by the co-bombardment of immature embryos by inserting 127 base pair single-stranded oligonucleotides as the sole DNA along with a pre-configured Cas9-gRNA complex.

7.2.2 THE PRESENT STATUS OF MAIZE GENE EDITING THROUGH CRISPR/CAS9

Maize is the 3rd most beneficial crop besides rice and wheat. The first maize gene knockout through CRISPR/Cas9 was successfully conducted by targeting the *ZmIPK* gene responsible for regulating phytic acid production. The researchers created two gRNAs for targeting that specific gene, resulting in 16–19% mutation frequency. This indicates that CRISPR/Cas9 is an effective tool for modifying maize genes (Liang et al., 2014).

Likewise, some other research used sgRNA to knock down the phytoene synthase gene under the control of the U6 promoter (Zhu et al., 2016). Using the CRISPR/Cas9 technology, researchers detected 10.67% cleavage efficiency of the maize phytoene synthase gene. Researchers also sequenced the altered gene to confirm the efficiency of mutation (Zhu et al., 2016). By utilizing the CRISPR/Cas9 technology for targeting the albino marker (Zmzb7) gene, 31% frequency of mutation in T0 genotypes was reported (Feng et al., 2016). The CRISPR/Cas9 technology was then used to the maize thermosensitive genic male-sterile 5 (ZmTMS5) gene as a target by creating sgRNAs that targeted the ZmTMS5 gene and induced mutations in protoplasts (Chen et al., 2018). The modified plants exhibited biallelic alteration, indicating that the CRISPR/Cas9 technology has an excellent ability to target maize mutagenesis (Chen et al., 2018). These findings indicate advancement in maize breeding techniques and crop improvement by using the CRISPR/Cas9 technology (Figure 7.2).

7.2.2.1 PROMOTER REPLACEMENT

One of the important characteristics of commercially available maize is tolerance to scarce water conditions (Cooper et al., 2014). The hormone ethylene has an important role in tolerance to scarce water levels. The higher the ethylene concentration, the lower the tolerance. The ARGOS8 gene codes for a protein that engages with the ethylene receptor signaling complex thereby, adversely controlling the ethylene response in maize (Shi et al., 2016). The gene expression levels of wild-type ARGOS8 are not high and almost non-existent in somatic cells. Research has shown that the overexpression of the ARGOS8 gene through moderate constitutive transgenesis can enhance tolerance to drought stress.

CRISPR/Cas is a genome-editing tool that is extensively being used for modifying the wild-type ARGOS8 gene to increase its expression. This is done by swapping the endogenous promoter with the GOS2 promoter which

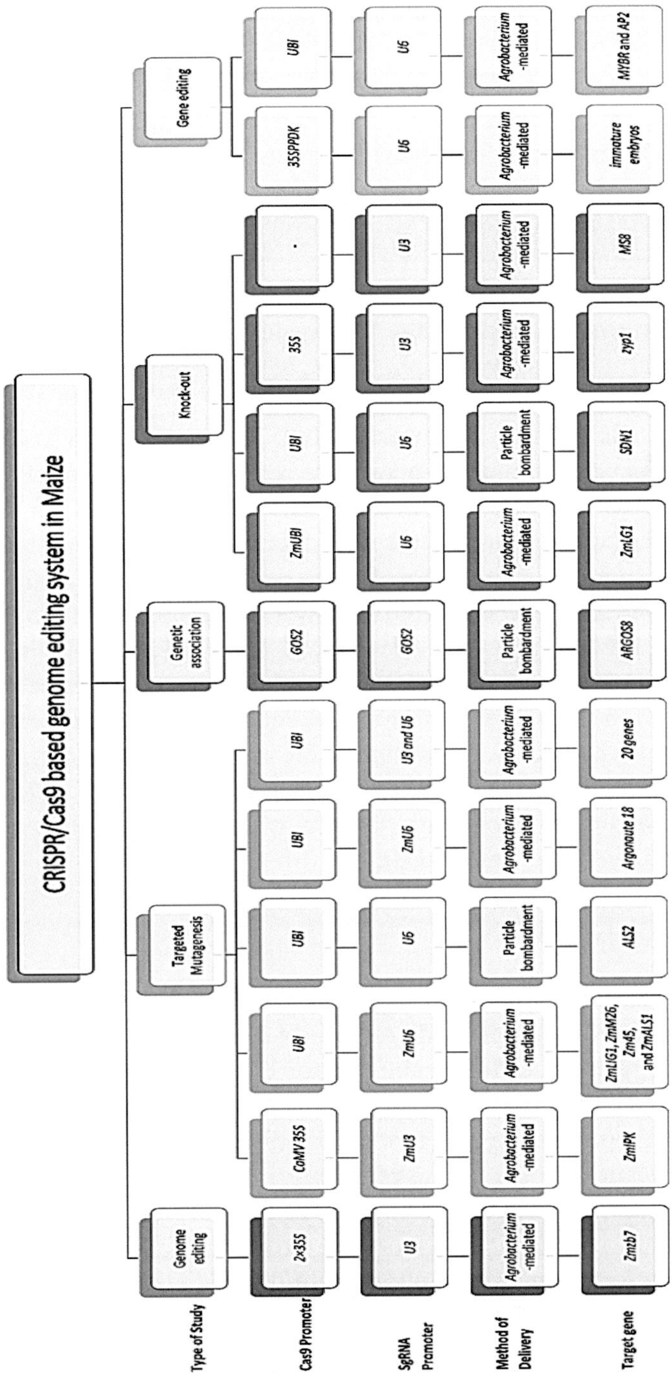

FIGURE 7.2 CRISPR/Cas9 based genome editing system in maize.

is originally involved in promoting SUI1 transcription factor expression in maize (de Pater et al., 1992).

The promoters are replaced by bombarding immature embryos with gold particles that have been coated with DNA repair templates carrying GOS2 promoters. This promoter is similarly surrounded by 400 bp DNA sequences that complement the sequences in the ARGOS8 locus that are right next to the Cas9 cleaving regions. This DNA repair template also includes five vectors containing the two gRNAs, *Streptococcus pyogenes* Cas9 gene, two vectors containing maize ODP2, maize WUS, and a PMI marker gene. The PMI selection system employs the PMI gene, which expresses PMI, which transforms harmful mannose-6-phosphate to harmless fructose-6-phosphate in maize plants (Negrotto et al., 2000).

After particle bombarding nearly 3,000 immature embryos, 334 plant shoots were put on mannose-6-phosphate regenerative media. Out of these shoots, 172 grew into plantlets. In the preliminary evaluation, 23 of the 172 plantlets tested positive for at least one junction. Three of them tested positive for both junctions, and two of them grew into fertile plants. One of the regenerated plants was devoid of any DNA material employed in the transforming process and was evaluated and analyzed in the fields.

By crossing the modified ARGOS8 variation to an inbred tester, the hybrid was evaluated in the wild in both watered and unwatered conditions against a similar wild-type hybrid. According to the earlier agronomic evaluation of the modified GOS2-ARGOS8, the altered hybrids behaved analogously to wild-type in an environment with plentiful water but had higher yield when cultivated under scarce water conditions (Shi et al., 2017).

7.2.2.2 GENE INSERTION

The large percentage of maize produced has several transgenic genes, generally consisting of herbicide resistance and several Bt transgenes that provide pest resistance. It is common for commercially available plants to include five distinct transgenes. These transgenes induce characteristics that were created through random modification and can be found all around the plant genome. Transgenes are introduced in inbreds through trait introgression when they are produced by breeders. This procedure entails mating a transgenic donor line with new elite inbred and then frequently backcrossing the resultant plant line with the elite line to restore the genome of the elite line as well as a short insertion carrying the gene of interest (Que et al., 2010).

Although trait introgression is successful, it has some drawbacks. It is expensive and time-intensive, needing numerous years and many generations. Some unwanted donor DNA is still retained accidentally next to the gene of interest in the result, typically containing multiple million nucleotides. The assembly of a huge number of characteristics is expensive and time-consuming rendering the introgression of over six unrelated loci unfeasible.

The apparent answer to this dilemma is to assemble genes of interest into a single genomic region. Since the limits of trait introgression seem to be largely linked to the number of introgressed loci, transgene colocation offers the ability to produce consumer goods with any number of recombinant characteristics. Colocation can be achieved by assembling molecular stacks that are typically constructs of DNA containing several genes of interest. For several years, this technique has been infrequent use. The TC1507 maize event trait, for example, has both a Bt gene that confers resistance to pests and an acetyltransferases gene that provides resistance to herbicides. This method is beneficial although it has many disadvantages. Firstly, when large constructs consisting of several transgenes are employed, transformation rates are poor. As a result, the number of combined transgenes is limited. Furthermore, neighboring genes within several kilobases may influence each other's ability to express proteins.

Complications arise in the designing and finding of optimal molecular stacks due to the possible interactions between genes (Eszterhas et al., 2002). Moreover, genetic crossing cannot change the individual components of molecular stacks if a specific function is not required anymore or an additional function is to be introduced in the molecular stack. The creation of complex trait loci is an alternate approach to these stacks. Complex trait loci are genetically collocated groups of genes of interest and are quite comparable to molecular stacks. The genes of interest are distanced at 50 kilobases to 1 megabase while remaining linked at a genetic level. The possibility of genes interacting with each other is reduced by the 50 kilobase distance between them. This enables the deletion and addition of new genes using recombinant techniques while still treating the transgenes as a single locus throughout the introgression of the traits.

CRISPR/Cas technology is frequently utilized to insert transgenes into specific genomic sequences by homology-directed repair to produce complex trait loci of maize. Unfortunately, this process is currently very inefficient. Site-specific gene integration by the FLP-FRT recombination system is comparatively very effective (Li et al., 2010). One of the factors that make site-specific integration effective is the use of a chromosomal promoter trap paired with a construct of DNA carrying a promoter-less gene marker. Because random integration of the DNA construct does not impart resistance to the gene marker in this scenario, the majority of the events retrieved are not the consequence of random integration; instead, they are true site-specific integration products. The main disadvantage of site-specific integration is that it needs a pre-integrated target site with a promoter trap and FRT recombination sites, while CRISPR/Cas can directly target the genome.

Researchers are integrating both CRISPR/Cas9 and Site-specific Integration for their functionalities. CRISPR/Cas for example can be used for inserting target construct at specific sites and site-specific integration is used for evaluating the transgenes present in those loci. The target sites were selected to be not too near or too far to be genetically separate at a distance of 0.1 centimorgan, but farther away from native genes to be genetically distinguishable. Many site-specific integration target lines were created by biolistically bombarding plasmids, Cas9 gene, maize ODP, containing gRNA, and *Streptococcus pyogenes*, maize WUS and a 400 base pair dsDNA template correlated to the sequences on either side of the cleavage location with NPT2 gene marker in the midst of bordering flippase recognition sequences. G418-containing media was used to regenerate events. With an average efficiency of around 0.1%, in other words for the bombardment of every 1,000 embryos, there was one perfect event, progenies now produced have the exact required sequence of DNA, as determined by styrene-butadiene-styrene. A selection of such lines has been described in terms of agronomics, genetic origin, transgenic expression level, and transformability. Complex trait loci may now be generated using these site-specific integration target lines.

7.2.3 THE FUTURE PROSPECTS OF MAIZE GENE EDITING THROUGH CRISPR/CAS9

The search for novel methods to extend the precision and scope of CRISPR/Cas9 mutagenesis is always an area of interest for studies. There are two nuclease enzymatic domains of Cas9, RuvC, and HNH. The RuvC domain cuts the non-complementary strand of the DNA while the HNH domain cuts the strand complementary to sgRNA. Cas9, in conjunction with its sgRNA, creates a double-strand DNA break when both nuclease domains are activated. In case of mutation in one domain, the enzymatic activity is lost and nCas9 or single-stranded nickase is formed. The mutation in only one domain entails that nCas9 can only cut one strand; however, gene alterations could still occur via the HDR or NHEJ pathways (Mali et al., 2013).

Cas9 nickases can be utilized in pairings or alone. The paired Cas9 nickase functions similarly to a ZFN pair, generating a dislocated double-strand DNA break (Sander and Joung, 2014). A pair of Cas9 nickases in conjunction with sgRNA was employed in rice and the precision was so accurate that even a nucleotide discrepancy with the target sequence was unable to cause mutations (Mikami et al., 2016). The combination of nCas9 with cytidine deaminase allowed for precise and targeted cytosine to thymine base editing in rice, wheat, and corn (Zong et al., 2017). The sgRNA/Cas9 nickase paired system might potentially be utilized for targeted gene insertions. However, in this instance, the length between the two nicks will hinder the versatility of the targeted gene substitution. Gene replacement in *Arabidopsis* has also been carried out by utilizing dual-sgRNA/Cas9 (Zhao et al., 2016).

This duality method formed DNA double-strand breaks and deleted the large intentional deletion length. The deletion length can be created from 1–300 kb in maize. Genomic tools that have numerous functions, such as deleting, knocking-in, and replacing genes, along with the ability to deal with huge sequences, are a game-changer for crop modification.

Cas9 that lacks the enzyme nuclease is called dead Cas9 (dCas9) as both mutated endonuclease domains cause a loss of enzymatic properties. dCas9 can target genomes without cleaving DNA through precise targeting based on the sequence of sgRNA. dCas9 can act as a gene-repression system or activator by fusing to different transcriptional domains. Many proteins can be localized on DNA by the attachment of dCas9. Molecular visualization of report genes, for example, can be achieved by fusing them with dCas9 (Shalem et al., 2015; Radzisheuskaya et al., 2016).

CRISPR/Cas9 technology has the potential to significantly benefit maize farmers as naturally produced variations are used in traditional breeding techniques. Plant breeders use comprehensive backcrossing to integrate desirable genes into the plant genome permanently. This process can be sped up by using gene editing to make accurate and calculated alterations to the plant genome. These genomes modified through CRISPR/Cas9 are not different from those produced by spontaneous or chemically-induced mutations and traditional breeding. This is why, in several nations, as long as the transgenic Cas9-sgRNA or other genetic material is separated from the inventory, crops produced by this technique are categorized as non-GM (Belhaj et al., 2013). Due to the regulation issues of *Agrobacterium*-mediated transformation, physical methods such as biolistic guns are being preferred for plant transformation as it does not require any disease-causing agents.

Maize that contains high concentrations of amylopectin may be the first CRISPR-edited variety cultivated at a commercial level (Waltz, 2016). The potential application of this editing technique on crop lines can be illustrated through the research recently being carried out on maize to develop variants for the ARGOS8 gene. During field testing, the variants of the ARGOS8 gene show a 314 kg per hectare increase in grain yield under scarce water conditions (Shi et al., 2017).

7.3 CONCLUSION

CRISPR/Cas9 shows promising results in maize quality and agronomic traits. A variety of genes containing naturally occurring allelic variations with minor effects are being used to improve crop yield and tolerance to stress since the possibility of hybrid maize became known. Ideal breeding lines do not contain these types of alleles in high frequencies. The study of genomics allows better grasping of variation at genetic levels, which may facilitate the designing of alleles through gene editing in the near future. Since there is no regulation price of utilizing these designed alleles, they can be pyramided and be of wide use even if they have minor effects. On the other hand, transgenics require the regulation prices of every gene to be carefully evaluated against the potential advantages. The seed industry of maize is well established as it is an economically important plant. New genetic techniques have been intensively and extensively applied to these plants in comparison to other plants. Significant advancement in genome editing in maize can herald a modern technology-based age, potentially setting the trend for more plants.

KEYWORDS

- ***agrobacterium***
- ***Arabidopsis***
- **clustered regularly interspaced short palindromic repeats**
- **homologous directed repair**
- **maize gene editing**
- **non-homologous end by joining**

REFERENCES

Ananiev, E. V., Wu, C., Chamberlin, M. A., et al., (2009). Artificial chromosome formation in maize (*Zea mays* L.). *Chromosoma., 118*(2), 157–177.

Belhaj, K., Chaparro-Garcia, A., Kamoun, S., & Nekrasov, V., (2013). Plant genome editing made easy: Targeted mutagenesis in model and crop plants using the CRISPR/Cas system. *Plant Methods, 9*, 39.

Beumer, K. J., Trautman, J. K., Christian, M., Dahlem, T. J., Lake, C. M., Hawley, R. S., Grunwald, D. J., et al., (2013). Comparing zinc finger nucleases and transcription activator-like effector nucleases for gene targeting in *drosophila. G3 Genes. Genomes Genet., 3*(10), 1717–1725.

Bortesi, L., & Fischer, R., (2015). The CRISPR/Cas9 system for plant genome editing and beyond. *Biotechnol. Adv., 33*, 41–52.

Brooks, C., Nekrasov, V., Lippman, Z. B., & Van, E. J., (2014). Efficient gene editing in tomato in the first generation using the clustered regularly interspaced short palindromic repeats/CRISPR-associated9 system. *Plant Physiol., 166*, 1292–1297.

Brutnell, T. P., (2002). Transposon tagging in maize. *Funct. Integr. Genomics, 2*, 4–12.

Burr, B., (1991). Recombinant in breds for molecular mapping in maize: Theoretical and practical considerations. *Trends Genet., 7*(2), 55–60.

Char, S. N., Neelakandan, A. K., & Nahampun, H., (2017). An agrobacterium-delivered CRISPR/Cas9 system for high-frequency targeted mutagenesis in maize. *Plant Biotechnol. J., 15*(2), 257–268.

Chen, R., Xu, Q., Liu, Y., et al., (2018). Generation of transgene-free maize male sterile lines using the CRISPR/Cas9 system. *Front Plant Sci., 9*, 1180.

Collins, G., (1909). *A New Type of Indian Corn from China* (Vol. 161, pp. 1–30). Washington: Government Printing Office. (Bulletin) Bureau of Plant Industry. http://library.umac.mo/ebooks/b28370405.pdf (accessed on 4 September 2022).

Cooper, M., Gho, C., Leafgren, R., Tang, T., & Messina, C., (2014). Breeding drought-tolerant maize hybrids for the US corn-belt: Discovery to product. *J. Exp. Bot., 65*(21), 6191–6204.

De Pater, B. S., Van, D. M. F., Rueb, S., et al., (1992). The promoter of the rice gene GOS_2 is active in various different monocot tissues and binds rice nuclear factor ASF-1. *Plant J., 2*(6), 837–844.

Doudna, J. A., & Charpentier, E., (2014). The new frontier of genome engineering with CRISPR–Cas9. *Science, 346*(6213), 1258096.

Eszterhas, S. K., Bouhassira, E. E., Martin, D. I. K., & Fiering, S., (2002). Transcriptional interference by independently regulated genes occurs in any relative arrangement of the genes and is influenced by chromosomal integration position. *Mol. Cell Biol., 22*(2), 469–479.

Feng, C., Yuan, J., Wang, R., Liu, Y., Birchler, J. A., & Han, F., (2016). Efficient targeted genome modification in maize using CRISPR/Cas9 system. *J. Genet. Genomics, 43*(1), 37–43.

Fraley, R. T., Rogers, S. G., Horsch, R. B., et al., (1983). Expression of bacterial genes in plant cells. *Proc. Natl. Acad. Sci. USA., 80*(15), 4803–4807.

Gao, J., Wang, G., Ma, S., Xie, X., Wu, X., Zhang, X., Wu, Y., et al., (2015). CRISPR/Cas9-mediated targeted mutagenesis in *Nicotiana tabacum. Plant Mol. Biol., 87*, 99–110.

Gilbert, L. A., Larson, M. H., Morsut, L., Liu, Z., Brar, G. A., Torres, S. E., Stern- Ginossar, N., et al., (2013). CRISPR-mediated modular RNA-guided regulation of transcription in eukaryotes. *Cell, 154*(2), 442–451.

Haughn, G. W., Smith, J., Mazur, B., & Somerville, C., (1988). Transformation with a mutant *Arabidopsis* acetolactate synthase gene renders tobacco resistant to sulfonylurea herbicides. *Mol. Gen. Genet., 211*(2), 266–271.

Hillary, E. V., & Antony, C. S., (2019). Application of CRISPR/Cas9 genome editing system in cereal crops. *The Open Biotechnology Journal, 13*, 173–179. doi: 10.2174/1874070701913010173.

Jia, H., & Wang, N., (2014). Targeted genome editing of sweet orange using Cas9/sgRNA. *PLoS One, 9*, e93806.

Jiang, W., Yang, B., & Weeks, D. P., (2014). Efficient CRISPR/Cas9- mediated gene editing in *Arabidopsis thaliana* and inheritance of modified genes in the T2 and T3 generations. *PLoS One, 9*, e99225.

Jinek, M., Chylinski, K., Fonfara, I., Hauer, M., Doudna, J. A., & Charpentier, E., (2012). A programmable dual-RNA-guided DNA endonuclease in adaptive bacterial immunity. *Science, 337*(6096), 816–821.

Joung, J. K., & Sander, J. D., (2013). TALENs: A widely applicable technology for targeted genome editing. *Nat. Rev. Mol. Cell Biol., 14*(1), 49–55. https://doi.org/10.1038/nrm3486.

Kim, H., & Kim, J. S., (2014). A guide to genome engineering with programmable nucleases. *Nat. Rev. Genet., 15,* 321–334.

Lee, K. Y., Townsend, J., Tepperman, J., et al., (1988). The molecular basis of sulfonylurea herbicide resistance in tobacco. *EMBO J., 7*(5), 1241–1248.

Li, Z., Moon, B. P., Xing, A., et al., (2010). Stacking multiple transgenes at a selected genomic site via repeated recombinases-mediated DNA cassette exchanges. *Plant Physiol., 154*(2), 622–631.

Liang, Z., Zhang, K., Chen, K., & Gao, C., (2014). Targeted mutagenesis in Zea mays using TALENs and the CRISPR/Cas system. *J. Genet. Genomics, 41*(2), 63–68.

Mali, P., Esvelt, K. M., & Church, G. M., (2013). Cas9 as a versatile tool for engineering biology. *Nat. Methods, 10,* 957–963.

Mazumdar, S., Quick, W. P., & Bandyopadhyay, A., (2016). CRISPR–Cas9 mediated genome editing in rice, advancements and future possibilities. *Indian J. Plant Physiol., 21*, 437–445.

Mikami, M., Toki, S., & Endo, M., (2016). Precision targeted mutagenesis via Cas9 paired nickases in rice. *Plant Cell Physiol., 57*, 1058–1068.

Murugan, K., Babu, K., Sundaresan, R., Rajan, R., & Sashital, D. G., (2017). The revolution continues: Newly discovered systems expand the CRISPR-Cas toolkit. *Mol. Cell, 68*(1), 15–25.

Negrotto, D., Jolley, M., Beer, S., Wenck, A. R., & Hansen, G., (2000). The use of phosphomannose isomerase as a selectable marker to recover transgenic maize plants (*Zea mays* L.) via Agrobacterium transformation. *Plant Cell Rep., 19*(8), 798–803.

Nelson, O. E., & Rines, H. W., (1962). The enzymatic deficiency in the waxy mutant of maize. *Biochem. Biophys. Res. Commun., 9*(4), 297–300.

Newman, M., & Ausubel, F. M., (2016). Introduction to gene editing and manipulation using CRISPR/Cas9 technology. *Curr. Protoc. Mol. Biol., 115*, 1–6.

O'Connell, M. R., Oakes, B. L., Sternberg, S. H., East-Seletsky, A., Kaplan, M., & Doudna, J. A., (2014). Programmable RNA recognition and cleavage by CRISPR/Cas9. *Nature, 516*(7530), 263–266.

Pegoraro, C., Mertz, L. M., Da Maia, L. C., Rombaldi, C. V., & De Oliveira, A. C., (2011). Importance of heat shock proteins in maize. *J. Crop Sci. Biotechnol., 14*, 85–95.

Puchta, H., Dujon, B., & Hohn, B., (1993). Homologous recombination in plant cells is enhanced by *in vivo* induction of double strand breaks into DNA by a site-specific endonuclease. *Nucl. Acids Res., 21,* 5034–5040.

Que, Q., Chilton, M. D., De Fontes, C. M., et al., (2010). Trait stacking in transgenic crops: Challenges and opportunities. *GM Crops, 1*(4), 220–229.

Radzisheuskaya, A., Shlyueva, D., Muller, I., & Helin, K., (2016). Optimizing sgRNA position markedly improves the efficiency of CRISPR/ dCas9-mediated transcriptional repression. *Nucl. Acids Res., 44*, e141.

Reis, A., Hornblower, B., Robb, B., & Tzertzinis, G., (2014) CRISPR/Cas9 and targeted genome editing: A new era in molecular biology. *NEB Expr., 1*, 3–6.

Sander, J. D., & Joung, J. K., (2014). CRISPR-Cas systems for editing, regulating and targeting genomes. *Nat. Biotechnol., 32*, 347–355.

Sandhya, D., Phanikanth, J., Venkateswar, R., Sadanandam, A., & Anshu, A., (2020). The present and potential future methods for delivering CRISPR/Cas9 components in plants. *Journal of Genetic Engineering and Biotechnology, 18*, 25. https://doi.org/10.1186/s43141-020-00036-8.

Shalem, O., Sanjana, N. E., & Zhang, F., (2015). High-throughput functional genomics using CRISPR–Cas9. *Nat. Rev. Genet., 16*, 299–311.

Shan, Q., Wang, Y., Li, J., & Gao, C., (2014). Genome editing in rice and wheat using the CRISPR/Cas system. *Nat. Protoc., 9*, 2395–2410.

Shi, J., Drummond, B. J., Wang, H., Archibald, R. L., & Habben, J. E., (2016). Maize and *Arabidopsis* ARGOS proteins interact with ethylene receptor signaling complex, supporting a regulatory role for ARGOS in ethylene signal transduction. *Plant Physiol., 171*(4), 2783–2797.

Shi, J., Gao, H., & Wang, H., (2017). ARGOS8 variants generated by CRISPR-Cas9 improve maize grain yield under field drought stress conditions. *Plant Biotechnol. J., 15*(2), 207–216.

Shi, J., Gao, H., Wang, H., Lafitte, H. R., Archibald, R. L., Yang, M., Hakimi, S. M., Mo, H., & Habben, J. E., (2017). ARGOS8 variants generated by CRISPR–Cas9 improve maize grain yield under field drought stress conditions. *Plant Biotechnol. J., 15*, 207–216.

Sprague, G. B. B., & Hixon, R., (1943). Some effects of the waxy gene in corn on properties of the endosperm starch1. *Agron. J., 35*(9), 817.

Sun, X., Hu, Z., Chen, R., Jiang, Q., Song, G., Zhang, H., & Xi, Y., (2015). Targeted mutagenesis in soybean using the CRISPR–Cas9 system. *Sci. Rep., 5*, 10342.

Svitashev, S., Schwartz, C., Lenderts, B., Young, J. K., & Cigan, A. M., (2016). Genome editing in maize directed by CRISPR–Cas9 ribonucleoprotein complexes. *Nat. Commun., 7*, 13274.

Svitashev, S., Schwartz, C., Lenderts, B., Young, J. K., & Mark, C. A., (2016). Genome editing in maize directed by CRISPR-Cas9 ribonucleoprotein complexes. *Nat. Commun., 7*, 13274.

Svitashev, S., Young, J. K., Schwartz, C., Gao, H., Falco, S. C., & Cigan, A. M., (2015). Targeted mutagenesis, precise gene editing, and site-specific gene insertion in maize using Cas9 and guide RNA. *Plant Physiol., 169*(2), 931–945.

Walbot, V., (2000). Saturation mutagenesis using maize transposons. *Curr. Opin. Plant Biol., 3*, 103–107.

Walter, K. L., Strachan, S. D., Ferry, N. M., Albert, H. H., Castle, L. A., & Sebastian, S. A., (2014). Molecular and phenotypic characterization of Als1 and Als2 mutations conferring tolerance to acetolactate synthase herbicides in soybean. *Pest Manag. Sci., 70*(12), 1831–1839.

Waltz, E., (2016). CRISPR-edited crops free to enter market, skip regulation. *Nat. Biotechnol., 34*, 582–583.

Wang, S., Zhang, S., Wang, W., Xiong, X., Meng, F., & Cui, X., (2015). Efficient targeted mutagenesis in potato by the CRISPR/Cas9 system. *Plant Cell Rep., 34*, 1473–1476.

Wessler, S., Tarpley, A., Purugganan, M., Spell, M., & Okagaki, R., (1990). Filler DNA is associated with spontaneous deletions in maize. *Proc. Natl. Acad. Sci. USA., 87*(22), 8731–8735.

Yu, Q., & Powles, S. B., (2014). Resistance to AHAS inhibitor herbicides: Current understanding. *Pest Manag. Sci., 70*(9), 1340–1350.

Zhang, Y., Liang, Z., Zong, Y., Wang, Y., Liu, J., Chen, K., Qiu, J. L., & Gao, C., (2016). Efficient and transgene-free genome editing in wheat through transient expression of CRISPR/Cas9 DNA or RNA. *Nat. Commun., 7*, 12617.

Zhao, Y., Zhang, C., Liu, W., Gao, W., Liu, C., Song, G., Li, W. X., et al., (2016). An alternative strategy for targeted gene replacement in plants using a dual-sgRNA/Cas9 design. *Sci. Rep., 6*, 23890.

Zhu, J., Song, N., Sun, S., et al., (2016). Efficiency and inheritance of targeted mutagenesis in maize using CRISPR-Cas9. *J. Genet. Genomics, 43*(1), 25–36.

Zong, Y., Wang, Y., Li, C., Zhang, R., Chen, K., Ran, Y., Qiu, J. L., et al., (2017). Precise base editing in rice, wheat and maize with a Cas9-cytidine deaminase fusion. *Nat. Biotechnol., 35*, 438–440.

CHAPTER 8

Genome Editing of Tomato Crop with CRISPR/Cas Systems: Past, Present, and Future

VINCENT EZIN[1] and RACHAEL C. SYMONDS[2*]

[1]*Faculty of Agricultural Sciences, University of Aboemy-Calavi, Benin*

[2]*School of Biological and Environmental Sciences, Liverpool John Moores University, UK*

[*]*Corresponding author. E-mail: r.c.symonds@ljmu.ac.uk*

ABSTRACT

Tomato is one of the world's most economically important crops, providing income for many rural economies. It is a significant source of nutrients, vitamins, and antioxidants; however, it is sensitive to many biotic and abiotic stresses. Additionally, as a soft fruit crop, there are major post-harvest losses. The impacts of climate change will intensify this situation, and new sources of breeding material are needed to produce breeding material with new and improved traits. Tomato has a very narrow genetic diversity due to intensive selection for selected traits during domestication, leaving few sources of new variation amongst the cultivated tomato. Traditional breeding methods have been successful, although time-consuming, and there have been successful breeding new traits from wild relatives. However, there are often deleterious effects associated with linkage drag, and many traits of interest are quantitative. Tomato is a model organism and benefited from the early availability of a completed genome sequence, and it was one of the very first crops to be genetically modified. The advent of CRISPR technology has allowed

CRISPR/Cas-Mediated Genome Editing in Plants, Deepu Pandita, Anu Pandita (Eds.)

the rapid expansion of genome editing in tomatoes for diverse traits. The advantages of the rapid production of transgene-free plants for both single and multiple traits, as well as the ease in which de novo domestication can be carried out has greatly expanded the material available for tomato breeding. In this chapter, we discuss the economic importance of tomatoes, the genome editing technologies used to date, and the future prospects and traits of interest.

8.1 INTRODUCTION

Tomato (*Solanum lycopersicum*) is the world's most important vegetable crop with approximately 180 million tons produced worldwide per annum approximately (FAO, 2019). Tomatoes are widely consumed as a culinary ingredient and contain high levels of vitamins C and A, potassium, and iron, as well as the red carotenoid lycopene (Bhowmik et al., 2012). Tomato consumption is inversely correlated with a range of health conditions and lower mortality from diseases, including heart disease, gastric, and prostate cancers, and cerebrovascular disease (Li et al., 2021). Furthermore, the pigment lycopene is a known antioxidant and dietary intake is inversely correlated with several diseases, including cardiovascular disease and stroke, cancer, and metabolic syndrome (Li et al., 2021).

Tomato is usually cultivated either in greenhouses or in field conditions (Lang, 2004). Open field conditions expose tomatoes to both abiotic and biotic stresses, which can reduce growth and yield. This is a major constraint on tomato production and results in many regions producing sub-optimal tomato yields (Silva et al., 2017). The impacts of climate change will only exacerbate this situation and require careful targeting of stress tolerance traits for further cultivar improvement.

Tomato was originally domesticated from the small red-fruited wild tomato species *Solanum pimpinellifolium* (Lin et al., 2014). From this wild species there have been major changes in the shape, size, color, and flavor of the fruit, changes to the plant architecture from indeterminant to determinant and changes to the biotic and abiotic stress tolerance (AT) during domestication and later breeding. Several traits, such as self-pruning (SP), fruit morphology, flowering time, and color have been specifically targeted for adaption to different growing conditions (Rodríguez et al., 2011).

The resulting cultivated tomato is a highly inbred species (Rick, 1987) and whilst domestication and then breeding has produced increases in yield and productivity, it has restricted genetic diversity (Zamir, 2001; Tam et al., 2005). This has resulted in a narrow genetic variation amongst commercial tomato cultivars of around only around 5% (Miller and Tanksley, 1990) greatly reducing the diversity needed for breeding new and improved cultivars. Wild tomatoes still harbor these resistance genes, and there has been some success developing improved germplasm using crosses and introgression (Bai and Lindhout, 2007). The tomato is a diploid model soft fruit plant, and tomato breeding and cultivar development has benefited from the early availability of the tomato reference genome (Ranjan et al., 2012; 100 Tomato Genome Sequencing Consortium et al., 2014).

With the advent of molecular breeding and genetic modification (GM) technology further gains in tomato improvements for new traits have been made. Tomato was the first transgenic fruit crop sold commercially, however, the difficulties with improving crops with a single gene insertion and lack of consumer acceptability have held back further gains (Schuch et al., 1991). A large number of studies followed using GM technologies to produce tomatoes with ripening, biotic, and abiotic resistance, and quality traits; however, to this day, commercial exploitation has lagged behind (Bai and Lindhout, 2007). The advent of genome editing techniques such as transcription activator-like effector nucleases (TALENS), ushered in a new dimension to the tomato breeding toolkit (Harwood et al., 2017), however, difficulties associated with this technique meant its use was not as widespread as could have been hoped for. The rapid paced expansion of CRISPR technologies has led to a new era in tomato genome editing. Within a few years, there have been tomatoes produced with traits including altered for architecture (Soyk et al., 2017), color (Cermak et al., 2015), biotic (Pramanik et al., 2021), and abiotic stress (Li et al., 2019). The present chapter focuses on the tomato crop and its production and health benefits, the genome editing technologies used to date, and the future of CRISPR and genome editing in tomatoes. A particular focus is given to CRISPR technologies associated with improvements in biotic and abiotic stress, improvements to genetic diversity through *de novo* domestication, metabolic engineering, and ripening and architecture traits (Figure 8.1).

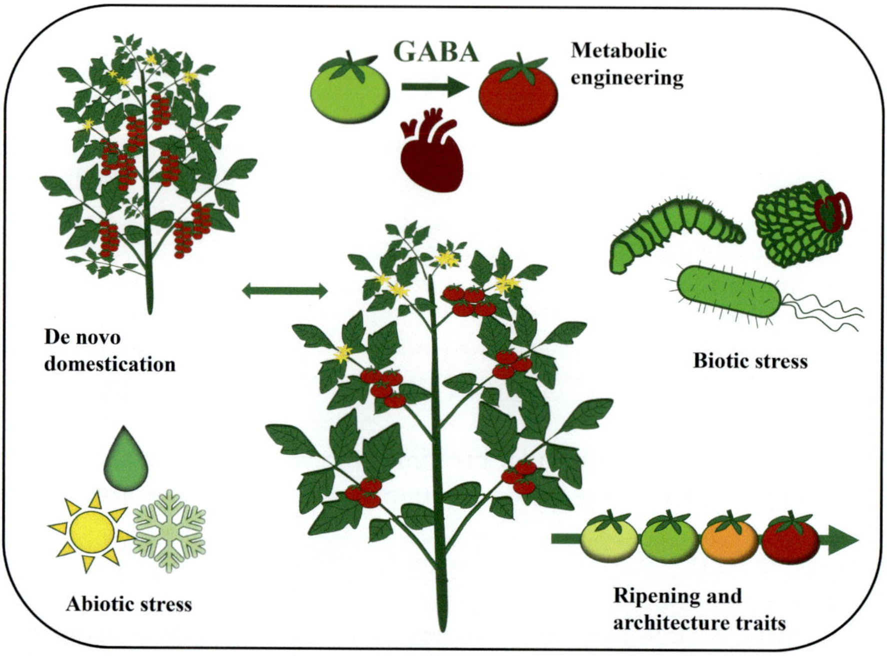

FIGURE 8.1 CRISPR/Cas system uses in tomato breeding for traits including *de novo* domestication, abiotic stress, biotic stress, metabolic engineering and ripening, and architecture traits.

8.2 BOTANICAL AND ECONOMIC IMPORTANCE OF TOMATO

Tomato (*S. lycopersicum*), native to South and Central America, is an herbaceous plant both annual and perennial with determinate and indeterminate growth (Bai and Lindhout, 2007). The flowers of cultivated tomatoes are of neutral days, thus indifferent to photoperiodism, which has allowed its adaptation to various latitudes. Tomato is a C3 plant, which produces carbohydrates through the first stable three carbon atoms. It is a self-pollinated crop based on its hermaphroditic flowers. The stamens and pistils are enclosed in the flowers which limits cross-pollination, although it is not a complete barrier (Welty et al., 2007). The wind and insects aid pollination by shaking the anthers, which release the pollen grains into the stigma. In 1694, tomato was classified in the genus *Lycopersicon* by French botanist Joseph Pitton and in 1753 it was changed to genus Solanum by Linnaeus, but in 1768, Philip Miller demonstrated that tomato is different from other species of the genus Solanum and classified it in the genus *Lycopersicon* with botanical name *Lycopersicon esculentum*. With the help of the modern techniques of molecular biology and genetics evidence, tomato has again been placed in the genus *Solanum*, thus proving Linnaeus correct (Peralta and Spooner, 2001; Peralta, 2006). The current name is therefore *Solanum lycopersicum*, although the name given by Miller is still used in many publications to date. Tomato was first domesticated in Mexico, then it was later introduced to Europe and other parts of the world (Harlan, 1987). Tomato names in some languages are: tima (Yoruba), tomate (Spain, France), tomat (Indonesia), faan ke'e (China), tomatl (Nahuatl), jitomate (Mexico), pomodoro (Italy), and nyanya (Swahili).

Tomato is one of the most popular and important fruit vegetables worldwide. It is grown in almost every country and is the most cultivated vegetable in the world. Its production provides money for many rural economies and significantly contributes to the GPD of every country in the world. Tomato fruits can be consumed fresh after harvest, and it can be processed and canned (in the forms of ketchup, tomato juice, tomato paste, and tomato sauces) in the industrial sector. It is rich in lycopene, antioxidants playing a role in the avoidance of many cancers and cardiovascular diseases, rich in chlorogenic acid, naringenin, vitamin C, potassium, minerals, folate, or vitamin B9, vitamin K, vitamins A, C, and E (Bhowmik et al., 2012). The tomato fruit contains chlorine which helps eliminate waste from the body, and sulfide protects the liver against certain blockages. Due to the presence of vitamin C, it helps in the assimilation of iron and calcium and reduces hypertension owing to the presence of potassium, which is very valuable for heart disease prevention and blood pressure control (D'Elia et al., 2011). Folate contributes to cell functions and tissue growth and is vital for pregnant woman (Fekete et al., 2012) while Vitamin K1 is vital for blood clotting and bone health (Bolton-Smith et al., 2007). Tomato skin contains naringin which possesses anti-inflammatory and antioxidant properties (Barthi et al., 2014). Chlorogenic acids are very effective in reducing blood pressure (Watanba et al., 2006). It has many benefits to human welfare, such as the prevention of stroke, high cholesterol, oxidative stress, and cancers. A 100-gram raw tomato contains calories (18), water (95%), protein (0.9 grams), carbs (3.9 grams), sugar (2.6 grams), fiber (1.2 grams), and fat (0.2 grams). High beta-carotene tomato varieties are available today to fight against malnutrition and vitamin A deficiency.

Tomato production has risen steadily and considerably over the past decades. It increased from 48 million tons in 1978, to 74 million in 1992, 89 million in 1998, and reached 124 million in 2006. Of the 16 countries that produced 1 million tons or more, 6 are well above 5 million tons (FAO,

2007). It is estimated that 30% of the tomatoes produced are processed. This percentage is very different from one country to another. About 1,77,118,248 tons of tomato fruits are harvested globally per year. China ranks first in the world with a production volume of 56,423,811 tons per year followed by India with an annual production of 18,399,000 tons and USA are third before Turkey in fourth position (FAOSTAT, 2013). Between 1961 and 2018, global tomato production increased 6.6-fold, from 27.6 million tons to 182.3 million tons. Most of this growth, however, can be attributed to Asia, whose tomato production has increased 14.7-fold, from 7.6 million tons to 111.7 million tons. In Africa, tomato production also increased sharply, from around 2 million tons to 20.8 million tons. The other continents have also seen their production increase, but in smaller proportions: x 3.7 for America, x 2.6 for Oceania and x 2.1 for Europe (FAOSTAT, 2019). Since 1961, China has usually been the first country in the world in terms of cultivated area. The United States were the first in terms of surface area in 1967 and 1970, then between 1972 and 1976. They are now in 3rd place with 1,30,000 hectares. Italy, which was third in 1961 with 1,27,000 hectares, dropped to 6th in 2018 with around 97,000 hectares. Similarly, in the course of the same years, Nigeria and India have significantly increased their area, rising from 50,000 hectares to 7,86,000 hectares and from 18,000 to 6,08,000 hectares, respectively. It should also be noted the rise of Cameroon, which rose from 6,500 hectares in 1961 to more than 93,000 hectares in 2018.

Despite all the above-mentioned benefits, tomato cultivation is not straightforwardly attained due to environmental and biotic constraints associated with crop management and input requirements. The limitations related to abiotic and biotic stress are variety and climatic condition dependent. To accomplish high yields and profits, know-how is needed in the field of agronomic practices, management of soils and its fertility, disease, and pest management, harvest, and post-harvest handling, conservation, and market availability for sale and adaptation and mitigation of climate change.

8.3 TOMATO AND CLIMATE CHANGE

Tomato is a vegetable crop that requires a cool, dry climate to provide a bountiful and quality harvest, however, tomato can adjust to a variety of climatic conditions ranging from the temperate climate to the hot and humid tropical climate. Tomato plants face multiple environmental stresses concomitantly, either in the form of multiple abiotic stresses or a combination of abiotic and biotic factors simultaneously (Havko et al., 2020). Generally speaking, climate change indicates the significant and lasting modification of climate parameters such as precipitation, temperature, wind, and sunshine. Tomato productivity is now threatened by persistent fluctuations in these climatic parameters. The rise in the average universal temperature has been known since the second half of the 20th century and is most likely attributable to the increase in concentrations of anthropogenic greenhouse gases (IPCC, 2007). On a wide scale, human activities are the primary causes leading to environmental problems, namely the atmospheric concentration of carbon dioxide (CO_2) and methane (CH_2). Air pollution from industry, transport, construction of buildings, energy production and agriculture from bush fires are among the major causes of this degradation. This warming is manifested mainly by the increase in extreme temperatures, which has significant impacts on rainfall and thus on every sector, including agriculture.

In general, agriculture will be negatively influenced by global warming (IPCC, 2007). Most plants, including tomatoes, would not be able to withstand the deleterious impact of climate change, whereas there is a vital need to increase crop production for food security reasons. The existing abiotic factors such as salinity, flooding, heat, heavy metal toxicity, frost, drought, and nutrient deficiency and biotic such as pests, nematodes, bacteria, fungi, and virus will be exacerbated and there will be more outbreak of new diseases and pests, e.g., *Spodoptera frugiperda* is ravaging maize crop in West Africa since their occurrence in 2016.

Global warming leads to increased stress on plants impeding growth and development, phenological parameters, abortion of flowers and sharp decline in fruit yields. The optimum temperature for most tomato varieties is between 21 and 24°C, as from 26°C, there is a significant decline in the tomato growth rate. Plants can overcome a certain temperature range, but below 10°C and above 38°C, plant tissue will be damaged (van Dam et al., 2005). For vegetative growth, tomatoes require above 18°C (Geisenberg and Stewart, 1986). Tomato requires a temperature discrepancy between day and night of 6 to 10°C in order to produce a maximum yield. Geisenberg and Stewart (1986) reported that beyond 31°C development and growth, fertilization, and the ripening of tomato diminishes. Flowering of tomato is very difficult in extreme temperatures and self-fertilization is almost impossible at 38°C. Fruiting is also difficult as from 30°C, and at 35°C fruit production is no longer possible. Ezin et al. (2018) reported that higher temperature associated with drought cause 100% abortion of flowers in sensitive tomato genotypes and hinder tomato fruit set. When temperatures are high, tomato plants infested by insects produce more jasmonates. However, this overproduction blocks the plant's ability to lower its internal temperature making it more sensitive to heat. In addition, when temperatures rise, the metabolism of insects' speeds up, causing them to feed more, leading to tomato plant destruction (Havko et al., 2020). Thus, climate change through high temperature will increase insect populations, range, and feeding rapaciousness (Raffa et al., 2013; Deutsch et al., 2018) increasing tomato susceptibility while lowering yields.

Globally, tomato production is frequently confronted with the rise in severe meteorological events owing to change in climate, which cause water deficit and flooding to be a serious menace to food security (FAO, 2017). Flooding or submergence, anoxia or hypoxia, soil saturation or waterlogging also affects all stages of crop production negatively from germination of seeds to flowering and fruit set (Pan et al., 2021) and causing serious economic losses. When flooding affects land or plots on tomato crops, there is a decline or complete inhibition of plant metabolism. The root systems of submerged tomato plants cease to function due to the absence of oxygen, thus inducing the closure of stomata (blocking photosynthesis), production of a high concentration of ethylene, and a cessation of essential nutrient absorption. Plants cannot grow normally, which leads to low yields or complete losses of yields (Ezin et al., 2010).

Soil salinity affects both the soil properties and structure and plant growth (Aubert, 1983). The ions such as Na^+ and Cl^- alter the physicochemical properties of soils and lead to structural and texture degradation which become averse to plant life. This scourge is becoming more serious with the advance of change in climate in Sahara and regions with irrigation systems for crop production. Salinity is a serious abiotic stress to tomato crop which impinge tomato seed germination, vegetative growth, and reproductive growth (Kayani et al., 1990). Salinity causes both ion toxicity and water deficit to tomato crop plants. The soil solution of salt ion is more intense than that of the tomato plant, which inhibits the plant from absorbing water from the soil. The up taking of salt ions by the root systems causes burning in the leaves, premature yellowing of the leaves, deterioration of cells, alteration of metabolic reactions such as photosynthesis, respiration, translocation of phyto-assimilates and biosynthesis of phytohormones (Brugnoli and Lauteri, 1991).

Water deficit is a major abiotic constraint in agricultural production and productivity, and drought is expected to be enhanced by climate change (Ximénez-Embún et al., 2018). First of all, by its negative impact on crop yield and product quality, and by its frequency due to global warming. Without a sufficient and accessible water resource at the root system level, the plant loses part of its water and the potential energy of water of the cells decreases. Drought can cause yield reduction of up to 50% or 71% depending on the intensity of drought (Boyer, 1982). Similarly, the loss of yield ranges from 30–90% depending on the crop stage and the degree and duration of drought (Pandit et al., 2018). The consequences are many and occur at different scales in the plant. Nowadays, 450 million people are subjected to severe water shortage and in 2025 this number may increase to about 2.7 billion (or 1/3 of the world population). Water deficit, among the various abiotic stress, is by far the most difficult and damaging globally (Pennisi, 2008). Agriculture is a main consumer of water worldwide and drought affects agriculture in 45% of the world's geographical area. The main responses of tomato plant to water deficit are a reduction in the pressure of turgor, the engine of cell growth and the growth of plant organs, and closure of the stomata which limits transpiration and decreases photosynthetic activity, decrease in morphological, physiological, and agronomical traits (Blum, 1997; Medyouni et al., 2021). Tomato plants are confronted with a kind of compromise when responding to water stress: maintaining a water balance while ensuring sufficient growth. Most compartments of the plant including the roots, the vegetative parts, and the reproductive organs, can be negatively impacted by the water deficit according to its severity.

Globally these extreme meteorological events are associated with economic losses, therefore it is a necessity to study abiotic factors' resistance related to tomato along with its morphological, physiological, anatomical, and molecular level mechanisms, so as to keep successfully producing tomato well adapted to climate change (Nishiuchi et al., 2012; Mondal et al., 2020; Pan et al., 2021).

8.4 HISTORY OF TOMATO BREEDING AND GENETIC MODIFICATION (GM)

Tomato improvement began with the domestication of the species by pre-Columbian civilizations, h in both temperate and tropical zones. Varietal improvement consists in improving the varieties of tomatoes (cultivars) by selecting numerous useful characters, including abiotic and biotic stress tolerance, yield, nutritional value, and conservation. The breeding techniques are domestication, varietal selection, varietal creation, hybridization, and genetically modified tomato or transgenic tomato.

Tomato domestication from wild species to cultivars seemingly took place in Mexico or in Peru, although up to date, there is no concrete evidence of it (Rick, 1969). Peralta and Spooner (2007) reported that the original place of domestication and all the activities involved are unclear. But at the same time, Larry and Joanne (2007) supposed that Mexico is the domestication site while Peru is the center of diversity of tomato wild species. The large fruits as a result of domestication already existed during the conquest of Mexico by the Spaniards in 1519 (Rick, 1969). Tomato is a diploid crop with 2n = 2x = 24 chromosomes. Selection breeding, cross-breeding, mutation breeding, and transgenic breeding have been successfully employed as breeding methods to develop new cultivars for human consumption. Varietal selection is the traditional method used and practiced by farmers since ancient times by preserving the most interesting individuals from one generation to another and by selecting the seeds on the most productive plants. In the 19th century, many developed varieties of tomato obtained either by the processes of domestication or selections were already available (Bai and Lindhout, 2007). The varietal selection also aims at purifying the variety through several generations of self-fertilization, resulting in a pure line to be used in varietal creation. Varietal creation hangs on the principles of predefined parental plants according to the criteria of abiotic and biotic tolerance, yield, nutritional value, conservation, shape, color, and performance of artificial cross-fertilization. The seeds obtained are therefore hybrids. The hybrid offspring are cultivated and the seeds obtained by self-pollination of the hybrids are harvested and reseeded. Among the population of plants of this 2nd generation, many different phenotypes coexist. The breeder selects among these plants those which have traits of interest, and then the process of varietal production begins by several generations of self-fertilization and selection of plants conforming to the desired variety. One of the efficient ways of producing new varieties in autogamy-pollinated crops especially tomato consists of crossing pure lines and selecting descendants with a set of interesting traits for disease and pest resistance, conservation, color, long shelf life, resistance to drought, extreme temperature, salinity, and flooding. Thus, the classical/conventional breeding relies on the principle of introducing genes from wild species into the cultivated species, the techniques have been used for years and the developed cultivars significantly contributed to human food security and welfare. *Solanum chmielewskii* genes have been introduced into the industrial tomato varieties genes which markedly improve the rate of soluble dry matter, an important criterion for the production of tomato paste. However, to develop a cultivar, it takes years before its release making the techniques time-consuming (Wang et al., 2019; Schouten et al., 2019).

In many countries in the world, *alc*, *nor*, and *rin* genes from tomato mutants have been successfully transferred to varieties to develop commercial hybrids with prolonged storage period in tomato (Mutschler, 1984; Lu et al., 1994; Seroczynska et al., 1998). Many hybrids have been developed in Australia using ripening mutants such as *rin-HRAS* 81–85 and *rin-795054-1*. The hybrids 'Red Center' and 'Juliette' were developed by crossing respectively HRAS 87–70 with *rin-HRAS* 81–85 and 79T-I x *rin-795054-1*. The two hybrids have 40 days' shelf life at 20°C temperature and have been commercialized (Nguyen et al., 1991; Nguyen, 1994). In Russia and China, 'Vasilisa' hybrid, and 'Changline' hybrid were respectively developed using *nor* gene and possess outstanding shelf life (Gavrish and Bogdanov, 1992; Lu et al., 1994). 'Rafal' hybrid with shelf life of between 77 and 99 days was developed in Poland using the same *nor* gene from the cross S 15 x *110r* (Seroczynska et al., 1998), 'Rafal' hybrid is of exceptional firmness and color and quality. The *alc* is the third gene involved delaying tomato fruit ripening and has been contributing greatly to extended long life, quality, and post-harvest preservation of tomato fruits. In the United Kingdom, "purple tomato" was created by the Center of John Innes. Similarly, B, Da, and F inbred lines and hybrids were obtained using the B, Da, and F GM tomato (Baranski et al., 2019). In this purple tomato, the high concentration of anthocyanin in charge of the purple color of the fruit came from of genes transferred from snapdragon, while the salt-tolerant tomato created from the variety 'Moneymaker' which received the AtNHX1 gene from *Arabidopsis thaliana* (Zuluaga et al., 2008). Moreover, tomato disease resistance, adaptation to heat, drought, and salinity have been released. In the tropics, research focuses on adaptation to climate and resistance to bacterial wilt and nematodes (Rick, 1990).

Contrary to classical breeding, the transfer of the genes of interest from one organism to another not closely related, which could be from microorganisms to plants has been possible thanks to recombinant DNA technology (RDT). The advent of molecular biology more 30 years ago made it possible to transfer virtually any genetic material from one species to another, irrespective of its origin or destination. The tomato resulting from RDT is called a "genetically modified" tomato. Many results of RDT have been proven positive such as the enhancement of tomato fruit yields and its quality, nutritional values, and shelf life (Wang et al., 2019). RTD is considered as the quickest technology in crop production (Parmar et al., 2017). A first set of "genetically modified" plants (GMP) emerged in the 1990s. Long-shelf-life tomatoes were the first GM crops to be marketed (1994), followed by insect resistant potatoes (1995), 1345-4 GM tomato in USA also called 'Endless Summer' (1995), soybeans resistant to the herbicide glyphosate (1996) and virus resistant papayas (1998). In 1994, after the risk assessments and the fulfillment of all the necessary conditions, the FDA (Food and Drug Administration) concurred in the GM "Flavr Savr™ tomato" for commercial growth in the USA (FDA, 1992; Bawa and Anilakumar, 2013) as the first tomato GM to be marketed but was prohibited for sale in 1997 (Baranski et al., 2019). "Flavr savr™ tomato" is a genetically modified tomato developed by the American company Calgene, via the antisense RNA technique with the objective of extending the life of the fruit after harvest and therefore the "quality" of the tomato for fresh consumption. In this tomato, they have succeeded in reducing/suppressing the expression of the gene responsible for the production of polygalacturonase, an enzyme responsible for the degradation of pectin in the cell walls during the ripening phase. In Canada, 1401F, H282F, 11013F, 70913F, TGT7F, and 7913F GM tomato were also developed to reduce pectin degradation by suppressing polygalacturonase activity (APHIS, 1994a, Meli et al., 2010). 1345-4 transgenic tomato was developed to reduce ethylene accumulation and therefore delay ripening. Monsanto created insect resistant (Lepidoptera) Bt-5345, the genetically modified tomato, which contains two new proteins: Cry1Ac provides Lepidoptera-resistance to tomato while NPTII gives kanamycin resistance to tomato. The British company Zeneca also created long-life Da (OECD UID: SYN-000DA-9), F (OECD UID: SYN-0000F-1) and B (OECD UID: SYN-0000B-6) GM tomatoes (APHIS, 1994a; Baranski et al., 2019). 35-1-N, 1345-4 and 8338 GM tomato were also developed to delay ripening and prolong shelf-life tomato through the change in ethylene biosynthesis produced by tomato as a climacteric fruit (Yokotani et al., 2009; Baranski et al., 2019). In China, GM tomatoes such as Da Dong No. 9 (for fruit quality), Huafan No. 1 (for fruit quality) and PK-TM8805R (for virus resistance), were developed (Chen and Yang, 1996; BiosafetyScanner, 2018). Baranski et al. (2019) reported that the three GM tomato are no longer produced in China today. Salt-tolerant GM tomato was also developed and it showed the build-up of salt in the leaves but not salt residues was observed in the fruits (Zhang and Blumwald, 2001) and drought and salt tolerant GM tomato were also created and it was shown that osmotin gene (BvHb2) is responsible for both salt and drought stress in transgenic tomato (Goel et al., 2010). Around 21 years after the introduction of the first GM crop was brought to market, these approved species are cultivated by approximately 8 million producers on 185 million hectares in 26 countries (ISAA, 2016).

The tomato genome consists of 12 pairs of chromosomes (2n = 24). Its size is estimated at 950 Mbp encoding around 35,000 genes. The majority of the gene sequences, representing 220 Mbp, are concentrated in contiguous euchromatic regions in the distal regions of each arm of the chromosomes. Tomato genome sequencing was initiated in 2005 by a consortium of 14 countries. It started with the genome of a cultivated tomato variety Heinz 1706, and the precise order of DNA nucleotides was determined and assembled using Sanger technology and then next-generation sequencing technologies. Likewise, the genome of the wild species *S. pimpinellifolium* was sequenced and assembled with Illumina short read sequencing technology. The comparison of these genomes (wild and cultivated) showed that there is a small difference between them (0.6% of SNP diverge) (Tomato Genome Consortium, 2012).

The sequencing made it possible to better understand the genome of the tomato. Indeed, the more precise identification of the modifications which occur during natural mutations could be carried out successfully. Various natural mutations have occurred in genes of interest during the evolution of the tomato. These are in particular mutations linked to the growth of the plant which results in a determined growth of the plant; mutations related to fruit coloring and fruit ripening mutations such as ripening inhibitor (rin), non-ripening, and alcobaca which block or slow the ripening of the fruits (Grassely et al., 2000). The sequencing facilitated the identification of polymorphic SNP-like molecular markers (Kissoudis et al., 2015). The techniques of detection of QTLs (quantitative trait locus) and genetics of association (GWAS: genome-wide association study) have allowed the discovery of many genes and QTLs involved in the quality of the tomato fruit, disease, pest resistance, drought, and salt resistance. A very large number of studies of QTLs (quantitative trait loci) responsible for the variation of different traits have been carried out (Mattoo and Razdan, 2007), in particular in response to abiotic stresses. For example, 54 QTLs were detected in the study of Diouf et al. (2018b), some of which are responsible for plant growth, fruit quality while others result from interaction of the plant with its environment. The precision and validity of QTL mapping might be enhanced by the type of population used. Indeed, the use of a MAGIC population, a population resulting from a mixed cross between eight parents, made it easier to detect causal SNPs or those linked to QTLs and to reduce the number of candidate genes and SNPs (Pascual et al., 2015a, b; Diouf et al., 2020). The use of these results in selection was first considered by marker-assisted selection (MAS). However, MAS has shortcomings by using a tiny part of the variation and relying on often overestimated effects. Genomic selection appears as an alternative. It allows with the aid of genomic prediction to save considerable time and resources and to overcome the shortcomings of previous methods.

8.5 PAST TECHNOLOGIES FOR GENOME EDITING

Genome editing is a new breeding method using only the genetic material of the plants unlike GMO techniques which utilize foreign genetic material to create new cultivars. And much more advantageous over the classical breeding techniques in the sense that there is no artificial crossing, and it only takes a few years to develop new cultivars (Rogowsky, 2016; Zhang et al., 2018).

Genome editing describes the process of inserting, removing, or replacing a specific gene in an organism's genome to introduce a new function or correct a mutation. Genome editing is a set of technologies allowing the targeted modification of genetic information by adding, deleting, or exchanging (replacing) nucleotides at a specific site in the sequence of the genome of a recipient organism. In the case of tomato plants, these technologies can deliver traits to the production chain, from farmer to consumer, faster than older methods. These very diverse traits include (i) resistance to biotic stresses, pathogens, and bio-aggressors, (ii) increased tolerance to abiotic stress such as flooding, drought or salinity or extreme temperature, and nutrient deficiency, (iii) improved health qualities, technological, and nutritional aspects of the harvested products (Haverkort et al., 2009; Haun et al., 2014; Clasen et al., 2016; Wang et al., 2019c; Xu et al., 2019; Kwon et al., 2020). Tomato is used as a model crop among the dicot crops due to the fact it presents numerous benefits (Martín-Pizarro and David Posé et al., 2018; Yuan et al., 2021): (a) its genome has been completely sequenced and of high quality; (b) it can easily be manipulated; (c) it is of economic importance as crop-generating incomes and as the most produced and consumed vegetable fruits worldwide.

Genetic engineering (GE) technique is based on the use of restriction endonucleases in order to modify the genome. The originality of this technique results in the high specificity of cleavage of the double strand of DNA by nucleases, unlike previous techniques where the modifications were largely random. Three past genome editing technologies based on the families of nucleases ("molecular scissors"), including the new one (CRISPR-Cas9 system) are used to cut DNA at the desired level, in order to add or remove a particular sequence. The modified DNA double strand is re-soldered by homologous recombination (HR) or by joining non-homologous ends. The insertion of genetic sequences by HR after double-stranded DNA (dsDNA) cleavage makes it possible to correct, modify or inhibit the target genetic sequence. The previous nucleases used are:

1. **Zinc Finger Nucleases (ZFNs):** They contain a zinc ion which form scissor-like shape. ZFNs possess two domains, including a specific DNA binding domain (which induces DNA recognition) and a non-specific catalytic domain (which induces the cleavage of the DNA double-strand). They work by binding specifically to a specific DNA sequence, leading to the targeted insertion or deletion of a nucleotide, possibly resulting in the appearance or suppression of a character (Pabo et al., 2001). In other words, ZFNs are hybrid proteins created by the fusion of genes encoding zinc finger proteins, interacting with a defined DNA sequence, and the gene encoding the enzyme restriction. Each zinc finger recognizes a specific short sequence (three nucleotides), the combination of several suitable zinc finger domains allows the sequence of interest to be uniquely targeted (Urnov et al., 2005). When L1L4 gene and LEC1-like4 protein gene were targeted in tomato using ZFN technology it resulted in creating phenotypic variability/heterochronic phenotypes during development such as plant height and number of florets, seed germination time, plant vigor, reduction in the number of locules per fruit, shifting in the time of flowering and ripening, change in fruit size and number of seeds and fruit weight (Hilioti et al., 2016). Traits related to herbicide tolerance (HT) were produced by replacing ALS (Aceto Lactate Synthase) gene in tobacco using ZFNs (Townsend et al., 2009).
2. **Transcription Activator-Like Effector Nucleases (TALENs):** These are artificial enzymes created by fusing a specific DNA binding domain (succession of TAL effectors) and the catalytic domain of the 'enzyme. TAL effectors are proteins secreted by the phytopathogenic bacterium Xanthomonas of the Pseudomonadaceae family. The protein sequence of an effector TAL varies across two amino acids involved in the specific recognition of a particular nucleotide in the genome. As a result, by combining different TAL

effectors, the DNA binding domain of TALENs can recognize a genomic sequence of interest (Christian et al., 2010). TALEN system can knock out some genes to create mutants in tomato. Tomato plant with PRO mutation presented extended internodes and smooth leaf margins, a phenotypic characteristic of procera mutant (Lor et al., 2014). There was high accumulation of anthocyanin when the 35S promoter replaced the promoter in tomato when SLANT1 gene was targeted and the phenotypic evaluation resulted to purple tomato (Čermák et al., 2015). Traits related to HT were produced by replacing ALS (Aceto Lactate Synthase) gene in rice using TALENs (Sun et al., 2016). Innovative traits linked to nutritional quality and processing ability have also been developed, for example in potatoes, where the simultaneous mutation by TALENs of the four alleles of the VInv gene (vacuolar invertase) has made it possible to decrease the accumulation of reducing sugars during cold storage of tubers and thus limit the production of acrylamide during frying (Clasen et al., 2016). Fragrant rice was created by mutation of the BADH2 gene using TALENs (Shan et al., 2015). Soybean lines with oil containing reduced saturated fatty acid c have been developed using TALENs targeting the gene encoding desaturase 2 (Haun et al., 2014). Tolerance to diseases is also a major issue for genome editing. It can be conferred by insertion of resistance alleles or by knock-out of susceptibility genes. In common wheat, mutation by TALENs of the three homologous MLO genes resulted in resistance to powdery mildew (Wang et al., 2014). Resistance to fire blight in rice could also be developed by TALENs targeting the binding site of pathogen TALE effectors (Li et al., 2012; Zhou et al., 2015).

3. **Meganucleases (EMRHE: Engineered Meganuclease Re-engineered Homin Endonucleases):** Unlike ZFN and TALEN, meganucleases recognize very long DNA sequences (12–45 bp) (Thierry and Dujon, 1992), which generate double-strand DNA breaks (DSBs). They split the DNA double-strand at highly specific sites, but these sequences are naturally defined and do not lend themselves to GE modification; they can therefore only be used in very special cases. Screens of random constructs of meganucleases were used to select the most specific of target genes (Smith et al., 2006). Puchta et al. (1999) was the first to use meganuclease tool (I-SceI) to modify tobacco plant and it resulted in the creation of DSBs. In 2008, D'Halluin et al. showed the successful insertion of the 35S promoter

at the site of I-SceI in maize crop thereby modifying the bar gene expression which give resistance to phosphinothricin (D'Halluin et al., 2008). The same authors (D'Halluin et al., 2013) made use of engineered meganucleases to insert two HT transgenes into cotton cells, fortunately 1,479 glyphosate-tolerant Calli have been generated and 27 Calli successful inserted the targeted genes. Fauser et al. (2012) used Arabidopsis line to generate a transgenic line which was then crossed with an I-SceI-expressing line and hybrid line was obtained. The F1 hybrid lines were selfing and the F2 seedlings obtained were screened, and nearly 1% of plants were identified and confirmed as the recombinant lines with the inserted genes using the engineered meganucleases.

4. **Other Past Genome Editing Tools Used in Tomato Improvement:** They are CBE, Target-AID, induced mutagenesis, oligonucleotide-directed mutagenesis (ODM), epigenome editing, transposable elements (Chaudhary et al., 2019; Vats et al., 2019). CBE was used as genome editing when SlDDB1, SlDET1 and SlCYC-B genes were targeted, and it led to enhanced carotenoid, lycopene, and β-carotene in tomato (Hunziker et al., 2020). Target-AID tool generated analogous phenotype to procera (allele of DELLA) as well as less-serrated leaflets in tomato; it also enhanced shelf-life in tomato (Shimatani et al., 2017).

The past three nucleases ZFN, Meganucleases, and TALEN bound and recognized the enzymes at specific sequences of DNA. Thus, the production of custom design nucleases for the genes of interest necessitates complex engineering processes (Martín-Pizarro and Posé, 2018). However, CRISPR-Cas9 system is the most recent technology, which has several advantages compared to Meganucleases, ZFN, and TALEN nucleases based on its simplicity, speed (development for a particular gene in less than two weeks) and cost (about 10 times less). However, both ZFN and TALEN, which rely on protein-DNA binding to achieve sequence specificity and are generated by fusing the DNA cleavage domain of the endonuclease FokI with zinc fingers or transcriptional activator-like effectors to achieve site-specific cleavage (Xu et al., 2019), Moreover, Xia et al. (2021) reported that CRISPR/Cas systems are the most used in tomato compared to the other genome editing technologies. It should be noted that gene editing using the CRISPR/Cas system has been used since 2014 to modified tomato genome and develop new tomato cultivars and mutants (Brooks et al., 2014; Lor et al., 2014).

8.6 CRISPR/CAS: WHERE ARE WE NOW?

Tomatoes have been improved using traditional breeding methods for decades and more recently using molecular breeding and genetic modification (GM) techniques. Genome editing techniques such as ZFNs, and TALENs, have had some success in trait and cultivar development but with the advent of CRISPR, there is now the opportunity to rapidly produce GM free genome edited crops.

CRISPR/Cas genome editing has many advantages over past genome editing technologies such as ZFNs and TALENs, including being cheaper, simpler, easy to multiplex and robust (Doudna and Charpentier, 2014). Using complexes of ribonucleoproteins (RNPs) to establish double stranded breaks means that edits can be redesigned multiple times by redesigning the specificity of the guide RNA (sgRNA). Currently, *Agrobacterium*-mediated stable transformation is the preferred delivery method for delivery of T-DNA integrated sgRNA and Cas in tomato and protocols are well established (Veillet et al., 2019). Early work in tomato reported by Ron et al. (2014), used hairy roots (*Rhizobium rhizogenes*) to transform tomato with CRISPR/Cas9 and disrupted GUS in a targeted and specific manner. This study was proof of concept that the CRISPR/Cas9 system could be successfully applied in tomato and the field of study has grown in the subsequent years, with multiple papers citing different applications of the CRISPR/Cas system for tomato improvement. The first reported stably inherited use of CRISPR/Cas9 in tomatoes was Brooks et al. (2014), who targeted the SlAGO7 gene, which resulted in a distinct phenotype. CRISPR/Cas9 has been used successfully in tomato cultivar research and improvement in various areas such as biotic and AT, *de novo* domestication, architecture, breeding traits, and quality improvement (Figure 8.1).

The CRISPR/Cas9 system has expanded the plant breeder's toolbox with simple gene knockout techniques, gene knock-in or replacement, base editing, and point mutations as well as gene regulation via mutations at the promoter site. Most CRISPR/Cas studies in tomato to date, have focused on single gene knockout and loss or gain of function mutants. However, CRISPR/Cas can also be easily modified to target multiple genes (Lowder et al., 2015). This rapidly increases the speed at which multiple traits can be stacked at one time.

Genetic variation is the foundation of any tomato breeding program and traditional cross-breeding can only work with material that is already available. CRISPR/Cas can be used to increase the genetic diversity available for cultivar improvement, either by *de novo* domestication of wild species, for example, *S. pimpinellifolium* (Rodrıguez-Leal et al., 2017), or by generation of new alleles such as the high throughput tomato mutagenesis strategy developed by Jacobs et al. (2017).

Compared to gene knockout, gene knock-in and precise editing is more complex. Plant mainly use the nonhomologous end-joining (NHEJ) pathway to repair breaks, compared to the homology-directed repair (HDR) pathway (Puchta, 2005). The precise modifications achieved by HDR are therefore challenging in plants, but the CRISPR/Cas9 GT-geminiviral replicon strategy has been successfully applied for large insertions and/or point mutations in tomato (Cermak et al., 2015). Further improvements to this system in tomato used a CRISPR/LbCpf1 geminiviral multi-replicon strategy, to produce salt tolerant plants due to a single amino acid change in the SlHKT1,2 gene (Vu et al., 2020). These plants were transgene-free and the HDR efficiency increased by 3-fold compared to existing Cas9 systems. A powerful addition to the CRISPR/Cas toolbox is base editing, and this has been successfully used in several crops, including tomato, and overcomes some of the challenges of using HDR in plants (Shimatani et al., 2017). Shimatani et al. (2017) utilized a CRISPR/Cas9-target-activation-induced cytidine deaminase system (Target-AID) to successfully induce point mutations into tomato, while Veillet et al. (2019) used this system to target the tomato acetolactate synthase (ALS) gene and produce transgene-free plants.

Base editing with CRISPR/Cas is still challenging, however the advent of new prime editing technologies in human cells has increased the applications for genome editing (Anzalone et al., 2019). In addition to targeting the genome in plants, the CRISPR/Cas system is now being used to target the epigenome (Papikian et al., 2019). The dCas9-SunTag system has been used to target the epigenome in Arabidopsis via DNA methylation and gene activation (Papikian et al., 2019). The rate of progress in CRISPR/Cas research is very rapid and so it should not be long before these kinds of studies are seen in tomato.

8.6.1 ABIOTIC STRESS

Plants are regularly subjected to multiple environmental stresses such as drought, salinity, heat, heavy metal toxicity, flooding, chilling, and UV light, to which they have developed adaptive mechanisms (He et al., 2018). Collectively these abiotic stresses can limit plant growth and in the case of crop plants can have a significant impact on yield and productivity (Kumar, 2020). The impacts of climate change are predicted to exacerbate abiotic stresses and crop productivity is projected to be reduced by approximately 50% due to stress related climate change (Mickelbart et al., 2015). High temperatures and sensitivity to limited soil moisture is one of the major causes of low vegetable yields, including tomato, in the tropics and this is likely to be exacerbated by climate change (Mattos et al., 2014).

Response and adaption to abiotic stress is a complex multigene process involving signaling, cellular, and regulatory processes (Mickelbart et al., 2015). Consequently, AT traits are mostly complex quantitative traits and so traditional breeding has relied on QTL analysis for many of these traits. The revolution in molecular techniques has led to GM crops tolerant to cold, drought, and salinity (Khare et al., 2010). However, there are often pleiotropic effects, and GM technology has not delivered the promised yield gains due to consumer backlash, length of time to develop new cultivars and regulations and red tape (Shukla et al., 2018). The CRISPR/Cas system has the advantage of being able to rapidly target single or multiple genes and produces transgene free plants.

Tomato is an economically important crop, and a large proportion is produced via open field cultivation. It has been predicted that large areas of tomato-growing land that are currently optimal for tomato growth are likely to become sub-optimal due to increasing abiotic pressures from climate change; however, some areas will become more suitable as temperatures in colder regions increase (Silva et al., 2017).

Plants are subjected to two broad categories of cold stresses, chilling 0–18°C and freezing >0°C, which limit plant distribution and productivity. Tomatoes are sensitive to cold stress and suffer damage to tissues below 10°C (Golem et al., 2012). The dehydration-responsive element-binding protein/C-repeat binding factors (DREB/CBFs) are an important family of transcription factors for plant responses to abiotic stress (Akhtar et al., 2012). The CBF gene family in tomato consists of 3 genes, SlCBF1, SlCBF2, and SlCBF3 (Zhang et al., 2004). Only SlCBF1 has been found to be cold-inducible in tomatoes, however SlCBF1 overexpression does not result in an increase in chilling tolerance in transgenic tomatoes (Zhang et al., 2004). In contrast to this, Li et al. (2018), generated SlCBF1 knockout mutants in tomato using the CRISPR-Cas9 system. These mutants had higher levels of chilling injury compared to WT plants and provided further insights into how chilling tolerance regulatory mechanisms work in tomato.

Tomato is of sub-tropical origin and can grow under diverse conditions with a wide range of optimum daytime temperatures 25–30°C but a much narrower night range of 20°C, with temperatures above 27°C starting to constrain growth and subject the tomato to heat stress (Camejo et al., 2005). The most critical stage for heat stress is the reproductive phase and tomato pollen is particularly susceptible to heat stress (Firon et al., 2006). High temperatures cause lack of pollen formation and effect fertility, a major cause of yield loss (Sangu et al., 2015). Parthenocarpy is the production of fruit without fertilization and is an important trait in agriculture when fertilization is prevented, e.g., due to elevated temperatures (Gorguet et al., 2005). Parthenocarpic tomato fruit has been produced by editing the Auxin response factor SlIAA9 resulting in plants with morphological changes in addition to the parthenocarpic fruit (Utea et al., 2017). Editing the MADS-box gene SlAGAMOUS-LIKE 6 (SlAGL6) with CRISPR/Cas9 partially solved this problem of heat stress-induced infertility by producing parthenocarpic fruits at high temperatures (Klap et al., 2017). In addition to producing parthenocarpic fruit at elevated temperatures, they showed improved yield under heat stress. Brassinazole resistant 1 (BZR1) controls brassinosteroid responses and the creation of a CRISPR-bzr1 mutant in tomato has established a role for this gene in heat tolerance in tomato via regulation of FERONIA (FER) homologs with mutants having reduced heat tolerance compared to the WT (Yin et al., 2018).

Drought is a major stress and causes untold damage to tomato as it is a fruit with lots of water. It is a major limit of crop production worldwide (Silva et al., 2017). The impact of climate change is likely to see increases in the frequency of droughts to accompany rising temperatures, which will further threaten food security (Piao et al., 2010). Mitogen-activated protein kinases (MAPKs) are involved in diverse stress responses in plants (Raja et al., 2017). Tomato SlMAPK3 knockout mutants were produced using the CRISPR/Cas system to investigate the function of this gene in tomato in response to drought stress (Wang et al., 2017). These plants were more susceptible to drought stress compared to WT plants with higher levels of wilting, hydrogen peroxide content, lower levels of antioxidants and higher levels of membrane damage. Knocking out this gene also resulted in changes of expression of drought responsive genes including SlDREB, indicating a role for SlMAPK3 in protecting cell membranes from damage.

Many genes are known to be responsive to abiotic stresses, however their precise function and role is often not known. The ease with which the CRISPR/Cas system can be employed to knockout single or multiple genes has opened many more possibilities for studying the function of these genes in response to environmental stress. One of the major genes involved in plant responses to pathogenesis is NPR1 however its role in abiotic stress is less clear, although it is strongly induced by drought (Li et al., 2019). This gene is a special receptor of salicylic acid (SA) and is considered part of a plant's systemic response to stress (Wu et al., 2012). Tomato NPR1 CRISPR/Cas9 knockout mutants have reduced stomatal conductance, higher drought stress and higher electrolyte leakage and anti-oxidant enzyme activities (Li et al., 2019). Major drought inducible genes such as DREB/CBF are also downregulated in these knockouts. Therefore, it appears that NPR1 does have a role to play in drought stress response in tomato (Li et al., 2019). A similar strategy was employed by Liu et al. (2020) to investigate the role of SlLBD40 which is involved in Jasmonic acid (JA) signaling. CRISPR/Cas9 knockout mutants showed enhanced drought tolerance tolerant compared to WT plants demonstrating a role for SlLBD40 as a negative regulator of drought tolerance in tomato (Liu et al., 2020). The role of the plant hormone Auxin is not well understood in plant stress however Bouzroud et al. (2020) demonstrated a role in drought, salinity, and osmotic stress by producing knockout mutants of the tomato auxin response factor gene SlARF4. Downregulation of this gene resulted in increased tolerance to both salinity and drought stress while also increasing the root system density.

8.6.2 BIOTIC STRESS

Plants are susceptible to various diseases which have a devastating effect on crops (Jones et al., 1991). Biotic stresses such as bacteria, viruses, nematodes, and fungal pathogens cause damage to crops, reduce yields and impact global food security (Savary et al., 2019). Various strategies employed by CRISPR include knocking out pathogen susceptibility genes in the host plant to confer resistance in the mutant (Acevedo-Garcia et al., 2014) or immunizing the plant against pathogen attack by targeting the pathogen genome (Pramanik et al., 2021). There are a large number of candidate immunity genes in tomato, but it has proved to be a technical challenge to screen them all for functional importance (Pombo et al., 2014). Many plant resistance genes are single genes and therefore many CRISPR/Cas studies have focused on single or double knockout mutants (Martínez et al., 2020; Nekrasov et al., 2017). The advantage of the CRISPR/Cas system is that it can rapidly knockout many genes at the same time. A large study produced 361 CRISPR mutants for 63 different candidate immunity genes and this valuable resource is now available for a more detailed search for immunity genes (Ahang et al., 2020). Most gene knockouts produced by this study were for single genes, with the exception of several multiple gene knockouts. The majority of these were stably inherited with some chimeric mutations observed and could be characterized into short deletions or insertions (Ahang et al., 2020). No off-target mutations were found despite the size of the study and demonstrating the broader effectiveness of using CRISPR for the characterization of immunity genes.

The plant growth regulator SA regulates resistance to biotrophs while ethylene and JA regulate resistance to necrotrophs (Glazebrook, 2005). These two pathways are antagonistic and therefore producing broad-spectrum resistance to multiple pathogens with resistance to necrotrophs leads to susceptibility to biotrophs and *vice versa*. Several different strategies have been used to combat bacterial infection. *Pseudomonas syringae* pv. tomato (Pst), which causes bacterial speck disease is responsible for huge tomato crop losses (Xin and He, 2013). Infection happens either through wounds or through stomatal openings (Melotto et al., 2017). This pathogen has evolved a sophisticated strategy to control plant hormones and bypass plant defenses, opening the stomata by producing coronatine which mimics bioactive JA (Fonseca et al., 2009). In *Arabidopsis thaliana* the Jasmonate-Zim domain protein 2 (AtJAZ2) is a major COR receptor (Gimenez-Ibanez et al., 2017). By editing the tomato ortholog of the COR, SlJAZ2, to produce a truncated form that lacked the C-terminal Jas domain (SlJAZ2Δjas), Ortigosa et al. (2019) produced gain of function mutants that did not open their stomata in response to JA. This mutation specifically targeted the stomata and so had no effect on wider JA signaling and therefore uncoupled the JA-SA antagonism in response to pathogen attack. This goes some way to providing broad-spectrum resistance.

Fungal pathogens are another major cause of crop losses in tomato with several different fungal pathogen species responsible for these losses. Tomato powdery mildew disease is caused by several different fungal species and covers the leaves in white fungal colonies and causes devastating crop losses (Kiss et al., 2001). The mildew resistant locus O (Mlo) confers susceptibility to powdery mildew and conversely the homozygous loss of function mutant confers resistance (Acevedo-Garcia et al., 2014). In tomato SlMlo1 is the major contributor to powdery mildew susceptibility

(Zheng et al., 2016). Naturally occurring loss of function mutants are available but introgression is a lengthy and complex process (Bai et al., 2005). Nekrasov et al. (2017) targeted an elite cultivar of tomato and created CRISPR/Cas9 knockout mutants targeting the SlMlo locus. These plants were resistant to powdery mildew and progeny were segregated for the transgene to produce plants with no evidence of foreign DNA. Additionally, no off-target mutations were detected. The only change was a 48 bp deletion that is similar to naturally occurring mutants and neither yield nor morphology were compromised. In a similar study, Pramanik et al. (2021) targeted (SlMlo1) in an elite tomato cultivar and produced plants that were completely resistant to powdery mildew. By using multiplex sgRNAs, Martínez et al. (2020) produced tomato plants with partial resistance to powdery mildew via mutation of the tomato callose synthase gene PMR4. The *pmr4* mutant plants had an enhanced hypersensitive response (Martínez et al., 2020).

Phytophthora infestans (*P. infestans*), the causal agent of late blight is another very damaging tomato pathogen (Nowicki et al., 2012). There are many single genes involved in host resistance/susceptibility, however micro RNAs are also known to play a role in plant defense against pathogens by repressing targeting genes (Li et al., 2012). Resistance to *P. infestans* in tomato plants was achieved using a multiplex CRISPR/Cas9 approach, which targeted two miRNAs, miR482b and miR482c simultaneously (Hong et al., 2020). Furthermore, targeting these miRNAs revealed cross regulation with other miRNAs.

There are two different ways of defending plants against viruses. The first is to design sgRNAs that will target a viral sequence if it infects the plant and the second is to confer resistance to the plant by the insertion of viral resistance genes (Tashkandi et al., 2018). The tomato yellow leaf curl virus (TYLCV), a geminivirus of the genus Begomovirus is transmitted by whiteflies and causes leaf yellowing and curling and plant stunting (Jones, 2003). Infection with this virus results in huge annual crop losses (Lefeuvre et al., 2010) and therefore both traditional breeding and transgenic methods of control have been employed to control TYLCV (Antignus et al., 2004; Dhaliwal et al., 2020). Traditional breeding incorporating resistance genes from wild tomato species such as *S. peruvianum* and *L. chilense* had some success and for many years was the mainstay of resistance to TYLCV (Pilowsky et al., 1990), however; it is a lengthy process in a crop like tomato with a longer lifecycle, and there are associated problems linkage between the resistance locus and poor fruit quality.

Transgenic modification has been successful in conferring resistance to TYLCV; for example, targeting of the viral replication-associated protein (Rep) genes resulted in tomato plants with enhanced resistance to TYLCV and no symptoms of infection were observed (Yang et al., 2004). However, this was accompanied by altered phenotypes in some plants. The transgene must be expressed continuously, and so low levels of consumer tolerance have delayed the progress of this approach. As an alternative strategy, CRISPR/Cas9 in planta viral interference has been used to target multiple TYLCV sequences at once, producing *Nicotiana benthamiana* virus-free plants (Ali et al., 2015). The plants showed reduced levels of virus compared to WT plants and significantly reduced symptoms. This system has the potential to target multiple DNA viruses at one time and is an exciting strategy for transferring this to both tomato and for use against other viruses. Tashkandi et al. (2018) designed sgRNAs to target either the TYLCV coat protein (CP) or replicase gene and showed a substantially reduced accumulation of TYLCV compared to WT plants. This was stably inherited, and viral resistance was active in the successive generations (Tashkandi et al., 2018). This strategy requires continuous CRISPR/Cas expression to maintain immunity against the virus and therefore these plants are classified as GMOs.

The CRISPR/Cas9 system has also been used to produce knockout mutants to identify key genes involved in viral resistance pathways, thereby identifying genes responsible for resistance. Tomato Dicer-like-2 (slDCL2) is a key component of tomato viral resistance pathways resistance (Wang et al., 2018a) and loss of function slDCL2 mutants have been produced using CRISPR/Cas9. These mutants were normal under normal conditions and resembled the WT, but when infected with tomato mosaic virus (ToMV) they showed developmental defects (Wang et al., 2018b). This was used to confirm the critical role of DCL2 in viral defense mechanisms. In contrast to studies which have targeted the viral genome, Pramanik et al. (2021) targeted host-virus susceptibility genes. A major QTL associated with TYLCY resistance includes the Ty-5 locus, which codes for the tomato mRNA surveillance factor Pelota (SlPelo) (Lapidot et al., 2015). Targeting the SlPelo gene to produce CRISPT/Cas9 knockouts produced mutants with restricted TYLCV proliferation resulting in viral resistance in these plants (Pramanik et al., 2021). Many of these targets are gene knockouts but many traits only have a single or multiple nucleotide substitution, so there is a need for precision targeting for CRISPR/Cas to be improved in tomato.

8.6.3 METABOLIC ENGINEERING AND NUTRITIONAL IMPROVEMENT (NI)

Consumption of tomato fruit has been associated with reduced incidence of many diseases, due to the high concentrations of nutrients and vitamins, and these properties extend to both raw and cooked tomatoes (Bhowmik et al., 2012).

Tomato is a major dietary source of the carotenoid lycopene, which has been associated with lower incidences of cancers and cardiovascular diseases (Li and Xu, 2014). There are different ways to manipulate the carotenoid biosynthesis pathway to increase lycopene levels. Various transgenic studies have manipulated single genes to produce elevated levels of lycopene in tomato, for example, lycopene β-cyclase 2 (LCY-B2) (Ronen et al., 2000), and phytoene synthase 1 gene (PSY1) (Fraser et al., 2007). However, it is hard to successfully target multiple genes at once using transgenic technologies. In contrast, CRISPR/Cas presents a robust mechanism for metabolic engineering as it can target multiple genes simultaneously, which are often a necessity for metabolic engineering. Levels of lycopene have been successfully increased fivefold using multiplex genome editing targeting 5 genes from the carotenoid biosynthetic pathway to produce knockout mutants (Li et al., 2018a). Compared to transgenics technologies this is an easier approach, has greater efficiency and in this study no off target effects were detected using CRISPR/Cas9.

Tomato contains a large amount of γ-aminobutyric acid (GABA) which behaves as an inhibitory neurotransmitter and has roles in controlling blood pressure in humans (Bachtiar et al., 2015). This beneficial nutritional component is very high in green fruit but declines rapidly once ripening occur (Rolin et al., 2000). CRISPR/Cas genome editing allows the possibility of creating new mutants related to metabolic engineering. By targeting glutamate decarboxylase (GAD), a key enzyme in GABA biosynthesis, Nonaka et al. (2017) were able to increase GABA concentration in red tomato fruits by up to 15-fold compared to WT plants. In plants, GABA is regulated by a shunt and multiplex genome editing has been used to knock out 5 key genes (GABA-TP1, GABA TP2, GABA-TP3, CAT9, SSADH) in this pathway (Li et al., 2018b). Several combinations of single as well as multisite knockouts mutants were produced and levels of GABA were significantly higher than that of the WT plants in both the leaves and in green and red fruit; however, this was associated with stunting and pale leaf color (Li et al., 2018b). Nevertheless, this study presents an exciting opportunity to demonstrate what multiplex CRISPR/Cas9 can do in precision metabolic engineering.

8.6.4 RIPENING, QUALITY, AND REPRODUCTIVE TRAITS

Major targets of tomato breeding are quality traits related to reproduction and tomato fruit growth. A fully ripe tomato is a soft fleshy fruit and ripening involves many genes associated with color, texture, and aroma (Wang et al., 2019). The soft nature of the tomato fruit presents problems for post-harvest storage and transport, and reducing post-harvest losses is a key strategy for ensuring global food security (Belik, 2018). Ripening, fruit size, color, aroma, nutritional content, shelf life, and associated quality traits are therefore ideal targets for genetic manipulation both to improve the quality of fruit reaching the market but also to understand the underlying mechanisms controlling these traits. Control of tomato fruit ripening was one of the first targets of GM technology (Sheehy et al., 1988). Early work including that of Picton et al. (1995), generated antisense tomato plants with a longer shelf life by first identifying and cloning ripening-related genes isolated from the RIN, tomato ripening mutant. The RIN gene encodes a MADS-box transcription factor that is the main fruit ripening regulator and master controller of color, firmness, and flavor (Fujisawa et al., 2013). This gene regulates and accelerates fruit ripening and tomatoes from plants that contain a *rin* mutation fail to ripen; however, there are compromises in texture and flavor (Vrebalov et al., 2002). A mutation was first reported at the RIN locus in 1968 (Robinson and Tomes, 1968) and has since been used to successfully produce F_1 hybrid tomato plants with improved shelf life, lower lycopene, and lower levels of fruit softening, however, these properties varied between plants (Kitagawa et al., 2005). CRISPR/Cas 9 has now been extensively used in fruit quality improvement, including in tomato (Xu et al., 2020). One of the earliest studies using CRISPR/Cas9 in tomato produced several stably inherited RIN knockout mutants by targeting three separate gene regions (Ito et al., 2015). These mutants produced fruit with incomplete ripening and had lower levels of the red pigment lycopene and these were novel mutants compared to other pre-existing ripening mutants (Ito et al., 2015). Li et al. (2020) produced CRISPR/Cas9 RIN knockout mutants with suppressed carotenoids and ethylene, volatiles, and ripening. These tomatoes could still partially ripen with reduced ethylene and in contrast to the naturally occurring *rin* mutants, fruit softening did still occur demonstrating the value of these knockout mutants in elucidating the exact role of this gene in ripening (Li et al., 2020). Although these studies have led to the increase in tomato shelf life this has been at the expense of color and flavor. The alcobaca (*alc*) mutant is a naturally occurring tomato mutant that has delayed ripening in the heterozygous form (Garg et al., 2008), and has been successfully used in breeding programs (Garg and Cheema, 2008). The majority of CRISPR/Cas studies in tomato ripening have generated knockout mutants via the NHEJ pathway (Ito et al., 2015; Li et al., 2020), however by using homology-directed repair (HDR), Yu et al. (2017) substituted the ALC allele with the *alc* to produce plants with longer shelf life with less impact on fruit quality. Not all ripening studies in tomato have concentrated on single gene locus. For example, CRISPR/Cas9 has been used to mutate the master ripening transcription factors SBP-CNR and NAC-NOR to produce tomato fruit with partial non-ripening or delayed ripening (Gao et al., 2019). Interestingly despite these mutations some ripening did occur, increasing the understanding of the roles of these two transcription factors in tomato fruit ripening.

The CRISPR/Cas system has also been used to elucidate the role of several post-transcriptional processes in tomato fruit ripening. A long coding RNA (lncRNA) related to ripening, lncRNA1459, loss of function mutant was created using CRISPR/Cas9 (Li et al., 2018c). Ripening, ethylene, and lycopene accumulation were repressed compared to WT plants demonstrating a role for this IncRNA in fruit ripening. Yang et al. (2017) identified and knocked out SlORRM4, an RNA editing factor related to ripening. Plants had delayed ripening, reduced respiration, and reduced ethylene production, shedding light on the relationship between RNA editing factors and fruit ripening.

Texture is an important part of tomato shelf life. Pectin plays a major role in intercellular adhesion and is modified during ripening, with pectate lyase (PL) being a particularly important enzyme in this process (Uluisik and Seymore, 2020). By targeting PL rather than RIN, Uluisik, et al. (2016) created CRISPER/Cas9 knockout mutants with altered texture but not flavor, color, or aroma. This is the first reported instance of fruit softening being manipulated to enhance firmness and shelf life without affecting other aspects of fruit ripening.

Another important target for tomato breeding is tomato size, however, as a quantitative trait, this has proved difficult for breeders. Tomato fruit size is dependent on a number of locules with as much as 50% of fruit size being dependent on a number of locules. Locule number is controlled by many different QTLs which contribute up to 50% of fruit size variance (Li et al., 2017). A study targeting the promoter region of the CLAVATA-WUSCHEL (CLV-WUS) gene (SlWUS CArG element) in both cultivated and the wild tomato *S. pimpinellifolium* generated mutants with increases in fruit size and yield, which resulted from changes in locule number (Rodrıguez-Leal et al., 2017). Novel QTLs was created demonstrating the use of cis-sequence mutagenesis in the generation of novel QTL traits.

Fruit color is a very important commercial trait for tomatoes and there are different preferences in different regions (Oltman et al., 2014). Pink colored fruit are preferred in Asia and the pink coloration comes from the absence of the yellow pigment flavonoid naringenin chalcone, which is controlled by the yellow (y) locus (Adato et al., 2009). There is evidence that the y locus may encode for SlMYB12, a transcription factor which is a key determinant in flavonoid accumulation (Ballester et al., 2010). Deng et al. (2018) used CRISPR Cas9 to knock out the SlMYB12 gene and produce pink tomatoes. They were able to produce transgene-free pink tomatoes within the year and more importantly without any resulting linkage drag. The problem of too little donor template being delivered to the plant was overcome by Cermak et al. (2015) who used geminivirus replicons rather than *agrobacterium* to create purple tomatoes by targeting the anthocyanin mutant (ANT1) which encodes for a Myb transcription factor and accumulates anthocyanin and results in a vivid purple color (Jin et al., 2003). This was stable in the next generation and had improved precision compared to *agrobacterium*-mediated transformation.

8.6.5 DE NOVO DOMESTICATION AND SHOOT ARCHITECTURE TRAITS: GENERATING GENETIC DIVERSITY FOR NEW BREEDING MATERIALS

Domestication and production of high yielding tomato cultivars with a focus on yield, fruit size, and plant architecture traits has resulted in a loss of genetic diversity (Williams et al., 1993). This has particularly affected the loss or narrowing of genetic diversity for traits associated with abiotic and biotic stress tolerance, taste, and nutritional value (Zsögön et al., 2018). This has resulted in little allelic variation for breeding amongst the cultivated tomato. Nevertheless, there are also reports of breeding increasing genetic diversity for certain traits, for example, genetic diversity and aroma in Dutch tomatoes has increased the 1960s, (Schouten et al., 2019).

CRISPR/Cas is an ideal tool for rapid production of new genetic variation in tomato. D'Ambrosio et al. (2018) created loss of function mutants for the carotenoid biosynthetic genes phytoene synthase 1 (Psy1) and the beta-carotene hydroxylase 2 (CrtR-b2). The resulting knockout mutants had yellow-fleshed fruit due to a lack of phytoene being knocked and 34 different alleles were produced, illustrating the rapid way in which CRISPR/Cas can generate new materials for breeding programs.

Inflorescence architecture plays an important role in yield and has been a highly selected trait during domestication and breeding improvement (Meyer and Purugganan, 2013). However, this trait has been underexploited in tomato breeding and inflorescence architecture has changed little from wild tomato ancestor species (Peralta and Spooner, 2005). Soyk et al. (2017) generated new alleles that were weakly expressed for branching inflorescence. These plants had moderate increases in branching with increased flower number. Other genes targeted by CRISPR/Cas genome editing for improving inflorescence architecture include blade on petiole (S1BOP), which causes pleiotropic effects in the knockout mutants, including modification of inflorescences into single flowers (Xu et al., 2016) and the MADS-box protein SlMBP21 which produced a jointless phenotype in the knockout mutants (Roldan et al., 2017). Tomatoes with a dwarf phenotype were produced by editing the tomato PROCERA gene, which encodes a DELLA protein (Tomlinson et al., 2019). These plants have a dominant and heritable dwarf phenotype with potential applications in the glasshouse industry and in conditions of strong winds.

To introduce new traits into cultivated tomatoes, breeders have looked to the wild tomato species, which harbor a treasure trove of abiotic, biotic, and nutritional traits (Rick, 1986). However, breeding this way is time consuming and there is the associated linkage drag from undesirable traits. The advent of CRISPR/Cas technology has allowed the rapid creation of *de novo* domesticated wild species of tomato (Østerberg et al., 2017; Zsögön et al., 2017). For *de novo* domestication to be successful, then the most useful traits must be identified, targeted, and selected for. An ideal tomato plant displaying the traits necessary for *de novo* domestication has been defined by Zsögön et al. (2017) who suggested several monogenic traits that could be targeted by genome editing. Targets such as the self-pruning gene (SP gene) (Pnueli et al., 1998) alter a tomato plant phenotype from an indeterminate phenotype common in wild tomato species to a determinant phenotype. The salt tolerant *S. pimpinellifolium* LA1589 has many suitable characteristics for cultivation; however, it requires multiple traits to be targeted to produce a productive crop (Li et al., 2018d). By targeting four genes for sensitivity to day length, shoot architecture and flower and fruit traits Li et al. (2018d) altered the plant architecture from indeterminant to determinant, biotic, and abiotic stress resistance.

The wild tomato relative *Physalis pruinose* has small, sweet fruit, however the fruit drop easily, and the plants have a sprawling weedy growth habit, presenting barriers to mass cultivation. Genome editing of the SP gene and the tomato CLAVATA gene produced plants with a compact architecture and higher yield (Lemmon et al., 2018). This was a rapid improvement of a wild tomato relative presenting new material for breeding programs.

Urban agriculture is becoming increasingly popular as a concept to help mitigate some effects of climate change (Langemeyer et al., 2021). However, this will require a change in architecture in shoots and shape to accommodate as presently it is not practical for most crops other than leafy vegetables and salad crops. The tomato regulator of stem length gene SlER, rapid flowering (SP5G) and precocious growth termination (SP) were stacked together using CRISPR/Cas 9 produced compact vine-like tomatoes suitable for urban agriculture (Kwon et al., 2020). Yield was not compromised, demonstrating the suitability of these plants for the urban agriculture environment.

8.7 FUTURE DIRECTIONS AND CONCLUDING REMARKS

The progress of CRISPR/Cas genome editing in tomato has been rapid, and in just a few years, there have been multiple advances in the technology (Figure 8.2). New developments in CRISPR/Cas mean that this is a rapidly evolving area, although there are still several gaps in its use in tomato. For example, heavy metal toxicity is an important abiotic stress, and there are major toxic metals polluting the worlds food supply (Manoj et al., 2020). There are currently no reported studies into heavy metal tolerance or mechanisms of phytoremediation in tomato using CRISPR, but the possible targets have been reviewed elsewhere in Sarma et al. (2021).

The majority of CRISPR/Cas studies in tomato involve knockout mutations, and knock-in insertions are much more challenging. The ease at which long sequences can be inserted into the genome has recently been improved thanks to a new study which used a combination of CRISPR/Cas9 with modified donor DNA to knock-in an insertion of over 2,000 bp in rice with an efficiency of 25% (Lu et al., 2020). This study further described methods for gene replacement using the HDR pathway for up to 130 bp

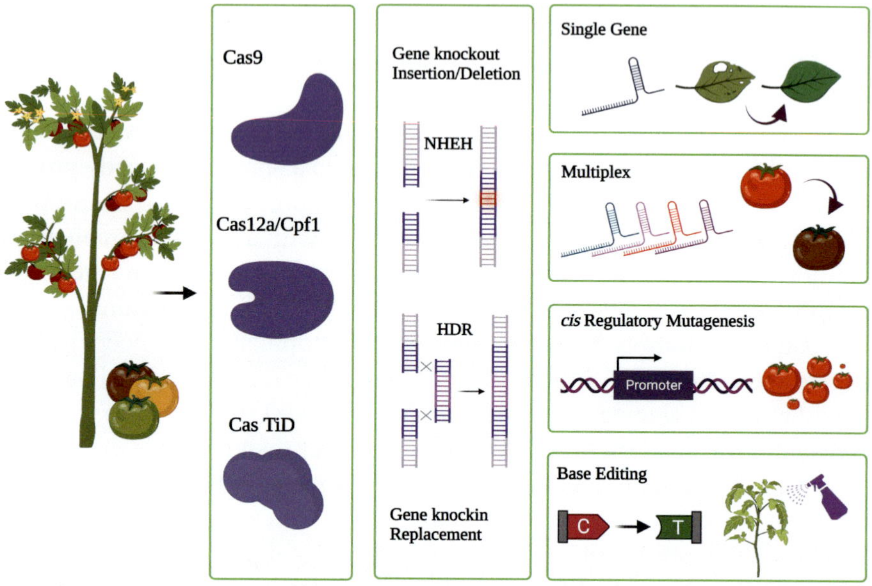

FIGURE 8.2 Examples of current CRISPR/Cas systems in use in tomato, including Cas9, Cas 12a/Cpf1 and Cas TiD. Pathways in tomato include both the NHEH and HDR pathways and have been used for applications including, single and multiple gene knockout, and single gene knock-in for biotic and abiotic traits, cis-regulatory mutagenesis for novel QTL development and increased genetic diversity, base editing for point mutation for herbicide resistance.

Source: Created with: Biorender.com.

with an efficiency of 6.1% (Lu et al., 2020). These new methods provide a leap forward in the development of targeted insertions and gene replacements in crop plants. Using a previously uncharacterized type I D rather than CRISPR/Cas system (TiD) Osakabe et al. (2020) targeted tomato plants and produced long-range and short indels deletions. These mutations were stably inherited but different from alternative Cas systems (Osakabe et al., 2020). This presents yet another novel CRISPR/Cas system for genome editing in tomato.

Many of the genetic changes targeted in early crop domestication were in *cis*-regulatory elements (Meyer and Purugganan, 2013; Swinnen et al., 2016). Unlike changes in coding sequences that interrupt or change a protein sequence, these mutations are involved in changing location, timing, and patterns of gene expression (Wittkopp et al., 2011). For example, changes in the expression of the LC/WUS gene controls locule number in tomato and increases floral meristem size, gives larger flatter fruit sizes (van der Knaap et al., 2014). So far, most CRISPR studies in tomatoes have targeted knock-outs in coding sequences with just a few exceptions (Rodríguez-Leal et al., 2017). Many novel and diverse cis-regulatory alleles were produced for fruit size, inflorescence branching, and plant architecture in tomato, by targeting CRISPR/Cas9 to cis-regulatory motifs (Rodríguez-Leal et al., 2017). Changes at the transcriptional level were a poor predictor of phenotype; however, a large increase in quantitative diversity was seen. This technique provides large amounts of new potential breeding material, and these changes often lack the pleiotropic effects seen in mutations in the coding region (Swinnen et al., 2016). The opportunities for increasing the limited genetic variation in cultivated tomato can therefore not be underestimated.

NHEJ mediated repair of DSB after Cas9 cleavage is error-prone and the resulting mutations produced are considered to be random. However, advances in machine learning and the availability of bioinformatics tools that can predict NHEJ repair outcomes allow the selection of suitable sgRNAs to introduce predictable mutations (Molla and Yang, 2020). Given that plants use the NHEJ pathway as their main repair mechanism, this should significantly improve the precision of mutations repaired by this pathway. The use of the HRD pathway for precision editing has been reported in tomato (Cermak et al., 2015; Vu et al., 2020) however efficiencies still need to be improved. Prime editing which has yet to be reported in tomato plants provides opportunities for improvements in precision editing (Anzalone et al., 2019). However, this is currently restricted to A.T to G.C and C.G to T.A substitutions (Chen et al., 2019). Methods for C.G to G.C have recently been reported in human cells raising the possibility of future applications in crop plants such as tomato (Kurt et al., 2021). As well as the development of protocols for crop plants the efficiencies of these methods must be improved.

CRISPR/Cas9 is the gold standard for genome editing of crops however CRISPR/Cas12a (CRISPR-Cpf1) has also been successfully used in tomato (Vu et al., 2020). While Cas9 generates small indels, Cas12a can generate larger deletions, has fewer off target edits compared to Cas9 and has the capability to use HDR greatly expanding the range of CRISPR/Cas applications available (Bernabe-Orts et al., 2019). A third CRISPR system CRISPR/Cas12b has recently been used successfully in rice and has promising potential for transcriptional repression and activation (Ming et al., 2020).

As well as being complex and multigene there are often barriers to tomato cultivar breeding and improvement because of whole genome duplication.

This means that single gene knockout does not always produce predictable or desired phenotypes. However, with CRISPR/Cas technology it is possible to precisely knock out more than one gene simultaneously. This has already been successfully carried out in crop plants such as Nicotiana and Arabidopsis (Li et al., 2013). There are additional problems with chromosomal rearrangements that present obstacles to successful cultivar breeding. An elegant solution was developed by Schmidt et al. (2020), who reversed an Arabidopsis 1.1 Mb inversion on chromosome 4 under egg-cell-specific expression of the Staphylococcus aureus Cas9 (SaCas9). With the rapid development of new CRISPR/Cas technologies, as well as the availability of transgene-free, genome-edited plants within one or two generations, it will surely only be a matter of time before CRISPR tomatoes are available on the supermarket shelf.

KEYWORDS

- **abiotic stress**
- **architecture traits**
- **biotic stress**
- **breeding new cultivars**
- **recombinant DNA technology**
- **tomato CRISPR/Cas**

REFERENCES

Acevedo-Garcia, J., Kusch, S., & Panstruga, R., (2014). Magical mystery tour: MLO proteins in plant immunity and beyond. *New Phytol., 204*, 273–281.

Adato, A., Mandel, T., Mintz-Oron, S., Venger, I., Levy, D., Yativ, M., Domínguez, E., et al., (2009). Fruit-surface flavonoid accumulation in tomato is controlled by a SlMYB12-regulated transcriptional network. *PLoS Genet., 5*(12), e1000777.

Aflitos, S., Schijlen, E., De Jong, H., De Ridder, D., Smit, S., Finkers, R., Wang, J., et al., (2014). Exploring genetic variation in the tomato (*Solanum* section Lycopersicon) clade by whole-genome sequencing. *Plant J., 80*(1), 136–148. doi: 10.1111/tpj.12616. E.

Akhtar, M., Jaiswal, A., Taj, G., Jaiswal, J. P., Qureshi, M. I., & Singh, N. K., (2012). DREB1/CBF transcription factors: Their structure, function and role in abiotic stress tolerance in plants. *J. Genet., 91*, 385–395.

Ali, Z., Abulfaraj, A., Idris, A., Ali, S., Tashkandi, M., & Mahfouz, M. M., (2015). CRISPR/Cas9-mediated viral interference in plants. *Genome Biol., 11*, 16–238.

Antignus, Y., Vunsh, R., Lachman, O., Pearlsman, M., Maslenin, L., Hananya, U., & Rosner, A., (2004). Truncated rep gene originated from tomato yellow leaf curl virus-Israel [Mild] confers strain-specific resistance in transgenic tomato. *Ann. Appl. Biol., 144*, 39–44.

Anzalone, A. V., Randolph, P. B., Davis, J. R., Sousa, A. A., Koblan, L. W., Levy, J. M., Chen, P. J., et al., (2019). Search-and-replace genome editing without double-strand breaks or donor DNA. *Nature, 576*, 149–157.

APHIS, (1994). *Petition for Release from Regulation for Modified Lines B, Da, F Derived from T7 Varieties of Processing Tomato*. USDA APHIS. https://www.aphis.usda.gov/brs/aphisdocs/94_29001p.pdf (accessed on 04 September 2022).

Bachtiar, V., Near, J., Johansen-Berg, H., & Stagg, C. J., (2015). Modulation of GABA and resting state functional connectivity by transcranial direct current stimulation. *Elife., 4*, e08789.

Bai, Y., & Lindhout, P., (2007). Domestication and breeding of tomatoes, what have we gained and what can we gain in the future? *Ann. Bot., 100*, 1085–1094. doi: 10.1093/aob/mcm150, available online at www.aob.oxfordjournals.org (accessed on 04 September 2022).

Bai, Y., Van, D. H. R., Bonnema, G., Marcel, T. C., Meijer-Dekens, F., Niks, R. E., & Lindhout, P., (2005). Tomato defense to *Oidium neolycopersici*: Dominant Ol genes confer isolate-dependent resistance via a different mechanism than recessive ol-2. Molecular plant-microbe interactions. *MPMI, 18*, 354–362.

Ballester, A. R., Molthoff, J., De Vos, R., Hekkert, B. T., Orzaez, D., Fernández-Moreno, J. P., Tripodi, P., et al., (2010). Biochemical and molecular analysis of pink tomatoes: Deregulated expression of the gene encoding transcription factor SlMYB12 leads to pink tomato fruit color. *Plant Physiol., 152*, 71–84.

Baranski, R., Klimek-Chodacka, M., & Lukasiewicz, A., (2019). Approved genetically modified (GM) horticultural plants: A 25-year perspective. *Folia Hort., 31*(1), 3–49. doi: 10.2478/fhort-2019-0001.

Bawa, A. S., & Anilakumar, K. R., (2013). Genetically modified foods: Safety, risks and public concerns—a review. *J. Food Sci. Technol., 50*, 1035–1046.

Belik, W., (2018). Impasses in transformation of the food system future of food. *J. Food Agric. Soc., 6*, 5–8.

Bernabe-Orts, J. M., Casas-Rodrigo, I., Minguet, E. G., Landolfi, V., Garcia-Carpintero, V., Gianoglio, S., Vazquez-Vilar, M., et al., (2019). Assessment of Cas12a-mediated gene editing efficiency in plants. *Plant Biotechnol.* J, *17*, 1971–1984.

Bharti, S., Rani, N., Krishnamurthy, B., & Arya, D. S., (2014). Preclinical evidence for the pharmacological actions of naringin: A review. *Planta Med., 80*(6), 437–451. doi: 10.1055/s-0034–1368351.

Bhowmik, D., Kumar, K. P. S., Paswan, S., & Srivastava, S., (2010). Tomato – a natural medicine and its health benefits. *J. Pharmacogn. Phytochem., 1*, 33–43.

Bhowmik, D., Paswan, S., & Srivastava, S., (2012). Tomato-a natural medicine and its health benefits. *J. Pharmacogn. Phytochem., 1*, 33–43.

Blum, A., (1997). Crop responses to drought and the interpretation of adaptation. In: Belhassen, I., (ed.), *Drought Tolerance in Higher Plants: Genetical, Physiological, and Molecular Biological Analysis* (pp. 57–70). Kluwer Academic Publ. Dordrecht. ISBN 0-7923-4123-6.

Bolton-Smith, C., McMurdo, E. T. M., Paterson, R. C., Mole, P. A., Harvey, M. J., Fenton, T. S., Prynne, C. J., et al., (2007). Two-year randomized controlled trial of vitamin k1 (phylloquinone) and vitamin D3 plus calcium on the bone health of older women. *Journal of Bone and Mineral Research, 22*(4). doi: 10.1359/JBMR.070116.

Bouzroud, S., Gasparini, K., Hu, G., Barbosa, M., Rosa, B. L., Fahr, M., Bendaou, N., et al., (2020). Down regulation and loss of auxin response factor 4 function using CRISPR/Cas9 alters plant growth, stomatal function and improves tomato tolerance to salinity and osmotic stress. *Genes, 11*, 272.

Brooks, C., Nekrasov, V., Lippman, Z. B., & Van, E. J., (2014). Efficient gene editing in tomato in the first generation using the clustered regularly interspaced short palindromic repeats/CRISPR-associated 9 system. *Plant Physiol., 166*, 1292–1297.

Brungnoli, E., & Lauteri, M., (1991). Effects of salinity on stomatal conductance, photosynthetic capacity, and carbon isotope discrimination of salt-tolerant (*Gossypium hirsutum* L.) and salt-sensitive (*Phaseolus vulgaris* L.) C3 non halophytes. *Plant Physiol., 95*, 628–935.

Camejo, D., Rodriguez, P., Morales, M. A., & Dell'Amico, J. M., (2005). High temperature effects on 450 photosynthetic activity of two tomato cultivars with different heat susceptibility. *J. Plant Physiol., 162*, 281–289.

Čermák, T., Baltes, N. J., Čegan, R., Zhang, Y., & Voytas, D. F., (2015). High-frequency, precise modification of the tomato genome. *Genome Biol., 16*, 232. http://dx.doi.org/10.1186/s13059-015-0796-9.

Chaudhary, J., Alisha, A., Bhatt, V., Chandanshive, S., Kumar, N., Mir, Z., Kumar, A., et al., (2019). Mutation breeding in tomato: Advances, applicability and challenges. *Plants, 8*, 128.

Chen, K., Wang, Y., Zhang, R., Zhang, H., & Gao, C., (2019). CRISPR/Cas genome editing and precision plant breeding in agriculture. *Annu. Rev. Plant Biol., 70*, 667–697.

Chen, Z., & Yang, R., (1996). *Method for Cultivating Tomato with Function of Anti Virus of Mosaic of Cucumber by Gene Engineering*. China patent CN 1054160, application by Beijing University, filed 4.07.1995, https://patentscope.wipo.int/search/en/detail.jsf?docId=CN82309046&redirected ID=true (accessed on 04 September 2022).

Christian, M., Cermak, T., Doyle, E. L., Schmidt, C., Zhang, F., Hummel, A., et al., (2010). Targeting DNA double-strand breaks with TAL effector nucleases. *Genetics, 186*, 757–761. doi: 10.1534/genetics.110.120717.

Clasen, B. M., Stoddard, T. J., Luo, S., Demorest, Z. L., Li, J., Cedrone, F., Demorest, Z. L., et al., (2015). Improving cold storage and processing traits in potato through targeted gene knockout. *Plant Biotechnology Journal, 14*(1), 169–176. doi: 10.1111/pbi.12370.

D'Ambrosio, C., Stigliani, A. L., & Giorio, G., (2018). CRISPR/Cas9 editing of carotenoid genes in tomato. *Transgenic Res., 27*, 367–378.

D'Elia, L., Barba, G., Cappuccio, F. P., & Strazzullo, P., (2011). Potassium intake, stroke, and cardiovascular disease a meta-analysis of prospective studies. *J. Am. Coll. Cardiol., 57*(10), 1210–1219. doi: 10.1016/j.jacc.2010.09.070.

D'Halluin, K., Vanderstraeten, C., Stals, E., Cornelissen, M., & Ruiter, R., (2008). Homologous recombination: A basis for targeted genome optimization in crop species such as maize. *Plant Biotechnol. J., 6*(1), 93–102.

D'Halluin, K., Vanderstraeten, C., Van, H. J., et al., (2013). Targeted molecular trait stacking in cotton through targeted double-strand break induction. *Plant Biotechnol. J., 11*(8), 933–941.

Deng, L., Wang, H., Sun, C., Li, Q., Jiang, H., Du, M., Li, C. B., & Li, C., (2018). Efficient generation of pink-fruited tomatoes using CRISPR/Cas9 system. *J. Genet. Genomics, 45*, 51–54.

Deutsch, C. A., et al., (2018). Increase in crop losses to insect pests in a warming climate. *Science, 361*, 916–919.

Dhaliwal, M. S., Jindal, S. K., Sharma, A., & Prasanna, H. C., (2020). Tomato yellow leaf curl virus disease of tomato and its management through resistance breeding: A review. *J. Hortic. Sc., 95*(4), 425–444.

Diouf, I. A., Derivot, L., Bitton, F., Pascual, L., & Causse, M., (2018b). Water deficit and salinity stress reveal many specific QTL for plant growth and fruit quality traits in tomato. *Frontiers in Plant Science, 9*, 279.

Diouf, I. A., Derivot, L., Koussevitzky, S., Carretero, Y., Bitton, F., Moreau, L., & Causse, M., (2020). *Genetic Basis of Phenotypic Plasticity and Genotype X Environment Interaction in a Multiparental Population*. bioRxiv 02.07.938456.

Doudna, J. A., & Charpentier, E., (2014). Genome editing. The new frontier of genome engineering with CRISPR-Cas9. *Science, 346*(6213), 1258096.

FDA, (1992). *Statement of Policy – foods Derived from New Plant Varieties*. Federal Register 57. Guidance to industry for food derived from new plant varieties. https://www.fda.gov/food/guidance-regulation-food-and-dietary-supplements/guidance-documents-regulatory-information-topic-food-and-dietary-supplements (accessed on 24 October 2022).

Fekete, K., Berti, C., Trovato, M., Lohner, S., Dullemeijer, C., Souverein, O. W., Cetin, I., & Decsi, T., (2012). Effect of folate intake on health outcomes in pregnancy: A systematic review and meta-analysis on birth weight, placental weight and length of gestation. *Nutrition Journal, 11*, 75.

Firon, N., Shaked, R., Peet, M. M., Pharr, D. M., Zamski, E., Rosenfeld, K., et al., (2006). Pollen grains of heat tolerant tomato cultivars retain higher carbohydrate concentration under heat stress conditions. *Sci. Hort., 109*, 212–217.

Fonseca, S., Chico, J. M., & Solano, R., (2009). The jasmonate pathway: The ligand, the receptor and the core signaling module. *Curr. Opin. Plant Biol., 12*, 539–547.

Food and Agriculture Organization of the United Nations [FAO], (2017). *The Impact of Disasters and Crises on Agriculture and Food Security* (p. 168). Rome: FAO.

Food and Agriculture Organization of the United Nations, (2019). *FAOSTAT Database*. Rome, Italy: FAO.

Fraser, P. D., Truesdale, M. R., Bird, C. R., Schuch, W., & Bramley, P. M., (1994). Carotenoid biosynthesis during tomato fruit development (evidence for tissue-specific gene expression). *Plant Physiol., 105*, 405–413.

Fujisawa, M., Nakano, T., Shima, Y., & Ito, Y., (2013). A large-scale identification of direct targets of the tomato MADS box transcription factor ripening inhibitor reveals the regulation of fruit ripening. *Plant Cell, 25*, 371–386.

Gao, Y., Zhu, N., Zhu, X., Wu, M., Jiang, C. Z., Grierson, D., Luo, Y., et al., (2019). Diversity and redundancy of the ripening regulatory networks revealed by the fruit ENCODE and the new CRISPR/Cas9 CNR and NOR mutants. *Hortic. Res., 6*, 39.

Garg, N., & Cheema, D. S., (2008). Genotype environment interactions for shelf life and yield attributes in tomato hybrids heterozygous at rin, nor, or alc loci. *J. Crop Improv., 22*, 17–30.

Garg, N., Cheema, D. S., & Dhatt, A. S., (2008). Utilization of rin, nor, and alc alleles to extend tomato fruit availability. *Int. J. Veg. Sci., 4*, 41–54.

Gavrish, S. F., & Korol, V. G., (1991). Some biological features of F1 tomato hybrids carrying the nor gene. *Izvestiya Timiryazevskoi Sel'skhozyaistvennoi Akademii, 1*, 118–132.

Gimenez-Ibanez, S., Boter, M., Ortigosa, A., García-Casado, G., Chini, A., Lewsey, M. G., Ecker, J. R., et al., (2017). JAZ2 controls stomata dynamics during bacterial invasion. *New Phytol., 213*, 1378–1392.

Glazebrook, J., (2005). Contrasting mechanisms of defense against biotrophic and necrotrophic pathogens. *Annu. Rev. Phytopathol., 43*, 205–227.

Goel, D., Singh, A. K., Yadav, V., Babbar, S. B., & Bansal, K. C., (2010). Overexpression of osmotin gene confers tolerance to salt and drought stresses in transgenic tomato (*Solanum lycopersicum* L.). *Protoplasma, 245*(1–4), 133–141. doi: 10.1007/s00709-010-0158-0.

Golam, F., Prodhan, Z. H., Nezhadahmadi, A., & Rahman, M., (2012). Heat tolerance in tomato. *Life Sci., 9*, 1936–1950.

Gorguet, B., Van, H. A. W., & Lindhout, P., (2005). Parthenocarpic fruit development in tomato. *Plant Biol., 7*, 131–139.

Grassely, D., Navez, B., & Letard M., (2000). *Tomato: For a Quality Product*. Paris: CTIFL Interprofessional Technical Center for Fruits and Vegetables.

Harlan, J. R., (1987). *Cultivated Plants and Man* (pp. 299, 300). ACCT/CILF/PUF, 10.1186/1475-2891-11-75.

Harwood, W., & Van, E. J., (2017). Gene editing in tomatoes. *Emerg Top Life Sci., 1*, 183–191.

Haun, W., Andrew, C., Benjamin, M. C., Zachary, L. D., Anita, L., Erin, R., Adam, R., et al., (2014). Improved soybean oil quality by targeted mutagenesis of the fatty acid desaturase 2 gene family. *Plant Biotechnology Journal, 12*(7), 934–940. doi: 10.1111/pbi.12201.

Haverkort, A. J., Struik, P. C., Visser, R. G. F., & Jacobsen, E., (2009). Applied biotechnology to combat late blight in potato caused by *Phytophthora infestans*. *Potato Research*, (52), 249–264 doi: 10.1007/s11540-009-9136-3.

Havko, N. E., Das, M. R., McClain, A. M., Kapali, G., Sharkey, T. D., Gregg, A., & Howe, G. A., (2020). Insect herbivory antagonizes leaf cooling responses to elevated temperature in tomato. *Proc. Natl. Acad Sci. U S A., 117*(4), 2211–2217. doi: 10.1073/pnas.1913885117.

He, M., He, C. Q., & Ding, N. Z., (2018). Abiotic stresses: General defenses of land plants and chances for engineering multistress tolerance. *Front. Plant Sci., 9*, 1771.

Hilioti, Z., Ganopoulos, I., Ajith, S., Bossis, I., & Tsaftaris, A., (2016). A novel arrangement of zinc finger nuclease system for *in vivo* targeted genome engineering: The tomato LEC1-LIKE4 gene case. *Plant Cell Reports, 35*(11), 2241–2255. doi:10.1007/s00299-016-2031-x.

Hong, Y., Meng, J., He, X., Zhang, Y., Liu, Y., Zhang, C., Qi, H., & Luan, Y., (2020). Editing miR482b and miR482c simultaneously by CRISPR/Cas9 enhanced tomato resistance to *Phytophthora infestans. Phytopathology*. 0.1094/PHYTO-08-20-0360-R.

ISAAA, (2016). *Global Status of Commercialized Biotech/GM Crops: 2016*. ISAAA Brief No. 52. ISAAA: Ithaca, NY.

Ito, Y., Nishizawa-Yokoi, A., Endo, M., Mikami, M., & Toki, S., (2015). CRISPR/Cas9-mediated mutagenesis of the RIN locus that regulates tomato fruit ripening. *Biochem. Biophys. Res. Commun., 467*, 76–82.

Jacobs, T. B., Zhang, N., Patel, D., & Martin, G. B., (2017). Generation of a collection of mutant tomato lines using pooled CRISPR libraries. *Plant Physiol., 174*, 2023–2037.

Jin, F., Li, S., Dang, L., Chai, W., Li, P., & Wang, N. N., (2012). PL1 fusion gene: A novel visual selectable marker gene that confers tolerance to multiple abiotic stresses in transgenic tomato. *Transgenic Res., 21*, 1057–1070.

Jones, D. R., (2003). Plant viruses transmitted by whiteflies. *Eur. J. Plant Pathol., 109*, 195–219.

Jones, J. B., Stall, R. E., & Zitter, T. A., (1991). *Compendium of Tomato Diseases*. APS Press, St Paul.

Khare, N., Goyary, D., Kumar, S. N., Shah, P., Rathore, M., Anandhan, S., Sharma, D., et al., (2010). Transgenic tomato cv. *Pusa uphar* expressing a bacterial mannitol-1-phosphate dehydrogenase gene confers abiotic stress tolerance. *Plant Cell, Tiss. Organ. Cult., 103*, 267–277.

Kiss, L., Cook, R. T. A., Saenz, G. S., Cunnington, J. H., Takamatsu, S., Pascoe, I., Bardin, M., et al., (2001). Identification of two powdery mildew, *Oidium neolycopersici* sp. Nov. and *Oidium lycopersici*, infecting tomato in different parts of the world. *Mycol. Res., 105*, 684–697.

Kissoudis, C., Chowdhury, R., Van, H. S., Van De, W. C., Finkers, R., Visser, R. G. F., Bai, Y., & Van, D. L. G., (2015). Combined biotic and abiotic stress resistance in tomato. *Euphytica, 202*, 317–332. doi: 10.1007/s10681-015-1363-x.

Kitagawa, M., Ito, H., Shiina, T., Nakamura, N., Inakuma, T., Kasumi, T., Ishiguro, Y., Yabe, K., & Ito, Y., (2004). Characterization of tomato fruit ripening and analysis of gene expression in F1 hybrids of the ripening inhibitor (rin) mutant. *Physiol. Plant, 123*, 331–338.

Klap, C., Yeshayahou, E., Bolger, A. M., Arazi, T., Gupta, S. K., Shabtai, S., Usadel, B., Salts, Y., & Barg, R., (2017). Tomato facultative parthenocarpy results from SlAGAMOUS-LIKE 6 loss of function. *Plant Biotechnol. J., 15*, 634–647.

Koyama, K. L., Levesley, A., Koebner, R. M. D., Flowers, T. J., & Yeo, A. R., (2001). Quantative trait loci for component physiological traits determining salt tolerance in rice. *Plant Physiol., 125*, 406–422.

Kumar, S., (2020). Abiotic stresses and their effects on plant growth, yield and nutritional quality of agricultural produce. *Int. J. Food Sci. Agric., 4*, 367–378.

Kurt, I. C., Zhou, R., Iyer, S., Garcia, S. P., Miller, B. R., Langner, L. M., Grünewald, J., & Joung, J. K., (2021). CRISPR C-to-G base editors for inducing targeted DNA transversions in human cells. *Nat Biotechnol., 39*, 41–46.

Kwon, C. T., Heo, J., Lemmon, Z. H., Capua, Y., Hutton, S. F., Van, E. J., Park, S. J., & Lippman, Z. B., (2020). Rapid customization of *Solanaceae* fruit crops for urban agriculture. *Nat Biotechnol., 38*, 182–188.

Kwon, C. T., Heo, J., Lemmon, Z., Capua, Y., Hutton, S., Eck, J., Park, S., & Lippman, Z., (2020). Rapid customization of *Solanaceae* fruit crops for urban agriculture. *Nat. Biotechnol., 38*, 1–7.

Lang, J., (2004). Exploring the tomato: Transformations of nature, society, and economy (review). *Technol. Cult., 45*, 222–224.

Langemeyer, J., Madrid, L. C., Mendoza, B. A., & Mendez, G., (2021). Urban agriculture — A necessary pathway towards urban resilience and global sustainability? *Landscape and Urban Planning, 210*, 104055. 10.1016/j.landurbplan.2021.104055.

Lapidot, M., Karniel, U., Gelbart, D., Fogel, D., Evenor, D., Kutsher, Y., Makhbash, Z., Nahon, S., Shlomo, H., Chen, L., et al., (2015). A Novel route controlling begomovirus resistance by the messenger RNA surveillance factor pelota. *PLoS Genet., 11*, e1005538.

Lefeuvre, P., Martin, D. P., Harkins, G., Lemey, P., Gray, A. J., Meredith, S., Lakay, F., et al., (2010). The spread of tomato yellow leaf curl virus from the Middle East to the world. *PLoS Pathog., 6*(10), e1001164. https://doi.org/10.1371/journal.ppat.1001164.

Lemmon, Z. H., Reem, N. T., Dalrymple, J., Soyk, S., Swartwood, K. E., Rodriguez-Leal, D., Van, E. J., & Lippman, Z. B., (2018). Rapid improvement of domestication traits in an orphan crop by genome editing. *Nature Plants., 4*, 766–770.

Li, F., Pignatta, D., Bendix, C., Brunkard, J. O., Cohn, M. M., Tung, J., Sun, H., Kumar, P., & Baker, B., (2012). MicroRNA regulation of plant innate immune receptors. *Proc. Natl. Acad. Sci. U.S.A, 109*, 1790–1795.

Li, H., Qi, M., Sun, M., Liu, Y., Liu, Y., Xu, T., Li, Y., & Li, T., (2017). Tomato transcription factor SlWUS plays an important role in tomato flower and locule development. *Front. Plant Sci., 8*, 457.

Li, J. F., Norville, J. E., Aach, J., McCormack, M., Zhang, D., Bush, J., Church, G. M., & Sheen, J., (2013). Multiplex and homologous recombination-mediated genome editing in Arabidopsis and Nicotiana benthamiana using guide RNA and Cas9. *Nat Biotechnol., 31*, 688–691.

Li, N., Wu, X., Zhuang, W., Xia, L., Chen, Y., Wu, C., Rao, Z., et al., (2021). Tomato and lycopene and multiple health outcomes: Umbrella review. *Food Chem., 343*, 128396.

Li, R., Fu, D., Zhu, B., Luo, Y., & Zhu, H., (2018c). CRISPR/Cas9-mediated mutagenesis of lncRNA1459 alters tomato fruit ripening. *Plant J., 94*, 513–524.

Li, R., Li, R., Li, X., Fu, D., Zhu, B., Tian, H., Luo, Y., & Zhu, H., (2018b). Multiplexed CRISPR/Cas9-mediated metabolic engineering of γ-aminobutyric acid levels in *Solanum lycopersicum*. *Plant Biotechnol. J., 16*, 415–427.

Li, R., Liu, C., Zhao, R., Wang, L., Chen, L., Yu, W., Zhang, S., Sheng, J., & Shen, L., (2019). CRISPR/Cas9-mediated SlNPR1 mutagenesis reduces tomato plant drought tolerance. *BMC Plant Biol., 19*, 38.

Li, R., Zhang, L., Wang, L., Chen, L., Zhao, R., Sheng, J., & Shen, L., (2018). Reduction of tomato-plant chilling tolerance by CRISPR-Cas9-mediated SlCBF1 mutagenesis. *J Agric Food Chem., 66*, 9042–9051.

Li, S., Zhu, B., Pirrello, J., Xu, C., Zhang, B., Bouzayen, M., Chen, K., & Grierson, D., (2020). Roles of RIN and ethylene in tomato fruit ripening and ripening-associated traits. *New Phytol., 226*, 460–475.

Li, T., Liu, B., Spalding, M. H., Weeks, D. P., & Yang, B., (2012). High-efficiency TALEN-based gene editing produces disease-resistant rice. *Nature Biotechnology, 30*(5), 390–392. doi: 10.1038/nbt.2199.

Li, T., Yang, X., Yu, Y., Si, X., Zhai, X., Zhang, H., Dong, W., et al., (2018d). Domestication of wild tomato is accelerated by genome editing. *Nature Biotechnol.*, 36. 10.1038/nbt.4273.

Li, X., & Xu, J., (2014). Meta-analysis of the association between dietary lycopene intake and ovarian cancer risk in postmenopausal women. *Sci. Rep., 4*, 4885.

Li, X., Wang, Y., Chen, S., Tian, H., Fu, D., Zhu, B., Luo, Y., & Zhu, H., (2018a). Lycopene is enriched in tomato fruit by CRISPR/Cas9-mediated multiplex genome editing. *Front. Plant Sci., 9*, 559.

Lin, T., Zhu, G., Zhang, J., Xu, X., Yu, Q., Zheng, Z., Zhang, Z., et al., (2014). Genomic analyses provide insights into the history of tomato breeding. *Nat. Genet, 46*, 1220–1226.

Liu, L., Zhang, J., Xu, J., Li, Y., Guo, L., Wang, Z., Zhang, X., et al., (2020). CRISPR/Cas9 targeted mutagenesis of SlLBD40, a lateral organ boundaries domain transcription factor, enhances drought tolerance in tomato. *Plant Sci., 301*, 110683.

Lowder, L. G., Zhang, D., Baltes, N. J., Paul, J. W., Tang, X., Zheng, X., Voytas, D. F., et al., (2015). A CRISPR/Cas9 toolbox for multiplexed plant genome editing and transcriptional regulation. *Plant Physiol., 169*, 971–985.

Lu, C. G., Xu, H. L., Yian, R. C., & Yu, W. G., (1994). Storage-linked physiological characters of tomato carrying fruit ripening mutant genes and their implications in breeding. *Jiangsu J. Agric. Sci., 10*, 5–10.

Lu, Y., Tian, Y., Shen, R., Yao, Q., Wang, M., Chen, M., Dong, J., et al., (2020). Targeted, efficient sequence insertion and replacement in rice. *Nature Biotechnol., 38*, 1402–1407.

Mageto, E. K., Makumbi, D., Njoroge, K., & Nyankanga, R., (2017). Genetic analysis of early-maturing maize (*Zea Mays* L.) inbred lines under stress and nonstress conditions. *Journal of Crop Improvement, 31*(4), 560–588. doi:10.1080/15427528.2017.131562.

Manoj, S. R., Karthik, C., Kadirvelu, K., Arulselvi, P. I., Shanmugasundaram, T., Bruno, B., & Rajkumar, M., (2020). Understanding the molecular mechanisms for the enhanced phytoremediation of heavy metals through plant growth promoting rhizobacteria: A review. *J. Environ. Manag., 254*, 109779.

Martín-Pizarro, C., & Posé, D., (2018). Genome editing as a tool for fruit ripening manipulation. *Front. Plant Sci., 9*, 1415. doi: 10.3389/fpls.2018.01415.

Mattoo, A., & Razdan, M. K., (2007). Genetic improvement of solanaceous crops (Vol. 2). Enfield, NH: Science Publishers.

Mattos, L. M., Moretti, C. L., Jan, S., Sargent, S. A., Lima, C. E. P., & Fontenelle, M. R., (2014). Climate changes and potential impacts on quality of fruit and vegetable crops. In: Ahmad, P., & Rasool, S., (eds.), *Emerging Technologies and Management of Crop Stress Tolerance: Biological Techniques* (Vol. 1, pp. 467–486). San Diego: Elsevier.

Medyouni, I., Zouaoui, R., Rubio, E., Serino, S., Ahmed, H. B., & Bertin, N., (2021). Effects of water deficit on leaves and fruit quality during the development period in tomato plant. *Food Sci. Nutr., 9*, 1949–1960.

Meli, V., Ghosh, S., Prabha, T., Chakraborty, N., Chakraborty, S., & Datta, A., (2010). Enhancement of fruit shelf life by suppressing N-glycan processing enzymes. *Proceedings of the National Academy of Sciences of the United States of America, 107*(6), 2413–2418. Bibcode:2010 PNAS.107.2413M. doi:10.1073/pnas.0909329107. PMC 2823905. PMID 20133661.

Melotto, M., Zhang, L., Oblessuc, P. R., & He, S. Y., (2017). Stomatal defense a decade later. *Plant Physiol., 174*, 561–571.

Meyer, R., & Purugganan, M., (2013). Evolution of crop species: Genetics of domestication and diversification. *Nat. Rev. Genet., 14*, 840–852.

Mickelbart, M. V., Hasegawa, P. M., & Bailey-Serres, J., (2015). Genetic mechanisms of abiotic stress tolerance that translate to crop yield stability. *Nat. Rev. Genet., 16*, 237–251.

Miller, J. C., & Tanksley, S. D., (1990). RFLP analysis of the phylogenetic relationships and genetic variation in the genus *Lycopersicon*. *Theor. Appl. Genet., 80*, 437–448.

Ming, M., Ren, Q., Pan, C., He, Y., Zhang, Y., Liu, S., Zhong, Z., et al., (2020). CRISPR-Cas12b enables efficient plant genome engineering. *Nat. Plants., 6*, 202–208.

Molla, K. A., & Yang, Y., (2020). Predicting CRISPR/Cas9-induced mutations for precise genome editing. *Trends Biotechnol., 38*, 136–141.

Mutschler, M. A., (1984b). Ripening and storage characteristic of alcobaca ripening mutant in tomato. *J. Am. Soc. Hortic. Sci., 109*, 504–507.

Nekrasov, V., Wang, C., Win, J., Lanz, C., Weigel, D., & Kamoun, S., (2017). Rapid generation of a transgene-free powdery mildew resistant tomato by genome deletion. *Sci. Rep., 7*, 482.

Nguyen, Y. Q., (1994). Juliette' fresh-market tomato. *Hort. Science, 29*(4), 332.

Nguyen, Y. Q., Ashcroft, W. J., Jones, K. J., & McGlasson, W. B., (1991). Evaluation of F1 hybrids incorporating the rin (ripening inhibitor) gene to improve the storage life and fruit quality of fresh market tomatoes (*Lycopersicon esculentum* Mill.). *Aust. J. Exp. Agric., 31*, 407–413.

Nishiuchi, S., Yamauchi, T., Takahashi, H., Kotula, L., & Nakazono, M., (2012). Mechanisms for coping with submergence and waterlogging in rice. *Rice, 5*, 2. doi: 10.1186/1939-8433-5-2.

Nonaka, S., Arai, C., Takayama, M., Matsukura, C., & Ezura, H., (2017). Efficient increase of γ-aminobutyric acid (GABA) content in tomato fruits by targeted mutagenesis. *Sci. Rep., 7*, 7057.

Nowicki, M., Foolad, M., Nowakowska, M., & Kozik, M., (2012). Potato and tomato late blight caused by *Phytophthora infestans*: An overview of pathology and resistance breeding. *Plant Dis., 96*. 4–17.

Oltman, A. E., Jervis, S. M., & Drake, M. A., (2014). Consumer attitudes and preferences for fresh market tomatoes. *J. Food Sci., 79*, S2091–S2097.

Ortigosa, A., Gimenez-Ibanez, S., Leonhardt, N., & Solano, R., (2019). Design of a bacterial speck resistant tomato by CRISPR/Cas9-mediated editing of SlJAZ2. *Plant Biotechnol. J., 17*, 665–673.

Osakabe, K., Wada, N., Miyaji, T., Murakami, E., Marui, K., Ueta, R., Hashimoto, R., et al., (2020). Genome editing in plants using CRISPR type I-D nuclease. *Commun Biol., 3*, 648. https://doi.org/10.1038/s42003-020-01366-6.

Østerberg, J. T., Xiang, W., Olsen, L. I., Edenbrandt, A. K., Vedel, S. E., Christiansen, A., Landes, X., et al., (2017). Accelerating the domestication of new crops: Feasibility and approaches. *Trends Plant Sci., 22*, 373–384.

Pabo, C. O., Peisach, E., & Grant, R. A., (2001). Design and selection of novel Cys2His2 zinc finger proteins. *Annual Review of Biochemistry, 70*, 313–340. doi: 10.1146/annurev.biochem.70.1.313.

Pan, J., Sharif, R., Xu, X., & Chen, X., (2021). Mechanisms of waterlogging tolerance in plants: Research progress and prospects. *Front. Plant Sci., 11*, 627331. doi: 10.3389/fpls.2020.627331.

Pandit, M., Sah, R. P., Chakraborty, M., Prasad, K., Chakraborty, M. K., Tudu, V., Narayan, S. C., et al., (2018). Gene action and combining ability for dual purpose traits in maize (*Zea mays* L.) under water deficit stress prevailing in eastern India. *Range Mgmt. & Agroforestry, 39*(1), 29–37. https://doi.org/10.1080/15427528.2017.1315625.

Papikian, A., Liu, W., Gallego-Bartolome, J., & Jacobsen, S., (2019). Site-specific manipulation of *Arabidopsis* loci using CRISPRCas9 SunTag systems. *Nat Commun., 10*, 729.

Parmar, N., Singh, K. H., Sharma, D., Singh, L., Kumar, P., Nanjundan, J., Khan, Y. J., et al., (2017). Genetic engineering strategies for biotic and abiotic stress tolerance and quality enhancement in horticultural crops: A comprehensive review. *3 Biotech, 7*(4). doi: 10.1007/s13205-017-0870-.

Pascual, L., Desplat, N., Huang, B. E., Desgroux, A., Bruguier, L., Bouchet, J. P., Le, Q. H., et al., (2015). Potential of a tomato MAGIC population to decipher the genetic control of quantitative traits and detect causal variants in the resequencing era. *Plant Biotechnology Journal, 13*, 565–577.

Pennisi, E., (2008). The blue revolution, drop by drop, gene by gene. *Science*, *320*, 171–173.

Peralta, I. E., & Spooner, D. M., (2001). Granule-bound starch synthase (GBSSI) gene phylogeny of wild tomatoes (*Solanum* L. section *Lycopersicon* (Mill.) Wettst. Subsection *Lycopersicon*). *American Journal of Botany, 88*(10), 1888–1902. doi:10.2307/3558365. JSTOR 3558365. PMID 21669622.

Peralta, I. E., & Spooner, D. M., (2005). Morphological characterization and relationships of wild tomatoes (*Solanum L.* section *Lycopersicon*). *Monogr. Syst. Bot. Missouri. Bot. Gard., 104*, 227–257.

Peralta, I. E., (2006). *Nomenclature for Wild and Cultivated Tomatoes* (No. 11). SOL Newslettre, (Consulted on the 30/07/2021.

Piao, S., Ciais, P., Huang, Y., Shen, Z., Peng, S., Li, J., Zhou, L., et al., (2010). The impacts of climate change on water resources and agriculture in China. *Nature, 467*, 43–51.

Picton, S., Gray, J. E., & Grierson, D., (1995). The manipulation and modification of tomato fruit ripening by expression of antisense RNA in transgenic plants. *Euphytica, 85*, 193–202.

Pilowsky, M., & Cohen, S., (1990). Tolerance to tomato yellow leaf curl virus derived from *Lycopersicon peruvianum*. *Plant Dis., 74*, 248–250.

Pnueli, L., Carmel-Goren, L., Hareven, D., Gutfinger, T., Alvarez, J., Ganal, M., Zamir, D., & Lifschitz, E., (1998). The SELF-PRUNING gene of tomato regulates vegetative to reproductive switching of sympodial meristems and is the ortholog of CEN and TFL1. *Development, 125*, 1979–1989.

Pombo, M. A., Zheng, Y., Fernandez-Pozo, N., Dunham, D. M., Fei, Z., & Martin, G. B., (2014). Transcriptomic analysis reveals tomato genes whose expression is induced specifically during effector-triggered immunity and identifies the Epk1 protein kinase which is required for the host response to three bacterial effector proteins. *Genome Biol., 15*, 492.

Pramanik, D., Shelake, R. M., Park, J., Kim, M. J., Hwang, I., Park, Y., & Kim, J. Y., (2021). CRISPR/Cas9-mediated generation of pathogen-resistant tomato against tomato yellow leaf curl virus and powdery mildew. *Int. J. Mol. Sci., 22,* 1878.

Puchta, H., (2005). The repair of double-strand breaks in plants: Mechanisms and consequences for genome evolution. *J. Exp. Bot., 56*, 1–14.

Raffa, F. K., Powell, E. N., & Townsend, P. A., (2013). Temperature-driven range expansion of an irruptive insect heightened by weakly coevolved plant defenses. *Proc. Natl. Acad. Sci. U.S.A, 110*, 2193–2198.

Raja, V., Majeed, U., Kang, H., Andrabi, K. I., & John, R., (2017). Abiotic stress: Interplay between ROS, hormones and MAPKs. *Environ. Exp. Bot., 137*, 142–157.

Ranjan, A., Ichihashi, Y., & Sinha, N. R., (2012). The tomato genome: Implications for plant breeding, genomics and evolution. *Genome Biol., 13*, 167.

Rick, C. M., (1986). Germplasm resources in the wild tomato species. *Acta Hortic., 190*, 39–48.

Rick, C. M., (1987). Genetic resources in *Lycopersicon*. In: Nevins, D. J., & Jones, R A., (eds.), *Tomato Biotechnology* (pp. 17–26). Alan Liss publishers, New York NY.

Rick, C. M., (1990). Perspectives from plant genetics: The tomato genetics stock center. In: *Genetic Resources at Risk: Scientific Issues, Technologies and Funding Policies.* Berkeley, Etatsunis, University of California.

Robinson, R., & Tomes, M., (1968). Ripening-inhibitor: A gene with multiple effects on ripening. *Rep. Tomato Genet. Coop., 18*, 36, 37.

Rodríguez, G. R., Muňos, S., Anderson, C., Sim, S. C., Michel, A., Causse, M., Gardener, B. B., et al., (2011). Distribution of SUN, OVATE, LC, and FAS in the tomato germplasm and the relationship to fruit shape diversity. *Plant Physiol., 156*, 275–285.

Rodríguez-Leal, D., Lemmon, Z. H., Man, J., Bartlett, M. E., & Lippman, Z. B., (2017). Engineering quantitative trait variation for crop improvement by genome editing. *Cell, 171*, 470–480.

Rogowsky, P., (2016). *Promises and Limits of New Plant Breeding Tools (NBT) [archive]*. presentation at the "Agriculture & Innovation 2025" meetings at the International Agricultural Show.

Roldan, M. V. G., P´erilleux, C., Morin, H., Huerga-Fernandez, S., Latrasse, D., Benhamed, M., & Bendahmane, A., (2017). Natural and induced loss of function mutations in SlMBP21 MADS-box gene led to jointless-2 phenotype in tomato. *Sci. Rep., 7*, 4402.

Rolin, D., Baldet, P., Just, D., Chevalier, C., Biran, M., & Raymond, P., (2000). NMR study of low subcellular pH during the development of cherry tomato fruit. *Aust. J. Plant Physiol., 27*, 61–69.

Ron, M., Kajala, K., Pauluzzi, G., Wang, D., Reynoso, M. A., Zumstein, K., Garcha, J., et al., (2014). Hairy root transformation using *Agrobacterium rhizogenes* as a tool for exploring cell type-specific gene expression and function using tomato as a model. *Plant Physiol., 166*, 455–469.

Ronen, G., Cohen, M., Zamir, D., & Hirschberg, J., (1999). Regulation of carotenoid biosynthesis during tomato fruit development: Expression of the gene for lycopene epsilon-cyclase is down-regulated during ripening and is elevated in the mutant delta. *Plant J., 17*, 341–351.

Sangu, E., Tibazarwa, F. I., Nyomora, A., & Symonds, R. C., (2015). Expression of genes for the biosynthesis of compatible solutes during pollen development under heat stress in tomato (*Solanum lycopersicum*). *J. Plant Physiol., 178*, 10–16.

Santillán, M. M. I., Bracuto, V., Koseoglou, E., Appiano, M., Jacobsen, E., Visser, R., Wolters, A. A., & Bai, Y., (2020). CRISPR/Cas9-targeted mutagenesis of the tomato susceptibility gene PMR4 for resistance against powdery mildew. *BMC Plant Biol., 20*, 284.

Sarma, H., Islam, N. F., Prasad, R., Prasad, M. N. V., Ma, L. Q., & Rinklebe, J., (2021). Enhancing phytoremediation of hazardous metal(loid)s using genome engineering CRISPR–Cas9 technology. *J. Hazard. Mater., 414*, 125493.

Savary, S., Willocquet, L., Pethybridge, S. J., Esker, P., McRoberts, N., & Nelson, A., (2019). The global burden of pathogens and pests on major food crops. *Nat. Ecol. Evol., 3*, 430–439.

Scheben, A., Wolter, F., Batley, J., Puchta, H., & Edwards, D., (2017). Towards CRISPR/Cas crops—bringing together genomics and genome editing. *New Phytol., 216*, 682–698.

Schmidt, C., Fransz, P., Rönspies, M., Dreissig, S., Fuchs, J., Heckmann, S., Houben, A., & Puchta, H., (2020). Changing local recombination patterns in *Arabidopsis* by CRISPR/Cas mediated chromosome engineering. *Nat Commun., 11*, 4418.

Schouten, H. J., Tikunov, Y., Verkerke, W., Finkers, R., Bovy, A., Bai, Y., & Visser, R. G. F., (2019). Breeding has increased the diversity of cultivated tomato in the Netherlands. *Front. Plant Sci., 10*, 1606.

Schuch, W., Kanczler, J., Robertson, D., Hobson, G., Tucker, G., Grierson, D., Bright, S., & Bird, C., (1991). Fruit quality characteristics of transgenic tomato fruit with altered polygalacturonase activity. *Hort. Science, 26*, 1517–1520.

Seroczynska, A., Niemirowicz, S. K., & Korrenicwsk, A., (1998). Utilizarion of three non~ripening mutants in tomato breeding for prolonged shelf lite. *Folia Hortic., 1*, 3–14.

Shan, Q., Zhang, Y., Chen, K., Zhang, K., & Gao, C., (2015). Creation of fragrant rice by targeted knockout of theOsBADH2gene using TALEN technology. *Plant Biotechnology Journal, 13*(6), 791–800. doi:10.1111/pbi.12312.

Sheehy, R. E., Kramer, M., & Hiatt, W. R., (1988). Reduction of polygalacturonase activity in tomato fruit by antisense RNA. *PNAS, 85*, 8805–8809.

Shimatani, Z., Kashojiya, S., Takayama, M., Terada, R., Arazoe, T., Ishii, H., Teramura, H., et al., (2017). Targeted base editing in rice and tomato using a CRISPR-Cas9 cytidine deaminase fusion. *Nat. Biotechnol., 35*, 441–443.

Shukla, M., Al-Busaidi, K. T., Trivedi, M., & Tiwari, R. K., (2018). Status of research, regulations and challenges for genetically modified crops in India. *GM Crops Food, 9*, 173–188.

Signaling in Tomato, (2018). *Plant Cell Physiol., 59*, 2239–2254.

Silva, R., Kumar, L., Shabani, F., & Picanço, M., (2017). Assessing the impact of global warming on worldwide open field tomato cultivation through CSIRO-Mk3·0 global climate model. *J. Agric. Sci., 155*, 407–420.

Smith, J., Grizot, S., Arnould, S., Duclert, A., Epinat, J. C., Chames, P., Prieto, J., et al., (2006). A combinatorial approach to create artificial homing endonucleases cleaving chosen sequences. *Nucleic Acids Res., 34*, e149. doi: 10.1093/nar/gkl720.

Soyk, S., Lemmon, Z., Oved, M., Fisher, J., Liberatore, K., Park, S., Goren, A., et al., (2017). Bypassing negative epistasis on yield in tomato imposed by a domestication gene. *Cell, 169*, 1142–1155.e12.

Sun, Y., Zhang, X., Wu, C., He, Y., Ma, Y., Hou, H., Guo, X., et al., (2016). Engineering herbicide-resistant rice plants through CRISPR/Cas9-mediated homologous recombination of acetolactate synthase. *Molecular Plant, 9*(4), 628–631. doi: 10.1016/j.molp.2016.01.001.

Swinnen, G., Goossens, A., & Pauwels, L., (2016). Lessons from domestication: Targeting Cis-regulatory elements for crop improvement. *Trends Plant Sci., 21*, 506–515.

Tam, S. M., Mhiri, C. A., Vogelaar, A., Kerkveld, M., Pearce, S. R., & Grandbastien, M. A., (2005). Comparative analysis of genetic diversification within tomato and pepper collections deleted by retrotransposon based SSAP, ALFP and SSR. *Theor. Appl. Genet., 7*, 583–597.

Tashkandi, M., Ali, Z., Aljedaani, F., Shami, A., & Mahfouz, M. M., (2018). Engineering resistance against tomato yellow leaf curl virus via the CRISPR/Cas9 system in tomato. *Plant Signal. Behav., 13*, e1525996.

Tomato, G. C., (2012). The tomato genome sequence provides insights into fleshy fruit evolution. *Nature, 485*, 635–641.

Tomlinson, L., Yang, Y., Emenecker, R., Smoker, M., Taylor, J., Perkins, S., Smith, J., et al., (2019). Using CRISPR/Cas9 genome editing in tomato to create a gibberellin-responsive dominant dwarf DELLA allele. *Plant Biotechnol. J., 17*, 132–140.

Townsend, J. A., Wright, D. A., Winfrey, R. J., Fu, F., Maeder, M. L., Joung, J. K., & Voytas, D. F., (2009). High-frequency modification of plant genes using engineered zinc-finger nucleases. *Nature, 459*(7245), 442–445. doi: 10.1038/nature07845.

Ueta, R., Abe, C., Watanabe, T., Sugano, S., Ishihara, R., Ezura, H., Osakabe, Y., & Osakabe, K., (2017). Rapid breeding of parthenocarpic tomato plants using CRISPR/Cas9. *Sci Rep., 7*, 507.

Uluisik, S., & Seymour, G. B., (2020). Pectate lyases: Their role in plants and importance in fruit ripening. *Food Chem., 309*, 125559.

Uluisik, S., Chapman, N. H., Smith, R., Poole, M., Adams, G., Gillis, R. B., Besong, T. M., et al., (2016). Genetic improvement of tomato by targeted control of fruit softening. *Nat Biotechnol., 34*, 950–952.

Urnov, F. D., Miller, J. C., Lee, Y. L., Beausejour, C. M., Rock, J. M., Augustus, S., Jamieson, A. C., et al., (2005). Highly efficient endogenous human gene correction using designed zinc-finger nucleases. *Nature, 435*, 646–651. doi: 10.1038/nature 03556.

Van, D. B., Goffau, M. D., Van, L. D. J. J., & Naika, S., (2005). *Cultivation of Tomato: Production, Processing and Marketing*. Agrodok; 17. Agromisa/CTA, Wageningen, The Netherlands.

Van, D. K. E., Chakrabarti, M., Chu, Y. H., Clevenger, J. P., Illa-Berenguer, E., Huang, Z., Keyhaninejad, N., et al., (2014). What lies beyond the eye: The molecular mechanisms regulating tomato fruit weight and shape. *Front Plant Sci., 5*, 227.

Vats, S., Kumawat, S., Kumar, V., Patil, G. B., Joshi, T., Sonah, H., Sharma, T. R., & Deshmukh, R., (2019). Genome editing in plants: Exploration of technological advancements and challenges. *Cells, 8*, 1386.

Veillet, F., Perrot, L., Chauvin, L., Kermarrec, M. P., Guyon-Debast, A., Chauvin, J. E., Nogué, F., & Mazier, M., (2019). Transgene-free genome editing in tomato and potato plants using agrobacterium-mediated delivery of a CRISPR/Cas9 cytidine base editor. *Int. J. Mol. Sci., 20*, 402.

Vrebalov, J., Ruezinsky, D., Padmanabhan, V., White, R., Medrano, D., Drake, R., Schuch, W., & Giovannoni, J., (2002). A MADS-box gene necessary for fruit ripening at the tomato ripening-inhibitor (rin) locus. *Science, 296*, 343e346.

Vu, T. V., Sivankalyani, V., Kim, E. J., Doan, D. T. H., Tran, M. T., Kim, J., Sung, Y. W., et al., (2020). Highly efficient homology-directed repair using CRISPR/Cpf1-geminiviral replicon in tomato. *Plant Biotechnol. J., 18*, 2133–2143.

Wang, L., Chen, L., Li, R., Zhao, R., Yang, M., Sheng, J., & Shen, L., (2017). *J. Agric. Food Chem., 65*, 8674–8682.

Wang, R., Tavano, E. C. D. R., Lammers, M., Martinelli, A. P., Angenent, G. C., & De Maagd, R. A., (2019). Re-evaluation of transcription factor function in tomato fruit development and ripening with CRISPR/Cas9-mutagenesis. *Sci Rep., 9*, 1696.

Wang, T., Deng, Z., Zhang, X., Wang, H., Wang, Y., Liu, X., Liu, S., et al., (2018). Tomato DCL2b is required for the biosynthesis of 22-nt small RNAs, the resulting secondary siRNAs, and the host defense against ToMV. *Hortic. Res., 5*, 62.

Wang, T., Zhang, H., & Zhu, H., (2019). CRISPR technology is revolutionizing the improvement of tomato and other fruit crops. *Horticulture Research, 6*, 77, https://doi.org/10.1038/s41438-019-0159-x.

Wang, Y., Cheng, X., Shan, Q., Zhang, Y., Liu, J., Gao, C., & Qiu, J. L., (2014). Simultaneous editing of three homoeoalleles in hexaploid bread wheat confers heritable resistance to powdery mildew. *Nature Biotechnology, 32*(9), 947–951. doi:10.1038/nbt.2969.

Wang, Z., Hardcastle, T. J., Canto, P. A., Yip, W. H., Tang, S., & Baulcombe, D. C., (2018a). A novel DCL2-dependent miRNA pathway in tomato affects susceptibility to RNA viruses. *Genes Dev., 32*, 17–18.

Welty, N., Radovich, C., Meulia, T., & Van, D. K. E., (2007). Inflorescence development in two tomato species. *Canadian Journal of Botany, 85*(1), 111–118.

Williams, C. E., & St. Clair, D. A., (1993). Phenetic relationships and levels of variability detected by restriction fragment length polymorphism and random amplified polymorphic DNA analysis of cultivated and wild accessions of *Lycopersicon esculentum*. *Genome, 36*, 619–630.

Wittkopp, P. J., & Kalay, G., (2011). Cis-regulatory elements: Molecular mechanisms and evolutionary processes underlying divergence. *Nat. Rev. Genet., 13*, 59–69.

Wu, Y., Zhang, D., Chu, J. Y., Boyle, P., Wang, Y., Brindle, I. D., De Luca, V., & Despres, C., (2012). The Arabidopsis NPR1 protein is a receptor for the plant defense hormone salicylic acid. *Cell Rep., 1*, 639–647.

Ximénez-Embún, M. G., González-Guzmán, M., Arbona, V., Gómez-Cadenas, A., Ortego, F., & Castañera, P., (2018). Plant-mediated effects of water deficit on the performance of *Tetranychus evansi* on tomato drought-adapted accessions. *Frontiers in Plant Science, 9*. doi: 10.3389/fpls.2018.01490.

Xin, X. F., & He, S. Y., (2013). *Pseudomonas syringae* pv. tomato DC3000: A model pathogen for probing disease susceptibility and hormone signaling in plants. *Annu. Rev. Phytopathol., 51*, 473–498.

Xin, Y., Yuan, Y., Feng, B., & Wei, D., (2020). CRISPR/Cas9-mediated gene-editing technology in fruit quality improvement. *J. Food Saf., 4,* 159–166.

Xu, C., Park, S. J., Van, E. J., & Lippman, Z. B., (2016). Control of inflorescence architecture in tomato by BTB/POZ transcriptional regulators. *Genes Dev., 30*, 2048–2061.

Xu, J. M., Hua, K., & Lang, Z. B., (2009). Genome editing for horticultural crop improvement. *Hortic Res, 6*, 113.

Yang, Y., Sherwood, T. A., Patte, C. P., Hiebert, E., & Polston, J. E., (2004). Use of tomato yellow leaf curl virus (TYLCV) Rep gene sequences to engineer TYLCV resistance in tomato. *Phytopathology, 94*, 490–496.

Yang, Y., Zhu, G., Li, R., Yan, S., Fu, D., Zhu, B., Tian, H., et al., (2017). The RNA editing factor SlORRM4 is required for normal fruit ripening in tomato. *Plant Physiol., 175*, 1690–1702.

Yin, Y., Qin, K., Song, X., Zhang, Q., Zhou, Y., Xia, X., & Yu, J., (2018). BZR1 transcription factor regulates heat stress tolerance through FERONIA receptor-like kinase-mediated reactive oxygen species signaling in tomato. *Plant and Cell Physiology,* 1–44. doi:10.1093/pcp/pcy146.

Yokotani, N., Nakano, R., Imanishi, S., Nagata, M., Inaba, A., & Kubo, Y., (2009). Ripening associated ethylene biosynthesis in tomato fruit is autocatalytically and developmentally regulated. *J. Exp. Bot., 60*(12), 3433–3442.

Yu, Q., Wang, B., Li, N., Tang, Y., Shengbao, Y., Yang, T., Xu, J., et al., (2017). CRISPR/Cas9-induced targeted mutagenesis and gene replacement to generate long-shelf life tomato lines. *Sci. Rep., 7*, 11874.

Yu, W., Wang, L., Zhao, R., Sheng, J., Zhang, S., Li, R., & Shen, L., (2019). Knockout of SlMAPK3 enhances tolerance to heat stress involving ROS homeostasis in tomato plants. *BMC Plant Biol., 19*, 354.

Yuan, S., Kawasaki, S., Abdellatif, I. M. Y., et al., (2021). Efficient base editing in tomato using a highly expressed transient system. *Plant Cell Rep., 40*, 667–676. https://doi.org/10.1007/s00299-021-02662-z.

Zafar, S. A., Zaidi, S. S., Gaba, Y., Singla-Pareek, S. L., Dhankher, O. P., Li, X., Mansoor, S., & Pareek, A., (2020). Engineering abiotic stress tolerance via CRISPR/ Cas-mediated genome editing. *J Exp Bot., 71*, 470–479.

Zamir, D., (2001). Improving plant breeding with exotic genetic libraries. *Nat. Rev. Gen, 12*, 983–989.

Zhang, H. X., & Blumwald, E., (2001). Transgenic salt-tolerant tomato plants accumulate salt in foliage but not in fruit. *Nature Biotechnology, 19*(8), 765–768. doi: 10.1038/90824. PMID 11479571. S2CID 1940765.

Zhang, N., Roberts, H. M., Van, E. J., & Martin, G. B., (2020). Generation and molecular characterization of CRISPR/Cas9-induced mutations in 63 immunity-associated genes in tomato reveals specificity and a range of gene modifications. *Front. Plant Sci., 11*, 10.

Zhang, X., Fowler, S. G., Cheng, H. M., Lou, Y. G., Rhee, S. Y., Stockinger, E. J., & Thomashow, M. F., (2004). Freezing-sensitive tomato has a functional CBF cold response pathway, but a CBF regulon that differs from that of freezing-tolerant *Arabidopsis*. *Plant J., 39*, 905–919.

Zheng, Z., Appiano, M., Pavan, S., Bracuto, V., Ricciardi, L., Visser, R. G., Wolters, A. M., & Bai, Y., (2016). Genome-wide study of the tomato SlMLO gene family and its functional characterization in response to the powdery mildew fungus *Oidium neolycopersici*. *Front. Plant Sci., 7*, 380.

Zhou, J., Zhao, P. Z., Long, J., Sosso, D., Liu, B., Eom, J. S., Huang, S., et al., (2015). *The Plant Journal, 82*(4), 632–643. doi: 10.1111/tpj.12838.

Zsögön, A., Čermák, T., Naves, E. R., Notini, M. M., Edel, K. H., Weinl, S., Freschi, L., et al., (2018). De novo domestication of wild tomato using genome editing. *Nat. Biotechnol., 36*, 1211–1216.

Zsögön, A., Cermak, T., Voytas, D., & Peres, L. E., (2017). Genome editing as a tool to achieve the crop ideotype and de novo domestication of wild relatives: Case study in tomato. *Plant Sci., 256*, 120–130.

Zuluaga, D. L., Gonzali, S., Loreti, E., Pucciariello, C., Degl'Innocenti, E., Guidi, L., Alpi, A., & Perata, P., (2008). *Arabidopsis thaliana* MYB75/PAP1transcription factor induces anthocyanin production in transgenic tomato plants. *Functional Plant Biology, 35*(7), 606–618. doi:10.1071/FP08021.PMID-32688816.

CHAPTER 9

Genome Editing Techniques and Their Applications in *Brassica* Crop Improvement

SANJAY J. JAMBHULKAR,[1*] ARCHANA RAI,[1] SHANKAR K. BHUJBAL,[1] and CHINMAYA DAS[2]

[1]*Nuclear Agriculture and Biotechnology Division, Bhabha Atomic Research Center, Mumbai – 400085, Maharashtra, India*

[2]*AICRP on Rapeseed-Mustard, Ranital (RRTTS, Ranital, OUAT), Bhubaneswar – 756111, Odisha, India*

[*]*Corresponding author. E-mail: jambhulkarsanjay52@gmail.com*

ABSTRACT

Genetic variations are the key to crop improvement. Mutation and recombination have played an important role in the evolution of crop plants and germplasm strengthening. Utilization of desirable variability in hybridization has resulted in the development of high-yielding varieties with a trait of choice. Climate change poses new threats to sustainable food security and needs the development of high-yielding variety with tolerance to biotic and abiotic stresses as well as enhanced nutritional quality in the shortest possible time. Therefore, tailoring crop plants with desirable traits need precision. CRISPR/Cas technology offers a suitable answer. Production of cereals and pulses has reached to near self-sufficiency in India but lacking in oilseed production. *Brassica* species have important oilseeds crops and genome editing in *Brassica* crops shall be useful in developing biotic and abiotic stress tolerance (AT) as well as enhanced nutritional quality varieties in lesser time.

CRISPR/Cas-Mediated Genome Editing in Plants, Deepu Pandita, Anu Pandita (Eds.)

9.1 INTRODUCTION

Plant breeding has immensely helped to overcome the problem food security through green revolution (Swaminathan, 2006) which is now accompanied by precision breeding (Hartung and Schiemann, 2014) and new plant technology (NBTs) (Lusser et al., 2011; Limera et al., 2017). However, crop improvement program needs to speed up to keep the pace with newly emerging problems engendered by climate change and the ever-increasing demand for food by burgeoning population. Here, the role of crop improvement plays an important role in designing the architecture of the genomic landscape of cultivars either manipulating existing natural genetic variations locked in different landraces, wild relatives or by creating novel variations through conventional hybridization-based breeding approaches or induced mutagenesis. Amalgamation of existing technology and newly emerged techniques such as genome editing (GE), artificial intelligence and deep learning (Mahood et al., 2020) shall show the rays of hope for sustainable food security. In India, near self-sufficiency has been observed in the production of cereal and pulse crops. However, oilseed production needs strengthening due to ever increasing demand of oil for which the government of India spends substantial money.

Oilseed *Brassica*s are one the major group of oilseed crops in India. This genus consists of 39 species and most of these species have enormous economic impact as oilseed, vegetables, forages, and condiments (Lv et al., 2020). *Brassica napus*, *B. juncea*, *B. rapa* are the major source of edible oil and as an animal feed (Jat et al., 2019). Due to increasing demand of vegetable oil, a major effort is being taken to enhance the production, productivity, and oil quality in a sustainable manner. But low variability among the agricultural crops restricted the sustainable crop improvement through hybridization technique (Ahmar et al., 2020). Genetic variation is the foundation of the crop improvement through plant breeding. Creation and use of mutations in crop improvements paved new possibilities and solutions for food security. Induced mutagenesis evolved as an important tool for the generation of variability among the various crop plants. Using physical (gamma rays, thermal neutron, X-rays) and chemical (EMS, NMU, NEU) mutagens, a large spectrum of mutations for morphological, biochemical, and yield contributing traits have been isolated (Raina et al., 2021). This technique has been extensively exploited for the generation of variability for traits like biotic and abiotic stress tolerance (AT), oil quality improvement, plant architecture, and phenology in *Brassica* species (Jambhulkar, 2007).

Till today, 41 varieties of have been developed using the mutation breeding approach (MVD/IAEA, 2021) in the crop *Brassica*s. Mutation breeding has become unpopular in recent years due to lack of precision, undesirable mutation, and advancement of GM techniques like transgenic and genome editing techniques (Mba, 2013).

CRISPR/Cas9 system is composed of a Cas9 nuclease and a single-guide RNA (sgRNA). Specific base-pairing between the target DNA sequence and sgRNA helps Cas9 in the recognition and cleavage of a specific target DNA sequence to generate double strand breaks (DSBs). Induction of DSBs triggers cellular-repair mechanisms and mutations during the process of repair are generated at or near DSBs (Bao et al., 2019). Now days it is possible to generate "null segregants," which are the transgene-free genome-edited plants in a short span of time (Wada et al., 2020). Further, site-directed mutagenesis at target sites of the genome and control of target-site gene expression pattern is feasible with CRISPR/Cas9 system which will not only promote molecular breeding with precision and ease, but also help in understanding basic design of biology for sustainable gain in crop productivity and enhanced stress tolerance in crop plants (Wada et al., 2020; Zaidi et al., 2020). An array of genome editing tools, including TALENs, zinc finger nucleases (ZFNs), and meganucleases have evolved for genetic manipulation of targeted genes in the genome of a crop plant (Razzaq et al., 2019). However, the feasibility of TALENs, ZFNs, and meganucleases is not sufficient to permit for successful genome editing with ease due to the highly expensive and complex and cumbersome procedure of such tools. All these problems were solved by the CRISPR/Cas9 system of genome editing, which is very easy to design, cost-effective and efficiently works in most of the organisms (Razzaq et al., 2019). It is a robust and powerful reverse genetics tool which enables to conduct targeted gene mutagenesis, single base substitution, gene knock-out, gene knock-in, gene replacement and transcriptional regulation of gene expression pattern in both single and multiplex-gene editing in plants (Hussain et al., 2018). Genome-edited crops are different from genetically modified crops as these don't contain foreign DNA sequence but consists of a precise edition of endogenous genes, which can also be achieved through mutagenesis using chemical or physical mutagenic agent. Genome editions have been classified into three types based on the length of sequence altered. SDN-1 for one or few base pair (bp) changes, SDN-2 for few bp changes and limited to 20 bps and SDN-3 gene replacements or long indels (Gao, 2018). Thus, the risk assessment should be done according to the degree of base edition for regulatory approval. Using

CRISPR/Cas9, various traits related to yield, biotic-, and abiotic-stress tolerance quality and hybrid-breeding techniques have been improvised in plant via knock out, knock-in, and replacement of the trait related genes in various crop plants like rice, wheat, maize, *Brassica* and tomato (Chen et al., 2019). In this view, mechanisms of genome editing, modifications in genome editing techniques, gene delivery systems and application of CRISPR-Cas technology in brassica crop improvement has been reviewed.

9.2 MECHANISMS OF GENOME EDITING IN BRIEF

9.2.1 SITE SPECIFIC NUCLEASES (SSNS)

Till now, multiple approaches have been developed for site-specific genetic alterations by using SSNs. The SSNs meganucleases, ZFNs, transcription activator-like effectors nucleases (TALENs), and recently popularized CRISPR/Cas9 system has opened new avenues in crop improvements. The experimental methodology of each technology evolved based upon the spectrum of the applications, complexity, and adaptability of the technique.

9.2.1.1 MEGANUCLEASES APPLICATION IN GENE EDITING

Meganucleases, also known as homing endonucleases (HEs) are the site-specific endonucleases. Meganucleases recognize >12 bp sequence and introduce a DSB at the central four base pairs (Jurica et al., 1998; Stoddard., 2005). Induction of double stranded breaks instigates DNA repair mechanisms and increases the efficiency of homologous recombination (HR). The ability of the meganucleases to make DSBs at the specific site makes them a useful tool for genome editing in plants and animals (Puchta et al., 1993). As it recognizes limited and central four bases of the target site, meganucleases limit the number of variable targets in DNA sequences. That restricts its usability in genome editing (Lambert et al., 2016). To overcome the barrier of the limited number of naturally existing DNA binding HEs, intense efforts are required in order to design, engineer, and select the suitable HE (Chames et al., 2005).

9.2.2 ZINC FINGER NUCLEASES (ZNFS) IN GENE EDITING

ZNFs have two major components; first is the sequence identifying domains, zinc finger proteins (ZNPs), each recognize 3 bp regions and are arranged in tandem repeats. The other component is the nuclease that introduces the double stranded breaks in the DNA, which are then repaired by using HDR offering a scope for alteration in the original sequence. In order to determine the precise location of the target sequence multiple zinc finger proteins against the longer target site are engineered together and linked with the endonuclease *FokI* (Bibikova et al., 2003; Kim et al., 1996; Bibikova et al., 2001; Urnov et al., 2010). Like meganucleases, ZNFs also have encountered a barrier of availability of specifically binding ZNFs and hindered the wide adoption of the technique as an efficient genome editing technology (Christian et al., 2010).

9.2.3 TRANSCRIPTION ACTIVATOR-LIKE EFFECTOR NUCLEASES (TALEN) IN GENE EDITING

TALEs are the transcription activator-like effectors known to bind to specific DNA sequences. TALEs identify the target DNA sequence by the protein that contains tandem repeats as many as 30 made of about 33–35 amino acids long motif. These proteins contain exceptional changes in amino acid sequence of the DNA binding motifs facilitating the recognition of variable DNA base pairs (Boch et al., 2009; Moscou et al., 2009). The TALEs are then engineered with endonuclease *FokI* that can successfully cleave the DNA, enabling further possibilities of gene editing (Miller et al., 2011). TALENs were found to be efficient in recognition of the target sequences and relatively easy for engineering and time saving as compared to ZNFs with lesser off-target effects (OTEs) (Hockemeyer et al., 2011; Wada et al., 2020).

9.2.4 CRISPR-BASED GENE EDITING

CRISPR-based gene editing is dependent upon the ability of the plant to repair the damaged DNA. CRISPR is mainly exploited for its ability to make RNA-guided cuts in the DNA. It introduces DSBs in plant genome thereby instigating DSB repair mechanisms mainly non-homologous end joining (NHEJ) and HR. NHEJ is a non-specific error-prone method of DNA repair resulting into multiple types of alteration in the repaired DNA and therefore used in gene knockout, insertion, and deletion of the DNA sequence (Belhaj et al., 2015; Liu, 2016). CRISPR-based genome editing has been adopted widely and modified to expand its functionality in plants. Different strategies are employed to introduce a desired genetic character in plants. The methodology involves multiple stages of gene manipulation from transferring the CRISPR/Cas components to the target plant to the selection of the desired plant type.

CRISPR/Cas9 system from the *Streptococcus pyogenes* has been adopted for the majority of plant genome editing applications (Jinek et al., 2012). CRISPR/Cas9-based genome editing relies on three components. The Cas9 protein, which is required to introduce a double-stranded break in the DNA and sgRNA guides the Cas9 protein to locate the target site. The third component is a target site located upstream of a protospacer-adjacent motif (PAM) sequence in the target genome (Jinek et al., 2012; Doudna and Charpentier, 2014). The sgRNA is engineered with protospacer-containing CRISPR RNA caller as crRNA and *trans*-activating tracer RNA with target site at the 5'end (Jinek et al., 2012). After successful recognition of the target sequence RucV and HNH nuclease domain of the Cas9 protein introduce precise DSB at three bases upstream of the PAM sequence (Ma et al., 2016). A subsequent repair of DSBs by NHEJ, which is an error prone repair mechanism introduces mutations in the target sequence. However, repair by HDR with the help of donor homologous sequence leads to knock in or replacement of the original gene sequence. By designing CRISPR/Cas9 system for multiple targets multiplex editing is now being practiced because of the wide applicability of the technique (Cong et al., 2013).

9.3 CRISPR/CAS SYSTEMS, MODIFICATIONS, AND APPLICATIONS

Many versions of the active CRISPR/Cas systems have been identified in bacteria since its discovery. There are two major classes of the CRISPR/Cas system based upon the number of effector proteins involved in the function. Class I CRISPR/Cas system requires multiprotein complex for the nuclease activity. Class-II CRISPR/Cas system uses single effector protein which is the most widely used in CRISPR/Cas system for biological applications. Depending upon the Cas enzyme is used, class II systems are further classified into three types – II, V, and VI. Type II Cas9 and type V Cas12 proteins have RNA-guided endonuclease activity while type VI Cas13 has preferentially RNA-targeting and cleavage activity.

9.3.1 CAS NUCLEASES

Most commonly used Cas nucleases is Cas9 and it forms a functional ribonucleoprotein (RNP) complex with CRISPR RNAs (crRNA) and *trans*-activating crRNAs (tracrRNA). Cas9 introduces DSB at the target site guided by single guide RNA (sgRNA), which is fused with crRNA and tracrRNA into a single longer RNA. Cas9 makes blunt end DSB after three nucleotides from the PAM sequence located towards 3'end of the protospacer (Ma et al., 2016; Jinek et al., 2012; Garneau et al., 2010). Cas12 is a frequently used Cas for genomic applications and works in a different fashion than Cas9. It possesses a single Ruv-C like nuclease domain which typically introduces a staggered DSB in the target site as compared to two nuclease domains in Cas9 (Makarova et al., 202).

9.3.1.1 USE OF CRISPR/CAS9 NUCLEASES IN MUTAGENESIS AND KNOCK OUTS

In *Brassica napus*, CRISPR/Cas9-based gene editing has been used for mutagenesis of fatty acid desaturase gene in order to increase the oleic acid content. Targeting four copies of *BnaFAD2* lead to 51.9% single base insertion or deletion mutation of total 74.1% short nucleotide change mutations. Knockdown resulted into the increase in the oleic acid content up to 80% as compared to the wild type which had oleic acid content of 66.43%. This increase in oleic acid content is attributed to the decreased levels of linoleic and linolenic acid content (Huang et al., 2020). Knock out of the gene using CRISPR/Cas9 gene editing led to decreased methionine derived glucosinolate in the field grown *B. oleracea* (Neequaye et al., 2021). Gene knocks out of BoPDS in *B. oleracea* could result into an albino phenotype. Thus, it is easier to screen the mutant plants with defined target and visible phenotype (Ma et al., 2019). Plants with high copy number and paralogous genes can be targeted by using a single sgRNA. In *B. napus* 12 genes were targeted with a single gRNA and frequency of mutations was recorded from 27.6% up to 96.6% (Yang et al., 2017).

9.3.1.2 MULTIPLEX GENE EDITING WITH MODIFIED T-RNA ENDOGENOUS PROCESSING

Multiplex gene editing involves the transfer of more than one sgRNA to the target plant. In order to increase the number of the sgRNA, multiple sgRNAs are linked with the tRNA in tandem repeat. Endogenous tRNA processing machinery can produce multiple copies of the sgRNAs. With this approach successful mutagenesis has been achieved in rice and *B.*

napus (Ma et al., 2019; Xie et al., 2015). In *B. napus*, *BnWRKY11 and BnWRKY70* genes involved in the pathogenic fungus *Sclerotinia sclerotiorum* resistance were mutated with multiplex gene editing. Mutation induction was precise and frequency observed is 54.5% for *BnWRKY11* and 50% for *BnWRKY70.* Transformants containing *BnWRKY11* showed increased resistance while the other showed no effect on the resistance to the pathogen (Sun et al., 2018).

9.3.2 DEAD CAS9 (DCAS9) APPLICATION IN TARGETING

Inactivated Cas9 is also termed as a dead Cas9 (dCas9) which is devoid of the Cas9 endonuclease activity and can be associated with numerous other DNA and protein modifying enzymes in order to carry out the desired changes. Different heterologous transcriptions factors can be used with dCas9 to reprogram the gene expression and regulation patterns (Sander and Joung, 2014). The catalytically inactive Cas nuclease fused with the corresponding deaminase enzyme which is used for the recognition and modification of the target site. Base editors are one of the modified forms of the CRISPR/Cas system that introduce a single base mutation in the targeted region. Based upon the type of nitrogenous base targeted, base editors are classified into two categories. Cytosine base editors (CBEs) and adenine base editors (ABEs) manifest the transition mutations leading to C→G and A→T respectively. This makes it possible to introduce desired point mutation at the target site which includes C→T, A →G, T→C, G→A transitions (Komor, 2016; Gaudelli, 2017). Inactivated Cas9 can also be fused with various epigenetic modifiers like histone acetylase and deacetylase. Engineering histone aceytalase with dCas9 efficiently located the target sequence and reprogramed expression of the secondary metabolite gene *breF* in *Aspergillus niger* (Li et al., 2021).

9.3.3 CASFISH

Traditional approaches of fluorescent in situ hybridization (FISH) require denaturation of the nuclear material thus distorting its natural organization of the nucleus. Fluorescently labeled Cas9 protein along with different guide RNAs could fluorescently label the specific genomic loci and will be useful for understanding the genome spatial organization (Deng et al., 2015).

9.3.4 CRISPR ASSOCIATED TRANSPOSON AND RECOMBINASES

Catalytically deficient Cas9 manifest RNA-guided site-specific integration of the mobile transposon in the target genome with the help of transposase. Targeted and site-specific insertion of the DNA fragments downstream to the protospacer sequence into *E. coli* genome has been achieved with a frequency up to 80% using Tn7-like transposases subunit and type V-K CRISPR (Klompe et al., 2019; Strecker et al., 2019). This system of targeted insertion can facilitate the DSB and donor DNA-free insertion. Targeted insertion of the large genomic fragments by using recombinases at user-decided loci is difficult. Thus, combining transposases with dCas9 can direct the efficient insertion of the DNA of interest at the single target locus (Chen and Wang, 2019).

9.3.5 PRIME EDITING

A small change in the desired target requires a high level of precision and HDR-mediated indels and base-edited mutations lead to undesirable changes in the genome. Moreover, base editors can only introduce the transition mutation and not the transversions. Prime editors are newly engineered protein complexes containing Cas9 nickase that lacks functional HNH domain fused with the reverse transcriptase (RT). The prime editing guide RNA (pegRNA) specifically locates the target site as well as codes for the desired change in the target sequence. The active nuclease, RuvC domain makes a nick in one strand of the target DNA making a 3'end free. Then pegRNA binds with complimentary region at free 3'end priming the extended DNA synthesis by RT using pegRNAas a template. Template pegRNA also contain the edited gene sequence/s and a region with homology in order to facilitate DNA repair after new DNA synthesis (Anzalone et al., 2019).

9.4 DELIVERY SYSTEMS FOR CRISPR/CAS9 EFFICIENT GENE EDITING

Most of the CRISPR based genome editing processes are carried out using *Agrobacterium*-mediated T-DNA transformation, biolistic, and PEG-mediated transformations. However, virus-induced gene silencing is frequently used in differential gene expression studies. Thus it can also be used to deliver the gRNA to the plants. The gRNA was successfully delivered in *Nicotiana benthamiana* using the *Tobacco rattle virus* (TRV) that replicates in the cytoplasm (Ali et al., 2015). Cabbage Leaf Curl virus (CaLCuV) is a DNA virus from the *Geminiviridae* family which replicates in the nucleus and is used to deliver the gRNA as like RNA viruses to induce specific mutations in the genome (Yin et al., 2015).

9.5 DNA FREE GENOME EDITING

All of the CRISPR-mediated successes were achieved by using mostly three transformation techniques viz. (i) *Agrobacterium tumefaciens*,(ii) biolistics, and (iii) PEG-mediated transformation of the protoplasts. The transgenic plants may target unpredictable sequences and prolonged exposure of plants to the CRISPR constructs causes unwanted OTEs and insertional mutations limiting the pertinence. Such constructs can be bred out by backcrossing but requires longer time for the process also it is not as useful in case the plants with self-incompatibility and in case of asexually producing plants (Murovec et al., 2018). The presence of the genome editing vectors, which are unwanted after the successful editing of the target DNA and raise the concerns of plants in question to be the GMO (Wolt et al., 2016). That's where the DNA free genome editing becomes significant.

Transfecting preassembled functional genome editing components work as precisely as they would when transformed by using routine transformation practices. A purified Cas9 protein mixed with excess gRNAs forms a functional ribonucelomprotein (RNPs) complex. RNA-guided endonuclease ribonucleoproteins (RGENs RNPs) were incubated with protoplasts derived from *A. Thaliana* and other species could produce indels with frequencies of 8.4% to 44%. Interestingly OTEs were also not detected in the homologous sequences differing by the 2–3 nt. Inventing such approaches would definitely speed up the CRISPR based genome editing research and overcoming the regulatory aspects related to the release of the GMOs as a crop variety (Woo et al., 2015). Nanoparticles have also been used to achieve DNA free genome editing and do not need any tissue culture intervene, thus applicable to all genotypes. Cas9 with sgRNA were transferred through. After the expression of the active Cas9 enzyme it formed a Cass9/gRNA complex and led to targeted gene editing (Tsanova et al., 2020). The nanoparticle material used in genome editing should be permeable and can be able to retain the payload for longer time (Blanco et al., 2016).

9.6 APPLICATION OF CRISPR/CAS TECHNOLOGY IN *BRASSICA* CROPS

Most of CRISPR/Cas technological applications have been reported in *B. napus* followed by *B. oleracea* and *B. compestris*. CRISPR/Cas based research activity in *Brassica* spp. is still in juvenile stage and most of the studies have been conducted to protocol standardization to efficiently genome editing (Cunfa et al., 2019; Murovec et al., 2018; Xiong et al., 2019; Yang et al., 2017; Sun et al., 2018). Phytoene desaturase gene (BoPDS) was used to study the efficiency of editing protocol since it gives albino plant. Dwarf and early plant type is one of the desirable characters in oil seed Brassica and successful gene editing have been reported to dwarf and early phenotype (Zheng et al., 2020; Lawrenson et al., 2015; Jeong et al., 2019; Jung et al., 2021). CRISPR/Cas9 system have effectively used in improving the pod shattering resistance *B. napus* (Braatz et al., 2017; Zhai et al., 2019; Zaman et al., 2019, 2021). Change in seed coat color (Zhai et al., 2020; Xie et al., 2020) and modification of fatty acid profiling to improve the oil quality (Huang et al., 2020; Okuzaki et al., 2018) have been reported using CRISPR/Cas9 system in oilseed *Brassica*. Yellow seed color and fatty acid profile change have also been achieved through mutagenesis, but these changes were found to be associated with other plant feature changes in oil seed crops (Jambhulkar et al., 2021). Besides, gene editing using CRISPR/Cas system have also been successfully utilized to provide resistance against fungal pathogen in *B. napus* (Sun et al., 2018; Pröbsting, 2020). These studies have been compiled in Table 9.1 and suggest the CRISPR/Cas system efficient application in desirable traits improvement in *Brassica* spp. Some of the challenging problems in oilseed *Brassica*s like Alternaria resistance, aphid tolerance, new sources of erucic acid, increased oil content and effective source-sink relationship could be resolved through genome editing.

TABLE 9.1 List of Genome Edited *Brassica* Crops for Different Characters

SL. No.	Crop	Objective of the Study/ Character Modified	Gene-Edited	Method Used in the Study	References
1.	*Brassica oleracea* var. Capitata	Precisely edition of cabbage genome	Phytoene desaturase gene (*BoPDS*)	CRISPR/Cas9 system	Cunfa et al. (2019)
2.	*Brassica spp.* (*B. oleracea, B. napus,* and *B. rapa*)	Develop a DNA-free protocol for site-directed mutagenesis RNP transfection	*FRI* and *PDS* genes	Delivery of preas-sembled ribonucleoprotein complexes (RNPs) composed of recombinant enzyme Cas9 and *in-vitro* transcribed sgRNA into protoplast.	Murovec et al. (2018)
3.	*B. campestris*	Efficiency of CRISPR/Cas9 system on single and multiplex genome editing	Pectin-methylesterase genes *Bra003491*, *Bra007665*, and *Bra014410*	Use of single-targeting vector and a multitargeting vector for CRISPR/Cas9 system	Xiong et al. (2019)
4.	*Brassica napus*	To examined the mutation efficiency of CRISPR/Cas9	Four genes of *BnaRGA* family, 3 genes of *BnaFUL* family, and paralogs of *BnaDA2* and *BnaDA1* family	CRISPR/Cas9 system	Yang et al. (2017)
5.	*B. oleracea var. alboglabra*	Mutagenesis through genome editing	Phytoene desaturase (PDS) gene	CRISPR/Cas9 system	Sun et al. (2018)

TABLE 9.1 *(Continued)*

SL. No.	Crop	Objective of the Study/ Character Modified	Gene-Edited	Method Used in the Study	References
6.	*Brassica napus*	To improve plant architecture	*BnaMAX1*	CRISPR/Cas9 system	Zheng et al. (2020)
7.	*Brassica oleracea* var. Capitata	Disrupting self-incompatibility and development of male sterile mutants	(i). phytoene desaturase gene (*BoPDS*) (ii) S-receptor kinase gene (*BoSRK*) (iii). male-sterility-associated gene (*BoMS1*)	CRISPR/Cas9 gene-editing system based on endogenous tRNA processing	Ma et al. (2019)
8.	*Brassica oleracea*	Dwarf phenotype	*BolC.GA4*	CRISPR/Cas9 system	Lawrenson et al. (2015)
9.	*Brassica rapa spp. pekinensis*	Early-flowering	*BraFLC2* and *BraFLC3*	CRISPR/Cas9 system	Jeong et al. (2019)
10.	*Brassica rapa L.*	Flowering time (Early)	Suppressor of overexpression of Constance 1 homologs (*BrSOC1s*)	CRISPR/Cas9 system	Jung et al. (2021)
11.	*Brassica napus*	Multilocular silique	*CLAVATA3* (*CLV3*) and its related receptors *CLV1* and *CLV2*	CRISPR/Cas9 system	Yang et al. (2018)
12.	*Brassica napus*	Pod shattering Resistance	*JAGGED* genes (*BnJAG.A02*, *BnJAG.C02*, *BnJAG.C06*, *BnJAG.A07*, and *BnJAG.A08*)	CRISPR/Cas9 system	Zaman et al. (2019)
13.	*Brassica napus*	Pod shattering Resistance	*ALCATRAZ* (*ALC*) genes	CRISPR/Cas9 system	Braatz et al. (2017)
14.	*Brassica napus*	Pod shattering Resistance	*BnSHP1* and *BnSHP2*	CRISPR/Cas9 system	Zaman et al. (2021)
15.	*Brassica napus*	Pod shatter Resistance	*Indehiscent* (*IND*) and *Alcatraz* (*ALC*) gene	CRISPR/Cas9 system	Zhai et al. (2019)
16.	*Brassica napus*	Yellow seed	*BnTT8*	CRISPR/Cas9 system	Zhai et al. (2020)
17.	*Brassica napus*	Yellow seed	*BnTT2*	CRISPR/Cas9 system	Xie et al. (2020)
18.	*Brassica. napus*	Identification of precise gene for oil production	Lysophosphatidic acid acyltransferase (*LPAT2* and *LPAT5*)	CRISPR-Cas9 with four single-gRNAs and two multi-gRNAs were deliberately designed from the conserved coding regions	Zhang et al. (2019)
19.	*Brassica napus*	Modifications of fatty acid profile	Fatty acid desaturase 2 (FAD2) gene	CRISPR/Cas9 system	Huang et al. (2020)
20.	*Brassica napus*	Modifications of fatty acid profile	Fatty acid desaturase 2 (FAD2) gene	CRISPR/Cas9 system	Okuzaki et al. (2018)

TABLE 9.1 *(Continued)*

SL. No.	Crop	Objective of the Study/ Character Modified	Gene-Edited	Method Used in the Study	References
21.	*Brassica oleracea*	Regulation of aliphatic glucosinolate (A-GSL) biosynthesis and associated sulfur metabolism	*MYB28*	CRISPR/Cas9 system	Neequaye et al. (2021)
22.	*Brassica napus*	Resistance against Verticillium wilt	*BnCRT1a* and *BnHVA22c*	CRISPR/Cas9 system	Pröbsting (2020)
23.	*Brassica napus*	JA- and SA-induced resistance to pathogenic fungus, *Sclerotinia sclerotiorum*	*BnWRKY11* and *BnWRKY70* genes	CRISPR/Cas9 system	Sun et al. (2018)

9.7 CONCLUSIONS

A large spectrum of variability achieved through mutation and recombination has helped to develop high-yielding varieties, which were one step closer to attend near food security. However, sustainable food security is a major concern in future in the arena of climate change and burgeoning population. Newer challenges in crop improvement program need to be tackled using advanced technology in the shortest possible time. Genome editing shall be one of the possible ways to tailor the crop of choice. Oilseed production and productivity, particularly in oilseed *Brassica* crops, could be enhanced by genome editing for plant type, resistance to biotic and abiotic stresses, and quality characters.

KEYWORDS

- **brassica**
- **CRISPR**
- **double-strand breaks**
- **food security**
- **genome editing**
- **homing endonucleases**

REFERENCES

Ahmar, S., Gill, R. A., Jung, K. H., Faheem, A., Qasim, M. U., Mubeen, M., & Zhou, W., (2020). Conventional and molecular techniques from simple breeding to speed breeding in crop plants: Recent advances and future outlook. *Int. J. Mol. Sci., 21*, 2590.

Ali, Z., Abul-faraj, A., Li, L., Gosh, N., Piatek, M., Mahjoub, A., Aouida, M., et al., (2015). Efficient virus-mediated genome editing in plants using the CRISPR/Cas9 system. *Mol. Plant, 8*, 1288–1291.

Anzalone, A. V., Randolph, P. B., Davis, J. R., Sousa, A. A., Koblan, L. W., Levy, J. M., Chen, P. Jet al., (2019). Search-and-replace genome editing without double-strand breaks or donor DNA. *Nature, 576*, 149–157.

Belhaj, K., Chaparro-Garcia, A., Kamoun, S., Patron, N., & Nekrasov, V., (2015). Editing plant genomes with CRISPR/Cas9.*Current Opinion in Biotechnology, 32*, 76–84.

Bibikova, M., Beumer, K., Trautman, J. K., & Carroll, D., (2003). Enhancing gene targeting with designed zinc finger nucleases. *Science, 300*, 764.

Bibikova, M., Caroll, D., Segal, D. J., Trautman, K., Smith, J., Kim, Y., & Chandrasegaran, S., (2001). Stimulation of homologous recombination through targeted cleavage by chimeric nucleases. *Mol. Cell Biol., 21*, 289–297.

Boch, J., Scholze, H., Schornack, S., Landgraf, A., Hahn, S., Kay, S., Lahaye, T., et al., (2009). Breaking the code of DNA binding specificity of TAL-type III effectors. *Science, 326*, 1509–1512.

Braatz, J., Harloff, H. J., Mascher, M., Stein, N., Himmelbach, A., & Jung, C., (2017). CRISPR-Cas9 targeted mutagenesis leads to simultaneous modification of different homoeologous gene copies in polyploid oilseed rape (*Brassica napus*). *Plant Physiol., 174*, 935–942.

Chames, P., Epinat, J. C., Guillier, S., Patin, A., Lacroix, E., & Paques, F., (2005). *In vivo* selection of engineered homing endonucleases using double-strand break-induced homologous recombination. *Nucleic Acids Research, 33*, 20.

Chen, K., Wang, Y., Zhang, R., Zhang, H., & Gao, C., (2019). CRISPR/Cas genome editing and precision plant breeding in agriculture. *Annu. Rev. Plant Biol., 70*, 667–697.

Chen, S. P., & Wang, H. H., (2019). An engineered Cas-transposon system for programmable and site-directed DNA transpositions. *CRISPR J., 2*, 376–394.

Christian, M., Cermak, T., Doyle, E. L., Schmidt, C., Zhang, F., Hummel, A., Bogdanove, A. J., & Voytas, D. F., (2010). Targeting DNA double-strand breaks with TAL effector nucleases. *Genetics, 186*, 757–761.

Cong, L., Ran, F. A., Cox, D., Lin, S., Baretto, R., Habib, N., Hasu, P. D., et al., (2013). Multiplex genome engineering using CRISPR/Cas systems. *Science, 339*, 819–823.

Deng, W., Shi, X., Robert, T., Lionnet, T., & Singer, R. H., (2015). CASFISH: CRISPR/Cas9-mediated in situ labeling of genomic loci in fixed cells. *PNAS, 112*, 11870–11875.

Doudna, J. A., & Charpentier, E., (2014). Genome editing. The new frontier of genome engineering with CRISPR-Cas9. *Science, 346*, 1258096.

Garneau, J. E., Dupuis, M., Villion, M., Romero, D. A., Patrick, B., Fremaux, C., Hovarth, P., et al., (2010). The CRISPR/Cas bacterial immune system cleaves bacteriophage and plasmid DNA. *Nature, 468*, 67–71.

Gaudelli, N. M., Komor, A. C., Rees, H. A., Packer, M. S., Badran, A. H., Bryson, D. I., & Liu, D. R., (2017). Programmable base editing of A•T to G•C in genomic DNA without DNA cleavage. *Nature, 551*, 464–471.

Goulet, B. E., Roda, F., & Hopkins, R., (2017). Hybridization in plants: Old ideas, new techniques. *Plant Physiology, 173*, 65–78.

Hajiahmadi, Z., Movahedi, A., Wei, H., Orooji, Y., Ruan, H., & Zhuge, Q., (2019). Strategies to increase on-target and reduce off-target effects of the CRISPR/Cas9 system in plants. *Int. J. Mol. Sci., 20*, 3719.

Hartung, F., & Schiemann, J., (2014). Precise plant breeding using new genome editing techniques: Opportunities, safety and regulation in the EU. *Plant J., 78*, 742–752.

Hockemeyer, D., Wang, H., Kiani, S., Lai, C. S., Gao, Q., Cassady, J. P., Cost, G. J., et al., (2011). Genetic engineering of human pluripotent cells using TALE nucleases. *Nat Biotechnol., 29*, 731–734.

Huang, H., Cui, T., Zhang, L., Yang, Q., Yang, Y., Xie, K., Fan, C., & Zhou, Y., (2020). Modifications of fatty acid profile through targeted mutation at BnaFAD2 gene with

CRISPR/Cas9-mediated gene editing in *Brassica napus*. *Theor. Appl. Genet., 133*, 2401–2411.

Jambhulkar, S. J., (2007). Mutagenesis: Generation and evaluation of induced mutations. In: Gupta, P. C., (ed.), *Advances in Botanical Research (Rapeseed Breeding)* (Vol. 45, pp. 418–428) Elsevier Publication Ltd. USA.

Jat, R. S., Singh, V. V., Sharma, P., & Rai, P. K., (2019). Oilseed brassica in India: Demand, supply, policy perspective and future potential. *OCL, 26*, 8.

Jeong, S. Y., Ahn, H., Ryu, J., Oh, Y., Sivanandhan, G., Won, K., Park, Y. A., et al., (2019). Generation of early-flowering Chinese cabbage (*Brassica rapa* spp. pekinensis) through CRISPR/Cas9-mediated genome editing. *Plant Biotechnol., 13*, 491–499.

Jinek, M., Chylinski, K., Fonfara, I., Hauer, M., Doudna, J. A., & Cherpntier, E., (2012). A programmable dual-RNA-guided endonuclease in adaptive bacterial immunity. *Science, 33*, 816–821.

Jung, H., Lee, A., Jo, S. H., Park, H. J., Jung, W. Y., Kim, H. S., Lee, H. J., et al., (2021). Nitrogen signaling genes and SOC1 determine the flowering time in a reciprocal negative feedback loop in Chinese cabbage (*Brassica rapa* L.) based on CRISPR/Cas9-mediated mutagenesis of multiple BrSOC1Homologs. *International Journal of Molecular Sciences, 22*, 4631.

Jurica, M. S., Monnat, R. J. Jr., & Stoddard, B. L., (1998). DNA recognition and cleavage by the LAGLIDADG homing endonuclease I-CreI. *Mol.Cell, 2*, 469–476.

Kim, Y. G., Cha, J., & Chandrasegaran, S., (1996). Hybrid restriction enzymes: Zinc finger fusions to FokI cleavage domain. *Proc. Natl Acad. Sci. USA, 93*, 1156–1160.

Klompe, S. E., Vo, P. L. H., Halpin-Healy, T. S., & Sternberg, S. H., (2019). Transposon-encoded CRISPR-Cas systems direct RNA-guided DNA integration. *Nature, 571*, 219–225.

Komor, A. C., Kim, Y. B., Packer, M. S., Zuris, J. A., & Liu, D. R., (2016). Programmable editing of a target base in genomic DNA without double-stranded DNA cleavage. *Nature, 533*, 420–424.

Lambert, A. R., Hallinan, J. P., Shen, B. W., Chik, J. K., Bolduc, J. M., Kulshina, N., Robins, L. I., et al., (2016). Indirect DNA sequence recognition and its impact on nuclease cleavage activity. *Structure, 2*, 862–873.

Lawrenson, T., Shorinola, O., Stacey, N., Li, C., Ostergaard, L., Patron, N., Uauy, C., & Harwood, W., (2015). Induction of targeted, heritable mutations in barley and *Brassica oleracea* using RNA-guided Cas9 nuclease. *Genome Biol., 16*, 258.

Li, X., Huang, L., Pan, L., Wang, B., & Pan, L., (2021). CRISPR/dCas9-mediated epigenetic modification reveals differential regulation of histone acetylation on *Aspergillus Niger* secondary metabolite. *Microbiological Research, 245*, 126694.

Limera, C., Sabbadini, S., Sweet, J. B., & Mezzetti, B., (2017). New biotechnological tools for the genetic improvement of major woody fruit species. *Front Pl. Sci., 8*, 1418.

Liu, D., Hu, R., Palla, K. J., Tuskan, G. A., & Yang, X., (2016). Advances and perspectives on the use of CRISPR/Cas9 systems in plant genomics research. *Current Opinion in Plant Biology, 30*, 70–77.

Lusser, M., Parisi, C., Plan, D., & Rodríguez-Cerezo, E., (2011). New plant breeding techniques. State of-the-art and prospects for commercial development. *JRC Scientific and Technical Reports.* EUR 24760 EN.

Lv, H., Fang, Z., Yang, L., Zhang, Y., & Wang, Y., (2020). An update on the arsenal: Mining resistance genes for disease management of brassica crops in the genomic era. *Horticulture Research, 7*, 34.

Ma, C., Liu, M., Li, Q., Si, J., Ren, X., & Song, H., (2019). Efficient BoPDS gene editing in cabbage by the CRISPR/Cas9 system. *Horticultural Plant Journal, 5*, 164–169.

Ma, C., Zhu, C., Zheng, M., Liu, M., Zhang, D., Liu, B., Li, Q., et al., (2019). CRISPR/Cas9-mediated multiple gene editing in *Brassica oleracea* var. capitata using the endogenous tRNA-processing system. *Horticulture Research, 6*, 20.

Ma, X., Zhu, Q., Chen, Y., & Liu, Y. G., (2016). CRISPR/Cas9 platforms for genome editing in plants: Developments and applications. *Molecular Plant, 9*, 961–974.

Mahood, E. H., Kruse, L. H., & Moghe, G. D., (2020). Machine learning: A powerful tool for gene function prediction in plants. *App. Plant Sci., 8*, e11376.

Makarova, K. S., Wolf, Y. I., Iranzo, J., Shmakov, S. A., Alkhnbashi, O. S., Brouns, S. J. J., Charpentier, E., et al., (2020). Evolutionary classification of CRISPR-Cas systems: A burst of class 2 and derived variants. *Nat. Rev. Microbiol., 18*, 67–83.

Miller, J. C., Tan, S., Qiao, G., Barlow, K. A., Wang, J., Xia, D. F., Meng, X., et al., (2011). A TALE nuclease architecture for efficient genome editing. *Nature Biotechnology, 29*, 2.

Moscou, M. J., & Bogdanove, A. J., (2009). A simple cipher governs DNA recognition by TAL effectors. *Science, 326*, 1501.

Murovec, J., Gucek, K., Bohanec, B., Avbelj, M., & Jerala, R., (2018). DNA-free genome editing of *Brassica oleracea* and B. rapa protoplasts using CRISPR-Cas9 Ribonucleoprotein complexes. *Front. Plant Sci., 9*, 1594.

Neequaye, M., Stavnstrup, S., Harwood, W., Lawrenson, W., Hundleby, P., Irwin, J., Troncoso-Rey, P., et al., (2021). CRISPR-Cas9-mediated gene editing of MYB28 genes impair glucoraphanin accumulation of *Brassica oleracea* in the field. *The CRISPR Journal, 4*, 3.

Paques, F., & Duchateau, P., (2007). Meganucleases and DNA double-strand break-induced recombination: Perspectives for gene therapy. *Current Gene. Therapy, 7*, 49–66.

Puchta, H., Dujon, A., & Hohn, B., (1993). Homologous recombination in plant cells is enhanced by *in vivo* induction of double strand breaks into DNA by a site-specific endonuclease. *Nucleic Acid Research, 21*, 22.

Sander, J. D., & Joung, J.K., (2014). CriSPr-Cas systems for editing, regulating and targeting genomes. *Nature Biotechnology, 32*, 4.

Strecker, J., Ladha, A., Garder, Z., Schmid-Burgk, J. L., Makarova, K. S., Koonin, E. V., & Zhang, F., (2019). RNA-guided DNA insertion with CRISPR-associated transposases. *Science, 365*, 48–53.

Sun, Q., Lin, L., Liu, D., Wu, D., Fang, Y., Wu, J., & Wang, Y., (2018). CRISPR/Cas9-mediated multiplex genome editing of the BnWRKY11 and BnWRKY70 genes in *Brassica napus* L. *Int. J. Mol. Sci., 19*, 2716.

Swaminathan, M. S., (2006). An evergreen revolution. *Crop Sci., 46*, 2293–2303.

Tsanova, T., Stefanova, L., Topalova, L., Atanasov, A., & Panctchev, I., (2020). DNA-free gene editing in plants: A brief overview. *Biotechnology and Biotechnological Equipments, 35*, 131–138.

Urnov, F. D., Rebar, E. J., Holmes, M. C., Zhang, H. S., & Gregory, P. D., (2010). Genome editing with engineered zinc finger nucleases. *Nat. Rev. Genet., 11*, 636–646.

Wada, N., Ueta, R., Osakabe, Y., & Osakabe, K., (2020). Precision genome editing in plants: State-of-the-art in CRISPR/Cas9-based genome engineering. *BMC Plant Biology, 20*, 234.

Wolt, J. D., Wang, K., & Yang, B., (2016). The regulatory status of genome-edited crops. *Plant Biotechnology Journal, 14*, 510–518.

Woo, J. W., Kim, J., Kwon, S. I., Corvalan, C., Cho, S. W., Kim, H., Kim, S., et al., (2015). DNA-free genome editing in plants with preassembled crisPr-cas9 ribonucleoproteins. *Nat Biotechnol., 33*, 11.

Xie, K. B., Minkenberg, B., & Yang, Y. N., (2015). Boosting CRISPR/Cas9 multiplex editing capability with the endogenous tRNA-processing system. *PNAS, 112*, 3570–3575.

Yang, H., Wu, J., Tang, T., Liu, K., & Dai, C., (2017). CRISPR/Cas9-mediated genome editing efficiently creates specific mutations at multiple loci using one sgRNA in *Brassica napus*. *Scientific Reports, 7*, 7489.

PART III

Potential Applications of CRISPR/Cas Systems for Climate Smart Crops

CHAPTER 10

CRISPR/Cas-Mediated Genome Editing for Abiotic Stress Tolerance

FARHANA HASSAN,[1] AMRINA SHAFI,[2] UMER MAJEED WANI,[1] and FIRDOUS A. KHANDAY[1*]

[1]*Department of Biotechnology, School of Biological Sciences, University of Kashmir, Hazratbal, Srinagar, Jammu and Kashmir – 190006, India*

[2]*Education Department, Government of J&K, Srinagar, Kashmir – 190002, Jammu and Kashmir, India*

[*]*Corresponding author. E-mail: khandayf@kashmiruniversity.ac.in*

ABSTRACT

Abiotic stresses, which include drought, salinity, light, temperature fluctuations, and others, have a huge impact on plants by adversely affecting their growth and development, thereby reducing overall plant productivity. Every year, abiotic stress alone causes huge damage to crop quality and productivity. The harmful effects of such factors are aggravated by climate change over the globe challenging food as well as nutritional security. Plants are equipped with an adequate defense system to respond against these negative environmental factors. However, routine exposure to one or the other harsh abiotic condition necessitates the installation of a system in plants that makes them resistant to such changes. Although the eco-friendly approach for such changes is achieved through conventional breeding of tolerant varieties, cumbersome screening and laborious crossing protocols restrict its utilization for overcoming abiotic stress and in obtaining the goal of sustainable food for the globe. Because of advancements in genomic engineering techniques, it is now possible to modify the genome at particular locations of plants to enhance adaptive capacity, increasing the yield and making plants more

CRISPR/Cas-Mediated Genome Editing in Plants, Deepu Pandita, Anu Pandita (Eds.)

resilient for stress factors. One such exciting genome editing approach is CRISPR/Cas-mediated editing, which is emerging as a game-changing technology and a flexible tool for modulating genomes in the desired way and in a manner that is specific to the target. In this chapter, the mechanistic approach of editing in plants via CRISPR/Cas technology is discussed. The main focus of the chapter is on CRISPR/Cas editing for enhancing plant tolerance for abiotic stress, which involves engineering plants for temperature (cold and heat), drought, and salinity tolerance. Further, the non-transgenic approach of the CRISPR/Cas system is also addressed in the chapter.

10.1 INTRODUCTION

Plants being sessile face multiple physical and chemical factors that lay a strong impact on overall plant productivity and is lethal for plant growth. Among many factors, drought, temperature, salinity, low or high light conditions, or heavy metals, these stresses have prominent consequences on growth and productivity of plants. These are collectively called abiotic stresses. Alteration of any of these stresses poses a great threat for plant survival. These conditions force plants to undergo changes which help them to adapt in harsh conditions to restore cellular homeostasis. Although abiotic stresses have been continuously shaping the evolution of plants, impaired molecular, biochemical, and physiological processes thrust serious consequences on growth, development, and overall plant productivity. The major challenge of the 21st century is environmental stress which presents unpredictable restrictions in meeting the demand for adequate food supply for the increasing population over the globe. These circumstances make it crucial to have better adaptive crop varieties which grow under varied environmental conditions without affecting the quality and quantity of the yield. To a significant extent, conventional breeding achieves the goal of higher production, but the loss of fitness and genetic variety limits its utilization. Also, its reliance on spontaneous allelic differences, as well as the fact that it takes too much of time, renders it ineffective for achieving sustainable crop production (Chen et al., 2019). Endonucleases like zinc finger nucleases (ZFNs) and transcription activator-like effector nucleases (TALENS) are used to expand the scope of molecular breeding. These nucleases are composed of programmable sequence-specific DNA-binding domains that are linked to the endonuclease FokI's nonspecific DNA cleaving domain. Despite its advantages in genome editing, ZFNs have a limited range of applications due to various intrinsic drawbacks, like it is costly and takes more time for optimizing ZF domain so that it binds DNA with strong affinity and it selects target sites with low frequency. It has high chances of off-target effects due to poor structure of the assembly and because the FokI cleavage domain binds nonspecifically. Also, the ZFNs can only be used after every 200 bps, these things limit its utilization. Similarly, the larger size of TALENs than ZFNs is a glaring drawback. A TALEN cDNA is about 3 kb in size, whereas a ZFN cDNA is normally about 1 kb in size. This makes delivering and expressing a pair of TALENs into cells more complicated than ZFNs. Further, some viral vectors have limited capacity to pack and distribute the TALENs due to their extremely repetitive structure (Holkers et al., 2013). Besides the difficulty in handling and laborious procedures, these methods can target only one single site at one time. Nodaway new techniques are discovered which overcome many restrictions associated with traditional breeding methods and new customized endonucleases in exposing the factors behind persistent loss in crop yield and in removing the roadblock in editing plants for preferable characters (Chen et al., 2019). One of such exciting advancement is the modulation of genome via CRISPR/Cas technology which is a promise in enhancing agricultural practices and is preferable to these procedures due to it being simple, target specific, manageable to multiplexing, and due to its strong efficiency and multidimensional applicability.

10.2 MECHANISTIC APPROACH OF CRISPR/CAS-MEDIATED GENOME EDITING IN PLANTS

The term CRISPR has been coined by Rudd Jansen and his colleagues to describe repeat sequences of approximately 21–37 bases in prokaryotes (Jansen et al., 2002). CRISPR is clustered regularly interspaced short palindromic repeats. CRISPR-Cas systems are complex in composition, architecture, and are broadly grouped into three types and all the three types use different mechanisms for recognizing and cleaving the nucleic acid. Type I involves the Cas3 enzyme, type II involves the Cas9 enzyme and Cas10 is used in type III system to recognize and cut the target sequences (Sharma et al., 2017). The system of CRISPR/Cas type II involves as follows: CRISPR RNA (crRNA), Cas9 endonuclease and trans-activating CRISPR RNA (tracrRNA) (Jinek et al., 2014; Biswas et al., 2021). The technology of CRISPR is originated out of the type II system, which is used as a powerful immune response for the plasmids and viruses which invade the bacteria (Doudna and Charpentier, 2014). Cas9 is taken to its target in the type II CRISPR/Cas system by two RNAs: the crRNA, which determines the Cas9 target, and the tracrRNA, which works as a scaffold attaching the crRNA to Cas9 to generate a ribonucleoprotein (RNP) complex/ternary complex. Trans-activating CRISPR-RNA pronounced as tracrRNA contains long chain of constant bases and giving rise to the structure resembling stem-loop which forms functional guide RNA (gRNA) by base pairing with crRNA. After hybridization, the two RNA components produce a mature gRNA that, depending on crRNA complementarity, programmably guides CRISPR nuclease to DNA regions. So, Cas9 uses tracrRNA to hold and the crRNA takes the complex towards the sequence targeted. In most cases of genome editing through CRISPR, tracrRNA, and crRNA are merged as one RNA sequence commonly called single guide RNA (sgRNA) which contains both the synthesized almost 20 bases long sequence to guide the Cas9 enzyme towards particular DNA sequence and second is scaffold sequence which is required to facilitate the binding of Cas9 enzyme. This specialized/synthesized sequence of crRNA also called gRNA, guides the RNP Complex towards the target DNA sequence by forming a complementary nucleotide pairing with its target sequence (Mehravar et al., 2019; Biswas et al., 2021). During this base pairing, DNA unwinds and forms a hybrid of DNA-RNA which leads to conformational change in Cas9, in turn the activation of HNH and Ruv C endonucleases (Jiang et al., 2016; Biswas et al., 2021). Cas9's positively charged grooves interact with the scaffold domain of gRNA inside the riboprotein complex, generating a conformational transition from a non-DNA-binding inactive conformation to a DNA-binding active conformation. Cas9 scans the whole genome for the signature of PAM by taking help from gRNA and once the homology match between the sequence to be targeted and the gRNA synthesized is achieved sufficiently, Cas9 binds to the target. After that, it goes through a second shift in conformation which leads to cleavage of opposing strands of target DNA by rightly oriented nuclease domains (Sharma et al., 2017) (Figure 10.1). Protospacer adjacent motif (PAM) sequences are 2–5 nucleotides and were reported as conserved 5'NGG3' motifs in Streptococcus thermophiles (Mojica et al., 2009). Cas9 recognizes the target as the sequence which is three base pair upstream of PAM and give blunt end cuts in both the strands of DNA (Biswas et al., 2021) via two different activities present at two different domains. The complementary DNA strand to the crRNA is nicked by the HNH nuclease domain, whereas the non-complementary DNA strand is nicked by the RuvC-like nuclease domain (Haurwitz et al., 2010; Majeed et al., 2018). As a result, there occurs insertion or deletion (indel) at the targeted location, and the same is repaired using several repair pathways depending on whether a repair template is available. When a homology sequence of DNA template (donor) is available, the damage gets repaired via HR (homologous recombination). In most cases the repair occurs via NHEJ wherein, the technique utilizes the error-prone tendency of cellular NHEJ DNA repair machinery (Zaidi et al., 2020).

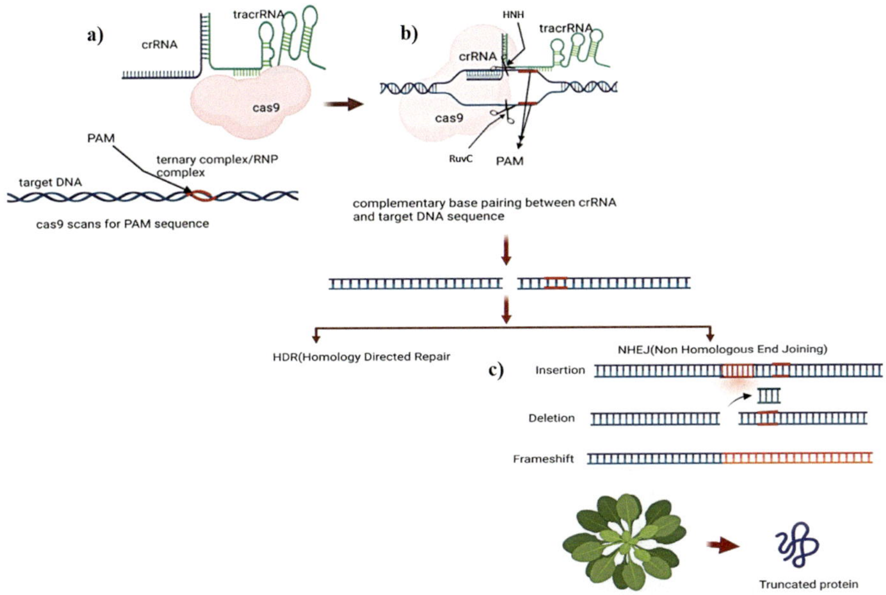

FIGURE 10.1 (a) Cas9 is guided to its target with the help of two RNAs; the crRNA which defines the target for Cas9 and the tracrRNA which acts as a scaffold linking crRNA to Cas9 to form a ribonucleoprotein (RNP) complex/ternary complex. (b) Guide RNA forms a complementary base pairing with its target. During this base pairing, DNA unwinds and Cas9 scans the genome for PAM sequence with the help of guide RNA. Once sufficient homology between gRNA and the target sequence is achieved Cas9 binds to the target and gives a double-stranded cut via RuvC and HNH. (c) NHEJ mechanism repairs the double-strand break without any homologous template and involves direct ligation of free ends. It produces random insertions or deletions (indels), which cause a frameshift in the coding sequence of a gene, disrupting its function.

Since abiotic stress is a complicated phenomenon controlled by numerous genes, also the presence of multigene families with functional redundancy for a large number of genes due to many duplication events in the genome of plants, it has become tough to find out the function of a gene by knocking out only a single gene since taking out only one gene does not always give the desired phenotype. CRISPR technology is an approach to overcome functional redundancy by allowing numerous genes to be targeted at the same time because of the ease of design and efficiency of sgRNA. The Cas9 endonuclease can be targeted to any of the individual DNA sequences having structure 5'N-(20)-NGG by merely changing the beginning 20-bp sequence in the gRNA. Cas9 sequences can be found in thousands of databases and a considerable number of orthologs for Cas9 are now available due to standard sequencing of microbial genome (Majeed et al., 2018). However, the PAM sequence necessary for such different Cas9 nucleases is limited. Nonetheless, multiple bioinformatic algorithms have speculated many sequences that are employed to recognize PAM for these novel Cas9 proteins close to the presumed proto-spacers in diverse viral genomes (Karvelis et al., 2015). There are several vectors available, such as pRGEB31, pRGE31, eSPCas9(1.1), lentiCRISPR v2 that contain all of the sequences required for manipulating genome, such as promoters, sites for restriction digestion, selection markers, multiple cloning site and the protein-coding sequence for Cas9 enzyme (Xie et al., 2013). These vectors have unique sites for restriction digestion which are used to clone gRNA for specific gene and subsequently utilizing the natural capacity of *agrobacterium* to transform the plants. Within plant cell, the Cas9 enzyme is instructed to cut the particular sequence and subsequently leads to the activation of NHEJ, the error prone repair pathway (Majeed et al., 2018). NHEJ mechanism repairs the double-strand break without any homologous template and involves direct ligation of free ends. It shifts the frame inside the coding region of a gene by producing random insertions or deletions (indels), disrupting its function.

10.3 CRISPR/CAS-MEDIATED EDITING OF GENOME FOR IMPROVING TOLERANCE TO ABIOTIC STRESS

According to the present scenario, it is thought that the yield of staple crops needs to be increased by >50% to achieve the United Nation's "zero hunger goal" up to the year 2030 (Wheeler and Braun, 2013). Increased environmental disturbance, such as extreme weather conditions, limits crop productivity, and increased urbanization and dwelling, which reduces the land available for agricultural purposes, adds to the urgent need for crop varieties that grow under various stress conditions to improve yield and satisfy the growing demand for food among the world's huge population. The CRISPR/Cas-mediated editing of genomes for different abiotic stresses in various plant species is as in subsections (Table 10.1).

TABLE 10.1 Abiotic Stress Tolerance Genes Targeted by the CRISPR/Cas System

Plant	Gene	Strategy	Improved Trait	References
Arabidopsis	Open stomata 2 (OST2)	*A. tumefaciens* mediated transformation produce site-directed modifications.	Drought tolerance	Osakabe et al. (2016)
Arabidopsis	(miRNA) miR169a	*Agrobacterium*-mediated transformation targeted deletion to create null mutations.	Resistance for drought	Zhao et al. (2016)
Arabidopsis	*Arabidopsis thaliana* vacuolar H^+-pyro phosphatase (AVP1)	*A. tumefaciens* mediated transformation to produce overexpression of the gene.	Resistance for drought	Park et al. (2017)
Arabidopsis	AREB1 (ABA-responsive element-binding protein-1)	*A. tumefaciens* mediated transformation to produce overexpression of the gene.	Resistance for drought	Paixão et al. (2019)
Maize	ARGOS8 (auxin-regulated gene involved in organ size 8)	Biolistic-mediated transformation/CRISPR-mediated DNA repair in untranslated regions of target genes to produce overexpression.	Drought tolerance	Shi et al. (2017)
Maize	WRKY40 (transcription factor with WRKY amino acid sequence at N terminus)	*A. tumefaciens*	Drought tolerance	Wang et al. (2018)
Maize	TMS5 (thermosensitive genic male sterile 5)	Mutational analysis	Heat tolerance	Li et al. (2017)
Tomato	BZR1 (brassinazole-resistant-1)	*A. tumefaciens*	Temperature tolerance	Yin et al. (2018)
Tomato	CBF1 (C-repeat binding factor)	*A. tumefaciens*	Cold tolerance	Li et al. (2018)
Tomato	HSA1(hybrid sterility-a1)	*A. tumefaciens* mediated deletion mutants	Heat tolerance	Qiu et al. (2018)
Tomato	SIAGL6 (siagamous-like 6)	*A. tumefaciens*-mediated editing	Heat tolerance	Klap et al. (2017)
Wheat	(Dehydration-responsive element-binding protein) DREB2, DREB3	Transformation using PEG	Resistance against drought	Kim et al. (2018)

TABLE 10.1 *(Continued)*

Plant	Gene	Strategy	Improved Trait	References
Rice	ANN3 (annexin coding gene)	CRISPR-mediated gene editing	Chill resistance	Shen et al. (2017)
Rice	Rice type-B response regulator (OsRR22)	*A. tumefaciens* mediated to generate knockdown plants.	Salt tolerance	Zhang et al. (2019)
Rice	RAV 2 (related to abI3 and vp1)	Mutational analysis	Resistance for salinity stress	Duan et al. (2016)
Rice	RR22 (rice type-B response regulator)	*Agrobacterium tumefaciens*	Salt resistance	Farhat et al. (2019)
Rice	OsMYB30 (a transcription factor)	*Agrobacterium*	Chill resistance	Zeng et al. (2020)
Rice	SAPK2 (Serine/threonine-protein kinase) SAPK2	*A. tumefaciens*	Resistance for drought	Lou et al. (2017)
Rice	DST (drought and salt tolerance)	*Agrobacterium*	Salt resistance	Kumar et al. (2020)

10.3.1 CRISPR/CAS SYSTEM IN TEMPERATURE TOLERANCE

Agriculture and food sustainability of globe are threatened by climate change. Just over an increase of 1°C in atmospheric temperature the wheat output is expected to decline by 6%, rice yield by 10%–20%, and maize yield by 21–31% (Zafar et al., 2020). Frequent temperature variations have a significant impact on agricultural plant adaptability and survival (Bita and Gerats, 2013). Heat stress of varying intensity and duration activates a variety of plant defense mechanisms, including the synthesis of enzymes that cause metabolites to accumulate, modulation in gene expression of scavenger proteins, accumulation of reactive oxygen species (ROS), heat shock proteins, and many other proteins (Fragkostefanakis et al., 2015). Genome editing through CRISPR technology is developed as a highly effective approach in manipulating important heat stress-related genes to produce heat-resistant crops (Biswal et al., 2019). Owing to the capability of tomato (*Solanum lycopersicum* L.) to undergo an efficient transformation and being an excellent model to study the modulation via system of CRISPR/Cas (Pan et al., 2016), different genes of the plants have been studied for heat stress tolerance. At around 20–32°C, tomato photosynthesizes maximum, but the further rise in temperature causes a decline in photosynthetic rate (Ruiz-Vera et al., 2015). Pollen viability has been depleted drastically (almost < 20%) when tomatoes were exposed to temperature of 43–45C (Frank et al., 2009). The function of HSA1 gene in tomato plant heat resistance was determined using CRISPR/Cas-mediated deletion mutants. Heat sensitivity was higher in heat-stress sensitive albino 1 (HSA1) deletion mutant varieties as compared to wild varieties (Qiu et al., 2018). Heat tolerance in tomatoes has also been accomplished by genome manipulation of SIAGAMOUS-LIKE 6 (SIAGL6), as a result of which fruit setting got enhanced in hot conditions (Klap et al., 2017). This also led to the production of fruits which were parthenocarpic (Klap et al., 2017), showing that Slagl6 also could be a competitive gene for exploring facultative parthenocarpy. Similarly, the role of BZR1 gene in thermotolerance was confirmed using CRISPR bzr1 mutant and overexpression lines in tomato (Yin et al., 2018). Mutant for bzr1 line had reduced H_2O_2 generation in the apoplast, as well as reduced heat resistance and decrease in the induction of Rboh1 gene in PM. Whereas increased expression increased H_2O_2 production and restored temperature tolerance. Suggesting that the gene could be involved in Rboh1 induction and Rboh-mediated ROS generation, which could be important in heat tolerance. Through mutational analysis, CRISPR/Cas was used to investigate the role of TMS5 gene in maize plants, resulting in the generation of seedlings which were thermosensitive (Li et al., 2017). Similarly, in rice, the PYL1/4/6 genes for abscisic acid (ABA) receptor were knockout using CRISPR technology, and mutants were found to be resistant to high temperature (Miao et al., 2018).

10.3.2 CRISPR/CAS SYSTEM IN COLD TOLERANCE

Tomato plants are susceptible to chilling, making their fruits vulnerable to cold stress. CRISPR/Cas-mediated C-repeat binding factor mutant generation 1 tomato gene was discovered to protect against chilling injury and aid in the prevention of electrolyte leakage (Li et al., 2018). The chilling tolerance for tomato plants was later discovered to be attributed to greater amounts of IAA and H_2O_2 (Li et al., 2018). Another gene coding for annexin which is a Ca^{2+} dependent phospholipid binding protein, was modified in one of the rice cultivars, which showed that the gene has a potent role in conferring cold tolerance to the plant (Shen et al., 2017). SAPK2 gene was found to attribute cold resistance to rice using CRISPR/Cas9 editing (Lou et al., 2017). CRISPR/Cas gene editing was found to have improved the physiological changes in mutant Osmyb30 with the efficiency of 63%, this gene was found to be highly relevant to cold stress tolerance (Zeng et al., 2020).

10.3.3 CISPR/CAS SYSTEM IN DROUGHT TOLERANCE

Drought is an unavoidable stress factor which threatens the process of growth and physiology in plants and puts their survival under danger (Xu et al., 2014). Plants that have undergone CRISPR/Cas editing have improved their quantitative and qualitative features, allowing them to achieve the needs of sustainable food for the globe. Wheat drought tolerance has been increased by CRISPR/Cas editing of ethylene-responsive factor 3 and dehydration responsive element binding protein 2 (Kim et al., 2018). Using the CRISPR/Cas technology, the rice genome was altered for varied abiotic stress tolerance (AT). Several members of the OsERF family (OsBIERF1, OsBIERF3, and OsBIERF4) were altered, resulting in enhanced development under a variety of abiotic stressors (Wang et al., 2016). Indica mega rice cultivar MTU1010 mutants of OsDST the gene involved in drought and salt tolerance had leaves which were broader but having lesser number of stomata, and the cultivar performed better under drought stress because its leaf water retention characteristic was improved (Ganie et al., 2021; Kumar et al., 2020).

SAPK2 is an important member in abscisic acid signaling. The gene was imparting tolerance against drought in rice mutants, where the mutant form of this gene brought about the plants that were more susceptible to oxidative stress and drought (Lou et al., 2017). Using the technology of CRISPR, the ZmWRKY40 gene was analyzed to be responding against abiotic stress and resultant CRISPR lines were shown to impart drought tolerance in maize (Wang et al., 2018). The ARGOS8 gene in maize has a role in the size of organs and is regulated by auxin. Its promoter sequence was replaced by GOS2 promoter via CRISPR/Cas editing. CRISPR developed lines had higher transcript levels of ARGOS8 and the production of grains under drought also got increased (Shi et al., 2017). In tomato, Slnpr 1 gene mutant developed by CRISPR/Cas9 system showed improved resistance against drought (Li et al., 2019). The technology of CRISPR was utilized to analyze the mechanism of regulation of the mitogen-activated protein kinases-3 (SIMAPK3) gene during drought stress. The gene was found to impart drought resistance and knockout lines displayed significant leaf wilting and bent stems, as well as oxidative stress and severe membrane damage. The gene was discovered to have certain significant motifs containing threonine-glutamic acid and tyrosine (TEY), which indicated to have an active part in response against drought (Wang et al., 2017). Since ABA is an important player in inhibiting the activity of proton ATPases in the plasma membrane and is responsible for closing stomata during drought conditions. When plasma membrane H^+-ATPase1 (AHA1) is active constitutively, it prevents ABA-mediated stomatal closure (Hyun, 2020). Suggesting that by downregulated H^+-ATPase of the plasma membrane, ABA-mediated stomatal closure can be induced. So, it becomes an attractive focus to develop drought resistance in crops having commercial importance. Using CRISPR technology OPEN STOMATA 2 gene (earlier represented as AHA1) has been edited using tissue-specific Cas9 cassette (Osakabe et al., 2016). Using this cassette, a 1 bp insertion at the target site of the gene has caused a frameshift which led to the introduction of a stop codon and a dominant-negative OST2 mutation enhanced the ABA-mediated stomatal closure, which resulted in lesser water loss via transpiration when measured against wild type (Osakabe et al., 2016). Nowadays, micro RNAs are also considered potent targets for modulation to enhance tolerance in crops for various abiotic stresses. The expression of several genes gets controlled by small non-coding RNA molecules via cleaving mRNA, modification after translation, DNA methylation or chromosomal remodeling (Shriram et al., 2016). Plants with a distinct miRNA expression pattern were shown to be more resistant or sensitive to certain abiotic stressors (Zhang, 2015). MiR169a gene was discovered to have a negative role in drought resistance and it was found to be mediating its role via an ABA-dependent pathway (Ni et al., 2013). The generated knock out line for the gene of this miRNA showed drought tolerance as against the wild type plants (Zhao et al., 2016). Results suggest CRISPR/Cas-based knockout of miRNA is a reliable strategy for improving abiotic stress. Among miRNA miR393 overexpression lines of rice were susceptible to drought (Xia et al., 2012) and in Arabidopsis, miR408 was found to be a negative player in drought resistance process (Ma et al., 2015). Suggesting that miRNAs may be considered a good choice to improve tolerance in plants against drought. The expression of many genes has been modulated by CRISPR/Cas through DNA repair in untranslated regions. By combining the trans-activating subunit of NFKappa B and activation domain of heat shock factor-1 with the activation domain of Herpes Simplex Virus VP64 using dCas9 system, the transcriptional level of vacuolar H-pyrophosphatase (AVP1) in Arabidopsis has been enhanced. The plants with higher AVP1 expression were found to have increased drought tolerance (Park et al., 2017). The gene expression of ABA-responsive element-binding protein-1 (Areb1) has also been enhanced using CRISPR/dCas 9 fusion with histone acetyltransferase. Drought resistance and chlorophyll content were improved in plants using the construct (Paixão et al., 2019). CRISPR/Cas gene mutants of the mitogen-activated protein kinase 3 (slmapk3) gene increased drought resistance in *Solanum lycopersicum* (Wang et al., 2017).

10.3.4 CISPR/CAS SYSTEM IN SALT TOLERANCE

Salinization alone is spreading annually at a pace of 10%, with over almost 50% of all arable land expected to be salinized by 2050 (Jamil et al., 2011). So, salt condition is a danger, particularly for coastal or semi-arid regions (Flowers, 2004; Bhat et al., 2021). Among most of the stresses, salinity is considered one of the serious constraints which affects germination and threatens plant survival. The salt stress is controlled by multiple physiological and molecular mechanisms which include tolerance at the osmotic level, ionic tolerance, and tolerance at the tissue level (Hyun, 2020). Plants try to minimize osmotic stress so they accumulate many organic solutes like proline, glycine betaine and some carbohydrates. All these physiological and biochemical activities are results of changes in gene expression and activation of many signaling pathways (Bhat et al., 2021). At cell signaling level plants respond by the generation of ROS, fluctuating Ca^{2+} levels, activating different kinases like Ca^{2+}/calmodulin-dependent kinases, etc., upregulation of many signaling pathways including SOS pathway (salt overly sensitive), synthesis of stress hormone, etc., (Zeng et al., 2015). At the molecular level, many genes have been found controlling the plant survival in salt stress conditions, e.g., OsRav2 gene imparts salt tolerance via regulation of GT-1 element was verified by CRISPR/Cas-based targeted mutation (Duan et al., 2016). SnRK2 and SAPK genes (Osmotic stress/ABA-activated protein kinases) were showing resistance against salinity stress in CRISPR-edited mutant lines of rice (Lou et al., 2017). Another rice gene discovered to be involved in response against salinity was OsRR22 that encode type-B response regulator. The gene-conferred salinity tolerance was confirmed by CRISPR/Cas-mediated knockdown by Zhang et al. (2019).

10.4 CRISPR/CAS AS NON-TRANSGENIC APPROACH

GM crops a term of common men comprehend, is used for the crops which are genetically modified and these initially came into market in 1996. Their consumption resulted in the decreased usage of pesticides on crops by almost 9% which is equal to 448 million kilograms, and the quotient of environmental impact has decreased by 17.9% which is a metric that quantifies the impact of pesticides and herbicides on the environment (Chen and Lin, 2013). The overall savings on carbon dioxide emissions linked with the approval of GM crop in 2010 were comparable to the removal of 8.6 million automobiles from the roads due to lower fuel consumption and greater absorption of carbon by soil (Chen and Lin, 2013). Up to very long period, the whole sector of GM was mostly focused on basic features, like, herbicide tolerance (HT) and insecticide resistance, that exclusively helped farmers in their profit. Aside from that, genetic engineering (GE) techniques in fruit crops have a wide range of uses, as they enable the development of critical agronomic traits such as quality of fruit and resistance against abiotic and biotic stress. Several fruit crops have been transformed using these strategies for the last two decades. Several future generation GM crops give profit to the end-users by supplying extra healthful supplements such as omega-3 fatty acids, many forms of vitamin A, and unsaturated fatty acids like oleic acid. As against traditional breeding, GE/recombinant DNA technology (RDT) permits for the insertion of beneficial genes of any microorganism or any useful plant in crops bearing the fruits, thereby, providing new genotypes with improved fruit quality and shelf life. This technology has been named the fastest-developing agricultural technology (Parmar et al., 2017) and organisms obtained via the technology are termed as "genetically modified" (GM). Flavr Savr tomato got approval for production commercially in 1994 by Food and Drug Administration (FDA) in America. Modification it possessed slow down its process of ripening and did not allow it to soften after harvesting. The GM papaya that has been approved for commercialization is resistant to the attacks of ringspot virus and has higher productivity. Similarly, genetically modified papaya accounts for 80% of Hawaiian papaya produced today (Bawa and Anilakumar, 2013). However, the regulatory approval processes have significant effect on progress of GM crops since the goal of setting such regulatory system is to avoid unnecessary damage to environment and health of humans and to avoid economic losses (Gould et al., 2017). These rules also serve to reassure consumers about the biosafety of genetically modified crops (Stirling et al., 2015). The consequences of this are that the expenses of acquiring permission for new genetically modified organisms (GMOs) are extremely high, and regulatory constraints also cause product marketing delays (Falck-Zepeda et al., 2012). These strict laws create unneeded obstacles for the new genetically modified crops being introduced.

The plant research and breeding fields were soon swept by editing of genome with Engineered Endonuclease Systems. The CRISPR technology is one such customized endonuclease system which offers a better option and solution to the difficulties associated with GM crops (Wang et al., 2019). Over 50 different plants have been utilized to test this genome editing technology, including model plants like Arabidopsis, major crops (Figure 10.2) like rice, wheat, maize, and crops which are not too important economically like cucumber, peanut, strawberry. In most of its uses, this approach has been employed for fundamental research as a trial to examine the function of genes. However, several market-oriented qualities, such as modified agronomic properties, enhanced quality of food, biotic, and abiotic resistance, or whether it be resistance against herbicides, all have been addressed. Genome editing via CRISPR technology is rapidly expanding and progressing at a very fast rate. Today CRISPR systems are obtained from a variety of prokaryotic organisms and also the existing system has been modified with several beneficial changes: including (a) changing PAM sequence, (b) increasing accuracy, (c) changing capability of cutting the DNA, (d) fusing Cas proteins with some other proteins. In addition to traditional ways of transformation where *Agrobacterium* or Rhizobium species were utilized, alternative transforming procedures are now available, and various new approaches are being tested in the sector of plant research. These methods rely exclusively on RNA-protein complexes or on proteins alone for transformation, allowing for genome manipulation without the introduction of recombinant DNA (Metje-Sprink et al., 2019). As a result, it is impossible for foreign DNA to be integrated inside the genome of the host cell and the resulting products are always transgene-free and labeled as "non-GM" products.

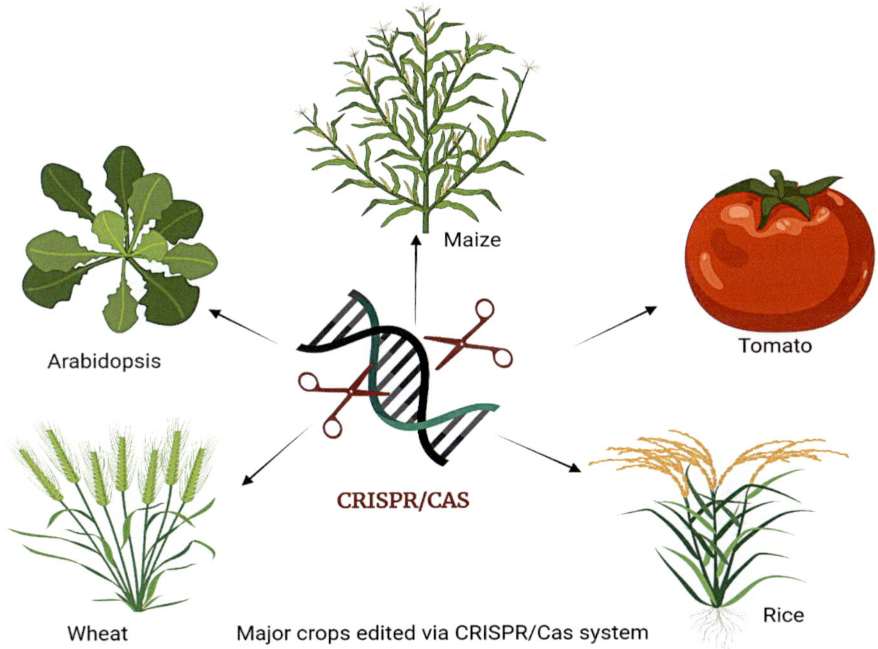

FIGURE 10.2 Representative image showing CRISPR involvement in editing major crops.

Nowadays, multiple vector methods are available for synthesizing the single guide RNA (sgRNA) in CRISPR-mediated editing for easier cloning and handling, making it cost-effective and allowing the generation of transgene-free plants with desired gene editing. This rapid improvement in genome editing technology allows for more precise genetic manipulation, reducing the risk of unintended consequences and increasing reliability. This technology contributes to promoting and facilitating the use of products in terms of public acceptance due to its correctness and "foreign DNA-free engineering" approach. A CRISPR-edited mushroom eluded US regulation in 2016 because it did not contain foreign DNA and hence did not fall under the GM organisms' statute (Kim and Kim, 2016). The FDA then allowed the marketing of a drought-tolerant soybean and enhanced oil content containing false flax seeds (Waltz, 2018), showing that crops edited via CRISPR do not come under such onerous rules which are faced by GM crops. This gives a hope that technology of CRISPR will undoubtedly be used in the future (Zaidi et al., 2018).

10.5 CONCLUSION

Abiotic stressors have major repercussions on the plant as a whole due to the challenges offered by climate change. With major concerns regarding yield decline, it frequently fails to provide sustainable food for the world's growing population. In these conditions, it seems critical to implement directional policies aimed at achieving sustainability in agriculture in order to meet the expanding population's need for food. Abiotic stress is a complex property regulated by several genes, with significant interactions in between many regulatory components, metabolic pathways and signaling molecules, leading to the abiotic stress response. Since last decade the modifications in basic CRISPR technology have made it a flexible and accurate gene editing tool for varied number of species which involve plants as well. Using this technology, many genes can be targeted at the same time since the designing of gRNA is very easy and it is highly accurate in specifying the target, thus the technology has the potential to solve the complicacy of functional redundancy in plants. The Cas9 protein acts as the scissors in this game, while an RNA scrap serves as the address mark, ensuring that the cut is made in the correct location. This genome editing tool allows one to introduce targeted mutations, insertion/deletion (indel), and exact sequence change in diverse organisms using customized nucleases. The use of CRISPR/Cas9 and related technologies in plant engineering would not only speed up basic research but is also becoming a widely established tool for next-generation breeding programs since it makes molecular breeding of crop plants easier. These technological discoveries have surely paved the way for the rapid increase of key and complex traits in crop plants, which involve particularly the salt tolerance and drought resistance, which is critical under situations of frequent changing climatic conditions and in the era of diminishing of suitable resources and lands for agriculture. The CRISPR-Cas9 system's rapid progress and developing applications offer a significant contribution to exploring the regulatory networks of genes driving response and acclimatization against abiotic stress to generate plants resistant to such stresses. The development of this technique with non-targeted safety assessment approaches has helped to lessen the potential risk of GM crops and to change the perception of the public regarding non-GM products.

KEYWORDS

- **abiotic stress**
- **CRISPR/Cas**
- **crop improvement**
- **drought**
- **non-transgenic**
- **salinity**
- **stress tolerance**
- **temperature stress**

REFERENCES

Bawa, A. S., & Anilakumar, K. R., (2013). Genetically modified foods: Safety, risks and public concerns—A review. *Food Sci. Technol., 50*(6), 1035–1046.

Bhat, M. A., Mir, R. A., Kumar, V., Shah, A. A., Zargar, S. M., Rahman, S., & Jan, A. T., (2021). Mechanistic insights of CRISPR/Cas-mediated genome editing towards enhancing abiotic stress tolerance in plants. *Physiol. Plant.*

Biswal, A. K., Mangrauthia, S. K., Reddy, M. R., & Yugandhar, P., (2019). CRISPR mediated genome engineering to develop climate-smart rice: Challenges and opportunities. *Semin. Cell Dev. Biol., 96*, 100–106. Academic Press.

Biswas, D., Saha, S. C., & Dey, A., (2021). CRISPR-Cas genome-editing tool in plant abiotic stress tolerance. *Plant Gene.,* 100286.

Bita, C., & Gerats, T., (2013). Plant tolerance to high temperature in a changing environment: Scientific fundamentals and production of heat stress-tolerant crops. *Front. Plant Sci., 4*, 273.

Chen, H., & Lin, Y., (2013). Promise and issues of genetically modified crops. *Curr. Opin. Plant Biol., 16*(2), 255–260.

Chen, K., Wang, Y., Zhang, R., Zhang, H., & Gao, C., (2019). CRISPR/Cas genome editing and precision plant breeding in agriculture. *Annu. Rev. Plant Biol., 70*, 667–697.

Doudna, J. A., & Charpentier, E., (2014). The new frontier of genome engineering with CRISPR-Cas9. *Science, 346*(6213).

Duan, Y. B., Li, J., Qin, R. Y., Xu, R. F., Li, H., Yang, Y. C., & Yang, J. B., (2016). Identification of a regulatory element responsible for salt induction of rice OsRAV2 through ex situ and in situ promoter analysis. *Plant Mol. Biol., 90*(1, 2), 49–62.

Falck-Zepeda, J., Yorobe, Jr. J., Husin, B. A., Manalo, A., Lokollo, E., Ramon, G., & Sutrisno, (2012). Estimates and implications of the costs of compliance with biosafety regulations in developing countries.*GM Crops Food, 3*(1), 52–59.

Farhat, S., Jain, N., Singh, N., Sreevathsa, R., Dash, P. K., Rai, R., & Rai, V., (2019). CRISPR-Cas9 directed genome engineering for enhancing salt stress tolerance in rice. *Semin. Cell Dev. Biol., 96*, 91–99. Academic Press.

Flowers, T. J., (2004). Improving crop salt tolerance. *J. Exp. Bot., 55*(396), 307–319.

Fragkostefanakis, S., Simm, S., Paul, P., Bublak, D., SCHARF, K. D., & Schleiff, E., (2015). Chaperone network composition in *Solanum lycopersicum* explored by transcriptome profiling and microarray meta-analysis. *Plant Cell Environ., 38*(4), 693–709.

Frank, G., Pressman, E., Ophir, R., Althan, L., Shaked, R., Freedman, M., & Firon, N., (2009). Transcriptional profiling of maturing tomato (*Solanum lycopersicum* L.) microspores reveals the involvement of heat shock proteins, ROS scavengers, hormones, and sugars in the heat stress response. *J. Exp. Bot., 60*(13), 3891–3908.

Ganie, S. A., Wani, S. H., Henry, R., & Hensel, G., (2021). Improving rice salt tolerance by precision breeding in a new era. *Curr. Opin. Plant Biol., 60*, 101996.

Gould, F., Amasino, R. M., Brossard, D., Buell, C. R., Dixon, R. A., Falck-Zepeda, J. B., & Whitaker, R. J., (2017). Elevating the conversation about GE crops. *Nat. Biotechnol., 35*(4), 302–304.

Haurwitz, R. E., Jinek, M., Wiedenheft, B., Zhou, K., & Doudna, J. A., (2010). Sequence-and structure-specific RNA processing by a CRISPR endonuclease. *Science, 329*(5997), 1355–1358.

Holkers, M., Maggio, I., Liu, J., Janssen, J. M., Miselli, F., Mussolino, C., & Goncalves, M. A., (2013). Differential integrity of TALE nuclease genes following adenoviral and lentiviral vector gene transfer into human cells. *Nucleic Acids Res., 41*(5), e63.

Hyun, T. K., (2020). CRISPR/Cas-based genome editing to improve abiotic stress tolerance in plants. *Bot. Serbica., 44*(2), 121–127.

Jamil, A., Riaz, S., Ashraf, M., & Foolad, M. R., (2011). Gene expression profiling of plants under salt stress. *CRC Crit Rev Plant Sci., 30*(5), 435–458.

Jansen, R., Embden, J. D. V., Gaastra, W., & Schouls, L. M., (2002). Identification of genes that are associated with DNA repeats in prokaryotes. *Mol. Microbiol., 43*(6), 1565–1575.

Jiang, F., Taylor, D. W., Chen, J. S., Kornfeld, J. E., Zhou, K., Thompson, A. J., & Doudna, J. A., (2016). Structures of a CRISPR-Cas9 R-loop complex primed for DNA cleavage. *Science, 351*(6275), 867–871.

Jinek, M., Jiang, F., Taylor, D. W., Sternberg, S. H., Kaya, E., Ma, E., & Doudna, J. A., (2014). Structures of Cas9 endonucleases reveal RNA-mediated conformational activation. *Science, 343*(6176).

Julkowska, M. M., & Testerink, C., (2015). Tuning plant signaling and growth to survive salt. *Trends Plant Sci., 20*(9), 586–594.

Karvelis, T., Gasiunas, G., Young, J., Bigelyte, G., Silanskas, A., Cigan, M., & Siksnys, V., (2015). Rapid characterization of CRISPR-Cas9 protospacer adjacent motif sequence elements. *Genome Biol., 16*(1), 1–13.

Khan, I., Khan, S., Zhang, Y., Zhou, J., Akhoundian, M., & Jan, S. A., (2021). CRISPR-Cas technology based genome editing for modification of salinity stress tolerance responses in rice (*Oryza sativa* L.). *Mol. Biol. Rep.,* 1–11.

Kim, D., Alptekin, B., & Budak, H., (2018). CRISPR/Cas9 genome editing in wheat. *Funct. Integr. Genomics, 18*(1), 31–41.

Kim, J., & Kim, J. S., (2016). Bypassing GMO regulations with CRISPR gene editing. *Nat. Biotechnol., 34*(10), 1014–1015.

Klap, C., Yeshayahou, E., Bolger, A. M., Arazi, T., Gupta, S. K., Shabtai, S., & Barg, R., (2017). Tomato facultative parthenocarpy results from SlAGAMOUS-LIKE 6 loss of function. *Plant Biotechnol. J., 15*(5), 634–647.

Kumar, V. S., Verma, R. K., Yadav, S. K., Yadav, P., Watts, A., Rao, M. V., & Chinnusamy, V., (2020). CRISPR-Cas9 mediated genome editing of drought and salt tolerance (OsDST) gene in indica mega rice cultivar MTU1010. *Physiol. Mol. Biol. Plants, 26*(6), 1099.

Li, J., Zhang, H., Si, X., Tian, Y., Chen, K., Liu, J., & Gao, C., (2017). Generation of thermosensitive male-sterile maize by targeted knockout of the ZmTMS5 gene. *J. Genet. Genomics, 44*(9), 465–468.

Li, R., Liu, C., Zhao, R., Wang, L., Chen, L., Yu, W., & Shen, L., (2019). CRISPR/Cas9-mediated SlNPR1 mutagenesis reduces tomato plant drought tolerance. *BMC Plant Biol., 19*(1), 1–13.

Li, T., Yang, X., Yu, Y., Si, X., Zhai, X., Zhang, H., & Xu, C., (2018). Domestication of wild tomato is accelerated by genome editing. *Nat. Biotechnol., 36*(12), 1160–1163.

Lou, D., Wang, H., Liang, G., & Yu, D., (2017). OsSAPK2 confers abscisic acid sensitivity and tolerance to drought stress in rice. *Front. Plant Sci., 8*, 993.

Ma, C., Burd, S., & Lers, A., (2015). mi R 408 is involved in abiotic stress responses in *Arabidopsis*. *Plant J., 84*(1), 169–187.

Majeed, U., Yaqoob, U., Qazi, H. A., Ahmad, S., & John, R., (2018). CRISPR/Cas system as an emerging technology to enhance plant viral immunity. *Physiol. Mol. Plant Path., 103*, 107–113.

Mehravar, M., Shirazi, A., Mehrazar, M. M., & Nazari, M., (2019). *In vitro* pre-validation of gene editing by CRISPR/Cas9 ribonucleoprotein. *Avicenna J. Med. Biotechnol., 11*(3), 259.

Metje-Sprink, J., Menz, J., Modrzejewski, D., & Sprink, T., (2019). DNA-free genome editing: Past, present and future. *Front. Plant Sci., 9*, 1957.

Miao, C., Xiao, L., Hua, K., Zou, C., Zhao, Y., Bressan, R. A., & Zhu, J. K., (2018). Mutations in a subfamily of abscisic acid receptor genes promote rice growth and productivity. *PNAS, 115*(23), 6058–6063.

Mojica, F. J., Díez-Villaseñor, C., García-Martínez, J., & Almendros, C., (2009). Short motif sequences determine the targets of the prokaryotic CRISPR defense system. *Microbiology., 155*(3), 733–740.

Ni, Z., Hu, Z., Jiang, Q., & Zhang, H., (2013). GmNFYA3, a target gene of miR169, is a positive regulator of plant tolerance to drought stress. *Plant Mol. Biol., 82*(1, 2), 113–129.

Osakabe, Y., Watanabe, T., Sugano, S. S., Ueta, R., Ishihara, R., Shinozaki, K., & Osakabe, K., (2016). Optimization of CRISPR/Cas9 genome editing to modify abiotic stress responses in plants. *Sci. Rep., 6*(1), 1–10.

Paixão, J. F. R., Gillet, F. X., Ribeiro, T. P., Bournaud, C., Lourenço-Tessutti, I. T., Noriega, D. D., & Grossi-de-Sa, M. F., (2019). Improved drought stress tolerance in *Arabidopsis* by CRISPR/dCas9 fusion with a histone acetyltransferase. *Sci. Rep., 9*(1), 1–9.

Pan, C., Ye, L., Qin, L., Liu, X., He, Y., Wang, J., & Lu, G., (2016). CRISPR/Cas9-mediated efficient and heritable targeted mutagenesis in tomato plants in the first and later generations. *Sci. Rep., 6*(1), 1–9.

Park, J. J., Dempewolf, E., Zhang, W., & Wang, Z. Y., (2017). RNA-guided transcriptional activation via CRISPR/dCas9 mimics overexpression phenotypes in *Arabidopsis*. *PLoS One, 12*(6), e0179410.

Parmar, N., Singh, K. H., Sharma, D., Singh, L., Kumar, P., Nanjundan, J., & Thakur, A. K., (2017). Genetic engineering strategies for biotic and abiotic stress tolerance and quality enhancement in horticultural crops: A comprehensive review. *3 Biotech, 7*(4), 1–35.

Qiu, Z., Kang, S., He, L., Zhao, J., Zhang, S., Hu, J., & Zhu, L., (2018). The newly identified heat-stress sensitive albino 1 gene affects chloroplast development in rice. *Plant Sci., 267*, 168–179.

Ruiz-Vera, U. M., Siebers, M. H., Drag, D. W., Ort, D. R., & Bernacchi, C. J., (2015). Canopy warming caused photosynthetic acclimation and reduced seed yield in maize grown at ambient and elevated [CO_2]. *Glob. Chang Biol., 21*(11), 4237–4249.

Sharma, S., Kaur, R., & Singh, A., (2017). Recent advances in CRISPR/Cas mediated genome editing for crop improvement. *Plant Biotechnol. Rep., 11*(4), 193–207.

Shen, C., Que, Z., Xia, Y., Tang, N., Li, D., He, R., & Cao, M., (2017). Knock out of the annexin gene OsAnn3 via CRISPR/Cas9-mediated genome editing decreased cold tolerance in rice. *J. Plant Biol., 60*(6), 539–547.

Shi, J., Gao, H., Wang, H., Lafitte, H. R., Archibald, R. L., Yang, M., & Habben, J. E., (2017). ARGOS 8 variants generated by CRISPR-Cas9 improve maize grain yield under field drought stress conditions. *Plant Biotechnol. J., 15*(2), 207–216.

Shriram, V., Kumar, V., Devarumath, R. M., Khare, T. S., & Wani, S. H., (2016). MicroRNAs as potential targets for abiotic stress tolerance in plants. *Front. Plant Sci., 7*, 817.

Shrivastava, P., & Kumar, R., (2015). Soil salinity: A serious environmental issue and plant growth promoting bacteria as one of the tools for its alleviation. *Saudi J. Biol. Sci., 22*(2), 123–131.

Stirling, A., Glover, D., & Millstone, E., (2015). Regulating Genetic Engineering: The Limits and Politics of Knowledge. *Issues Sci Technol., 31*, 4.

Waltz, E., (2018). With a free pass, CRISPR-edited plants reach market in record time. *Nat. Biotechnol., 36*(1), 6–8.

Wang, C. T., Ru, J. N., Liu, Y. W., Yang, J. F., Li, M., Xu, Z. S., & Fu, J. D., (2018). The maize WRKY transcription factor ZmWRKY40 confers drought resistance in transgenic *Arabidopsis*. *Int. J. Mol. Sci., 19*(9), 2580.

Wang, F., Wang, C., Liu, P., Lei, C., Hao, W., Gao, Y., & Zhao, K., (2016). Enhanced rice blast resistance by CRISPR/Cas9-targeted mutagenesis of the ERF transcription factor gene OsERF922. *PLoS One, 11*(4), e0154027.

Wang, L., Chen, L., Li, R., Zhao, R., Yang, M., Sheng, J., & Shen, L., (2017). Reduced drought tolerance by CRISPR/Cas9-mediated SlMAPK3 mutagenesis in tomato plants. *J. Agric. Food Chem., 65*(39), 8674–8682.

Wang, T., Zhang, H., & Zhu, H., (2019). CRISPR technology is revolutionizing the improvement of tomato and other fruit crops. *Hortic. Res., 6*(1), 1–13.

Wheeler, T., & Von, B. J., (2013). Climate change impacts on global food security. *Science., 341*(6145), 508–513.

Xia, K., Wang, R., Ou, X., Fang, Z., Tian, C., Duan, J., & Zhang, M., (2012). OsTIR1 and OsAFB2 downregulation via OsmiR393 overexpression leads to more tillers, early flowering and less tolerance to salt and drought in rice. *PLoS One, 7*(1), e30039.

Xie, K., & Yang, Y., (2013). RNA-guided genome editing in plants using a CRISPR–Cas system. *Mol Plant., 6*(6), 1975–1983.

Xu, J., Yuan, Y., Xu, Y., Zhang, G., Guo, X., Wu, F., & Lu, Y., (2014). Identification of candidate genes for drought tolerance by whole-genome resequencing in maize. *BMC Plant Biol., 14*(1), 1–15.

Yin, Y., Qin, K., Song, X., Zhang, Q., Zhou, Y., Xia, X., & Yu, J., (2018). BZR1 transcription factor regulates heat stress tolerance through FERONIA receptor-like kinase-mediated reactive oxygen species signaling in tomato. *Plant Cell Physiol., 59*(11), 2239–2254.

Zafar, S. A., Zaidi, S. S. E. A., Gaba, Y., Singla-Pareek, S. L., Dhankher, O. P., Li, X., & Pareek, A., (2020). Engineering abiotic stress tolerance via CRISPR/Cas-mediated genome editing. *J. Exp. Bot., 71*(2), 470–479.

Zaidi, S. S. E. A., Mahas, A., Vanderschuren, H., & Mahfouz, M. M., (2020). Engineering crops of the future: CRISPR approaches to develop climate-resilient and disease-resistant plants. *Genome Biol., 21*(1), 1–19.

Zaidi, S. S. E. A., Mukhtar, M. S., & Mansoor, S., (2018). Genome editing: Targeting susceptibility genes for plant disease resistance. *Trends Biotechnol., 36*(9), 898–906.

Zeng, H., Xu, L., Singh, A., Wang, H., Du, L., & Poovaiah, B. W., (2015). Involvement of calmodulin and calmodulin-like proteins in plant responses to abiotic stresses. *Front. Plant Sci., 6*, 600.

Zeng, Y., Wen, J., Zhao, W., Wang, Q., & Huang, W., (2020). Rational improvement of rice yield and cold tolerance by editing the three genes OsPIN5b, GS3, and OsMYB30 with the CRISPR–Cas9 system. *Front. Plant Sci., 10*, 1663.

Zhang, A., Liu, Y., Wang, F., Li, T., Chen, Z., Kong, D., & Luo, L., (2019). Enhanced rice salinity tolerance via CRISPR/Cas9-targeted mutagenesis of the OsRR22 gene. *Mol. Breed., 39*(3), 1–10.

Zhang, B., (2015). MicroRNA: A new target for improving plant tolerance to abiotic stress. *J. Exp. Bot., 66*(7), 1749–1761.

Zhao, Y., Zhang, C., Liu, W., Gao, W., Liu, C., Song, G., & Xie, C., (2016). An alternative strategy for targeted gene replacement in plants using a dual-sgRNA/Cas9 design. *Sci. Rep., 6*(1), 1–11.

CHAPTER 11

CRISPR/Cas-Mediated Genome Editing for Biotic Stress Tolerance

PRASHANT SHINGOTE,[1*] RAVINDRA KALE,[1] SHOBHA SURBHAIYYA,[1] DHIRAJ WASULE,[1] VASHNAVI PARMA,[1] PRATIK PUSADKAR,[1] PRASHANT KALE,[1] PRASHANT KAWAR,[2] and NARSING PARLAWAR[1]

[1]*Dr. Panjabrao Deshmukh Krishi Veedyapeeth, Akola, Maharashtra – 444104, India*

[2]*ICAR–Directorate of Floriculture Research, Shivaji Nagar, Pune, Maharashtra – 411005, India*

[*]*Corresponding author. E-mail: shingoteprashant@pdkv.ac.in*

ABSTRACT

Present cropping systems are threatened by increasing pathogen burden. Molecular understanding of plant-pathogen interactions and genomic information, leads to the emergence of genome editing as disease resistance in crops. Genome editing can be utilized in numerous ways for effective crop improvement. Clustered regularly interspaced short palindromic repeats (CRISPR) associated endonuclease Cas9 (CRISPR/Cas) as genome editing method permits to explore of the genetic modification (GM) potential of different crops for resilience. The present chapter emphasizes on novel approaches to identify novel targets for resistance breeding with a focus on CRISPR/Cas-mediated genome editing tool. The present chapter is focused on major biotic stress challenges with the necessity of CRISPR/Cas gene editing, origin, and development in genome editing tools to combat abiotic stresses. The chapter further emphasizes on CRISPR/Cas-mediated abiotic stress tolerance (AT) studies for bacterial and fungal diseases. CRISPR/Cas tools

CRISPR/Cas-Mediated Genome Editing in Plants, Deepu Pandita, Anu Pandita (Eds.)

with improved and broader specificity will allow for practically any desired alteration. CRISPR technology will become an imperative tool to overcome the beneficial resistance traits in plants to meet upcoming challenges.

11.1 INTRODUCTION

Modification of genome using genome editing has become an indispensable tool in biological research, which is done traditionally using random mutagenesis or by low-efficiency gene targeting. The rapid development in genome editing research helped to overcome the limitations of conventional breeding methods. The use of engineered site-specific nucleases (SSNs) for genome editing has revolutionized the field of plant agriculture research and seems to be powerful tools for the crop improvement. The different SSNs, such as CRISPR/Cas9, transcriptional activator-like effector nucleases (TALENs), and zinc finger nucleases (ZFNs) works via breakage in DNA strand and repairing (Miglani, 2017).

Due to it is simple, robust, and highly efficient features, CRISPR/Cas9 is thought to be the most effective SSN, (Xu et al., 2016). CRISPR was first identified in *Escherichia coli* which encompasses two components, the first is short 20–40 bp regularly interspaced DNA repeats, and the second part is hypervariable spacer DNA, which is unique in nature and matches up perfectly with the viral DNA especially bacteriophages. Precise modification of genome can be generated using CRISPR-Cas9 genome-editing technique (Knott and Doudna, 2018). With various advanced application in biological research, CRISPR/Cas9 with modified versions such as CRISPR/Cpf1 system and base editors become popular genome editing technique (Doudna and Charpentier, 2014; Miglani, 2017; Mishra et al., 2018).

11.2 GENETIC MODIFICATION (GM) STRATEGIES FOR BIOTIC STRESS TOLERANCE IN THE PRE-EDITING AGE

Crops yields can be improved by controlling damage caused by certain biotic stresses, particularly pathogens (viruses, bacteria, and fungi), insect pests, nematode parasites, protists, and weeds. The genetic modification (GM) can be stepping stone in understanding possible role of different genes influencing activity of pathogens like fungi, bacteria, and viruses damaging the crops. As many resistant varieties developed by conventional breeding need to be further improved as pathogenicity of pathogen altered with mutation which confers variation with the time for success of controlling pests. The transgenic crops with multiple stacked genes development should be followed as a strategy for genetically engineered crops and have to be moved to the field as varieties resistant to various biotic stresses (James, 2013). The transgenic varieties including number of genes discovered earlier containing biotic stress tolerance factors such as natural and synthetic *Cry* genes, *protease* inhibitors (PI), cystatin genes aryloxy alkanoate, dioxygenase, *nitrilase*, GTgene, AFP (anti-freezing protein gene), Chitinase II and III gene, *glucanase*, osmotin, defensin, and Rps1-k found to be the factors conferring resistivity of crop to certain disease-producing pathogens.

The protection of plants against insect attacks can be settled by boosting the natural defense system of plants with PI genes which confers to the proteolytic activity of insect digestive gut. Plant PI such as serine and cysteine PIs are usually abundant in seed as well as storage tissues but sometimes can also be expressed in other aerial tissues against insect attacks on plants by interfering with the proteolytic activity of insect's digestive gut (Reeck et al., 1997; Hilder et al., 1987). There is another approach called effector-triggered immunity, in this plant immune system enhancement can be done for the production of resistive products of genes by host plants through processes such as GE for conferring disease resistance in crops. Also, the plant possesses innate immunity as damage-associated molecular patterns (DAMPs) can induce downstream responses that are partially overlapping, including the aggregation of plant hormones, and expression of ROS, which confers activation in the defense system of plants known as triggered immunity (Howe & Jander, 2008; Dodds & Rathjen, 2010; Tang et al., 2017).

Enzyme triggered response also followed, such as laccase lignin polymerization enzyme expression in cotton as GhLac1, which modulates increased lignification leads to broad-spectrum biotic stress tolerance response via manipulating Phenylpropanoid pathway and Jasmonic acid (JA) synthesis helpful to control insects like bollworms and aphids (Hu, 2018). Phenylpropanoid metabolism was found to be the crucial secondary metabolic pathway involved in plant defense against biotic stress. The Lectins and Alpha-*amylase* inhibitors, which found abundant in the seeds are effective against the sap sucking Hemiptera and seed weevils (Powell et al., 1995). Plant cell walls with higher lignin are difficult to suck or pierce by herbivores and hard to digest by leaf-eating herbivores (Naoumkina et al., 2010; Mottiar et al., 2016). Enhancing the presence of lignin in some plant tissues may have an insect tolerant effect. Polyamines (PA) metabolism well known to garble in plant cells from long time in response to intuitive alteration in plants by interacting with fungal diseases (Asthir et al., 2004; Greenland and Lewis,

1984), viral pathogens (Torrigiani et al., 1997; Walters, 2003). Even so, the progress regarding PA and its relation to plant diseases remained unperceptive (Walters, 2000, 2003). The overexpression of spermine is a spermidine-derived biogenic polyamine responsible for overexpression of PA having important role as a mediator in defines signaling against pathogen enabling resistance to virus infection (Takahashi et al., 2003, 2004; Yamakawa et al., 1998). PA trigger plant defense against different biotic pathogens like fungi and viruses through molecular biology and modern biochemical perspective (Mehta et al., 2002; Walters, 2003).

11.3 MAJOR BIOTIC STRESS CHALLENGES WITH NECESSITY OF GENE EDITING TOOLS

In the current scenario, the most pressing concern facing humanity is to ensure the security of food for an expanding population. The world's population will have reached 10 billion people by 2050, so the food supply needs to be increased up to a 60–100% increased globally (FAOSTAT, 2016). In addition to the growing population rate, extreme climate, limited farmland availability, increased biotic and abiotic pressure are major barriers to agriculture and food production. Aside from drought and flooding, biotic stress poses a significant threat to agricultural output stability. Savary et al. (2019) observed that from 2010 to 2014, global pest production losses were 21% for wheat, 30% for rice, 22% for maize, 17% for potato, and 21% for soybean. Moreover, the largest losses were recorded in countries with food shortages and fast-growing populations.

In the past, pesticides were used to proficiently control the diseases, but their adverse effect on biodiversity demands alternative strategies. Plants with inherent disease resistance can help reduce pesticide use and increase yields. Disease defenses that are constantly active, on the other hand, typically result in a growth trade-off, resulting in reduced growth or dwarfism in plants (Butselaar and Ackerveken, 2020; Gao et al., 2020). Alternative resistance types that find a balance among resistance, growth, and yield are considered desirable. Pathogens evolved to their hosts throughout co-evolution and produced unique effectors to regulate the plant's defense response. Effector research and target identification enable not only the development of novel biotechnological tools (Doron et al., 2018), but also the disruption of this connection by target mutation in order to avoid host manipulation and, as a result to boost resistance (Doron et al., 2018). The development of technology that can aid in crop enhancement can help to boost output to some level.

Chemical, physical, and biological (transposons/T-DNA) mutagenesis methods have made important contributions to understanding the significance of genes and revealing biological pathways for crop improvement in the previous several decades (Ma et al., 2018). Transgenic approaches have been utilized to learn about basic plant biology and to improve crops for the past three decades. However, when it comes to edible crop species, transgene integration in host plant is random, often unstable, and under biosafety regulations (Stephens and Barakate, 2017). Recently SSNs based genome editing approaches confirmed precise editing in plants. The SSNs produce double-stranded breaks (DSBs) in the specific DNA. To repair DSB, nonhomologous end joining (NHEJ) or homology-directed recombination (HDR) are employed, resulting in INDELS and substitution mutations in the target region(s) (Jinek et al., 2012).

Genome editing methods, unlike transgenic procedures, which produce random insertions and, in many cases, random phenotypes, produce specific mutants, making them valuable modern crop improvement approaches. Genome-edited crops have an advantage over transgenic plants in that they can contain the changed DNA of the desired characteristic (Malzahn et al., 2017). When compared to traditional genetically modified (GM) crops, such enhanced crops can be utilized in breeding programs, and the resultant varieties can be used immediately with less acceptability/consumption concerns and fewer regulatory procedures (Waltz, 2018). Accumulation of increasing information on the molecular processes governing plant-pathogen interactions, helps researchers to test a variety of ideas regarding possible breeding targets and see if manipulating them improves resistance.

11.4 CRISPR/CAS SYSTEM AS GENOME EDITING TOOL

Genetic engineering (GE) is a part of modern breeding techniques, such as ZFN, TALEN, oligo-directed mutagenesis (ODM) and CRISPR/Cas, which allow for the replacement of random mutagenesis and significantly reduce the background of unwanted off target mutations. Crops with complex genome, such as tetraploid potato (Johansen et al., 2019), hexaploid wheat (Wang et al., 2014), octoploid strawberry (Wilson et al., 2019), and allotetraploid cotton species can be precisely edited. Precise GE can be used to target multiple alleles at the same time to produce desired trait (Braatz et al.,

2017; Wang et al., 2018). Use of two TALEN gives very precise targeting of a genomic site, but programming TALEN is quite tedious and challenging, as is delivering them jointly into the plant's genome (Langner et al., 2018). As a result, TALEN was rapidly superseded by alternate feasible method CRISPR/Cas, with ease to formulate, comparatively affordable, and well-suited multiplexing (Thompson et al., 2018).

There are only two components of this system Cas protein and single guide RNA (sgRNA) with respective roles of causing DSB in DNA and guiding the Cas protein to the genomic target area of interest. The system's specificity is provided approximate 20 bp changeable target complementary guide sequence inside the sgRNA. This makes reprogramming sgRNAs relatively simple, such as by a succession of PCR procedures (Li et al., 2013). *Streptococcus pyogenes*-derived Cas9 is the most often utilized Cas protein which needs a 3-nt-long PAM sequence NGG (Doudna and Charpentier, 2014). In the CRISPR/Cas system, the PAM is a changeable sequence (Leenay and Beisel, 2017). With available genome sequence organisms, it is feasible to design specific sgRNA by choosing a conserved region of a particular allele than the other alleles. Other CRISPR/Cas systems have been identified in the meanwhile, greatly expanding the GE toolbox.

Khatodia et al. (2016) recently published a paper that includes many bioinformatics tools designed to generate appropriate sgRNAs and detect potential off-target locations by for CRISPR/Cas system. Following DSB, the cells must repair it in one of two methods. Insertions or deletions may occur when a DSB is repaired using the error-prone NHEJ process (InDels). A simple random mutation would arise, causing a frameshift and maybe a loss-of-function phenotype. CRISPR/Cas has the benefit of continuing until all target sites have been changed to the point where they are no longer recognized by the sgRNA/Cas combination. On the other hand, InDels of nucleotides in multiples three, will remain in frame and may not cause the desired loss of function (Figure 11.1(a)). HDR permits the insertion of a desired sequence when a complementary flanking template is given (Figure 11.1(b)). This transforms a knocking into a "safe harbor," removing the possibility of inactivating a critical gene or element for plant growth. Because HDR allows for alteration of specific alleles at the DSB, selecting euchromatic areas with strong transcriptional activity can assist boost transgenic expression even further (Figure 11.1(c)). This type of rewriting in the targeted gene sequence by introducing precise alterations may be utilized to customize enzyme activity or substrate selectivity by changing a particular amino acid. In addition to CRISPR/Cas system, HDR necessitates an adequate number of templates DNA to be transported into the genome. As late S or G2 phases are typically stages of HDR, therefore Nocodazole, CCND1 are used to improve GE efficiency (Zhang et al., 2017).

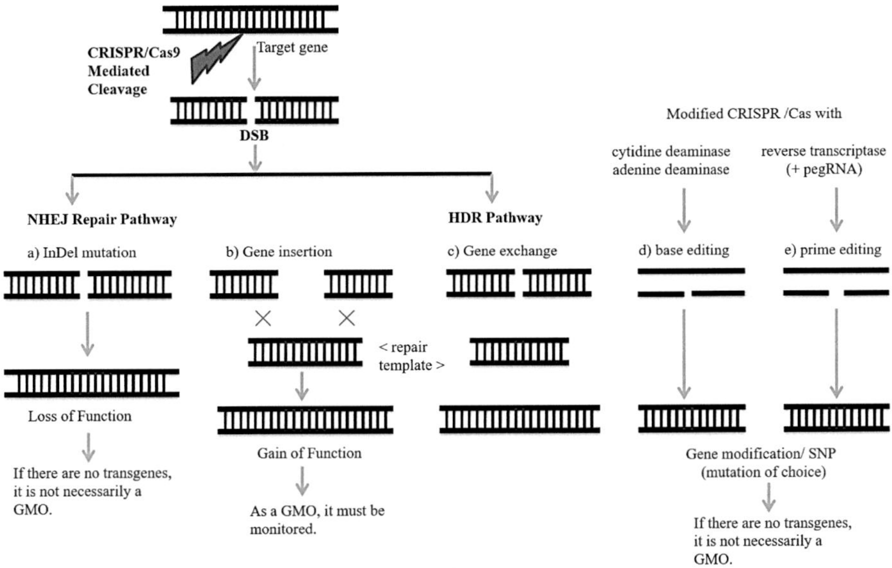

FIGURE 11.1 CRISPR/Cas system to improve disease resistance.

Note: (a) Depending on the repair mechanism, CRISPR/Cas-induced DSBs can result in simple mutations of a specific sequence (through NHEJ pathway), resulting in small InDels, causing frameshift mutations. (b) Instead, if a repair template is available, the DSB can be repaired via HDR, which allows for the insertion of bigger extraneous DNA into a desired genomic region. (c) Injecting particular mutations into a designated region. (d) Cytosine base (CB) editors with cytidine or adenine base editors (ABEs) can be used to produce single-nucleotide mutations. (e) Prime editing can result in more specified alterations; it is a combination of nCas9 along with reverse transcriptase (RT) and uses specific gRNA than the conventional sgRNAs.

Virus-derived vectors seem to be beneficial, by providing a large number of repair templates and therefore boosting the likelihood of effective HDR events (Baltes et al., 2014). Targeted base editing has previously been done in cotton, rice, and tomato utilizing dCas9 and nCas9 combined with cytidine deaminase and adenine deaminase (Li et al., 2017; Shimatani et al., 2017; Molla et al., 2020). CB editors are nCas9 fusions used for converting C–G bp to T–A bp and ABEs for converting an A–T bp to a G–C bp (Marzec and Hensel, 2018).

CRISPR can be used to induce point mutations in plants, for example, have resulted in herbicide-tolerant rice (Shimatani et al., 2017). When single changes, such as altering 20 bp sgRNA target site are insufficient, then prime editing, as shown in Figure 11.1(e), may be the preferred editing system for accurate GE. Such new method does not require a "repair template," as HDR editing does (Anzalone et al., 2019). A designed Cas9 nickase coupled to RT is utilized in combination of specific gRNA (prime editing) programming the desired edit in prime editing (Marzec and Hensel, 2020). Furthermore, this approach has previously demonstrated potential in crops such as wheat and rice (Lin et al., 2020), and it promises a handy tool in advanced resistance breeding programs. Careful designing of sg RNA in the CRISPR/Cas system allows targeting multiple genes of the same or different alleles through multiplexed sgRNAs and having least with minimal off-target effects (OTEs) (Hahn and Nekrasov, 2019). The amounts of off-target events in plants were reported to be relatively lesser and can be further decreased using optimized Cas proteins (Langner et al., 2018). Even a few off-target mutations are acceptable to breeders when contrasted to the massive number of undesired modifications generated by undirected ethyl methane sulfonate (EMS) or X-ray mutagenesis. Prior to the discoveries of TALEN and CRISPR/Cas, it appeared that crops with complex genomes would be unable to create a recognizable phenotype from recessive resistance genes. Many crops are experiencing bottlenecks due to technical issues with genetic transformation methods like *Agrobacterium*-mediated or biolistic, etc., for transporting the CRISPR/Cas system to plant. However, novel approaches like as de novo meristem induction may one day assist in overcoming these constraints, allowing all plants to be GE-friendly (Table 11.1) (Maher et al., 2020).

TABLE 11.1 CRISPR/Cas-Methods Targeting of Viral Genomes to Overcome Resistance

Plant Species	Resistance to Virus	Target Region	Cas System	References
Arabidopsis/*N. benthamiana*	Beet severe curly top virus (BSCTV)	IR, CP, Rep	Cas9	Ji et al. (2015)
Arabidopsis/*N. benthamiana*	Cucumber mosaic virus (CMV)/tobacco mosaic virus (TMV)	1A, CP, 3'UTR-A	FnCas9 > RNA	Zhang et al. (2018a)
Tomato	Tomato yellow leaf curl virus (TYLCV)	CP, Rep	Cas9	Tashkandi et al. (2018)
Barley	Wheat dwarf virus (WDV)	MP/CP, Rep/RepA LIR	Cas9	Kis et al. (2019)
Potato	Potato virus Y (PVY)	P3, CI, Nib, CP	Cas13a > RNA	Zhan et al. (2019)

11.5 GENOME-EDITING STRATEGIES FOR BIOTIC STRESS RESISTANCE

The biotic stress of plants has always been a subject of concern to the fraternity of plant research and breeders. During the entire growth cycle, plants are relentlessly vulnerable to various biotic factors like a pathogen, virus, pests, insects, etc. As a consequence, crops are substantially lost which threatens/negatively impacted agriculture and global food security. To overcome such challenges and to improve productivity and sustainability in agriculture, researchers are in a continuous effort to develop a smart crop that can tolerate such baleful/menacing biotic stress. Genetic improvement in crop plants, a chief strategy has been adopted to rescue such crop plants for decades now. Traditionally, crop modification was focused on the use of conventional and mutant-induced genetic variation to obtain a favorable genetic combination. The conventional plant breeding approaches involved natural mutation, inter-genic crosses, hybridization, and mutagenesis by physical or chemical mutagen. However, such an approach has given stress-tolerant and elite varieties but also possess major limitation in terms of its specificity, cumbersome work process, time consumption and overall difficulty in screening and identification of desired genetic change. Meanwhile, genetic manipulations by physical, chemical, and biological means came into the limelight for the development of transgene or genetically modified organisms (GMOs). Although, GE has contributed significantly towards crop improvement for biotic stress tolerance, but faced stringent regulatory policies for risk assessment and lacked social acceptance. With the advancement of technologies, more efficient and precise techniques were adopted by researchers over the past few decades. One such advanced technology that greatly impacted the research field is the use of genome editing tools.

11.5.1 GENOME EDITING: A JOURNEY FROM MEGA NUCLEASE TO CRISPR

It is all rooted in the concept of target modification of an organism carried out by the term Genome editing. Genome editing enables a modification and modulation of the functioning DNA sequences of an endogenous genome of an organism. Earlier, gene editing was initiated by the integration of a small single segment of DNA into the host genome by homologous recombination (HR). However, the output of before said leads to random integration and was inefficient to achieve target modification (Adli, 2018). The emergence of the era began with the introduction of double-strand break (DSB) at the genomic locus that needs to be altered (Carroll, 2007). Furthermore, the DSB assisted by a single specific nuclease (SSN) transitioned the area of genetic and molecular research. The SSN is a customizable engineered restriction enzyme that creates site-specific dsDNA breaks that stimulate HR, gene insertion, and gene inactivation. The DSBs at target site induced by nuclease is repaired by two mechanism-nonhomologous ends joining (NHEJ) and homology direct repair (HDR). NHEJ is an efficient, imprecise, error-prone repair mechanism that introduces small insertions or deletions (indels) at the site of the break, which disrupts gene function and is responsible for knock-out of the gene whereas, HDR repairs DNA break by using a homologous DNA sequence as the template.

During early studies of DNA sequences modification, naturally occurring endonucleases were engineered and employed as the first editing tool. Such homing endonuclease or meganuclease was capable of recognizing and cleaving off long stretches up to 14–40 bp of target DNA (Sander et al., 2014). Though, it often faces limitations in its catalytic activity whenever the target site specificity changes and also were unable to manipulate complex genome. A key technical breakthrough in genome editing has been the development of engineered specific DNA binding protein that acts as a nuclease and cleave target site, such protein-based molecular tools are ZFN and transcription activator-like effector nucleases (TALENs). ZFN is a chimeric restriction nuclease, customized by ligating zinc finger domain, typically recognizes trinucleotide motif on DNA with a nonspecific catalytic domain of type II endonuclease FokI. To improve the efficiency and off-target DNA cleavage, FokI dimers were employed. ZFN turns out to be a successful nuclease for target-specific DSB, but also have a downside as not all DNA sequence are bound by ZFN domain (Carroll, 2017).

In the interim, a modular approach was realized with the introduction of transcription activator protein. TAL effector nuclease follows the same principal design of ZFN, having two TAL proteins fused with catalytic domain FokI. Each method has its shortfalls, even though TALENs are preferred over ZFNs, in terms of its simple modulation and cost-effectiveness, however, when we talk about specificity, simplicity, efficiency, versatility, robustness, and preciseness in genome editing CRISPR (clustered regularly interspaced short palindromic repeats)/Cas9 holds a great promise.

11.5.2 ORIGIN OF CRISPR TECHNOLOGY

A remarkable and revolutionary journey of CRISPR has begun long back in the year 1987 when for the first time, a tandem array of DNA repeats with a short interspaced variable repeat were encountered in one strain of *E. coli*. Following the year till 2000, the same peculiar loci of the genome were observed in archaea and different other bacteria. During initial discovery, the origin and function of these clustered repeats were unknown. Thanks to the advent of genome sequencing technology, numerous, and diverse genome sequences of bacteria and archaea were available, which elucidated (Mojica et al., 2016) to realize tandem sequences are functionally related. This eventually cleared the vision that sequences are partially palindromic and occurred in clusters, regularly interspersed by unique spacer sequences of constant length, similar to that of the repeats helping the researcher to term this hypervariable locus as CRISPR. Aforesaid, until 2005, the biological function of a unique spacer (32 bp) was unknown when three different research groups revealed that unique spacer show sequence homology with viruses or plasmids (Barrangaou and Horvath, 2017). This observation provides an intimation of the acquired bacterial immune system. In the same year, a PAM was discovered as a conserved DNA flanking protospacer (an original sequence of viruses and plasmids from which the spacers were derived). Later, in 2007 the postulate associated with CRISPR in bacterial defense was proved and proclaimed that bacteria acquired memory/immune function as a resistance against the exogenous genome. The bacterial defense system is associated with a cluster of arrays called CRISPR-associated or Cas gene. It is encoded with a series of Cas proteins located just next to the CRISPR locus. It has a domain that functions similar to endonuclease, helicase, polymerase, and nucleotide-binding proteins (Sontheimer et al., 2015; Ishino et al., 2018). After the discovery of defensive CRISPR immunity in bacteria and archaea, the following years revealed the detailed mechanism of CRISPR/Cas. Amongst the year in 2008, the role of CRISPR small RNAs (crRNA) as a guide RNA (gRNA) was determined. Also, during the same year, the importance of conserved sequences present next to protospacers for CRISPR-mediated phage resistance was described. 2010 a year, where CRISPR-Cas interference mechanism and Cas cleaving DNA at target sites were determined. The next three consecutive years were a progressive year as genome editing tools. During these years, trans-activating crRNAs (tracrRNAs) were identified. Following with the in-vitro reconstruction of CRISPR-Cas systems, 2013 was a year of demonstration of CRISPR-Cas tools for efficient genome editing in mammalian cells (Mojica et al., 2016).

11.6 CRISPR-CAS SYSTEM

The discovery of the CRISPR/Cas system evolved naturally from acquired bacterial or archaeal immunity against viral genome has revolutionized genome editing technology. Unlike ZFN and TALEN which is a protein-based engineered nuclease, CRISPR/Cas editing approach targets DNA by RNA-guided engineered nuclease. CRISPR/Cas till now groups in two classes (I&II) classified on the constitution of effector protein, six types and over 30 subtypes based on presence/absence of signature genes, protein sequence conservation, and organization of the respective genomic loci. Class, I CRISPR/Cas system is characterized based on the presence of multiple Cas protein whereas class II system has single Cas protein akin to class 1 multiple complexes. Although, amongst all class II CRISPR/Cas9 systems derived from *Streptococcus pyogenes* is a widely adopted editing technique because of its simple construction.

11.7 MOLECULAR MECHANISM OF CRISPR-CAS SYSTEM

The CRISPR loci consist of Cas operon, protospacer, and spacer sequences and follow three major steps: The first step is an acquisition, where a new spacer, a short segment of viral genome introduced to bacterial cell by the exogenous viral genome, is incorporated with the help of Cas1 and Cas2 protein near leader end of CRISPR array. Followed by the expression of CRISPR array into pre-CRISPR RNA (pre-crRNA), consequently transcribe into small, mature CRISPR RNA (crRNAs) upon reinfection. Each crRNAs have a target segment obtained from a unique CRISPR spacer. Lastly, during interference, crRNAs proceed further and binds Cas effector nuclease along with a complementary nucleic acid segment of a target sequence and perform specific cleavage (Figure 11.2) (Sontheimer et al., 2015; Barrangaou and Horvath, 2017).

11.8 WORLDWIDE PERSPECTIVES/CURRENT STATUS FOR GENOME-EDITED CROPS

Agricultural crops are important components in nutritional security. Additionally, the higher productivity of industrial plants is having an economic role in human development, leading to improved living. Thus, GMs for high-yielding plants have played a major role in the global food security, nutritional self-sufficiency, health sector, and ultimately the economic development. The genetic improvement in crops and plants is aimed to meet the demand of industries, consumers, and farmers. From conventional time to

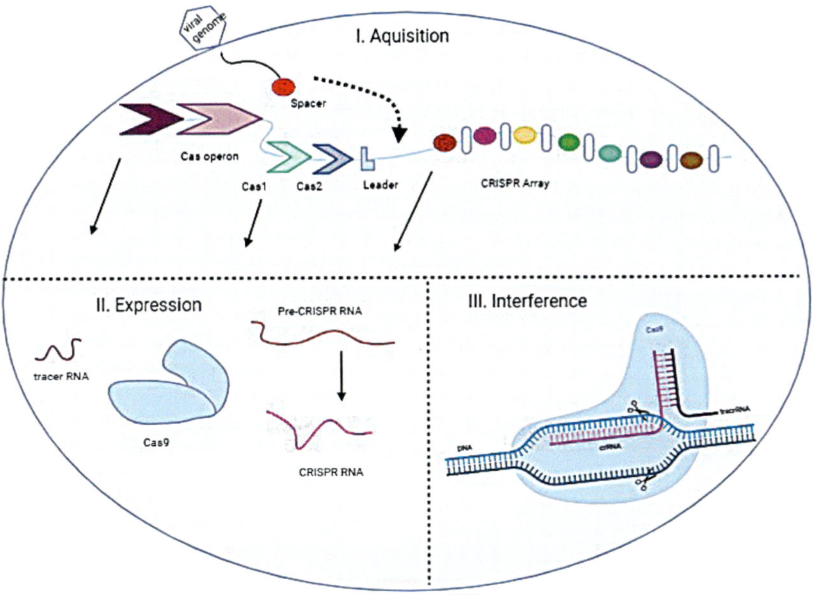

FIGURE 11.2 Development of CRISPR-Cas immunity in *E. coli*.

Note: (i) Acquisition of foreign DNA as CRISPR spacer: Protospacers (shown as different colors circles) of double-stranded DNA from invading virus or pathogen is adapted at the leader end of a CRISPR array on *E. coli* genome (host). A CRISPR array involves of unique spacers interspaced between repeats. Acquisition ensued by Cas1 and Cas2 proteins. (ii) Expression of CRISPR locus: RNA polymerase synthesized Pre-CRISPR RNA (Pre-crRNA) on the leader region further cleaved by Cas proteins into smaller crRNAs that contain a single spacer and a partial repeat. (iii) Interference: crRNA comprising a spacer with a perfect complementarity to incoming foreign DNA initiates a cleavage mechanism through Cas proteins. This cleavage event interferes with virus replication or plasmid activity and governs immunity to the host bacteria.

the recent era, the wider target of this development is quality and uniformity in farm produce, higher yield, tolerance to diseases, pests, abiotic stresses, climate resilience, etc. Along with this, traits like appearance, shelf-life, taste, and nutritional value are important that is being studied. Very few crops and cultivars are moderately resistant or resistant to biotic and abiotic stresses and pathogens and pests are also co-evolving and affecting non-host plants. The transgenics commonly referred to as genetically modified (GM) crops offered a suitable, sustainable, and economic alternative to conventional breeding methods aiming to achieve full genetic potential and overall improvement.

As per the recent data from the GM approval database of ISAAA (https://www.isaaa.org); till now 32 crops/plants have been reported with successful

GMs. This includes different agriculturally and industrially important crops (Figures 11.3 and 11.4). As per the data from the same reports; up-till now, 44 different traits/pathways have been genetically altered to achieve 428 successful events (Graphs). Worldwide, about 44 countries have adopted transgenic crop events (altogether 2,397 events) for agricultural or industrial purposes.

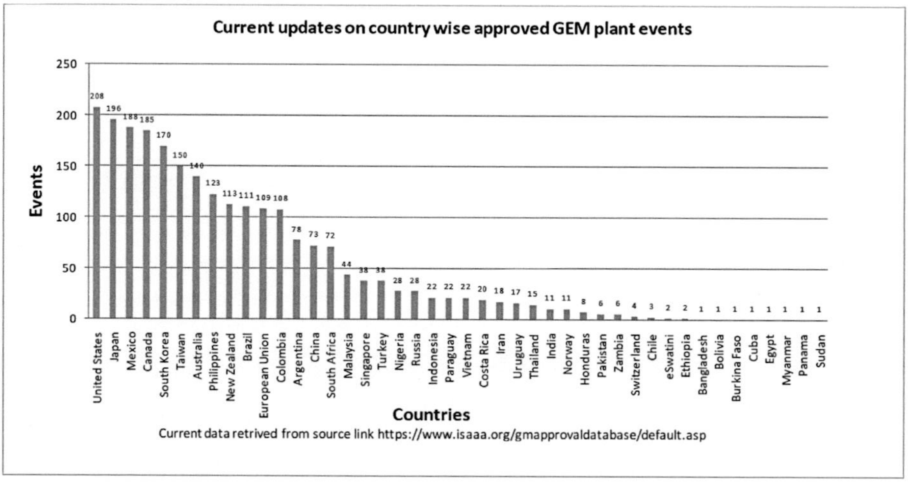

FIGURE 11.3 Country-wise approved genome editing plant events.

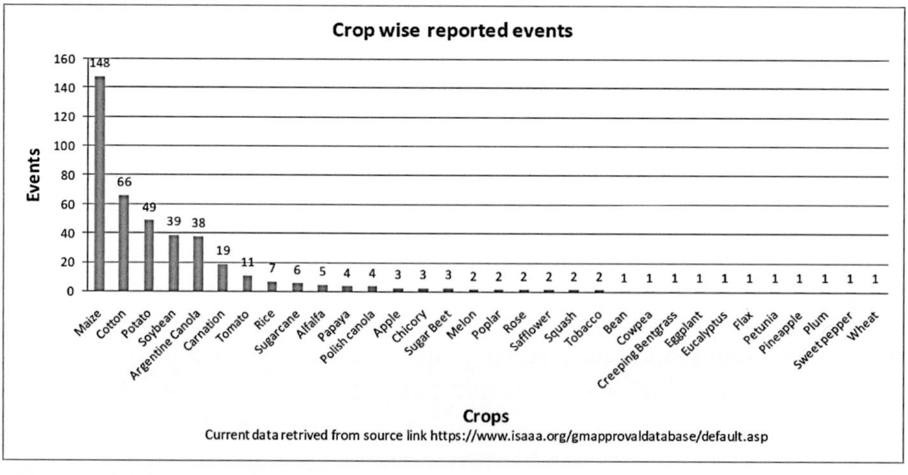

FIGURE 11.4 Crop-wise reported genome editing events.

11.9 CRISPR/CAS9 AS A GENOMIC TOOL

CRISPR/Cas9 genomic editing tool consists of Cas9 protein and sgRNA. sgRNA is a chimeric mature RNA, form from crRNA-tracrRNA duplex. Trans-activating crRNA (tracrRNA) a novel type of RNA (25 bp), complementary to CRISPR repeats, is needed for the processing of crRNAs. crRNA-tracrRNA complex or sgRNA acts as a reference that activates Cas9 and guides it to the target site for specific cleavage. Cas9 endonuclease recognized PAM (a 2–5 nucleotides bp) in dsDNA and prompts a dsDNA break at a specific target site just next to a recognition sequence.

The CRISPR/Cas9 protein interferes endonucleolytic cleavage at the specific site because of the presence of Cas9 protein. The Cas9 is structurally bi-lobed with recognition (REC) and a nuclease (NUC) lobe. The NUC lobe integrates two nuclease domains, RuvC and HNH, and a PAM-interacting domain (PI). Cas9 (Rec lobe) and gRNA look for a PAM (the trinucleotide NGG) of target DNA and binds to it. Anchored PAMs, activates Cas9 HNH and RuvC nickase activity cleaving RNA–DNA hybrid and RuvC ds DNA to form a double-strand break (DSB) (Selle et al., 2015).

11.10 MANAGEMENT OF FUNGAL DISEASES THROUGH CRISPR

11.10.1 FUNGAL INFECTIONS A SERIOUS THREAT FOR CROP PRODUCTION

With continuous change in environmental factors, it is difficult to identify new plant diseases. One of the major fast-growing diseases in plants caused by fungi and closely related organisms and this majorly affect life at developing country where staple food crops are important part of people life causing economical loss and impacting global poverty. Fungal infection indirectly affects agricultural production by causing diseases in invertebrate hosts. Various approaches such as biological control, genetic transformation techniques, fungicides or resistant crop varieties represent a valid and promising tool to control fungal infection (Singh, 2016). But these problems have some limitations, such as restricted use of genetically modified crops in field, due to continuous evolution in fungi, resistance varieties are not sustaining for longer time and thus it needs new strategies to improve their management.

11.10.2 CRISPR/CAS TOOLS TO CONFER MILDEW RESISTANCE

Tomato yellow leaf curl virus (TYLCV) and powdery mildew are the two important diseases of tomato caused heavy yield loss resulted into severe economic set back. To address these issues to develop disease resistant crop varieties, the promising valuable approach is the (CRISPR)/CRISPR-associated protein (Cas). In tomato to develop pathogen resistance varieties, CRISPR-Cas9 tool used to introduce targeted mutagenesis, where instead of the viral genome, host plant genome susceptibility factors encoded to achieve pathogen resistance. Knockout lines generated using CRISPR/Cas in SlMlo1 gene of tomato which resulted in achieving resistance for powdery mildew pathogen while knockout lines for SlPelo gene plants showed resistance to TYLCV infection (Pramanik et al., 2021).

In tomato gene silencing of, powdery mildew resistance 4 (PMR4) susceptibility (S) gene using RNAi has resulted in improved powdery mildew pathogen resistance. In another study tomato PMR4 gene fully knock-out using CRISPR/Cas9 to produce PMR4 CRISPR mutants. Total five diverse mutation events identified and all showed minimum susceptibility to powdery mildew but not complete protection from the pathogen susceptibility. CRISPR/Cas9 found promising to study S-genes (Martínez et al., 2020). Powdery mildew disease is an important disease in Grapevine (*Vitis vinifera*). VvMLO3 and VvMLO4 genes which belong to MLO family are supposed to serve as S factors for powdery mildew pathogen targeted with CRISPR/Cas9 technology for generating mutation in the Thompson Seedless powdery mildew-susceptible cultivar. Almost 20 VvMLO3/4-edited lines received with various mutations status out of which 6 resulted in normal growth, whereas remaining showed disease symptoms. CRISPR/Cas9-enabled targeted mutagenesis of VvMLO3 essential for the powdery mildew disease development in a grapevine cultivar (Wan et al., 2020).

The CRISPR/Cas9 technology used to generate a non-transgenic powdery mildew resistant variety of tomato which confirmed by absence of any foreign DNA in edited variety. Also, there was no any off-target effect in gene-edited variety (Nekrasov et al., 2017).

11.10.3 CRISPR/CAS-APPROACHES ENHANCES VERTICILLIUM LONGISPORUM RESISTANCE

Verticillium longisporum (Vl43) fungus is one of the major threats for brassica napus with no effective genetic resistance. Study of gene regulation in brassica after V143 infection identified major genes that are upregulated after infection. The mutant lines of Arabidopsis with knock out of gene results in drastically minimizing susceptibility of cultivar to Vl43. These results are validated with the help of CRISPR/Cas-induced BnCRT1a mutants that showed overall reduced susceptibility in gene-edited lines (Probsting et al., 2020).

11.10.4 CRISPR TO CONFER BLAST RESISTANCE

Magnaporthe oryzae is the most devastating pathogen that affects wheat and rice. In rice for non-race-specific resistance to blast a natural allele of an Ethylene-type transcription factor found to play an important role. The said allele resulted in a single nucleotide change in the promoter of the bsr-d1 gene, causes in reduction of expression of that gene which resulted in an inhibition of H_2O_2 degradation and ultimately improve in disease resistance. The native Bsr-d1 gene removed using CRISPR/Cas9 to establish functional validation of this gene for blast resistance (Li et al., 2017). In addition to single modification at OsERF922, Cas9/Multi-target-sgRNAs used to target multiple sites within gene modification found useful (Wang et al., 2016). The fast neutron-generated mutant line with two base pair deletions within the Ptr coding region produced a truncated protein, causing susceptibility to *M. oryzae*. CRISPR/Cas9 used to produce resistant cultivar that causes reduced blast susceptibility which confirm its resistance function (Zhao et al., 2018).

11.10.5 CRISPR/CAS TO CONFER PHYTOPHTHORA RESISTANCE

Phytophthora caused severe damage and huge losses, and it was the cause of the Irish potato famine (Fry et al., 2015). Various diseases caused by phytophthora such as phytophthora root rot (PRR), Phytophthora blight, *Phytophthora capsici* Leon causes significant yield loss in the major agricultural crops (Barbosa and Bosland, 2010; Kieu et al., 2021). With the help of gene editing, a homozygous mutant PpalEPIC8 were generated to tackle the problem of *Phytophthora palmivora* infection (Gumtow et al., 2018). In a study alternative selection marker to improve the efficiency of studies in Phytophthora transformation by using a fungicide resistance gene, PcMuORP1 found promising (Wang et al., 2019).

An effector gene Avr4/6 edited with CRISPR/Cas9 demonstrated that the function of specific genes in *P. sojae* can be achieved efficiently using CRISPR/Cas9-mediated gene disruption and gene deletion strategy (Fang and Tyler, 2016). In another study, results showed that gene editing in PcORP1 gene confer resistance to oxathiapiprolin in *P. capsica* (Miao et al., 2018). Total seven putative S-genes in potatoes used for mutation screening using a CRISPR/Cas9 system, that involves the use of two gRNAs, which generated knockout mutants of StDND1, StCHL1, and DMG400000582 (StDMR6-1) results in improve resistance against late blight. This is the first kind of report in potato that uses CRISPR/Cas9 system, and the product is ready for field trials (Kieu et al., 2021).

11.11 MANAGEMENT OF BACTERIAL DISEASES

In the bacterial infection process type III effectors system primarily interrupts the host's defense pathways and/or activate the S genes to develop disease in plant (Zaidi et al., 2018). CRISPR/Cas9-mediated gene editing will exhibit plant resistance to S genes and negative regulators of plant innate immune response which are important target sites. CRISPRCas9 has been newly engineered a powerful gene editing tool for disease resistance development (Jinek et al., 2013; Cho et al., 2013). This CRISPR-Cas9 system involves Cas9 nuclease and a single chimeric guide RNA (sgRNA), playing a role in the introduction of precise DNA double-strand breaks (DSBs), were used efficiently to stimulate gene targeting. The system of targeting of the RNA-guided Cas9 nuclease to a desired DNA sequence is done by the protospacer sequence of the sgRNA. Management of bacterial plant disease is done by the avoidance, exclusion, eradication, and protection of the pathogen, genetic resistance, agronomic practices, and biocontrol agents (Kerr, 2016). OsSWEET13 is a susceptibility (S) gene encoding to perform function of a sucrose transporter responsible in interaction of plant-pathogen which was used in rice for achieving disease resistance against bacterial blight pathogen g-proteobacterium *Xanthomonas oryzae* pv. *oryzae* (Zhou et al., 2015). *X. oryzae* released an effectors protein, PthXo2 that resulted in OsSWEET13 expression in the host and the following state of susceptibility. Use of a null mutation was done in OsSWEET13 in order to better identify PthXo2-dependent disease susceptibility, and the resultant mutants were resistant to bacterial blight pathogen (Zhou et al., 2015). A significant improvement in the resistance to *X. oryzae* pv. *oryzae* (Xoo) was observed without altering the pollen development by the removal of the endogenous OsSWEET11 gene in rice (Kim et al., 2019). In rice varieties multiplex genome editing of the promoter regions of OsSWEET11, OsSWEET13, and OsSWEET14 was done to converse broad-spectrum disease resistance to Xoo (Oliva et al., 2019). By using TALEN approach in OsSWEET14 promoter mutagenesis the disease resistance was developed, and the disruption of their targeted gene rendered the *X. oryzae* effector unable to bind OsSWEET14 (Li et al., 2012). The PthA4 effector binding elements in the promoter of the Lateral Organ Boundaries 1 (CsLOB1) gene editing in Duncan grapefruit resulted in canker resistant mutants with minimum symptoms (Jia et al., 2016). Soft-rot and blackleg disease of potato caused by *Pectobacterium atrosepticum* has been utilized to explore protein-to-protein interactions and also in complex formation in the subtype I-F CRISPR/Cas system which includes Cas1, Cas3, and the four subtype-specific proteins namely Csy1, Csy2, Csy3, and Cas6f (Richter et al., 2012). The Cas9/sgRNA is a molecular tool used for the study of citrus gene function and for targeting the genetic manipulation for desire traits. In the citrus disease resistance to citrus canker conferred by genome editing tool using the promoter region of the S gene CsLOB1, that targets the TALE from *X. citri* pv. *citri* (Xcc), (Jia et al., 2016, 2017; Peng et al., 2017). In sweet orange, the CRISPR/Cas9 technique and synthetic sgRNA tools were used to target the CsPDS gene which was introduced into leaves via agroinfiltration technique facilitated by the pre-treatment with Xanthomonas *citri* subsp. *citri* (Xcc). Conformation results were obtained by DNA sequencing that the CsPDS gene was mutated without any off-target mutagenesis, (Jia and Wang, 2014). CRISPR-Cas9 tool was used to nullify the DMR6 ortholog in tomato and found that dmr6 mutants achieve resistance to different bacterial pathogens which include *Pseudomonas syringae*, Phytophthora capsica, and Xanthomonas spp., without many negative effects (Thomazella et al., 2016).

11.12 PROMISES AND LIMITATIONS OF CRISPR/CAS TECHNOLOGIES

A drastic revolution can be achieved in plant breeding with the help of genome editing tool which helps to carry out precise mutation in the genome. Genome editing helps to achieve trait improvement in plants as well as do functional characterization of important genes. Another advantage is the production of non-transgenic plants that use transient expression plasmid. In addition to single gene editing, genome editing is also applicable for modification of biological pathway by targeting multiple genes involved in the pathway. The system has the advantage of changing the target gene by changing the sgRNA spacer sequence (Hsu et al., 2015). Genome editing found a promising tool in fungal resistance research area which is very difficult to develop using conventional breeding approach (Munoz et al., 2019). Compared with TALENs, CRISPR/Cas9-expressing vectors are much easier to construct and can be competed in just two or three days and induces a much higher mutation rate than TALENs.

Whereas the requirement for a PAM (-NGG) sequence and the possibility of OTEs are limitations of CRISPR/Cas9 system. As there are chances of the release of a mutant strain in the field, it should be evaluated for their safety and also for public acceptance.

Regardless of GMO regulation difficulties, given the ever-expanding CRISPR toolkit, it seems evident for speeding up plant breeding for improved traits. Genome editing is found applicable in various research areas and any desired alteration.

KEYWORDS

- **biotic stress tolerance**
- **CRISPR**
- **genome editing**
- **polyamines**
- **protease inhibitors**
- **site-specific nucleases**
- **susceptibility factors**

REFERENCES

Adli, M., (2018). The CRISPR tool kit for genome editing and beyond. *Nat Commun., 9*, 1911.

Anzalone, A. V., Randolph, P. B., Davis, J. R., Sousa, A. A., Koblan, L. W., Levy, J. M., Chen, P. J., et al., (2019). Search-and-replace genome editing without double-strand breaks or donor DNA. *Nature, 576*, 149–157.

Asthir, B., Spoor, W., & Duffus, C. M., (2004). Involvement of polyamines, diamine oxidase and polyamine oxidase in resistance of barley to *Blumeria graminis* f. sp. hordei. *Euphytica, 136*(3), 307–312.

Baltes, N. J., Gil-Humanes, J., Cermak, T., Atkins, P. A., & Voytas, D. F., (2014). DNA replicons for plant genome engineering. *Plant Cell, 26*, 151–163.

Barrangou, R., & Horvath, P., (2017). A decade of discovery: CRISPR functions and applications. *Nature Microbiology, 2*(7), 1–9.

Braatz, J., Harloff, H. J., Mascher, M., Stein, N., Himmelbach, A., & Jung, C., (2017). CRISPR-Cas9 targeted mutagenesis leads to simultaneous modification of different homoeologous gene copies in polyploid oilseed rape (*Brassica napus*). *Plant Physiol., 174*, 935–942.

Carroll, D., (2017). Focus: Genome editing: Genome editing: Past, present, and future. *The Yale Journal of Biology and Medicine, 90*(4), 653.

Daley, J. M., Kwon, Y., Niu, H., & Sung, P., (2013). Focus: 50 years of DNA repair: The yale symposium reports: Investigations of homologous recombination pathways and their regulation. *The Yale Journal of Biology and Medicine, 86*(4), 453.

Dodds, P. N., & Rathjen, J. P., (2010). Plant immunity: Towards an integrated view of plant-pathogen interactions. *Nature Reviews Genetics, 11*(8), 539–548.

Doron, S., Melamed, S., Ofir, G., Leavitt, A., Lopatina, A., Keren, M., Amitai, G., & Sorek, R., (2018). Systematic discovery of antiphase defense systems in the microbial pangenome. *Science, 359*, eaar4120.

Doudna, J. A., & Charpentier, E., (2014). The new frontier of genome engineering with CRISPR-Cas9. *Science, 346*(6213).

Fang, Y., & Tyler, B. M., (2016). Efficient disruption and replacement of an effector gene in the oomycete *Phytophthora sojae* using CRISPR/C as9. *Molecular Plant Pathology, 17*(1), 127–139.

Fry, W. E., Birch, P. R. J., Judelson, H. S., Grünwald, N. J., Danies, G., Everts, K. L., Gevens, A. J., et al., (2015). Five reasons to consider *Phytophthora infestans* a reemerging pathogen. *Phytopathology, 105*(7), 966–981.

Gao, Z., Liu, Q., Zhang, Y., Chen, D., Zhan, X., Deng, C., Cheng, S., & Cao, L., (2020). OsCUL3a-associated molecular switches have functions in cell metabolism, cell death, and disease resistance. *J. Agric. Food Chem., 68*, 5471–5482.

Greenland, A. J., & Lewis, D. H., (1984). Amines in barley leaves infected by brown rust and their possible relevance to the formation of 'green islands.' *New Phytologist, 96*(2), 283–291.

Gumtow, R., Wu, D., Uchida, J., & Tian, M., (2018). A *Phytophthora palmivora* extracellular cystatin-like protease inhibitor targets papain to contribute to virulence on papaya. *Molecular Plant-Microbe. Interactions, 31*(3), 363–373.

Hahn, F., & Nekrasov, V., (2019). CRISPR/Cas precision: Do we need to worry about off-targeting in plants? *Plant Cell Rep., 38*, 437–441.

Hilder, V. A., Gatehouse, A. M., Sheerman, S. E., Barker, R. F., & Boulter, D., (1987). A novel mechanism of insect resistance engineered into tobacco. *Nature, 330*(6144), 160–163.

Howe, G. A., & Jander, G., (2008). Plant immunity to insect herbivores. *Annu. Rev. Plant Biol., 59*, 41–66.

Hsu, P. D., Lander, E. S., & Zhang, F., (2014). Development and applications of CRISPR-Cas9 for genome engineering. *Cell, 157*(6), 1262–1278.

Hu, Q., Min, L., Yang, X., Jin, S., Zhang, L., Li, Y., Ma, Y., et al., (2018). Laccase GhLac1 modulates broad-spectrum biotic stress tolerance via manipulating phenylpropanoid pathway and jasmonic acid synthesis. *Plant Physiology, 176*(2), 1808–1823.

Ishino, Y., Krupovic, M., & Forterre, P., (2018). History of CRISPR-Cas from encounter with a mysterious repeated sequence to genome editing technology. *Journal of Bacteriology, 200*(7), e00580–17.

James, C., (2013). Global status of commercialized biotech/ GM crops: 2012. *China Biotechnol., 33*, 1–8.

Jeffry, D. S., & Keith, J. J., (2014). CRISPR-Cas systems for editing, regulating and targeting genomes. *Nature Biotechnology, 32*(4).

Jia, H., & Wang, N., (2014). Targeted genome editing of sweet orange using Cas9/sgRNA. *PloS One, 9*(4), e93806.

Jia, H., Orbovic, V., Jones, J. B., & Wang, N., (2016). Modification of the PthA4 effector binding elements in type I Cs LOB 1 promoter using Cas9/sg RNA to produce transgenic Duncan grapefruit alleviating XccΔpthA4: dCs LOB 1.3 infection. *Plant Biotechnology Journal, 14*(5), 1291–1301.

Jinek, M., Chylinski, K., Fonfara, I., Hauer, M., Doudna, J. A., & Charpentier, E., (2012). A programmable dual-RNA–guided DNA endonuclease in adaptive bacterial immunity. *Science, 337*(6096), 816–821.

Johansen, I. E., Liu, Y., Jørgensen, B., Bennett, E. P., Andreasson, E., Nielsen, K. L., Blennow, A., & Petersen, B. L., (2019). High efficacy full allelic CRISPR/Cas9 gene editing in tetraploid potato. *Sci. Rep., 9*, 17715.

Kerr, A., (2016). Biological control of crown gall. *Australasian Plant Pathology, 45*(1), 15–18.

Khatodia, S., Bhatotia, K., Passricha, N., Khurana, S. M. P., & Tuteja, N., (2016). The CRISPR/Cas genome-editing tool: Application in improvement of crops. *Front. Plant Sci., 7*, 506.

Kieu, N. P., Lenman, M., Wang, E. S., Petersen, B. L., & Andreasson, E., (2021). Mutations introduced in susceptibility genes through CRISPR/Cas9 genome editing confer increased late blight resistance in potatoes. *Scientific Reports, 11*(1), 1–12.

Kleinstiver, B. P., Pattanayak, V., Prew, M. S., Tsai, S. Q., Nguyen, N. T., Zheng, Z., & Joung, J. K., (2016). High-fidelity CRISPR-Cas9 nucleases with no detectable genome-wide off-target effects. *Nature, 529*, 490–495.

Knott, G. J., & Doudna, J. A., (2018). CRISPR-Cas guides the future of genetic engineering. *Science, 361*(6405), 866–869.

La Camera, S., Gouzerh, G., Dhondt, S., Hoffmann, L., Fritig, B., Legrand, M., & Heitz, T., (2004). Metabolic reprogramming in plant innate immunity: The contributions of phenylpropanoid and oxylipin pathways. *Immunological Reviews, 198*(1), 267–284.

Langner, T., Kamoun, S., & Belhaj, K., (2018). CRISPR crops: Plant genome editing toward disease resistance. *Annu. Rev. Phytopathol., 56*, 479–512.

Leenay, R. T., & Beisel, C. L., (2017). Deciphering, communicating, and engineering the CRISPR PAM. *J. Mol. Biol., 429*, 177–191.

Li, J. F., Norville, J. E., Aach, J., McCormack, M., Zhang, D., Bush, J., Church, G. M., & Sheen, J., (2013). Multiplex and homologous recombination-mediated genome editing in *Arabidopsis* and *Nicotiana benthamiana* using guide RNA and Cas9. *Nat. Biotechnol., 31*, 688–691.

Li, T., Liu, B., Spalding, M. H., Weeks, D. P., & Yang, B., (2012). High-efficiency TALEN-based gene editing produces disease-resistant rice. *Nature Biotechnology, 30*(5), 390–392.

Li, W., Zhu, Z., Chern, M., Yin, J., Yang, C., Ran, L., Cheng, M., et al., (2017). A natural allele of a transcription factor in rice confers broad-spectrum blast resistance. *Cell, 170*(1), 114–126.

Lin, Q., Zong, Y., Xue, C., Wang, S., Jin, S., Zhu, Z., Wang, Y., et al., (2020). Prime genome editing in rice and wheat. *Nat. Biotechnol., 38*, 582–585.

Ma, J., Chen, J., Wang, M., Ren, Y., Wang, S., Lei, C., & Cheng, Z., (2018). Disruption of OsSEC3A increases the content of salicylic acid and induces plant defense responses in rice. *Journal of Experimental Botany, 69*(5), 1051–1064.

Maher, M. F., Nasti, R. A., Vollbrecht, M., Starker, C. G., Clark, M. D., & Voytas, D. F., (2020). Plant gene editing through de novo induction of meristems. *Nat. Biotechnol., 38*, 84–89.

Malzahn, A., Lowder, L., & Qi, Y., (2017). Plant genome editing with TALEN and CRISPR. *Cell Biosci., 7*, 21. doi: 10.1186/s13578-017-0148-4.

Martínez, M. I. S., Bracuto, V., Koseoglou, E., Appiano, M., Jacobsen, E., Visser, R. G., Wolters, A. M. A., & Bai, Y., (2020). CRISPR/Cas9-targeted mutagenesis of the tomato susceptibility gene PMR4 for resistance against powdery mildew. *BMC Plant Biology, 20*(1), 1–13.

Marzec, M., & Hensel, G., (2018). Targeted base editing systems are available for plants. *Trends Plant Sci., 23*, 955–957.

Marzec, M., & Hensel, G., (2020). Prime editing: Game changer for modifying plant genomes. *Trends Plant Sci., 25*, 722–724.

Mehta, R. A., Cassol, T., Li, N., Ali, N., Handa, A. K., & Mattoo, A. K., (2002). Engineered polyamine accumulation in tomato enhances phytonutrient content, juice quality, and vine life. *Nature Biotechnology, 20*(6), 613–618.

Miao, J., Chi, Y., Lin, D., Tyler, B. M., & Liu, X., (2018). Mutations in ORP1 conferring oxathiapiprolin resistance confirmed by genome editing using CRISPR/Cas9 in *Phytophthora capsici* and *P. sojae*. *Phytopathology, 108*(12), 1412–1419.

Miglani, G. S., (2017). Genome editing in crop improvement: Present scenario and future prospects. *Journal of Crop Improvement, 31*(4), 453–559.

Mikami, M., Toki, S., & Endo, M., (2016). Precision targeted mutagenesis via Cas9 paired nickases in rice. *Plant Cell Physiol., 57*, 1058–1068.

Mishra, R., Joshi, R. K., & Zhao, K., (2018). Genome editing in rice: Recent advances, challenges, and future implications. *Frontiers in Plant Science, 9*, 1361.

Mojica, F. J., & Montoliu, L., (2016). On the origin of CRISPR-Cas technology: From prokaryotes to mammals. *Trends in Microbiology, 24*(10), 811–820.

Molla, K. A., Shih, J., & Yang, Y., (2020). Single-nucleotide editing for zebra3 and wsl5 phenotypes in rice using CRISPR/Cas9-mediated adenine base editors. *3Biotech, 1*, 106–118.

Monroy-Barbosa, A., & Bosland, P. W., (2010). A rapid technique for multiple-race disease screening of phytophthora foliar blight on single *Capsicum annuum* L. plants. *HortScience, 45*(10), 1563–1566.

Mottiar, Y., Vanholme, R., Boerjan, W., Ralph, J., & Mansfield, S. D., (2016). Designer lignins: Harnessing the plasticity of lignification. *Current Opinion in Biotechnology, 37*, 190–200.

Muñoz, I. V., Sarrocco, S., Malfatti, L., Baroncelli, R., & Vannacci, G., (2019). CRISPR-Cas for fungal genome editing: A new tool for the management of plant diseases. *Frontiers in Plant Science, 10*, 135.

Naoumkina, M. A., Zhao, Q., Gallego-Giraldo, L. I. N. A., Dai, X., Zhao, P. X., & Dixon, R. A., (2010). Genome-wide analysis of phenylpropanoid defense pathways. *Molecular Plant Pathology, 11*(6), 829–846.

Nekrasov, V., Wang, C., Win, J., Lanz, C., Weigel, D., & Kamoun, S., (2017). Rapid generation of a transgene-free powdery mildew resistant tomato by genome deletion. *Scientific Reports, 7*(1), 1–6.

Oliva, R., Ji, C., Atienza-Grande, G., Huguet-Tapia, J. C., Perez-Quintero, A., Li, T., Eom, J. S., et al., (2019). Broad-spectrum resistance to bacterial blight in rice using genome editing. *Nature Biotechnology, 37*(11), 1344–1350.

Peng, A., Chen, S., Lei, T., Xu, L., He, Y., Wu, L., Yao, L., & Zou, X., (2017). Engineering canker-resistant plants through CRISPR/Cas9-targeted editing of the susceptibility gene Cs LOB 1 promoter in citrus. *Plant Biotechnology Journal, 15*(12), 1509–1519.

Powell, K. S., Gatehouse, A. M. R., Hilder, V. A., & Gatehouse, J. A., (1995). Antifeedant effects of plant lectins and an enzyme on the adult stage of the rice brown planthopper, *Nilaparvata lugens*. *Entomologia Experimentalis et Applicata, 75*(1), 51–59.

Pramanik, D., Shelake, R. M., Park, J., Kim, M. J., Hwang, I., Park, Y., & Kim, J. Y., (2021). CRISPR/Cas9-mediated generation of pathogen-resistant tomato against tomato yellow leaf curl virus and powdery mildew. *International Journal of Molecular Sciences, 22*(4), 1878.

Pröbsting, M., Schenke, D., Hossain, R., Häder, C., Thurau, T., Wighardt, L., Schuster, A., et al., (2020). Loss of function of CRT1a (calreticulin) reduces plant susceptibility to *Verticillium longisporum* in both *Arabidopsis thaliana* and oilseed rape (*Brassica napus*). *Plant Biotechnology Journal, 18*(11), 2328–2344.

Reeck, G. R., (1997). Proteinase inhibitors and resistance of transgenic plants to insects. *Advances in Insect Control: The Role of Transgenic Plants,* 157–183.

Richter, C., Gristwood, T., Clulow, J. S., & Fineran, P. C., (2012). *In vivo* protein interactions and complex formation in the *Pectobacterium atrosepticum* subtype IF CRISPR/Cas system. *PloS One, 7*(12), e49549.

Savary, S., Willocquet, L., Pethybridge, S. J., Esker, P., McRoberts, N., & Nelson, A., (2019). The global burden of pathogens and pests on major food crops. *Nat. Ecol. Evol., 3*, 430–439.

Schiml, S., Fauser, F., & Puchta, H., (2016). Repair of adjacent single-strand breaks is often accompanied by the formation of tandem sequence duplications in plant genomes. *Proc. Natl. Acad. Sci. U S A, 113*, 7266–7271.

Selle, K., & Barrangou, R., (2015). Harnessing CRISPR–Cas systems for bacterial genome editing. *Trends in Microbiology, 23*(4), 225–232.

Seshadri, R., Leahy, S. C., Attwood, G. T., Teh, K. H., Lambie, S. C., Cookson, A. L., Eloe-Fadrosh, E. A., et al., (2018). Cultivation and sequencing of rumen microbiome members from the hungate1000 collection. *Nature Biotechnology, 36*(4), 359–367.

Shimatani, Z., Kashojiya, S., Takayama, M., Terada, R., Arazoe, T., Ishii, H., Teramura, H., et al., (2017). Targeted base editing in rice and tomato using a CRISPR-Cas9 cytidine deaminase fusion. *Nat. Biotechnol., 35*, 441–443.

Singh, S., (2016). Role of nonpathogenic fungi in inducing systemic resistance in crop plants against phytopathogens. In: *Microbial Inoculants in Sustainable Agricultural Productivity,* (pp. 69–83). Springer, New Delhi.

Stephens, J., & Barakate, A., (2017). Gene editing technologies – ZFNs, TALENs, and CRISPR/Cas9. In: Thomas, B., Murray, B. G., & Murphyp, D. J., (eds.), *Encyclopedia of Applied Plant Sciences* (2nd edn., pp. 157–161) Cambridge, MA: Academic Press. doi: 10.1016/B978-0-12-394807-6.00242-2.

Takahashi, Y., Berberich, T., Miyazaki, A., Seo, S., Ohashi, Y., & Kusano, T., (2003). Spermine signaling in tobacco: Activation of mitogen-activated protein kinases by spermine is mediated through mitochondrial dysfunction. *The Plant Journal, 36*(6), 820–829.

Tang, D., Wang, G., & Zhou, J. M., (2017). Receptor kinases in plant-pathogen interactions: More than pattern recognition. *The Plant Cell, 29*(4), 618–637.

Thompson, D. B., Aboulhouda, S., Hysolli, E., Smith, C. J., Wang, S., Castanon, O., & Church, G. M., (2018). The future of multiplexed eukaryotic genome engineering. *ACS Chem. Biol., 13*, 313–325.

Torrigiani, P., Rabiti, A. L., Bortolotti, C., Betti, L., Marani, F., Canova, A., & Bagni, N., (1997). Polyamine synthesis and accumulation in the hypersensitive response to TMV in *Nicotiana tabacum. The New Phytologist, 135*(3), 467–473.

Vakulskas, C. A., Dever, D. P., Rettig, G. R., Turk, R., Jacobi, A. M., Collingwood, M. A., Bode, N. M., et al., (2018). A high-fidelity Cas9 mutant delivered as a ribonucleoprotein complex enables efficient gene editing in human hematopoietic stem and progenitor cells. *Nat. Med., 24*, 1216–1224.

Van, B. T., & Van, D. A. G., (2020). Salicylic acid steers the growth–immunity tradeoff. *Trends Plant Sci., 25*, 566–576.

Walters, D. R., (2000). Polyamines in plant–microbe interactions. *Physiological and Molecular Plant Pathology, 57*(4), 137–146.

Walters, D. R., (2003). Polyamines and plant disease. *Phytochemistry, 64*(1), 97–107.

Waltz, E., (2018). With a free pass, CRISPR-edited plants reach market in record time. *Nat. Biotechnol., 36*, 6, 7. doi: 10.1038/nbt0118-6b.

Wan, D. Y., Guo, Y., Cheng, Y., Hu, Y., Xiao, S., Wang, Y., & Wen, Y. Q., (2020). CRISPR/Cas9-mediated mutagenesis of VvMLO3 results in enhanced resistance to powdery mildew in grapevine (*Vitis vinifera*). *Horticulture Research, 7*(1), 1–14.

Wang, F., Wang, C., Liu, P., Lei, C., Hao, W., & Gao, Y., (2016). Enhanced rice blast resistance by CRISPR/Cas9-targeted mutagenesis of the ERF transcription factor gene OsERF922. *PLoS One, 11*, e0154027. doi: 10.1371/journal.pone. 0154027.

Wang, P., Zhang, J., Sun, L., Ma, Y., Xu, J., Liang, S., Deng, J., et al., (2018). High efficient multisites genome editing in allotetraploid cotton (*Gossypium hirsutum*) using CRISPR/Cas9 system. *Plant Biotechnol. J., 16*, 137–150.

Wang, W., Xue, Z., Miao, J., Cai, M., Zhang, C., Li, T., Zhang, B., Tyler, B. M., & Liu, X., (2019). PcMuORP1, an oxathiapiprolin-resistance gene, functions as a novel selection marker for phytophthora transformation and CRISPR/Cas9 mediated genome editing. *Frontiers in Microbiology, 10*, 2402.

Wang, Y., Cheng, X., Shan, Q., Zhang, Y., Liu, J., Gao, C., & Qiu, J. L., (2014). Simultaneous editing of three homoeoalleles in hexaploid bread wheat confers heritable resistance to powdery mildew. *Nature Biotechnology, 32*(9), 947–951.

Wilson, F. M., Harrison, K., Armitage, A. D., Simkin, A. J., & Harrison, R. J., (2019). CRISPR/Cas9-mediated mutagenesis of phytoene desaturase in diploid and octoploid strawberry. *Plant Methods, 15*, 45.

Xu, R., Yang, Y., Qin, R., Li, H., Qiu, C., Li, L., Wei, P., & Yang, J., (2016). Rapid improvement of grain weight via highly efficient CRISPR/Cas9-mediated multiplex genome editing in rice. *J. Genet. Genomics, 43*, 529–532.

Zaidi, S. S. E. A., Mukhtar, M. S., & Mansoor, S., (2018). Genome editing: Targeting susceptibility genes for plant disease resistance. *Trends in Biotechnology, 36*(9), 898–906.

Zhang, Y., Bai, Y., Wu, G., Zou, S., Chen, Y., Gao, C., & Tang, D., (2017). Simultaneous modification of three homoeologs of TaEDR1 by genome editing enhances powdery mildew resistance in wheat. *Plant J., 91*, 714–724.

Zhao, H., Wang, X., Jia, Y., Minkenberg, B., Wheatley, M., Fan, J., Jia, M. H., et al., (2018). The rice blast resistance gene Ptr encodes an atypical protein required for broad-spectrum disease resistance. *Nature Communications, 9*(1), 1–12.

Zhou, X., Jacobs, T. B., Xue, L. J., Harding, S. A., & Tsai, C. J., (2015). Exploiting SNP s for biallelic CRISPR mutations in the outcrossing woody perennial Populus reveals 4-coumarate: CoA ligase specificity and redundancy. *New Phytologist, 208*(2), 298–301.

CHAPTER 12

CRISPR/Cas-Mediated Genome Editing for Improvement of Plant Traits

POOJA SARASWAT, AMBIKA CHATURVEDI, and RAJIV RANJAN*

Department of Botany, Faculty of Science, Dayalbagh Educational Institute, Dayalbagh, Agra – 282005, Uttar Pradesh, India

**Corresponding author. E-mail: rajivranjanbt@gmail.com*

ABSTRACT

Crop development has always been a major area of research for plant breeders. The present stress conditions can bring changes in the plant physiology and metabolism which thereby affect the yield and other possible traits. So far, the conventional methods of breeding were the only option, but now new innovations have made ways for efficient modifications in plant genomes in a shorter time period comparatively. The present scenario of plant research is focused on the development of sustainable, better, and efficient crop system. Considering this, many genome editing techniques have developed like ZFNs, TALENS, CRISPR/Cas, etc. The recently developed CRISPR/Cas system of plant genome editing is a technique used to obtain desired traits and minimizing the barriers in the field of genome editing. Different characteristics of plant in terms of its growth, yield, quality, etc., are well modified using CRISPR/Cas. This chapter focuses on the recent researches done and how CRISPR is well utilized for improvement in plant traits.

12.1 INTRODUCTION

The changes in the environment and climate have raised issues related to crop development. The different crops being important for human consumption

CRISPR/Cas-Mediated Genome Editing in Plants, Deepu Pandita, Anu Pandita (Eds.)

are highly affected by these changes, which may include biotic and abiotic stresses. So, there is an obvious need to improve crops against different forms of stresses with consideration to not affect the quality and yield. In addition, the focus is on increasing the nutritional content by using various techniques. Since, past decades there has been a wide use of various technologies to improve agriculture in a significant manner. Growing crops with high quality nutritive content is of great concern because it affects human health by supplying all the essential nutrients like carbohydrates, proteins, vitamins, minerals, fibers, and other compounds (Slavin and Lloyd, 2012). Even the scientific community is now more focused on the development of such crops with improved quality and for which different strategies have been applied in the improvement of different crop traits. It is observed that the crop plants generally lose their genetic diversity which brings changes in the phenotype as well. The genetic erosion can be seen both at species (selection during domestication) and genetic level (breeding) (Asano et al., 2011; Avni et al., 2017). Such variations in the genetic level makes the modern plants more exposed to stress, when compared to their wild types (Budak et al., 2015). With a thought to increase genetic diversity, plant breeders are more into finding ways to create new genetic variants. These variants can further be selected for better adaptability and yield. But the conventional strategies used are time taking and labor intensive and may also result in some unwanted mutations (Hussain et al., 2015). Recently, the development of genome editing (GE) technology has made genome modifications much easier with great advantages to the crop breeders.

The synthetic nucleases such as zinc-finger nucleases (ZFNs) (Miller et al., 2007; Sander et al., 2011) Transcription activator-like effector nucleases (TALENs) (Christian et al., 2010; Sanjana et al., 2012) and very recent RNA-guided CRISPR-Cas nuclease system (Cong et al., 2013; Horvath et al., 2010) have been so far known for their effective mutagenic applications. The ZFNs and TALENs use endonuclease catalytic domain and DNA binding proteins that result in target double-stranded breaks (DSBs) in the DNA. The ZFNs were the first synthetic nucleases for gene editing formed by combination of FokI endonuclease domain and zinc finger DNA-binding domain (Kim et al., 1996). Whereas, TALENs perform DNA cleavage by using two domains, one is a *FokI* cleavage domain and the other is a TALE proteins-based DNA-binding domain. Though when compared with the ZFNs, TALENs have higher target specificity with low off-targets (Joung and Sander, 2013). Hence, it is used in various applications as an effective tool (Shan et al., 2013; Wang et al., 2014). But due to their complex construct have limited their use in plants. While, CRISPR system is an RNA-based nuclease which is based on Watson-Crick base pairing to target the complementary DNA (Garneau et al., 2010; Jinek et al., 2012) with high efficiency, ease, specificity, and broad range of applications in various cell types and organisms. Like other techniques, CRISPR/Cas nucleases also promote gene editing at the target location.

12.2 CRISPR/CAS SYSTEMS FOR PLANTS

In 1987, CRISPR was first reported in *Escherichia coli* and recognized as an immune response in bacteria to fight against viral and other foreign DNA (Ishino et al., 1987) and later identified as an immune system response that identifies and cleaves foreign DNA (Jinek et al., 2012). Based on the Cas proteins CRISPR/Cas system has been classified into two classes (Class 1 and Class 2) in which each of them have three subtypes (I, II, III, IV, V, and VI). Among this type II system is widely used in gene editing. This system consists of Cas9 nuclease (*Streptococcus pyogenes*), crRNA array (guide RNAs and trans-activating crRNA-tracrRNA) and a Cas9 gene (Jinek et al., 12). As per the mechanism followed during gene editing precursor CRISPR RNA (pre-crRNAs) are processed into mature crRNAs in the presence of different Cas proteins as per different classes (Li et al., 2017). Like, CRISPR/Cas9 system uses Cas9 protein during maturation and interference steps. In this system the Cas9 nuclease identifies the protospacer-bearing DNA and cleaves at target position complementary to the mature crRNA. The PAM (Jinek et al., 2012) is the sequence in the target gene, which identifies the target locus and may vary on the basis of the CRISPR system. For example, *Streptococcus pyogenes*-based CRISPR system has PAM sequence in the form of 5′-NGG immediately after the target gene. Similarly, other orthologs may have different PAM sequence. After the cleavage at the target site near PAM sequence, the cell undergoes a DNA repair system which is either (NHEJ) or (HDR) pathway. The both pathways enable to achieve desired editing outcome. When the repair template is not present, DSBs are repaired by NHEJ mechanism which forms insertion/deletion type mutations (Wyman and Kanaar, 2006). And therefore, the CRISPR/Cas system can be used to knockdown, knock-in, and knock-out genes. Additionally, this is also employed in targeting of multiple genes which is eventually contributing to understanding different pathways. The CRISPR/Cas9 system for genome engineering has emerged as a burning topic amongst the plant scientists preferable for desired mutations in plants system. The chapter later discusses about the recent applications of CRISPR/Cas system in the improvement of plant traits.

12.3 CRISPR/CAS SYSTEMS IN PLANT TRAIT IMPROVEMENT

Crop quality and improved traits play an important role in determining the crop's market value at a commercial scale. In general, we see quality of crop is based on the internal and external traits. The external qualities largely focus on the physical aspects of the plant, including color, size, texture, etc. But the internal qualities usually cover the bioactive compounds (carotenoids, flavonoids, etc.), available in the plant. The CRISPR/Cas9 system is largely used in the development of plants with enhanced traits, which may improve their physical appearances, nutritive quality, etc. (Table 12.1), discussed further in the chapter. The diagram showing the process flow of the CRISPR/Cas technique in the plants is shown in Figure 12.1.

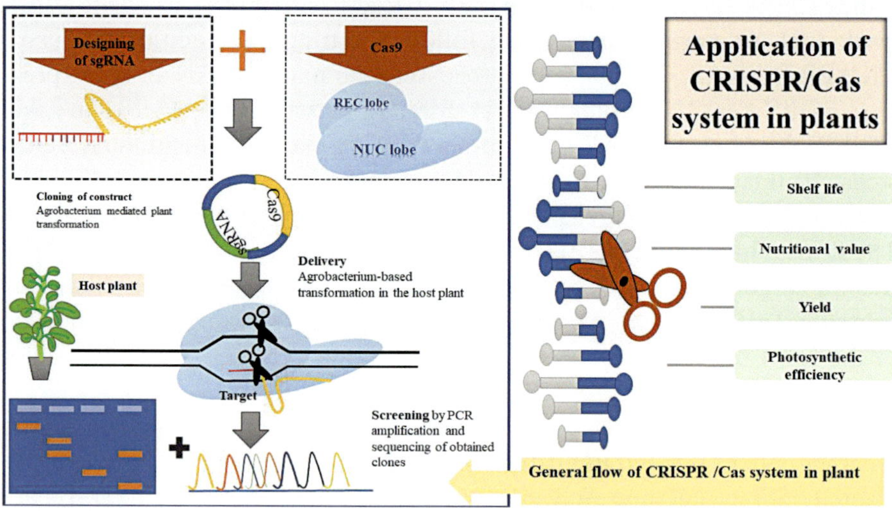

FIGURE 12.1 The diagrammatic representation of the process involved in gene-editing in plants using CRISPR/Cas9: sgRNA and Cas9 together forms the construct and is transferred into the plant through *Agrobacterium*-mediated transformation. The edited plants are checked and identified for their genotypes and further validated by sequencing.

12.3.1 NUTRITIONAL QUALITY

The market value and acceptance among consumers for any crop is determined by its eating and cooking quality (ECQ). Rice with low amylose content, i.e., 7–10% makes its texture sticky. A Waxy (Wx) gene is known for amylose synthesis and codes for granule-bound starch synthase I (GBSSI). There have been several studies done to improve amylose content by mutating the Wx gene in rice. In a study, CRISPR/Cas9 was used to mutate the waxy gene in two varieties of *Japonica*. The results showed that mutations occurred in the Waxy genes have caused a reduction in the amylose content and increased gluten with no changes found in the other desirable traits, which in turn offers an effective way to improve gluten rich rice cultivars (Zhang et al., 2017). Similarly, GABA is an inhibitory neurotransmitter, which functions in controlling anxiety and blood pressure (Nuss et al., 2015). Therefore, the development of GABA-rich plants can be good for food industries. In this respect, researchers have tried to focus on creating GABA enriched rice by knocking out *OsGAD3* by CRISPR/Cas9, which resulted in a seven-fold increase in GABA content (Akama et al., 2020). To improve GABA levels in tomato, multiplex gene editing was employed using CRISPR/Cas9 to knock-out SlGABA-Ts and SlSSADH by which GABA increased by 20-fold but had to compromise in the size and yield (Li et al., 2018). Raffinose family oligosaccharides (RFOs) are some soluble carbohydrates present in soybean which are hard to digest by humans and other animals. Therefore, it becomes necessary to minimize RFO levels as required in soyabean to enhance its nutritive quality. In a study, two galactinol synthase (GOLS) genes namely *GmGOLS1A* and *GmGOLS1B* were truncated and the findings suggested that the two genes are important in RFO pathway which can be used in further improvements related to efforts (Le et al., 2020). As saturated fatty acids are not good for health (linked to cardiovascular problems), they need to be removed or reduced in any crop system to enhance its nutritive value. Similarly, two genes coding for FATB known as GmFATB1a and GmFATB1b were knocked down using CRISPR/Cas9 system. The single fatb1a and fatb1b mutant resulted in the decrease of stearic acid and palmitic acid in the leaves. While double mutant (fatb1a:1b) showed more decrease in the fatty acid composition in comparison to single mutants. After the seed oil profile analysis, it was observed that the two mutants have a significant effect on lowering the steric acid and palmitic acid content without compromising the content of unsaturated fatty acids. In addition, the linoleic acid content was enhanced by significant levels, i.e., 1–3% approximately (Ma et al., 2021).

Aroma of any food is one of the qualities that are preferred after ECQ. Rice with good fragrance is popular in both continents like Asia and Europe which also increases its marketability and acceptance. In researches, it has been reported that most of the aromatic rice contains 2-acetyl-1-pyrroline (2AP) compound (Buttery et al., 1983). Studies have also suggested that BADH2 (encoding a betaine aldehyde dehydrogenase) is associated with aroma production (Singh et al., 2010). A synthesized cassette of 5.2 kb carotenoid was inserted in the rice genome using CRISPR/Cas9 system. Marker-free plants with enhanced carotenoid content were obtained without affecting the yield and other traits (Dong et al., 2020). Similarly, in another study, a non-aromatic rice variety has been introduced with aroma by developing new alleles of *OsBADH2* (Ashokkumar et al., 2020).

Carotenoids are well known for their antioxidant property and helpful in eye-related problems. But as humans are not able to synthesize carotenoids, they have to gain it from diet. Also, other pigments like phytoene and lycopene reduces chances of cancer and heart related diseases. Banana being an important crop is worldwide famous for its nutrition content. But due to the triploid genome and parthenocarpic nature, it becomes difficult to manipulate it at genetic level easily. In a work presented by Kaur and colleagues, mutation has been introduced in the *lycopene epsilon-cyclase* (*LCYε*) gene to develop banana cultivars enriched with β-carotene. The result of the metabolic profiling done on the mutated line showed an increase in the β-carotene content to a significant level (up to 6-fold) as compared to the non-edited ones (Kaur et al., 2020). Two genes named as IbGBSSI (encoding granule-bound starch synthase I) and IbSBEII (encoding starch branching enzyme II) important in the starch biosynthetic pathway have been identified and targeted using CRISPR/Cas binary vector in two cultivars of sweet potato (*Ipomoea batatas*) rich in carotenoid (Taizhong6) and starch (Xushu22). The mutation performed by CRISPR/Cas system generated about 72 and 35 transgenic lines of Xushu22 and Taizhong6 respectively which were further used for analysis. The results concluded that transgenic lines obtained did not show any significant change in the starch content when compared to the wild type. But, IbGBSSI showed decrease while the knock-out of IbSBEII increased the amylose content in the allopolyploid sweet potato (Wang et al., 2019).

12.3.2 *YIELD*

The crop improvement is significant when the yield is also considered as an important part of the improvement process. As many crop plants face major yield losses due to different factors. Like, "Easy-to-shatter" is one of

TABLE 12.1 Recent Work Done for the Improvement of Plant Traits Using CRISPR/Cas System

Plant	Gene Name	Method of Mutagenesis	Result Obtained	References
Tomato	ALC gene	Gene replacement	Long shelf life	Yu et al. (2017)
Rice (*Oryza sativa*)	OsAAP3	Knock in	Grain yield	Lu et al. (2018)
Rice (*Oryza sativa*)	*qSH1* gene	–	Reduced seed shattering	Sheng et al. (2020)
Rice (*Oryza sativa*)	*OsGRF4*	Knock in	Grain size	Li et al. (2016)
Rice (*Oryza sativa*)	*MADS78 and MADS79*	–	Early seed development	Paul et al. (2020)
Rice (*Japonica*)	*GS3* and *GL3.1*	Gene knock-out	Improve grain size	Yuyu et al. (2020)
Rice	GLW2; GS9; TMS5; GW2, GW5, and TGW6; Gn1a, DEP1, GS3, IPA1	Gene knock-out/gene knock-in	Yield improvement	Li et al. (2016); Zhao et al. (2018); Xu et al. (2016); Zhou et al. (2016)
Solanum lycopersicum	lncRNA1459	Gene knock-out	Long shelf life	Soyk et al. (2017)
Triticum aestivum	GASR7; α-gliadin	Gene knock-out	Kernel weight; low gluten	Zhang et al. (2018); Sánchez-León et al. (2018)
Rice	SBEI and SBEIIb	Gene knock-out	High-amylose content	Sun et al. (2017)
Soybean	*GmGOLS1A* and *GmGOLS1B*	Gene knock-out	Raffinose family oligosaccharides (RFO)	Le et al. (2020)
Glycine max	*FAD2-1A*, *FAD2-1B*	Gene knock-out	Improved oil quality	Kim et al. (2017)
Zea mays	*PPR*, *RPL*	Knock-out	Reduced zein protein	Qi et al. (2016)
Solanum tuberosum	*GBSS*	Gene knock-out	Increase amylopectin/amylose	Andersson et al. (2017)
Soybean	*GmFATB1a* and *GmFATB1b*	Gene knock-out	Decrease in palmitic and stearic acid content	Ma et al. (2021)
Banana	*Lycopene epsilon-cyclase* (*LCYε*) gene	Gene knock-out	Enhanced β-carotene	Kaur et al. (2020)

TABLE 12.1 *(Continued)*

Plant	Gene Name	Method of Mutagenesis	Result Obtained	References
Rapeseed (*Brassica napus*)	*BnaMAX1*	Gene knock-out	Increased yield	Zheng et al. (2020)
Rice	*Os03g0603100*, *Os03g0568400*, and *GL3*.	Gene knock-out	Improved aroma and yield	Usman et al. (2020)
Rice	*OsHXK1*	Gene knock-out	Photosynthetic efficiency and yield	Zheng et al. (2021)
Banana	MA-ACO1	Gene knock-in	Shelf life	Hu et al. (2020)

the traits responsible for yield loss in the rice plant which requires a need to emphasize more on the development of such rice cultivars which shows low seed shattering. Keeping this in mind, research has been conducted in *Indica* rice in which mutation in the qSH1 gene was done using CRISPR/Cas9 to develop less seed shattering and more yield. In the study, it was seen that the homozygous mutant lines of *qsh1* show significant reduction in seed shattering relative to wild type (HR1128-WT) (Sheng et al., 2020). In studies, it is known that *ARGOS8* negatively regulates the ethylene responses as its overexpression reduces ethylene sensitivity and improves yield. New *ARGOS8* variants have been developed using CRISPR/Cas-mediated technology. The field observation concluded that *ATGOS8* variants shows increase in yield under flowering stress conditions and no effect under normal water conditions (Shi et al., 2017). Two *BnaMAX1* genes in rapeseed are similar to *Arabidopsis MAX1* gene which is responsible in regulation of axillary bud outgrowth and plant height. Two sgRNAs has been designed to edit *BnaMAX1* homologs in rapeseed with efficient mutations (56.30%–67.38%) in T_0 plants. The results of knock-out of four *BnaMAX1* alleles simultaneously developed semi-dwarf plants with an increase in yield when compared to their wild types. Such mutation provides development of desired germplasm for more breeding practices for better yield in rapeseed (Zheng et al., 2020).

The better grain yield and quality does not often develop together as they are antagonistic to each other but of great demand to breeders and consumers. The function of grain yielding gene belonging to the cytochrome P450 family is still not known in rice. In a study performed by Usman and coworkers (2020), three genes related to cytochrome p450 (Os03g0603100, Os03g0568400, and GL3.2) and OsBADH2 have been edited all together using CRISPR/Cas9 system for high yield and aroma. The caused mutations resulted in the increase of 2AP content, grain size, and grain quantity and no effect on other traits. The proteomic and deep RNA sequencing of the edited plants revealed that the regulation of Deep RNA sequencing of the edited plants revealed and proteomic results revealed the regulation of grain size and cell cycle controlled by cytochrome P450 family (Usman et al., 2020). Also, the grain size was increased by editing OsSPL16/qGW8 in rice via CRISPR/Cas9 and the variation obtained from mutations were analyzed by proteomics. Like many studies, the desired trait (yield) was improved with no changes in the agronomic trait (Usman et al., 2021). The mutant gene of *abnormal cytokinin response1 repressor1* (*are1*) in rice shows delay in senescence and therefore increases grain yield under low nitrogen availability. But, the role of *ARE1* was not much known in the case of wheat. A wheat cultivar (ZhengMai

7698) has been used to isolate and characterize three *TaARE1* homoeologs. Later, these were targeted by CRISPR/Cas9 to obtain lines with enhanced grain yield under nitrogen-deficient conditions in fields (Zhang et al., 2021).

The development and production of micronutrient-rich crops is a sustainable approach, as biofortification of any crop enhances its nutritional value and increases its demand in the commercial market. The Fe (iron) content in the grain has been increased/improved in rice, by knocking vacuolar iron transporter (VIT) genes (OsVIT2) using CRISPR/Cas9. The mutations observed in OsVIT2 caused the distribution of Fe to various parts of the grain, including embryo and endosperm, with no negative effect on its yield (Che et al., 2021). Also, in a study the arsenite tolerant 1 (astol1) mutant has been gained in rice. This helped in increasing the level of selenium (Se), which has antioxidant properties to the human body and therefore is an essential micronutrient. The genome editing can also help in the regulation of gene expression in the homeostasis of ion in both rice and wheat to develop micronutrient-enriched varieties (Sun et al., 2021).

12.3.3 SHELF LIFE

Maintaining the fruit texture is another important property for the production of any crop. So, the modification to be done in the traits involved in long shelf life is a continual process for breeders. The use of CRISPR/Cas9 technology is showing its great potential in increasing the shelf life of many fruits including bananas and tomatoes. One study suggested that mutations in the alc genes affect the increase in shelf life and does not affect other traits like color and fragrance of the fruit the mutation not only prolonged shelf life but also kept fruit color and fragrance (Casals et al., 2011). Mutations in *the ALC* gene have been done successfully using HDR-mediated gene replacement in tomato plant, and the resulting T1 generation with alc homozygous mutants showed great storage ability (Yu et al., 2017). Ethylene is studied as a factor influencing the preservation and shelf life of banana. A gene called MA-ACO1 is associated with ethylene synthesis which promotes fruit ripening of banana (Elitzur et al., 2016).

12.3.4 OTHER PLANT TRAITS

CRISPR/Cas9 technology has also been employed to optimize other traits like the shape and size of any crop as per the needs. Many genes or quantitative trait loci (QTLs) are often seen involved in the appearance related characters for any crop. Like two QTLs *GS3* and *GL3.1* are known for their negative regulation in grain size. Based on this, CRISPR/Cas9 system has been used in *japonica* rice in targeting of multiple genes *GS3* and *GL3.1* and their knock-out improved the grain size in rice (Yuyu et al., 2020). The color of the plant or fruit is determined by the pigments present like anthocyanin, carotenoids, and polyphenols. In studies, it has been reported that the pink color of plants is obtained due to the absence of flavonoids in peel. Therefore, manipulations can be done through CRISPR/Cas9 in the fruit color by targeting such genes involved in respective pigment synthesis. A MYB12 gene is a transcription factor for regulation of flavonoid biosynthetic pathway which therefore determines the flavonoids accumulation and results in pink color phenotype. Pink colored tomatoes have been developed by using the CRISPR/Cas system by the knock-out of SlMYB12 gene (Yang et al., 2019).

12.4 CONCLUSION

The advanced technique of genome editing, CRISPR/Cas9 is doing wonders to the plant research community because of its ability to modify plant traits efficiently. It can develop new varieties or lines with better and improved trait features of desired function. There are many quality traits that have been improved successfully using CRISPR/Cas system. Even many edited crops have been launched in the commercial markets with good efficacy. However, some factors limit the use of CRISPR/Cas system, which requires attention to develop the better application of this technique. Apart from the technical problems, there are society concerns related to genome-edited crop utilization, which are under government regulations which becomes another factor limiting its application. But still, we researchers see it as the beginning of the CRISPR revolution as new opportunities are developing day by day, which ultimately lead to a bright future.

KEYWORDS

- **CRISPR/Cas9**
- **double-stranded breaks**
- **eating and cooking quality**
- **galactinol synthase**
- **granule-bound starch synthase I**
- **knock-in**
- **knock-out**
- **shelf-life**
- **trait improvement**
- **yield**

REFERENCES

Akama, K., Akter, N., Endo, H., Kanesaki, M., Endo, M., & Toki, S., (2020). An *in vivo* targeted deletion of the calmodulin-binding domain from rice glutamate decarboxylase 3 (OsGAD3) increases γ-aminobutyric acid content in grains. *Rice, 13*, 20.

Andersson, M., Turesson, H., Nicolia, A., Fält, A. S., Samuelsson, M., & Hofvander, P., (2017). Efficient targeted multiallelic mutagenesis in tetraploid potato (*Solanum tuberosum*) by transient CRISPR-Cas9 expression in protoplasts. *Plant Cell Rep., 36*, 117–128.

Asano, K., Yamasaki, M., Takuno, S., Miura, K., Katagiri, S., Ito, T., Doi, K., et al., (2011). Artificial selection for a green revolution gene during japonica rice domestication. *PNAS, 108*, 11034–11039.

Ashokkumar, S., Jaganathan, D., Ramanathan, V., Rahman, H., Palaniswamy, R., Kambale, R., & Muthurajan, R., (2020). Creation of novel alleles of fragrance gene OsBADH2 in rice through CRISPR/Cas9 mediated gene editing. *PloS One, 15*, e0237018.

Avni, R., Nave, M., Barad, O., Baruch, K., Twardziok, S. O., Gundlach, H., & Distelfeld, A., (2017). Wild emmer genome architecture and diversity elucidate wheat evolution and domestication. *Science, 357*, 93–97.

Budak, H., Hussain, B., Khan, Z., Ozturk, N. Z., & Ullah, N., (2015). From genetics to functional genomics: Improvement in drought signaling and tolerance in wheat. *Front. Plant Sci., 6*, 1012.

Buttery, R. G., Ling, L. C., Juliano, B. O., & Turnbaugh, J. G., (1983). Cooked rice aroma and 2-acetyl-1-pyrroline. *J. Agric. Food Chem., 31*, 823–826.

Casals, J., Cebolla-Cornejo, J., Roselló, S., Beltrán, J., Casañas, F., & Nuez, F., (2011). Long-term postharvest aroma evolution of tomatoes with the alcobaça (alc) mutation. *Eur. Food Res. Technol., 233*, 331–342.

Che, J., Yamaji, N., & Ma, J. F., (2021). Role of a vacuolar iron transporter OsVIT2 in the distribution of iron to rice grains. *New Phytol., 230*, 1049–1062.

Christian, M., Cermak, T., Doyle, E. L., Schmidt, C., Zhang, F., Hummel, A., & Voytas, D. F., (2010). Targeting DNA double-strand breaks with TAL effector nucleases. *Genetics., 186*, 757–761.

Cong, L., Ran, F. A., Cox, D., Lin, S., Barretto, R., Habib, N., & Zhang, F., (2013). Multiplex genome engineering using CRISPR/Cas systems. *Science, 339*, 819–823.

Dong, O. X., Yu, S., Jain, R., Zhang, N., Duong, P. Q., Butler, C., Li, Y., et al., (2020). Marker-free carotenoid-enriched rice generated through targeted gene insertion using CRISPR-Cas9. *Nat. Commun., 11*, 1–10.

Elitzur, T., Yakir, E., Quansah, L., Zhangjun, F., Vrebalov, J., Khayat, E., Giovannoni, J. J., & Friedman, H., (2016). Banana MaMADS transcription factors are necessary for fruit ripening and molecular tools to promote shelf-life and food security. *Plant Physiol., 171*, 380–391.

Garneau, J. E., Dupuis, M. È., Villion, M., Romero, D. A., Barrangou, R., Boyaval, P., Fremaux, C., et al., (2010). The CRISPR/Cas bacterial immune system cleaves bacteriophage and plasmid DNA. *Nature, 468*, 67–71.

Horvath, P., & Barrangou, R., (2010). CRISPR/Cas, the immune system of bacteria and archaea. *Science, 327*, 167–170.

Hu, C., Sheng, O., Deng, G., He, W., Dong, T., Yang, Q., Dou, T., et al., (2021). CRISPR/Cas9-mediated genome editing of MaACO1 (aminocyclopropane-1-carboxylate oxidase-1) promotes the shelf life of banana fruit. *Plant Biotechnol. J., 19*, 654.

Hussain, B., (2015). Modernization in plant breeding approaches for improving biotic stress resistance in crop plants. *Turk. J. Agric. For., 39*, 515–530.

Ishino, Y., Shinagawa, H., Makino, K., Amemura, M., & Nakata, A., (1987). Nucleotide sequence of the iap gene, responsible for alkaline phosphatase isozyme conversion in *Escherichia coli*, and identification of the gene product. *J. Bacteriol., 169*, 5429–5433.

Jinek, M., Chylinski, K., Fonfara, I., Hauer, M., Doudna, J. A., & Charpentier, E., (2012). A programmable dual-RNA–guided DNA endonuclease in adaptive bacterial immunity. *Science, 337*, 816–821.

Joung, J. K., & Sander, J. D., (2013). TALENs: A widely applicable technology for targeted genome editing. *Nat. Rev. Mol. Cell Biol., 14*, 49–55.

Kaur, N., Alok, A., Kumar, P., Kaur, N., Awasthi, P., Chaturvedi, S., Pandey, P., et al., (2020). CRISPR/Cas9 directed editing of lycopene epsilon-cyclase modulates metabolic flux for β-carotene biosynthesis in banana fruit. *Metab. Eng., 59*, 76–86.

Kim, H., Kim, S. T., Ryu, J., Kang, B. C., Kim, J. S., & Kim, S. G., (2017). CRISPR/Cpf1-mediated DNA-free plant genome editing. *Nat. Comm., 8*, 1–7.

Kim, Y. G., Cha, J., & Chandrasegaran, S., (1996). Hybrid restriction enzymes: Zinc finger fusions to FokI cleavage domain. *Proc. Natl. Acad. Sci. USA, 93*, 1156–1160.

Le, H., Nguyen, N. H., Ta, D. T., Le, T. N. T., Bui, T. P., Le, N. T., & Chu, H. H., (2020). CRISPR/Cas9-mediated knock-out of galactinol synthase-encoding genes reduces raffinose family oligosaccharide levels in soybean seeds. *Front. Plant Sci., 11*, 2033.

Li, C., Unver, T., & Zhang, B., (2017). A high-efficiency CRISPR/Cas9 system for targeted mutagenesis in cotton (*Gossypium hirsutum* L.). *Sci. Rep., 7*, 43902.

Li, M., Li, X., Zhou, Z., Wu, P., Fang, M., Pan, X., Lin, Q., et al., (2016). Reassessment of the four yield-related genes *Gn1a*, *DEP1*, *GS3*, and *IPA1* in rice using a CRISPR/Cas9 system. *Front. Plant Sci., 7*, 377.

Li, R., Li, R., Li, X., Fu, D., Zhu, B., Tian, H., Luo, Y., & Zhu, H., (2018). Multiplexed CRISPR/Cas9-mediated metabolic engineering of γ-aminobutyric acid levels in *Solanum lycopersicum*. *Plant Biotechnol. J., 16*, 415–427.

Li, S., Gao, F., Xie, K., Zeng, X., Cao, Y., Zeng, J., He, Z., et al., (2016). The OsmiR396c-OsGRF4-OsGIF1 regulatory module determines grain size and yield in rice. *Plant Biotechnol. J., 14*, 2134–2146.

Lu, K., Wu, B., Wang, J., Zhu, W., Nie, H., Qian, J., Huang, W., & Fang, Z., (2018). Blocking amino acid transporter Os AAP 3 improves grain yield by promoting outgrowth buds and increasing tiller number in rice. *Plant Biotech. J., 16*, 1710–1722.

Ma, J., Sun, S., Whelan, J., & Shou, H., (2021). CRISPR/Cas9-mediated knock-out of GmFATB1 significantly reduced the amount of saturated fatty acids in soybean seeds. *Inter. J. Mol. Sci., 22*, 3877.

Miller, J. C., Holmes, M. C., Wang, J., Guschin, D. Y., Lee, Y. L., Rupniewski, I., Beausejour, C. M., et al., (2007). An improved zinc-finger nuclease architecture for highly specific genome editing. *Nature Biotechnol., 25*, 778–785.

Nuss, P., (2015). Anxiety disorders and GABA neurotransmission: A disturbance of modulation. *Neuropsychiatr. Dis. Treat., 11*, 165.

Paul, P., Dhatt, B. K., Miller, M., Folsom, J. J., Wang, Z., Krassovskaya, I., Liu, Ket al., (2020). MADS78 and MADS79 are essential regulators of early seed development in rice. *Plant Physiol., 182*, 933–948.

Qi, W., Zhu, T., Tian, Z., Li, C., Zhang, W., & Song, R., (2016). High-efficiency CRISPR/Cas9 multiplex gene editing using the glycine tRNA processing system-based strategy in maize. *BMC Biotech., 16*, 58.

Sánchez-León, S., Gil-Humanes, J., Ozuna, C. V., Giménez, M. J., Sousa, C., Voytas, D. F., & Barro, F., (2018). Low-gluten, nontransgenic wheat engineered with CRISPR/Cas9. *Plant Biotech. J., 16*, 902–910.

Sander, J. D., Dahlborg, E. J., Goodwin, M. J., Cade, L., Zhang, F., Cifuentes, D., Curtin, S. J., et al., (2011). Selection-free zinc-finger-nuclease engineering by context-dependent assembly (CoDA). *Nature Methods, 8*, 67–69.

Sanjana, N. E., Cong, L., Zhou, Y., Cunniff, M. M., Feng, G., & Zhang, F., (2012). A transcription activator-like effector toolbox for genome engineering. *Nature Prot., 7*, 171–192.

Shan, Q., Wang, Y., Chen, K., Liang, Z., Li, J., Zhang, Y., Zhang, K., et al., (2013). Rapid and efficient gene modification in rice and *Brachypodium* using TALENs. *Mol. Plant, 6*, 1365–1368.

Sheng, X., Sun, Z., Wang, X., Tan, Y., Yu, D., Yuan, G., Yuan, D., & Duan, M., (2020). Improvement of the rice "easy-to-shatter" trait via CRISPR/Cas9-mediated mutagenesis of the qSH1 gene. *Front. Plant Sci., 11*, 619.

Shi, J., Gao, H., Wang, H., Lafitte, H. R., Archibald, R. L., Yang, M., Hakimi, S. M., et al., (2017). ARGOS 8 variants generated by CRISPR-Cas9 improve maize grain yield under field drought stress conditions. *Plant Biotechnol. J., 15*, 207–216.

Singh, A., Singh, P. K., Singh, R., Pandit, A., Mahato, A. K., Gupta, D. K., Tyagi, K., et al., (2010). SNP haplotypes of the BADH1 gene and their association with aroma in rice (*Oryza sativa* L.). *Mol. Breed. New Strategy. Plant Improv., 26*, 325–338.

Slavin, J. L., & Lloyd, B., (2012). Health benefits of fruits and vegetables. *Adv. Nutr., 3*, 506–516.

Soyk, S., Müller, N. A., Park, S. J., Schmalenbach, I., Jiang, K., Hayama, R., Zhang, L., et al., (2017). Variation in the flowering gene SELF PRUNING 5G promotes day neutrality and early yield in tomato. *Nat Genet., 49*, 162.

Sun, S. K., Xu, X. J., Tang, Z., Tang, Z., Huang, X. Y., Wirtz, M., Hell, R., & Zhao, F. J., (2021). A molecular switch in sulfur metabolism to reduce arsenic and enrich selenium in rice grain. *Nat. Commun., 12*, 1392.

Usman, B., Nawaz, G., Zhao, N., Liao, S., Qin, B., Liu, F., Liu, Y., & Li, R., (2021). Programmed editing of rice (*Oryza sativa* L.) OsSPL16 gene using CRISPR/Cas9 improves grain yield by modulating the expression of pyruvate enzymes and cell cycle proteins. *Intern J. Mol. Sci., 22*, 249.

Usman, B., Nawaz, G., Zhao, N., Liu, Y., .Li, R., (2020). Generation of high yielding and fragrant rice (*Oryza sativa* L.) lines by CRISPR/Cas9 targeted mutagenesis of three homoeologs of cytochrome P450 gene family and OsBADH2 and transcriptome and proteome profiling of revealed changes triggered by mutations. *Plants, 9*, 788.

Wang, H., Wu, Y., Zhang, Y., Yang, J., Fan, W., Zhang, H., Zhao, S., et al., (2019). CRISPR/Cas9-based mutagenesis of starch biosynthetic genes in sweet potato (Ipomoea Batatas) for the improvement of starch quality. *Inter. J. Mol. Sci., 20*, 4702.

Wang, Y., Cheng, X., Shan, Q., Zhang, Y., Liu, J., Gao, C., & Qiu, J. L., (2014). Simultaneous editing of three homoeoalleles in hexaploid bread wheat confers heritable resistance to powdery mildew. *Nat. Biotechnol., 32*, 947–951.

Wyman, C., & Kanaar, R., (2006). DNA double-strand break repair: All's well that ends well. *Annu. Rev. Genet., 40*, 363–383.

Xu, R., Yang, Y., Qin, R., Li, H., Qiu, C., Li, L., Wei, P., & Yang, J., (2016). Rapid improvement of grain weight via highly efficient CRISPR/Cas9-mediated multiplex genome editing in rice. *J. Genet. Genom., 43*, 529–532.

Yang, T., Deng, L., Zhao, W., Zhang, R., Jiang, H., Ye, Z., Li, C. B., & Li, C., (2019). Rapid breeding of pink-fruited tomato hybrids using the CRISPR/Cas9 system. *J. Genet. Genom., 46*, 505.

Yu, Q. H., Wang, B., Li, N., Tang, Y., Yang, S., Yang, T., Xu, J., et al., (2017). CRISPR/Cas9-induced targeted mutagenesis and gene replacement to generate long-shelf life tomato lines. *Sci. Rep., 7*, 1–9.

Yuyu, C., Aike, Z., Pao, X., Xiaoxia, W., Yongrun, C., Beifang, W., & Yingxin, Z., (2020). Effects of GS3 and GL3. 1 for grain size editing by CRISPR/Cas9 in rice. *Rice Sci., 27*, 405–413.

Zhang, J., Zhang, H., Botella, J. R., & Zhu, J. K., (2018). Generation of new glutinous rice by CRISPR/Cas9-targeted mutagenesis of the waxy gene in elite rice varieties. *J. Integr. Plant Biol., 60*, 369–375.

Zhang, J., Zhang, H., Li, S., Li, J., Yan, L., & Xia, L., (2021). Increasing yield potential through manipulating of an ARE1 ortholog related to nitrogen use efficiency in wheat by CRISPR/Cas9. *J. Integr. Plant Biol.*

Zhao, D. S., Li, Q. F., Zhang, C. Q., Zhang, C., Yang, Q. Q., Pan, L. X., Ren, X. Y., et al., (2018). *GS9* acts as a transcriptional activator to regulate rice grain shape and appearance quality. *Nat. Commun., 9*, 1240.

Zheng, M., Zhang, L., Tang, M., Liu, J., Liu, H., Yang, H., Fan, S., et al., (2020). Knock-out of two Bna MAX 1 homologs by CRISPR/Cas9-targeted mutagenesis improves plant architecture and increases yield in rapeseed (*Brassica napus* L.). *Plant Biotechnol. J., 18*, 644–654.

Zheng, S., Ye, C., Lu, J., Liufu, J., Lin, L., Dong, Z., Li, J., & Zhuang, C., (2021). Improving the rice photosynthetic efficiency and yield by editing OsHXK1 via CRISPR/Cas9 system. *Int. J. Mol. Sci., 22*, 9554.

Zhou, H., He, M., Li, J., Chen, L., Huang, Z., Zheng, S., Zhu, L., et al., (2016). Development of commercial thermo-sensitive genic male sterile rice accelerates hybrid rice breeding using the CRISPR/Cas9-mediated *TMS5* editing system. *Sci. Rep., 6*, 37395.

CHAPTER 13

Investigation of Plant-Microbe Interactions by CRISPR/Cas for Sustainable Agriculture

POOJA BHADRECHA,[1*] SHILPY SINGH,[2] and S. RUTH ASSUMI[3]

[1]*University Institute of Biotechnology, Chandigarh University, Mohali, Punjab, India*

[2]*Department of Biotechnology, Noida International University, Noida, Uttar Pradesh, India*

[3]*Division of System Research and Engineering, ICAR Research Complex of NEH Region, Umiam, Meghalaya, India*

[*]*Corresponding author. E-mail: pbhadrechapb@gmail.com*

ABSTRACT

The profiling of the plant microbiome is a growing idea to investigate relationships between plants and microbes. Microbiota expands its plant capacity through different methods to adapt to environmental circumstances. Modern technologies like next-generation sequencing, omics (including transcriptomics, proteomics, metagenomics, and metabolomics), and computing techniques help to analyze the communities-level molecular features for plant-microbe interactions regulating the characteristics of plants. Numerous researches have studied various plant microbiome components and the impact of host genotyping on distinct microbiome features. To investigate the spectrum of various mutualistic and/or harmful relationships within a complex ecosystem involving various species, by removing particular species or by just deleting some targeted alleles, and analyze its consequences on the

CRISPR/Cas-Mediated Genome Editing in Plants, Deepu Pandita, Anu Pandita (Eds.)

whole microbiome is possible with CRISPR/Cas9 system. By targeting the processes of central dogma, well-programmed editing of target genome is becoming very popular, and it relies on alterations caused by gene silencing and insertion of desired genes in target site at the RNA, epigenome editing. Editing plant genome by procuring CRISPR technology has been proved to be efficient in enhancing agronomic traits of the crops, e.g., nutrients, yields, tolerance to biotic and abiotic stress.

13.1 INTRODUCTION

In the environment, flora and fauna interact continually to many microbes throughout their lifetime. Early evolutionary period's people can be exposed to diverse microbial environment, which enhances their potential for adaptation to healthy living (Kirjavainen et al., 2019). Likewise, plants live with microorganisms, including archaeans, protista, bacteria, and fungi, known as microbiota (Bulgarelli et al., 2013). The microbial life began at the very beginning of life, indicating that during a long period prior to the adaption for plants to terrestrial life that is prior 450 million of year's microbes and microbes' interactions developed and diversified (Hassani et al., 2018). Higher plants or algae have incorporated *cyanobacteria*l endosymbionts, now known as chloroplasts or plastids, throughout the evolution (Keeling, 2010). Therefore, the development of plants and microorganisms share common roots and interdependent survival. As a result, increased focus has paid to "plant microbiota" that occurs on or around plant components (Waghunde et al., 2017). Profiling the genomic composition of all the microorganisms associated with plant systems has been gaining popularity. Microbiome explores all aspects via distinct methods to adapt to various environmental circumstances. Some of the interactions are beneficial to both, some are beneficial to one but neutral to the other one and some may be beneficial to one but harmful to the other. Plant-microbe's (PM) beneficial interactions include stimulating plant growth, biotic or abiotic stress protection by promoting plant immunity or plant defense induction, adapting to fluctuated environment, mycorrhizal symbiosis, nutrient absorption, and the processing of unused forms of nutrients into plant accessible forms (Hassani et al., 2018). PM interactions are two-way, as microorganisms also produce nutrients from their host plants. The exchange among plants and microorganisms may grow into a partnership based over its effect upon plant health, that is mutually (each parties' benefit, symbiotic), neutrally (one-partner benefit, commensalistic) and noxious (deleterious for host plant, pathogenic) (Thrall et al., 2007; Rodriguez et al., 2019). The PM interactions in sustainable agricultural practices and the environment are essential for food supply and community health, respectively. A study of host plant in conjunction with its associated microbiomes (also termed holobiont) reveals coevolution of interactions between plant-microbe, plant-plant, and microbe-microbe (Rosenberg and Zilber-Rosenberg, 2018).

For certain crops and related microbes, genetic data concerning PM interactions became accessible. Insights into molecular component of PM interaction at the genetic level are one critical step towards improved utilization of microbiome in farming (Bulgarelli et al., 2015; Cavicchioli et al., 2019). In this respect, breakthrough technologies like CRISPR, technology for genome editing, capable of generating precise genetic modifications (GMs) (Knott and Doudna, 2018) are the optimal platform for the knowledge and accuracy of the basic PM interaction to increase crop efficiency and the resistance of disease (Zaidi et al., 2018). In this study, we put up a critical evaluation of current updates on PM interactions concerning the diversity, structure, and determinants affecting plant microbiota development. We explore furthermore the CRISPR technologies and its uses for useful symbiotic or pathogenic PM interactions in sustainable agriculture practice development. We also discuss the limits, regulatory problems, and potential routes for the future by using CRISPR-mediated agricultural techniques in either plants or microorganisms.

Unveiling of antimicrobial characteristics offered by the microbiome associated with plant systems, at the molecular level, has been rendered by knock-down tools and techniques like CRISPR/Cas9 (Bikard et al., 2014). Various mutualistic and harmful relationships in a complex ecosystem require removal of target species or deletion of target alleles. This level of specificity has made the technique popular among researchers aiming at exploring natural/environmental complex microbiomes. The codes for intruding nucleic acids are well utilized in preserved form, in case of technique CRISPR-Cas, which conserves DNA segments with short and repetitive bases (CRISPR RNA, crRNA). Genes like the Cas9 and spacer sequences which are complementary to specific (target) sequences in the genome and are present in genomic locus, producing CRISPR array with crRNAs and tracrRNA (i.e., trans-activating crRNA) are content of this system. Activated crRNA/tracrRNA complexes are produced by the activity of RNA polymerase III, and direct the Cas proteins for targeting particular gene loci, resulting in broken double strands on the target sequences (Hille et al., 2018). The option for selecting and hence targeting specific sites on the genome is possible by using the guide RNA (sgRNA) and utilizing it with specific endonucleases (Cas9) gives this privilege to researchers and breeders (Jinek et al., 2012). CRISPR techniques were instantly accepted by researchers worldwide because they offer highly efficient editing, enable editing of multiplexes and are user friendly. The plant mutants produced via conventional breeding and natural mutations are recorded to be similar to mutants generated by implementations of genome-editing technologies and are devoid of foreign genes. Entities like transgenic organisms and RNAi lines always fall under the category of genetically modified organisms (GMOs), but the same doesn't apply to CRISPR lines because here the foreign DNA segment doesn't inevitably integrate in host genome to be expressed in upcoming progenies. Still, in some places, e.g., Europe, such practices have been prohibited from application to plants.

Researchers have successfully created herbicide-resistant crops and enhanced crop-domestication by utilizing various techniques to edit plant genome and bring about desired changes (Li et al., 2018; Sun et al., 2016). Several issues that may diminish the efficiency of CRISPR-Cas technology are off-target effects (OTEs), lesser homologous recombinants and restrictive PAM (protospacer adjacent motif); but researchers have successfully modified the techniques to surpass such limitations; e.g., new Cas9 orthologs like Cpf1 and Cas13, and also successful integration of prime editing received by merging Cas9 to reverse transcriptase (RT) which provide provisions to widen the applications of genome editing (Anzalone et al., 2019). Enzymes like cytidine deaminase and adenosine deaminase can convert bases, i.e., they can change one nucleotide to another, and these enzymes are combined together with nickase Cas9 (nCas9) to generate point mutations and hence offer a new genome editing method (Komor et al., 2016; Gaudelli et al., 2017). Dead Cas9 (dCas9), which is also known as dCas9, has inactivated catalytic domains. These dCas9 are combined with activator or repressor elements of transcription which forms an engineered Cas9 and this is further directional for recognition and attachment to 'target promoter region' (Qi et al., 2013). Several targeted sites can be edited with the help of CRISPR (Xie et al., 2015). CRISPR-Cas ribonucleoproteins (RNPs) and transient expression systems have been successfully employed to transfer DNA (T-DNA) cassettes which code for components utilized in editing the plant genomes, and hence researchers have succeeded to achieve plant genomes edited with a DNA-free approach (Woo et al., 2015). Central dogma is basically targeted to procure a well-planned editing of the gene of interest and insertion of desired genes at the RNA (Cricks, 1970). Hence researchers worldwide have been focusing on refining genetic codes' flow throughout central dogma along with interacting with other factors which determine the epigenetic memory of plant's interactions with its vicinity and environment (Pramanik et al., 2021).

13.2 UNDERSTANDING THE KEY ELEMENTS OF PLANT-MICROBE INTERACTIONS

PM interactions are based on both microbiota and host plants genetics (Levy et al., 2018). Several mechanisms of plant immune system have been developed to detect microbiota released molecules as signals for the host to defend against phytopathogens (Netea and van der Meer, 2017). Plant hormones include ethylene, jasmonic acid (JA) and salicylic acid (SA) are also markers of the interactions amongst PMs (Pieterse et al., 2014). In order to increase our knowledge of their fundamental developmental principles, our comprehension of the "Red Queen" dynamics of PM interactions (coevolutionary cycles sustained because of conflicts between pathogens and plants) (Han, 2019). Standard methods including 16S RNA sequencing allow the makeup of microbial population easier to differentiate. Moreover, several omics-tools (genomics, transcriptomics, proteomics, and also metabolomics) also highlight plant-related gene functions and pathways at a community level. But these technologies cannot identify detrimental, neutral, or favorable contact with host plant. In addition, omics data generally highlights single points in time during PM interactions, so that spatiotemporal dynamic connections cannot be identified (Pinski et al., 2019). Due to its potential function of microorganisms in plant health, pinpointing key genes for PM interactions that regulate the agronomic characteristics would enhance specific plant characteristics for industry and agriculture. One of the major uses of CRISPR-based GE technologies is the investigation of genetic alteration activities among plants and microorganisms. The overall knockdown of desired gene is indeed distinctive benefit for CRISPR-based technique in comparison with RNA silencing mechanism that generates partial phenotype (Bisht et al., 2019). Overall, utilizing CRISPR-based technologies, genetic information can be gained more precisely, especially genetic function within PM molecular interaction. The research of model microorganisms, such as root-rhizobium and *P. syringae*, has been examined mechanically to understand genetic mechanisms causing mutual and harmful interactions with hosts, respectively (Xin et al., 2018). CRISPR/Cas method has recently been developed using single-stranded recombinants for genetic changes includes gene deletional, insertional, substitutional, and transcriptional suppression of rhizospheric bacteria *P. putida* KT2440 (Sun et al., 2018; Aparicio et al., 2018). However, mechanistic investigations using non-modeled microbiological isolates needed to investigate gene-level linkages in noxious or harmful PM interaction. Hence, GE for non-model organisms with powerful CRISPR/Cas technologies allows study to identify connections among genes and functionalities. Furthermore, a new pattern of use of biomaterials as a form of DNA/RNA or protein provides a unique option of CRISPR/Cas delivery into microorganisms which cannot be achieved by conventional techniques (Eoh and Gu, 2019).

13.3 PLANT DEVELOPMENT AND UPTAKE OF NUTRIENTS

The widely studied plants growth promoting microorganisms (PGPM) include nitrogen-fixing rhizobium and arbuscular mycorrhizal (AM) fungi. For instance, the plant root-based AM fungus occupy 80%–90% of land invasive plants and mobilizing soil nutrients to plant (Bonfante and Genre, 2010). Similarly, *Arabidopsis* isolated root endophytic fungi *C. tofieldiae* shows PGP activities and mobilizes P in host plant in contexts with phosphate deficiencies (Hiruma et al., 2016). Although pathogenic PM interactions are being intensively studied to produce disease resistant plants, some plant microbiota PGPMs that interact with favorable PM interactions are being less recognized. But the favorable PM interactions generally rely on a certain species or cultivar. Pathogenic PM interactions seem to be much similar to plant signaling pathways with beneficial effects. However, the outcomes of specific PM interactions are yet unclear. The effective PM interaction results in complicated mechanisms such as increasing accessibility of nutrients by iron absorption, nitrogen-fixing, and by solubilizing potassium or phosphorus. This includes active plant defense mechanisms against biotic stress via signaling molecules, particularly systemic inherited and systemic generated resistance (Rodriguez et al., 2019). Some bacteria and archaea linked with the legumes have been able to transform the unavailable (atmospheric) N in the accessible (ammonium) form. This symbiotic relation for plant-rhizobium includes food trading in return for fixing Nitrogen. A systemic investigation of symbiotic relationships on nodules organogenesis can provide strategies to host N-fixing bacteria for non-legume plants (Geurts et al., 2016). In instance, GE techniques already have been developed for certain model plant species like, *L. japonicus* (Wang et al., 2016), *M. truncatula* (Curtin, 2018), *G. max* (Li et al., 2015) and *V. unguiculata* (Ji et al., 2019). Furthermore, CRISPR procedures for certain rhizosphere PGPMs rhizosphere also intended for genetics including *B. mycoides* and *B. subtilis* (Yi et al., 2018). Therefore, the expansion of CRISPR toolkit might change food legume and non-legume engineering to achieve effective N-fixation rhizobium and Phosphate-solubilizing microorganisms. Detailed bimolecular study on PM interactions with GE allows field application to improve agricultural production for non-host plants, organic, and unfertilized land. The use of these beneficial GE-modified microorganisms for agriculture is also the alternative to the notoriously ineffective agrochemicals (Avellan et al., 2019). Up to 95% and 99.9% of the agriculturally applied micronutrients and pesticides are wasted and are never targeted at polluting the environment (Castro et al., 2014; Monreal et al., 2016). Therefore, PGPM might thus be better choices for reducing cost in an environmentally responsible manner. The adoption of these genetically modified microbial inoculants can prevent the quick fall of the microbial population and subsequent benefit of crops (Qiu et al., 2019). This also makes it possible to remediate polluted soils via plant microbiota (Basu et al., 2018).

13.4 MODERN TOOLS FOR EXPLORING PLANT-MICROBE INTERACTIONS

To understand the underlying processes of plant-specific microbiome, an appropriate approach in agriculture, as plant microbiota substantially affects the host's phenotyping. More concisely, a study of PM interactions between microbes and plant genes is crucial for future usage of plant microbiota. Molecular biology, all omics plus NGS (next-generation sequencing) techniques have significantly enhanced our knowledge of plant microbiota, including PGP microorganisms and phytopathogens (Jansson and Hofmockel, 2018). For example, there was a change in the makeup of the favorable microbiota population in plant-pathogen interactions (Niu et al., 2017; Kwak et al., 2018). Even though some meta and proteo-genomic publications give data from an enlarged functional category, alternative techniques are necessary for genes and protein functional understanding.

The utilization of microbial consortiums (a collection of species) is a study issue on agricultural PM applicability, whereas a different research topic involves a precise GM for plants and microbe. GM techniques have been commonly employed in the past, coupled with gene silence, to research genetic functions to improve characteristics. Transgene technology is one successful technique to get rapid results; however, the addition of external genetic material restricts its broad application owing to regulatory problems

(Munoz et al., 2019). In this respect, GE technology allows scientists to modify genomic sequences more accurately without integrating a foreign gene (Knott and Doudna, 2018). Genome editing technique uses endonucleases designed to induce double-stranded breaks (DSB) that trigger DNA repair by endogenous processes and produce several kinds of abnormalities (Gaj et al., 2013). DSB repair processes have two main routes, non-homologous end-joining (NHEJ) and homology-directed repair (HDR). HDR are accurate but far less frequent than NHEJ or can be used for specific replacement of donor-dependent genes. Although targeted genetic changes may be performed in a number of ways, three mega nucleases (MNs) are the most commonly used in the last few years, including transcriptional activator-like effector nuclease (TALEN), zinc-finger nuclease (ZFN) and CRISPR/Cas technique (Brandt and Barrangou, 2019). ZNF and TALEN GM is focused on the capacity of a DNA-binding domain to identify nearly any target DNA sequence. The GE abilities of ZNF/TALEN thus largely depend on the DNA-bound affinity and specificity for zinc-finger or TALE protein assembled (Gaj et al., 2013). Compared to ZFN and TALEN, CRISPR/Cas systems are adventitious for easy design, flexibility, economical, better efficiency, multiplexing, and sensitivity (Brandt and Barrangou, 2019). Since CRISPR/Cas mechanisms have now been adapted to GE for eukaryotes (Cong et al., 2013), it became as a prominent GE tool, also referred as the toolbox of nature and 'magic wall' of genomic engineering (Sakuma and Yamamoto, 2017). Therefore, CRISPR-mediated GE techniques and applications are mostly explored in PM research.

13.5 METABOLIC ENGINEERING

Secondary metabolites (SMs) are generally characterized as natural organic molecules which are not essential for the promotion of life and growth. The plant and the accompanying microorganisms produce several SMs, mostly via certain chemical routes. Whilst these SMs seem crucial among the pathways of plant protection or microbial defense, several have been used as nutrition, medications, repellents, aroma, flavor, and dye (Pyne et al., 2019). Many techniques for the creation of native or heterologous SMs in plants (Rybakova et al., 2017) and microorganisms were adopted (Lv et al., 2016; Podolsky et al., 2019). Various plant microbiome researches give an intriguing overview of microbial taxa that supports plant SM synthesis activity. In particular, the seeding microbiome of medicinal plants, for example, *S. miltiorrhiza* (Chen et al., 2018), is particularly overlapping in a set of fungal and bacterial species with maize (Liu et al., 2013), bean (Klaedtke et al., 2016), rice (Midha et al., 2016), rapeseed, etc. (Rybakova et al., 2017). PM-associated microorganisms also have similar metabolic terpenoids to host plants, which indicates their potential of them as an SM-related gene repository. In addition, plant SM synthesis changes with geography owing in part to the change in the microbiota, which indicates that the plant metabolome and the associated microbiome must be directly linked (Koberl et al., 2013). CRISPR-based alteration of PM-related genes that engage in SM route presents a new and appealing method to achieve more persistent and bioactive SM synthesis. In recent studies, CRISPR-associated metabolic engineering experiments were conducted in either microorganisms or plants to understand fundamental SM pathways or to improve the synthesis of SMs (Shi et al., 2017; Shanmugam et al., 2019). Some microbial researches involve *B. bassiana* (uridine biosynthesis) (Chen et al., 2017), *T. reesei* (uridine biosynthesis) (Liu et al., 2015), *E. coli* (flavonoid synthesis) (Wu et al., 2015), *Myceliophthora* (cellulase synthesis) (Liu et al., 2017), *A. niger* (mucic acid production) (Kuivanen et al., 2016), *A. oryzae* (pigment production) (Katayama et al., 2016), *S. bambusicola* (hypocrellin production) (Deng et al., 2017), and *A. fumigatus* (trypacidin synthesis) (Weber et al., 2017). Furthermore, some instances of CRISPR-associated metabolic engineering among plants includes *P. somniferum* (morphine biosynthesis) (Alagoz et al., 2016), *S. lycopersicum* (GABA, i.e., γ-aminobutyric acid synthesis) (Li et al., 2018), *S. miltiorrhiza* (tanshinone production) (Li et al., 2017), *Dendrobium* sp. (lignocellulose synthesis) (Kui et al., 2017), *C. sativa* (fatty acid biosynthesis) (Aznar-Moreno and Durrett, 2017), *N. tabacum* (N-glycan biosynthesis) (Mercx et al., 2017), and *S. miltiorrhiza* (phenolic compounds) (Zhou et al., 2018). Therefore, it is interesting to employ CRISPR technology to alter PM interaction to discover new features of SM pathways and regulation of SM production routes. In general, the utilization of CRISPR technique offers a huge potential to recognize the key problems and uses of future PM interactions on sustainable agriculture.

13.6 MAJOR ELEMENTS OF THE CLUSTERED REGULARLY INTERSPACED SHORT PALINDROMIC REPEATS (CRISPR)/CAS SYSTEM

CRISPR/Cas is bacterial Type II immune response present across various prokaryotes, like archaea and bacteria. Ishino et al. (1987) initially described describe some CRISPR components in *E. coli*, although their function was that time unknown. Around 10 years ago, CRISPR arrays were defined as a programmable, bacterial defense mechanism against phages (Brouns et al., 2008). CRISPR immune response components were modified soon towards GE mediated CRISPR. Single gRNA (sgRNA) and Cas endonuclease are two primary components of the CRISPR technology (Jinek et al., 2012). The chimeric sgRNA include guide RNA (crRNA) and trans-activating RNA (tracrRNA). One main component for CRISPR application is its protospacer adjacent motifs (PAMs), a short region of recognition near gRNA. SgRNA design for target genes is simple and numerous companies offer sgRNA synthetization services, including plasmid construct (Cas endonuclease) are accessible at Addgene library (https://www.addgene.org/) that provide CRISPR technology for researchers at affordable pricing. Moreover, multiplexing utilizing CRISPR techniques are also a simple method for editing in a single cell for several target locations (Cong et al., 2013). *S. pyogenes* Cas9 is first Cas enzyme utilized in GE as well as the most widely employed in CRISPR-based tools for the creation and deployment of robust efficiency and availability of PAMs (Knott and Doudna, 2018). Table 13.1 enlists distinct characteristics of several CRISPR/Cas systems, which can be utilized in studies focused on PM interactions (Gosavi et al., 2020).

TABLE 13.1 Characteristics of Various CRISPR/Cas Systems Available for Plant-Pathogen Interactions

CRISPR System	Cas9	Cas12a	Cas12b	Cas13	Cas14
Type and Class	Type II Class 2	Type III Class 2	Type IV Class 2	Type V Class 2	Type VI Class 2
Pre-crRNA processing	No	Yes	No	Yes	Yes
tracrRNA	Yes	No	Yes	No	Yes
PAM/PFS	3,'G-rich	5,' T-rich	5,' T-rich	3,' non-G PFS	Independent-PAM
Nuclease domain	RuvC, HNH	RuvC	RuvC	HPEN	RuvC
Substrate	dsDNA	dsDNA	dsDNA	RNA	ssDNA

13.7 CRISPR-MEDIATED TECHNIQUES AND ITS APPLICATIONS

Besides using CRISPR/Cas technique as GE technologies, modified tools with updated Cas9 variants are created (like nCas9 and dCas9). Such CRISPR-based techniques have repeatedly proven their utility in simple base editing, genetic control, epigenetic modifications, chromosome technology, imaging, etc. (Xu and Qi, 2019; Adli, 2018). The CRISPR-based techniques of next-generation extended over DSB-based GE and provided the capacity for the exact targeting of the DNA sequence.

In fundamental research the endonuclease capacity of SpCas9 was studied to develop knock-out and knock-in mutants for gene function research purposes (Jinek et al., 2012; Cong et al., 2013) or to enhance HR-based donor DNA incorporation at certain locations (Bassett et al., 2014). The multiplex CRISPR technique can be more efficient and aim at numerous locations as needed in order to comprehend the complicated regulatory network (Jiang et al., 2015). EvolvR is now a new targeted technique to integrate semi-random mutations in plasmids or prokaryotic genomes in dCas9/gRNA-targeted location (Halperin et al., 2018). David Liu's group used the combination of Cas9 impaired (nCas9 or dCas9) with cytidine or adenine deaminase to develop cytidine base editing (CBE) and adenine base editing (ABE), respectively (Komor et al., 2016). CBE or ABE generate C to T and A to G nucleotide alterations already utilized to plants (Li et al., 2017) and microorganisms (Zheng et al., 2018). Furthermore, RNA-targeted Cas9 (RCas9) mRNA trafficking was seen in live cells which might allow *in vivo* RNA monitoring of desired transcripts (Nelles et al., 2016).

CRISPR-mediated screening techniques like detection of nucleic acid (Gootenberg et al., 2017, 2018), molecular tagging (Roberts et al., 2017), bar-coding (Kalhor et al., 2018), and pedigree tracing (McKenna et al., 2016), together with specific functional genome libraries (Miles et al., 2016) allows the target to be screened concisely from huge population. Imaging live cells CRISPR techniques are extremely effective for understanding the structure of chromatin (Qin et al., 2017), chromatin dynamics (Dreissig et al., 2017) and the topology of manipulation of chromatin links between regulating genomic sites (Morgan et al., 2017). The system CRISPR/Cas9 has reprogramed for altering gene expression through transcription, translation, and regulation of epigenome. The dCas9 was coupled with activator proteins (Domains VP64) in the CRISPR activation tool (CRISPRa), to activate transcripts which are effective for the usage in plants or bacterial species, whereas CRISPRi system based on dCas9 employs the repressor complex to inhibit gene transcription (Piatek et al., 2015). The Cas13 has shown that the mRNA of the target gene is eliminated and gene function is thereby eliminated (Cox et al., 2017). Furthermore, dCas9 epigenetic factors are efficient in activating repressed genes/transposons by histone remodeling and DNA demethylation catalysis of plants (Gallego-Bartolome et al., 2018). In all, the CRISPR has designed a consumer-friendly GE tool for expanding the CRISPR toolbox, enabling a secure and rapid solution for GM technology as well as for time-intensive classical breeding methods.

13.7.1 GENOME EDITING BY CRISPR/CAS9

Wheat plant is one of the most utilized and most explored communities because of its high demand for human consumption. This drew researchers' attention towards efforts to diminish infections caused by plant pathogens like *Blumeria graminis Tritici* which leads to deteriorating powdery mildew in wheat plants. So, successful mutations were brought about in wheat plants by employing CRISPR/Cas9 in one (TaMLO-A1) in one study, and by developing resistance to the pathogen by knocking out TaEDR1 by utilizing CRISPR/Cas9 technology (Wang et al., 2014; Zhang et al., 2017). Gene TaHRC codes for a putative protein that is rich in histidine and binds to calcium. The nuclear protein confers susceptibility to *Fusarium* head blight (FHB) disease and deleting the start codon of the gene produces FHB resistance, with the help of CRISPR-Cas9-mediated gene editing (Su et al., 2019). The procured mutant plants show diminished severity of FHB as compared to the wild type and this proposes the effectiveness of TaHRC in FHB susceptibility and hence proves that Fhb1 resistance is conferred by inactivating/editing TaHRC. Figure 13.1 describes the events involved in procuring disease resistant crops through CRISPR/Cas9 system (Tyagi et al., 2020).

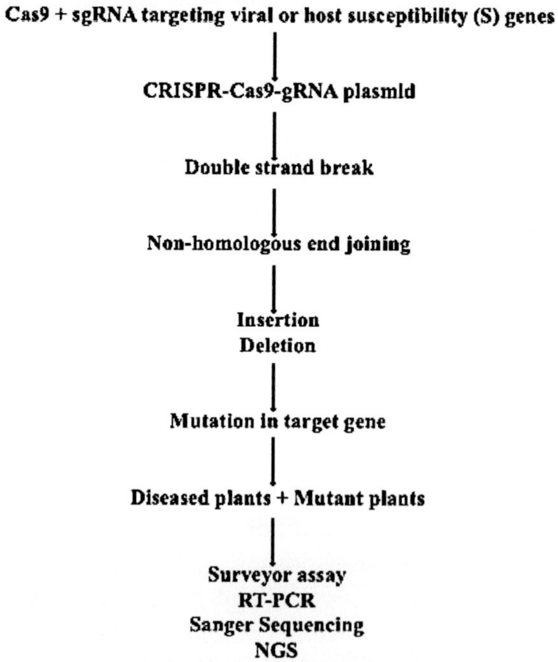

FIGURE 13.1 Generating disease resistant crops through CRISPR/Cas9 system.

Rice plants resistant to blast diseases have also been produced employing CRISPR/Cas9-mediated editing and disruption of OsERF922 and OsSEC3A genes in the wild type plants. Other mutant reported have better resistance against *Magnaporthe oryzae* and have heightened SA content with the help of Ossec3 which induces exocytosis, but produces dwarf plants (Wang et al., 2016; Ma et al., 2018). Whereas, transcription factors ERF922 (ET responsive factor), which are related to various stress responses, produce mutants that don't represent any editing in agronomic traits though the T1 and T2 transgene-free mutants. In fact, lesser blast lesions were recorded in seedling and tillering stages in the mutants (Ma et al., 2018). But there are very few records of utilizing CRISP/Cas techniques against bacterial infections in plants. One example is editing the OsSWEET13 gene of rice plants, which procured plants resistant against *Xanthomonas oryzae* pv. *oryzae*, responsive for bacterial blight disease (Zhou et al., 2015). The CRISPR-associated protein 9 (Cas9) technology has been successfully used to block the synthesis pathway for abscisic acid (ABA) by editing the *ZEP* gene. The system provides high efficiency of up to 85.4% where the CRISPR/Cas9 gene-editing vector system offers easier and faster genome editing of multiple monocot and dicot species (Li et al., 2021; Ma et al., 2016; Mao et al., 2019). Table 13.2 highlights some examples of gene editing techniques utilized to enhance disease resistance in cereals.

TABLE 13.2 Some Examples of Gene Editing Techniques to Increase Disease Resistance in Cereals

Molecular Technique	Biotechnological Intervention	Species	Gene	Enhanced Resistance to	References
CRISPR/Cas9	Silencing of host genes	Wheat	*TaMlo-A1*	*Blumeria graminis* f. sp. *tritici*	Wang et al. (2014)
			TaEDR1	*Blumeria graminis* f. sp. *tritici*	Zhang et al. (2017)
			TaLpx-1	*Fusarium graminearum*	Wang et al. (2018)
			TaHRC	*Fusarium graminearum*	Su et al. (2019)
		Rice	*OsSWEET13*	*Xanthomonas oryzae* pv. *oryzae*	Zhou et al. (2015)
			OsERF922	*Magnaporthe oryzae*	Wang et al. (2016)
			OsSEC3A	*Magnaporthe oryzae*	Ma et al. (2018)

13.7.2 CRISPR-BASED PM APPLICABILITY IN SUSTAINABLE AGRICULTURE

Numerous publications emphasize the utilization of CRISPR-based technologies for plants that focus particularly on plant-pathogen interactions for disease resistance development. Here we highlight the uses of CRISPR to react sustainable to the growing population's food demand by discovering key pathways for PM interaction, developing disease-resistant cultivars, and improving PGP activities.

The average global output loss due to pathogenic PM and plant/insect interactions are between 11 and 30% mostly in food-scarce areas (Savary et al., 2019). Thus, the production of resistant cultivars against phytopathogens was a continual job for scientists. Conventional techniques have been implemented for inducing GMs for better disease-resistant plant types, such as selective breeding, natural abnormalities, hybridization, radiation/chemical mutation, and biological mutation (Langner et al., 2018). Therefore, these approaches nonetheless create numerous non-target alterations, and screening is still a long and tedious task. The common modes of phytopathogenic resistance in the GE technologies include blocking pathogenic entry, altering plant protection, regulation of recessive/susceptibility-genes (S-genes), stimulation of the dominant resistance genes (R-genes), antimicrobial peptide expression and silencing application (Van Esse et al., 2019). As previously mentioned, GM technologies-based transgene insertion is seen as a problem. Also, an integral method for combating phytopathogens, for instance, has limitations by use of GM technologies together with bio-controls, chemical fertilizers, and pesticides, and after pathogen development several pesticides don't really effective and higher doses can lead to harm for human beings and for the environment (Berger et al., 2017). In the previous decade, an alternative for the conventional plant-breeding and GM tool has been created by a novel collection of techniques jointly labeled as novel plant breeding strategies (NPBTs or NBTs), and also CRISPR/Cas (Langner et al., 2018; Zaidi et al., 2018). CRISPR mediated genomic editing of plants and pathogens, including bacteria, fungi, and viruses, in order to maximize disease resistance. The main step among effective GE application is to comprehend the molecular detail of a plant or pathogen target gene. Whole-genome sequencing many plant species and related microbiome is accessible. CRISPR-based disease resistance could be attained either by targeting a gene either from host plant or from the pathogen.

Unlike humans, the plants may not have an adaptive immune response or mobile immune cells, and rely on the two-tier innate immune response to deal with pathogenic interactions (Spoel and Dong, 2012; Jones et al., 2016). One aspect utilizes pattern-recognition receptors (PRRs) to differentiate "non-self" and "self," i.e., pathogen-associated molecular patterns (PAMPs) and damage-associated molecular patterns (DAMPs) derived by plant, and activate PAMP-triggered immunity (PTI) (Boller and Felix, 2009). The other employs resistant (R) proteins to handle microbes-derived effectors to inhibit the pathogen entry into the host cell and activate effector-triggered immunity (ETI) (Upson et al., 2018). Receptor-like proteins and kinases are reported to function as PRRs, whereas R-genes encode intracellular nucleotide-binding leucine-rich repeat receptor (NLRs) proteins. These important genes in plant innate immunity are engaged in host susceptibility (S-genes) and resistance (R-genes) generated against the pathogenic responses are widely studied. The GE strategies targeted the above-mentioned plant genes (PTI or ETI) or the pathogen genes to develop resistance towards bacteria (Li et al., 2012; Jiang et al., 2013; Zhou et al., 2015; Jia et al., 2017; Peng et al., 2017; Malnoy et al., 2016; Ortigosa et al., 2019), fungi and oocytes (Kong et al., 2019; Li et al., 2018), and also DNA/RNA viruses (Aman et al., 2018; Tripathi et al., 2019). In addition, CRISPR/Cas system is inserted in several microbes for targeted GE acting as biological control agents, such as insect-pathogenic fungi *B. bassiana* and filamentous fungi such as Trichoderma species (Jiao et al., 2019; Chen et al., 2017).

13.8 PROMISING HOST SUSCEPTIBILITY GENES FOR ANTIVIRAL ENGINEERING IN PLANTS

The efficiency of virus vector-based CRISPR/Cas9 system has been strongly proved to be efficient and promising for generating plant mutants that possess all the craved traits, and this is achieved following précised methodologies that produce exclusive results in no time. Still, the need of the hour has been convenient and more optimized CRISPR/Cas9 technology which can provide significant antiviral engineering.

DNA viruses like BeYDV, cabbage leaf curl virus and WDV; and the RNA viruses like barley stripe mosaic virus (BSMV), beet necrotic yellow vein virus, PEBV, TMV, and TRV have been targeted by CRISPR/Cas9 system for gene targeting frequency in model plants like *A. thaliana* and *N. Benthamiana*; and common crops like potato, tomato, rice, wheat, and maize (Cody et al., 2017; Zaida and Mansoor, 2017; Hameed et al., 2018; Hu et al., 2019; Jing et al., 2019). Two distinct kinds of virus-induced gene editing vectors were prepared, which had pC86 and pV86 promoters of BSCTV, which were later trans-activated via co-infection with BSCTV. This strategy was planned for repressing OTEs of antiviral phenomenon liaised via CRISPR/Cas9 systems in plants. The prepared vectors which could be induced in BSCTV were to some extent, responding, and could positively suppress BSCTV accumulation in transgenic *A. thaliana* and *N. benthamiana*, with no effects on the off-target (Ji et al., 2018). Time and effort can be saved by utilization of transgene-free CRISPR/Cas9 system delivered by SYNV (Sonchus yellow net virus). But this approach can only be applied in *N. bethamiana* plants, limiting its utility in crops (Ma et al., 2020). Exclusive gene editing with accurate results can also be achieved by distinct CRISPR/Cas editing agents like base editors and primer editors (Anzalone et al., 2020).

13.8.1 TARGETING HOST FACTORS/PLANT-VIRUS INTERACTIONS BY CRISPR/CAS TO CREATE VIRUS RESISTANT PLANTS

Phytopathogenic viruses inscribe into the translation machinery of the host and recruit their transcription factors for translation of viral RNA and proceed with their infection pathway. Hence, researchers termed these translation-related factors of the host as 'pro-viral factors.' Recessive mutation in gene is confabulated to be behind 'recessive resistance' resulting in expression of host factors which are believed to be essential for viral infection and the life cycle of the virus; or the mutation may lead to negative regulation of plant defense responses (Hashimoto et al., 2016). Hence, procuring recessive gene mutants by CRISPR/Cas systems stands out as an ideal methodology. Various species of crops demand recessive resistant genes against translation mechanism of plant RNA virus and the most employed eukaryotic initiation factors for translation for the same are eIF4E, eIF4G, eIF (iso) 4E and eIF (iso) 4G. They successfully defend the host plant from viral pathogens (Hashimotot et al., 2016; Lellis et al., 2002; Robaglia and Caranta, 2006; Sanfacon, 2015; Wang and Krishnaswamy, 2012). Absolute resistance was developed against ZYMV (zucchini yellow mosaic virus), PRSV-W (papaya ringspot mosaic virus-W) and CVYV (cucumber vein yellowing virus) using CRISPR/Cas9-edited eIF4E-knockout cucumber plants (Chandrasekaran et al., 2016). In another study CRISPR/Cas9-targeted mutagenesis procured eIF4G (novel allele of rice plant) which was completely resistant to RTSV (rice tungro spherical virus) (Baltes et al., 2015). Absolute resistance against CBSV (cassava brown streak virus) in cassava plants and TuMV-GFP in *A. thaliana* plants was achieved by inducing mutations in CRISPR/Cas9-mediated eIF(iso)4E (Pyott et al., 2016; Gomez et al., 2019). Few alternative eIF4E-like proteins, e.g., nCBP-1 (novel cap-binding protein-1) and n CBP-2, are expressed by cassava and when mutant plants were procured employing CRISPR/Cas9 targeting the nCBP-1 and nCBP-2, these mutants were found to be resistant against CBSV (Gomez et al., 2019). The Arabidopsis allele which makes the plant susceptible to ClYVV (clover yellow fever vein virus) was altered at the bases by CRISPR-Cas9n-cytidine base editor, and the resultant mutants were discovered with resistance against the same virus, by the intrusion of N176K mutations (Bastet et al., 2019). CRIPSR/Cas technique has also been successfully utilized for the production of soybean with resistance to the SMV (soybean mosaic virus). For this, a decoy protein of soybean, i.e., the PBS1 protein was altered and the new modified protein possessed cleavage site for NIa protease of the SMV, which induces cell death in soybean protoplast (Helm et al., 2019).

Phytopathogen *Xanthomonas citri* causes citrus canker diseases in citrus plants and CRISPR-Cas9 has been utilized to achieve resistance against it by modifying a target gene involved in ETI. The effector binding element PthA4 for the gene CsLOB1 (i.e., lateral organ boundaries-1) is involved in host susceptibility. This gene was targeted and mutated, and the mutant gene was found to have lost recognition and response to effector elements for the bacteria and therefore, the mutants were confirmed to possess heightened resistance against the phytopathogen (Jia et al., 2016). Tomato plants are susceptible to various infections like *Pseudomonas syringae* pv. *tomato* and *Phytophthora capsica*. To produce some resistant plants, tomato ortholog SlDMR6-1 was removed using CRISPR-Cas9 which resulted in resistant plants with no deteriorating impacts on plant's overall growth and yields. In a similar way, the *DMR6* gene of *A. thaliana* was mutated with the help of CRISPR/Cas9 which resulted in higher levels of phytohormone SA and the plants were found to be resistant against a wide spectrum of phytopathogens (Zeilmaker et al., 2015). Tomato plant resistant against *P. syringae* pv. *tomato* was also procured by mutating *SlJAZ2* (*Jasmonate ZIM-domain-2*) ortholog, with the help of CRISPR-Cas9 (Ortigosa, 2019). Plant's susceptibility to fire blight disease is a joint effort of interaction of *dspE* protein with the *DIPM* (DspE-interacting proteins of malus) genes (viz. *DIPM 1, 2, 3, 4*). CRISPR-Cas9 and S gene were utilized to abbreviate DIPM genes in the protoplast of apple plant, and this mutation can be applied to both short-lived plants and plants with longer life span (Das et al., 2019). An advanced and modified version of CRISPR-Cas9 is the 'EvolvR' which has been successfully utilized to generate novel alleles in rice plants, which enables the plant to detect all the strains and ligands of phytopathogen *Xanthomonas* (Luu et al., 2019). 'Begomoviruses' refer to heterogenous group of viruses, enlisting CLCuRaV (cotton leaf curl Rajasthan virus), CLCuAlV (cotton leaf curl Alabad virus), CLCuKoV (cotton leaf curl Kokhran virus), CLCuMuV (cotton leaf curl Multan virus) and CLCuBaV (cotton leaf curl Bangalore virus) (Iqbal et al., 2016). CRISPR techniques were also utilized to inactivate sequences of eBSV (endogenous banana streak virus) which infects banana plants (Tripathi et al., 2019). The isoforms *eif4e* (i.e., novel cap-binding protein-1 and 2) are colligated with viral genome proteins. These were targeted by researchers to produce mutants with diminished disease symptoms of CBSV, using the CRISPR-Cas9 system (Gomez et al., 2019). The gene *eIF4E1* has also been targeted using CRISPR-Cas9 and this resulted in resistant plants of *A. thaliana* against ClYVV (*clover yellow vein virus*) (Bastet et al., 2016).

13.8.2 RESISTANCE TO PLANT DNA AND RNA VIRUSES USING CRISPR/CAS9

There are several other technologies that offer DNA editing, e.g., ZFN and TALEN, but since CRISPR/Cas presents even better vantage of editing, especially in the case of engineering antiviral plants, it is more promising and hence has gained popularity. Mutations have been brought about in plants *Nicotiana benthamiana* and *Arabidopsis thaliana* using CRISPR/Cas9 techniques employing sgRNA (single guide RNA). In these transgenic plants the intergenic regions (IRs) that allow for viral replication-associated region is targeted to procure effective DNA interference and hence they are resistant to BSCTV (curly top virus), CLCuMuV and BeYDV (bean yellow dwarf virus) (Baltas et al., 2015; Ji et al., 2015; Yin et al., 2019). Transgenic plants of tomato and *N. benthamiana* produced with the help of Cas9-single guide RNA reported less accumulation of TYLCV (tomato yellow leaf curl virus) DNA. The effective changes were brought about by targeting the sequences that encode coat protein (CP) or replicase (Tashkandi et al., 2018). Geminiviruses like bipartite geminivirus, CLCuKoV, Merremia mosaic virus (MeMV), monopartite geminivirus and various strains of TYLCV geminivirus have been successfully edited using sgRNAs. The two reported approaches for the same include targeting the stem-loop sequence and the viral CP region and replication-associated region in-between the IR (Ali et al., 2016). Researchers have also targeted proteins responsible for movement and CP region for procuring WDV (wheat dwarf virus) and BSV (banana streak virus) resistance, respectively. These were achieved by duplexing CRISPR/Cas9 system with sgRNAs (Kis et al., 2019; Tripathi et al., 2019). CRISPR/Cas9 has also proven to be efficient in defending plants against dsDNA viruses. The CaMV (cauliflower mosaic virus) from the *Caulimoviridae* family has significantly been targeted at the CP region by joint expression of Cas9 protein and sgRNAs (Liu et al., 2018). Similarly, victorious efforts to produce transgenes with resistance to geminivirus have been produced by targeting the virus's DNA, using CRISPR/Cas9 (Kalinina et al., 2020). Figure 13.2 represents general steps followed to procure DNA and RNA variants of viruses generated by employing the CRISPR/Cas system (Cao et al., 2020).

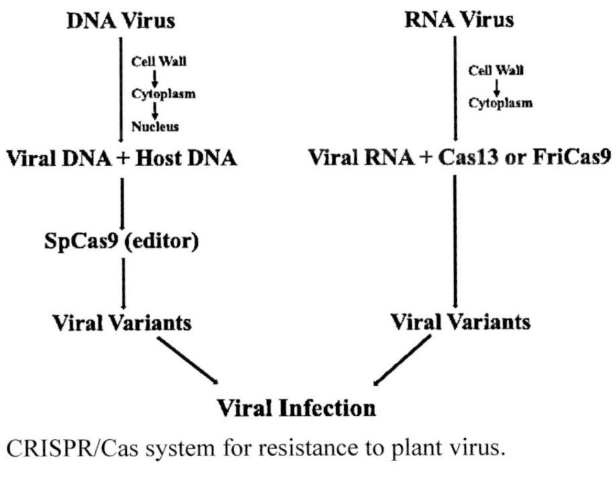

FIGURE 13.2 CRISPR/Cas system for resistance to plant virus.

Lately few targeting and editing CRISPR-associated proteins have been reported to be quite effective, e.g., FnCas9 from *Francisella novicida* and Cas13a from *Leptotrichia shahii* (Price et al., 2015; Shmakov et al., 2015). The application of CRISPR/Cas techniques were spread out from DNA to RNA by controlled and directed RNA-targeting and LshCas13a stands out as the very first Cas13 ortholog for this (Abudayyeh et al., 2016). Targeted bacterial mRNAs can be knocked down using a fusion of LshCas13a protein with protospacers (Abudayyeh et al., 2017), providing the researchers with a substitute CRISPR/Cas system against pathogenic RNA viruses. New

Cas effector proteins promisingly manipulate the target sites in RNA virus genomes applying CRISPR/Cas techniques. Resistant *N. benthamiana* and *A. thaliana* plants against CMV (cucumber mosaic virus) and TMV (tobacco mosaic virus), respectively, were achieved via a plant codon-optimized version of FnCas9 fused with rgRNA, a combination which proved to significantly target plant ssRNA viruses (Zhang et al., 2018). Another example of an effective entity against ssRNA viruses in plants has been reported, which was prepared by using class 2 type VI-A CRISPR/Cas effector 'LshCas13a' which has sgRNA targeting viral RNA sequences. The result was fast and efficient RNA interference (RNAi) against TuMV (turnip mosaic virus) in which the LshCas13a transgenic plants *N. benthamiana* and *A. thaliana* had sgRNA directed by PEBV (pea early browning virus) promoter in TRV (tobacco rattle virus)-based vector (Aman et al., 2018a, b). Likewise, tobacco and rice transgenic plants which had LshCas13a were found to be resistant against RSMV (rice stripe mosaic virus) and SRBSDV (Southern rice black-streaked dwarf virus). Significance of CRISPR/Cas13a system was also proven via focusing P3-, Nib- or CP-coding sequences in PVY (potato virus Y) which resulted in inhibition of PVY infection (Zhan et al., 2019). But, in another study, a conserved single nucleotide mutation was recorded to confer resistance to CRISPR/Cas9 cleavage, finally leading to no resistance to geminivirus during glasshouse inoculations. This was recorded between 33 and 48% of edited viral genomes (Mehta et al., 2019).

13.9 CONCLUSION AND FUTURE RECOMMENDATIONS

Comprehension of the basic concepts of PM interactions, threat causing phytopathogens, and utilization of exclusive gene-editing techniques like CRISPR/Cas technology in sustainable agriculture is the perfect method to meet future food demands. In the past, only a few representative species were considered as plant growth-promoting microorganisms and were utilized in plant protection microbiological investigations. Therefore, considerable molecular research on the advantage of microbes-mediated plants has been carried out in the recent few years to expand the field of agricultural PM biotechnology. CRISPR/Cas technique offers a huge opportunity for researchers to study the principles of PM interactions, to conquer the phytopathogenic traits and design optimal plants/microbes for agricultural use. Therefore, more considerable plant microbiome sequencing analysis and metatranscriptomic data are additional to a study of a wider number of plant species to comprehend the community-level molecular processes under field settings. CRISPR-based applications in sustainable agriculture practices will allow the selection of certain agricultural characteristics, plants, and/or microbial genes.

KEYWORDS

- **arbuscular mycorrhizal**
- **CRISPR/Cas**
- **double-stranded breaks**
- **genome editing**
- **metabolic engineering**
- **metabolomic**
- **microbiota**
- **next-generation sequencing**

REFERENCES

Abudayyeh, O., Gootenberg, J., Essletzbichler, P., Han, S., Joung, J., Belanto, J. J., Verdine, V., et al., (2017). RNA targeting with CRISPR-Cas13. *Nature, 550*(7675), 280–284.

Abudayyeh, O., Gootenberg, J., Konermann, S., Joung, J., Slaymaker, I., Cox, D., Shmakov, S., et al., (2016). C2c2 is a single-component programmable RNA-guided RNA-targeting CRISPR effector. *Science, 353*(6299), aaf5573.

Adli, M., (2018). The CRISPR tool kit for genome editing and beyond. *Nature Communications, 9*, 1911. https://doi.org/10.1038/s41467-018-04252-2.

Alagoz, Y., Gurkok, T., Zhang, B., & Unver, T., (2016). Manipulating the biosynthesis of bioactive compound alkaloids for next-generation metabolic engineering in opium poppy using CRISPR-Cas 9 genome editing technology. *Scientific Reports, 6*, 30910. https://doi.org/10.1038/srep30910.

Ali, Z., Ali, S., Tashkandi, M., Zaidi, S. S., & Mahfouz, M., (2016). CRISPR/Cas9- mediated immunity to geminiviruses: Differential interference and evasion. *Scientific Reports, 6*(1), 26912.

Aman, R., Ali, Z., Butt, H., Mahas, A., Aljedaani, F., Khan, M. Z., Ding, S., & Mahfouz, M., (2018a). RNA virus interference via CRISPR/Cas13a system in plants. *Genome Biology, 19*(1), 1–9. https://doi.org/10.1186/s13059-017-1381-1.

Aman, R., Mahas, A., Butt, H., Aljedaani, F., & Mahfouz, M., (2018b). Engineering RNA virus interference via the CRISPR/Cas13 machinery in *Arabidopsis*. *Viruses, 10*(2), 732.

Anzalone, A. V., Randolph, P. B., Davis, J. R., Sousa, A. A., Koblan, L. W., Levy, J. M., Chen, P., et al., (2019). Search-and-replace genome editing without double-strand breaks or donor DNA. *Nature Cell Biology, 576*(7785), 149–157.

Anzalone, A., Koblan, L., & Liu, D., (2020). Genome editing with CRISPR-Cas nucleases, base editors, transposases and prime editors. *Nature Biotechnology, 38*(7), 824–844.

Aparicio, T., De Lorenzo, V., & Martínez-Garcia, E., (2018). CRISPR/Cas9-based counterselection boosts recombineering efficiency in *Pseudomonas putida*. *Biotechnology Journal, 13*(5), e1700161. https://doi.org/10.1002/biot.201700161.

Avellan, A., Yun, J., Zhang, Y., Spielman-Sun, E., Unrine, J. M., Thieme, J., Li, J., et al., (2019). Nanoparticle size and coating chemistry control foliar uptake pathways, translocation, and leaf-to-rhizosphere transport in wheat. *ACS Nano, 13*(5), 5291–5305. https://doi.org/10.1021/acsnano.8b09781.

Aznar-Moreno, J. A., & Durrett, T. P., (2017). Simultaneous targeting of multiple gene homeologs to alter seed oil production in *Camelina sativa. Plant & Cell Physiology, 58*(7), 1260–1267. https://doi.org/10.1093/pcp/pcx058.

Baltes, N., Hummel, A., Konecna, E., Cegan, R., Bruns, A., Bisaro, D., & Voytas, D. F., (2015). Conferring resistance to geminiviruses with the CRISPR-Cas prokaryotic immune system. *Nature Plants, 1*(10), 15145.

Bassett, A. R., Tibbit, C., Ponting, C. P., & Liu, J. L., (2014). Mutagenesis and homologous recombination in *Drosophila* cell lines using CRISPR/Cas9. *Biology Open, 3*(1), 42–49. https://doi.org/10.1242/bio.20137120.

Bastet, A., Lederer, B., Giovinazzo, N., Arnoux, X., German-Retana, S., Reinbold, C., Brault, V., et al., (2016). Trans-species synthetic gene design allows resistance pyramiding and broad-spectrum engineering of virus resistance in plants. *Plant Biotechnology Journal, 16*(9), 1569–1581.

Bastet, A., Zafirov, D., Giovinazzo, N., Guyon-Debast, A., Nogue, F., Robaglia, C., & Gallois, J. L., (2019). Mimicking natural polymorphism in eIF4E by CRISPR-Cas9 base editing is associated with resistance to potyviruses. *Plant Biotechnology Journal, 17*, 1736–1750.

Basu, S., Rabara, R. C., Negi, S., & Shukla, P., (2018). Engineering PGPMOs through gene editing and systems biology: A solution for phytoremediation? *Trends in Biotechnology, 36*(5), 499–510. https://doi.org/10.1016/j.tibtech.2018.01.011.

Berger, S., El Chazli, Y., Babu, A. F., & Coste, A. T., (2017). Azole resistance in *Aspergillus fumigatus*: A consequence of antifungal use in agriculture? *Frontiers in Microbiology, 8*, 1024. https://doi.org/10.3389/fmicb.2017.01024.

Bikard, D., Euler, C. W., Jiang, W., Nussenzweig, P. M., Goldberg, G. W., Duportet, X., Fischetti, V. A., & Marrafinni, L. A., (2014). Exploiting CRISPR-Cas nucleases to produce sequence-specific antimicrobials. *Nature Biotechnology, 32*(11), 1146–1150.

Bisht, D. S., Bhatia, V., & Bhattacharya, R., (2019). Improving plant-resistance to insect-pests and pathogens: The new opportunities through targeted genome editing. *Seminars in Cell & Developmental Biology, 96*, 65–76. https://doi.org/10.1016/j.semcdb.2019.04.008.

Boller, T., & Felix, G., (2009). A renaissance of elicitors: Perception of microbe-associated molecular patterns and danger signals by pattern-recognition receptors. *Annual Review of Plant Biology, 60*, 379–406. https://doi.org/10.1146/annurev.arplant.57.032905.105346.

Bonfante, P., & Genre, A., (2010). Mechanisms underlying beneficial plant–fungus interactions in mycorrhizal symbiosis. *Nature Communications, 1*(1), 1–11. https://doi.org/10.1038/ncomms1046.

Brandt, K., & Barrangou, R., (2019). Applications of CRISPR technologies across the food supply chain. *Annual Review of Food Science and Technology., 10*, 133–150. https://doi.org/10.1146/annurev-food-032818-121204.

Brouns, S. J., Jore, M. M., Lundgren, M., Westra, E. R., Slijkhuis, R. J., Snijders, A. P., Dickman, M. J., et al., (2008). Small CRISPR RNAs guide antiviral defense in prokaryotes. *Science, 321*(5891), 960–964. https://doi.org/10.1126/science.1159689.

Bulgarelli, D., Garrido-Oter, R., Munch, P. C., Weiman, A., Droge, J., Pan, Y., McHardy, A. C., & Schulze-Lefert, P., (2015). Structure and function of the bacterial root microbiota in wild and domesticated barley. *Cell Host & Microbe., 17*(3), 392–403. https://doi.org/10.1016/j.chom.2015.01.011.

Bulgarelli, D., Schlaeppi, K., Spaepen, S., Ver, L. V. T. E., & Schulze-Lefert, P., (2013). Structure and functions of the bacterial microbiota of plants. *Annual Review of Plant Biology, 64*, 807–838. https://doi.org/10.1146/annurev-arplant-050312–120106.

Cao, Y., Zhou, H., Zhou, X., & Li, F., (2020). Control of plant viruses by CRISPR/Cas system-mediated adaptive immunity. *Frontiers in Microbiology, 11*, 593700.

Castro, M. J., Ojeda, C., & Cirelli, A. F., (2014). Advances in surfactants for agrochemicals. *Environmental Chemistry Letters, 12*(1), 85–95. https://doi.org/10.1007/s10311-013-0432-4.

Cavicchioli, R., Ripple, W. J., Timmis, K. N., Azam, F., Bakken, L. R., Baylis, M., Behrenfeld, M. J., et al., (2019). Scientists' warning to humanity: Microorganisms and climate change. *Nature Reviews Microbiology, 17*(9), 569–586. https://doi.org/10.1038/s41579-019-0222-5.

Chandrasekaran, J., Brumin, M., Wolf, D., Leibman, D., Klap, C., Pearlsman, M., Sherman, A., et al., (2016). Development of broad virus resistance in non-transgenic cucumber using CRISPR/Cas9 technology. *Molecular Plant Pathology, 17*, 1140–1153.

Chen, H., Wu, H., Yan, B., Zhao, H., Liu, F., Zhang, H., Sheng, Q., Miao, F., & Liang, Z., (2018). Core microbiome of medicinal plant *Salvia miltiorrhiza* seed: A rich reservoir of beneficial microbes for secondary metabolism?. *International Journal of Molecular Sciences, 19*(3), 672. https://doi.org/10.3390/ijms19030672.

Chen, J., Lai, Y., Wang, L., Zhai, S., Zou, G., Zhou, Z., Cui, C., & Wang, S., (2017). CRISPR/Cas9-mediated efficient genome editing via blastospore-based transformation in entomopathogenic fungus *Beauveria bassiana. Scientific Reports, 7*(1), 1–10. https://doi.org/10.1038/srep45763.

Cody, W. B., Scholthof, H. B., & Mirkov, T. E., (2017). Multiplexed gene editing and protein overexpression using a tobacco mosaic virus viral vector. *Plant Physiology, 175*(1), 23–35.

Cong, L., Ran, F. A., Cox, D., Lin, S., Barretto, R., Habib, N., Hsu, P. D., et al., (2013). Multiplex genome engineering using CRISPR/Cas systems. *Science, 339*(6121), 819–823. https://doi.org/10.1126/science.1231143.

Cox, D., Gootenberg, J. S., Abudayyeh, O. O., Franklin, B., Kellner, M. J., Joung, J., & Zhang, F., (2017). RNA editing with CRISPR-Cas13. *Science (New York, N.Y.), 358*(6366), 1019–1027. https://doi.org/10.1126/science.aaq0180.

Crick, F., (1970). Central dogma of molecular biology. *Nature Cell Biology, 227*(5258), 561–563.

Curtin, S. J., (2018). Editing the *Medicago truncatula* genome: Targeted mutagenesis using the CRISPR-Cas9 reagent. In: *Functional Genomics in Medicago Truncatula* (pp. 161–174), Humana Press, New York, NY. https://doi.org/10.1007/978-1-4939-8633-0_12.

Das, A., Sharma, N., & Prasad, M., (2019). CRISPR/Cas9: A novel weapon in the arsenal to combat plant diseases. *Frontiers in Plant Science., 9*, 2008.

Deng, H., Gao, R., Liao, X., & Cai, Y., (2017). Genome editing in *Shiraia bambusicola* using CRISPR-Cas9 system. *Journal of Biotechnology., 259*, 228–234. https://doi.org/10.1016/j.jbiotec.2017.06.1204.

Dreissig, S., Schiml, S., Schindele, P., Weiss, O., Rutten, T., Schubert, V., Gladilin, E., et al., (2017). Live-cell CRISPR imaging in plants reveals dynamic telomere movements. *The Plant Journal: For Cell and Molecular Biology, 91*(4), 565573. https://doi.org/10.1111/tpj.13601.

Eoh, J., & Gu, L., (2019). Biomaterials as vectors for the delivery of CRISPR-Cas9. *Biomaterials Science, 7*(4), 1240–1261. https://doi.org/10.1039/c8bm01310a.

Gaj, T., Gersbach, C. A., & Barbas, III. C. F., (2013). ZFN, TALEN, and CRISPR/Cas-based methods for genome engineering. *Trends in Biotechnology, 31*(7), 397–405. https://doi.org/10.1016/j.tibtech.2013.04.004.

Gallego-Bartolome, J., Gardiner, J., Liu, W., Papikian, A., Ghoshal, B., Kuo, H. Y., et al., (2018). Targeted DNA demethylation of the *Arabidopsis* genome using the human TET1 catalytic domain. *Proceedings of the National Academy of Sciences of the United States of America, 115*(9), E2125–E2134. https://doi.org/10.1073/pnas.1716945115.

Gaudelli, N. M., Komor, A. C., Rees, H. A., Packer, M. S., Badran, A. H., Bryson, D. I., & Liu, D. R., (2017). Programmable base editing of A_T to G_C in genomic DNA without DNA cleavage. *Nature, 551*(7681), 464–471.

Geurts, R., Xiao, T. T., & Reinhold-Hurek, B., (2016). What does it take to evolve a nitrogen-fixing endosymbiosis? *Trends in Plant Science, 21*(3), 199–208. https://doi.org/10.1016/j.tplants.2016.01.012.

Gomez, M., Lin, Z., Moll, T., Chauhan, R., Hayden, L., Renninger, K., Beyene, G., et al., (2019). Simultaneous CRISPR/Cas9-mediated editing of cassava eIF4E isoforms nCBP-1 and nCBP-2 reduces cassava brown streak disease symptom severity and incidence. *Plant Biotechnology Journal, 17*, 421–434.

Gootenberg, J. S., Abudayyeh, O. O., Kellner, M. J., Joung, J., Collins, J. J., & Zhang, F., (2018). Multiplexed and portable nucleic acid detection platform with Cas13, Cas12a, and Csm6. *Science (New York, N.Y.), 360*(6387), 439–444. https://doi.org/10.1126/science.aaq0179.

Gootenberg, J. S., Abudayyeh, O. O., Lee, J. W., Essletzbichler, P., Dy, A. J., Joung, J., Verdine, V., et al., (2017). Nucleic acid detection with CRISPR-Cas13a/C2c2. *Science (New York, N.Y.), 356*(6336), 438–442. https://doi.org/10.1126/science.aam9321.

Gosavi, G., Yan, F., Ren, B., Kuang, Y., Yan, D., Zhou, X., & Zhou, H., (2020). Applications of CRISPR technology in studying plant-pathogen interactions: Overview and perspective. *Phytopathology Research, 2*, 21.

Halperin, S. O., Tou, C. J., Wong, E. B., Modavi, C., Schaffer, D. V., & Dueber, J. E., (2018). CRISPR-guided DNA polymerases enable diversification of all nucleotides in a tunable window. *Nature, 560*(7717), 248–252. https://doi.org/10.1038/s41586-018-0384-8.

Hameed, A., Zaidi, S. S., Shakir, S., & Mansoor, S., (2018). Applications of new breeding technologies for potato improvement. *Frontiers in Plant Science., 9*, 925.

Han, G. Z., (2019). Origin and evolution of the plant immune system. *New Phytologist, 222*(1), 70–83. https://doi.org/10.1111/nph.15596.

Hashimoto, M., Neriya, Y., Yamaji, Y., & Namba, S., (2016). Recessive resistance to plant viruses: Potential resistance genes beyond translation initiation factors. *Frontiers in Microbiology, 7*, 1695.

Hassani, M. A., Duran, P., & Hacquard, S., (2018). Microbial interactions within the plant holobiont. *Microbiome, 6*(1), 58. https://doi.org/10.1186/s40168-018-0445-0.

Helm, M., Qi, M., Sarkar, S., Yu, H., Whitham, S., & Innes, R., (2019). Engineering a decoy substrate in soybean to enable recognition of the soybean mosaic virus NIa protease. *Molecular Plant-Microbe Interactions, 32*(6), 760–769.

Hille, F., Richter, H., Wong, S. P., Bratovic, M., Ressel, S., & Charpentier, E., (2018). The biology of CRISPR-Cas: Backward and forward. *Cell, 172*(6), 1239–1259.

Hiruma, K., Gerlach, N., Sacristan, S., Nakano, R. T., Hacquard, S., Kracher, B., Neumann, U., et al., (2016). Root endophyte *Colletotrichum tofieldiae* confers plant fitness benefits that are phosphate status dependent. *Cell, 165*(2), 464–474. https://doi.org/10.1016/j.cell.2016.02.028.

Hu, J., Li, S., Li, Z., Li, H., Song, W., Zhao, H., Lai, J., et al., (2019). A barley stripe mosaic virus-based guide RNA delivery system for targeted mutagenesis in wheat and maize. *Molecular Plant Pathology, 20*, 1463–1474.

Iqbal, Z., Sattar, M. N., & Shafiq, M., (2016). CRISPR/Cas9: A tool to circumscribe cotton leaf curl disease. *Frontiers in Plant Sciences, 7*, 475.

Ishino, Y., Shinagawa, H., Makino, K., Amemura, M., & Nakata, A., (1987). Nucleotide sequence of the *iap* gene, responsible for alkaline phosphatase isozyme conversion in *Escherichia coli*, and identification of the gene product. *Journal of Bacteriology, 169*(12), 5429–5433. https://doi.org/10.1128/jb.169.12.5429-5433.1987.

Jansson, J. K., & Hofmockel, K. S., (2018). The soil microbiome-from metagenomics to metaphenomics. *Current Opinion in Microbiology, 43*, 162–168. https://doi.org/10.1016/j.mib.2018.01.013.

Ji, J., Zhang, C., Sun, Z., Wang, L., Duanmu, D., & Fan, Q., (2019). Genome editing in cowpea *Vigna unguiculata* Using CRISPR-Cas9. *International Journal of Molecular Sciences, 20*(10), 2471. https://doi.org/10.3390/ijms20102471.

Ji, X., Zhang, H., Zhang, Y., Wang, Y., & Gao, C., (2015). Establishing a CRISPR Cas-like immune system conferring DNA virus resistance in plants. *Nature Plants, 1*(10), 15144.

Jia, H., Orbovic, V., Jones, J. B., & Wang, N., (2016). Modification of the PthA4 effector binding elements in type I CsLOB1 promoter using Cas9/sgRNA to produce transgenic Duncan grapefruit alleviating XccΔpthA4: dCsLOB1.3infection. *Plant Biotechnology Journal, 14*(5), 1291–1301.

Jia, H., Zhang, Y., Orbovic, V., Xu, J., White, F. F., Jones, J. B., & Wang, N., (2017). Genome editing of the disease susceptibility gene CsLOB1 in citrus confers resistance to citrus canker. *Plant Biotechnology Journal, 15*(7), 817–823. https://doi.org/10.1111/pbi.12677.

Jiang, W., Zhou, H., Bi, H., Fromm, M., Yang, B., & Weeks, D. P., (2013). Demonstration of CRISPR/Cas9/sgRNA-mediated targeted gene modification in *Arabidopsis*, tobacco, sorghum and rice. *Nucleic Acids Research, 41*(20), e188. https://doi.org/10.1093/nar/gkt780.

Jiang, Y., Chen, B., Duan, C., Sun, B., Yang, J., & Yang, S., (2015). Multigene editing in the *Escherichia coli* genome via the CRISPR-Cas9 system. *Applied and Environmental Microbiology, 81*(7), 2506–2514. https://doi.org/10.1128/AEM.04023-14.

Jiao, Y., Li, Y., Li, Y., Cao, H., Mao, Z., Ling, J., Yang, Y., & Xie, B., (2019). Functional genetic analysis of the leucinostatin biosynthesis transcription regulator lcsL in *Purpureocillium lilacinum* using CRISPR-Cas9 technology. *Applied Microbiology and Biotechnology, 103*(15), 6187–6194. https://doi.org/10.1007/s00253-019-09945-2.

Jinek, M., Chylinski, K., Fonfara, I., Hauer, M., Doudna, J. A., & Charpentier, E., (2012). A Programmable dual-RNA-guided DNA endonuclease in adaptive bacterial immunity. *Science, 337*(6096), 816–821.

Jones, J. D., Vance, R. E., & Dangl, J. L., (2016). Intracellular innate immune surveillance devices in plants and animals. *Science (New York, N.Y.), 354*(6316), aaf6395. https://doi.org/10.1126/science.aaf6395.

Kalhor, R., Kalhor, K., Mejia, L., Leeper, K., Graveline, A., Mali, P., & Church, G. M., (2018). Developmental barcoding of whole mouse via homing CRISPR. *Science (New York, N.Y.), 361*(6405), eaat9804. https://doi.org/10.1126/science.aat9804.

Kalinina, N. O., Khromov, A., Love, A. J., & Taliansky, M. E., (2020). CRISPR applications in plant virology: Virus resistance and beyond. *Phytopathology, 110*(1), 18–28.

Katayama, T., Tanaka, Y., Okabe, T., Nakamura, H., Fujii, W., Kitamoto, K., & Maruyama, J., (2016). Development of a genome editing technique using the CRISPR/Cas9 system in the industrial filamentous fungus *Aspergillus oryzae*. *Biotechnology Letters, 38*(4), 637–642. https://doi.org/10.1007/s10529-015-2015-x.

Keeling, P. J., (2010). The endosymbiotic origin, diversification and fate of plastids. *Philosophical Transactions of the Royal Society of London. Series B, Biological Sciences, 365*(1541), 729–748. https://doi.org/10.1098/rstb.2009.0103.

Kirjavainen, P. V., Karvonen, A. M., Adams, R. I., Taubel, M., Roponen, M., Tuoresmaki, P., Loss, G., et al., (2019). Farm-like indoor microbiota in non-farm homes protects children from asthma development. *Nature Medicine, 25*(7), 1089–1095. https://doi.org/10.1038/s41591-019-0469-4.

Kis, A., Hamar, E., Tholt, G., Ban, R., & Havelda, Z., (2019). Creating highly efficient resistance against wheat dwarf virus in barley by employing CRISPR/Cas9 system. *Plant Biotechnology Journal, 17*(6), 1004–1006.

Klaedtke, S., Jacques, M. A., Raggi, L., Preveaux, A., Bonneau, S., Negri, V., Chable, V., & Barret, M., (2016). Terroir is a key driver of seed-associated microbial assemblages. *Environmental Microbiology, 18*(6), 1792–1804. https://doi.org/10.1111/1462-2920.12977.

Knott, G. J., & Doudna, J. A., (2018). CRISPR-Cas guides the future of genetic engineering. *Science, 361*(6405), 866–869. https://doi.org/10.1126/science.aat5011.

Kong, G., Wan, L., Deng, Y. Z., Yang, W., Li, W., Jiang, L., et al., (2019). Pectin acetylesterase PAE5 is associated with the virulence of plant pathogenic oomycete *Peronophythora litchii*. *Physiological and Molecular Plant Pathology, 106*, 16–22. https://doi.org/10.1016/j.pmpp.2018.11.006.

Kui, L., Chen, H., Zhang, W., He, S., Xiong, Z., Zhang, Y., Yan, L., et al., (2017). Building a genetic manipulation toolbox for orchid biology: Identification of constitutive promoters and application of CRISPR/Cas9 in the orchid, *Dendrobium officinale*. *Frontiers in Plant Science, 7*, 2036. https://doi.org/10.3389/fpls.2016.02036.

Kuivanen, J., Wang, Y. M. J., & Richard, P., (2016). Engineering *Aspergillus Niger* for galactaric acid production: Elimination of galactaric acid catabolism by using RNA sequencing and CRISPR/Cas9. *Microbial Cell Factories, 15*(1), 1–9. https://doi.org/10.1186/s12934-016-0613-5.

Kwak, M. J., Kong, H. G., Choi, K., Kwon, S. K., Song, J. Y., Lee, J., Lee, P. A., et al., (2018). Rhizosphere microbiome structure alters to enable wilt resistance in tomato. *Nature Biotechnology, 36*(11), 1100–1109. https://doi.org/10.1038/nbt.4232.

Langner, T., Kamoun, S., & Belhaj, K., (2018). CRISPR crops: Plant genome editing toward disease resistance. *Annual Review of Phytopathology, 56*, 479–512. https://doi.org/10.1146/annurev-phyto-080417-050158.

Lellis, A., Kasschau, K., Whitham, S., & Carrington, J., (2002). Loss-of-susceptibility mutants of *Arabidopsis thaliana* reveal an essential role for eIF(iso)4E during potyvirus infection. *Current Biology, 12*(12), 1046–1051.

Levy, A., Conway, J. M., Dangl, J. L., & Woyke, T., (2018). Elucidating bacterial gene functions in the plant microbiome. *Cell Host & Microbe, 24*(4), 475–485. https://doi.org/10.1016/j.chom.2018.09.005.

Li, B., Cui, G., Shen, G., Zhan, Z., Huang, L., Chen, J., & Qi, X., (2017). Targeted mutagenesis in the medicinal plant *Salvia miltiorrhiza*. *Scientific Reports, 7*(1), 1–9. https://doi.org/10.1038/srep43320.

Li, J., Sun, Y., Du, J., Zhao, Y., & Xia, L., (2017). Generation of targeted point mutations in rice by a modified CRISPR/Cas9 system. *Molecular Plant, 10*(3), 526–529. https://doi.org/10.1016/j.molp.2016.12.001.

Li, J., Zhang, Y., Zhang, Y., Yu, P. L., Pan, H., & Rollins, J. A., (2018). Introduction of large sequence inserts by CRISPR-Cas9 to create pathogenicity mutants in the multinucleate filamentous pathogen *Sclerotinia sclerotiorum. M. Bio., 9*(3), e00567–18. https://doi.org/10.1128/mbio.00567-18.

Li, R., Li, R., Li, X., Fu, D., Zhu, B., Tian, H., Luo, Y., & Zhu, H., (2018). Multiplexed CRISPR/Cas9-mediated metabolic engineering of γ-aminobutyric acid levels in *Solanum lycopersicum. Plant Biotechnology Journal, 16*(2), 415–427. https://doi.org/10.1111/pbi.12781.

Li, T., Liu, B., Spalding, M. H., Weeks, D. P., & Yang, B., (2012). High-efficiency TALEN-based gene editing produces disease-resistant rice. *Nature Biotechnology, 30*(5), 390–392. https://doi.org/10.1038/nbt.2199.

Li, T., Yang, X., Yu, Y., Si, X., Zhai, X., Zhang, H., Dong, W., et al., (2018). Domestication of wild tomato is accelerated by genome editing. *Nature Biotechnology, 36*(12), 1160–1163.

Li, Y., Jiao, M., Li, Y., Zhong, Y., Li, X., Chen, Z., Chen, S., & Wang, J., (2021). *Penicillium chrysogenum* polypeptide extract protects tobacco plants from tobacco mosaic virus infection through modulation of ABA biosynthesis and callose priming. *Journal of Experimental Botany, 72*(10), 3526–3539.

Li, Z., Liu, Z. B., Xing, A., Moon, B. P., Koellhoffer, J. P., Huang, L., Ward, R. T., et al., (2015). Cas9-guide RNA directed genome editing in soybean. *Plant Physiology, 169*(2), 960–970. https://doi.org/10.1104/pp.15.00783.

Liu, Q., Gao, R., Li, J., Lin, L., Zhao, J., Sun, W., & Tian, C., (2017). Development of a genome-editing CRISPR/Cas9 system in thermophilic fungal *Myceliophthora* species and its application to hyper-cellulase production strain engineering. *Biotechnology for Biofuels, 10*(1), 1–14. https://doi.org/10.1186/s13068-016-0693-9.

Liu, R., Chen, L., Jiang, Y., Zhou, Z., & Zou, G., (2015). Efficient genome editing in filamentous fungus *Trichoderma reesei* using the CRISPR/Cas9 system. *Cell Discovery, 1*(1), 1–11. https://doi.org/10.1038/celldisc.2015.7.

Liu, Y., Zuo, S., Zou, Y., Wang, J., & Song, W., (2013). Investigation on diversity and population succession dynamics of endophytic bacteria from seeds of maize (*Zea mays* L., Nongda108) at different growth stages. *Annals of Microbiology, 63*(1), 71–79. https://doi.org/10.1007/s13213-012-0446-3.

Luu, D. D., Joe, A., Chen, Y., Parys, K., Bahar, O., Pruitt, R., Chan, L. J. G., et al., Biosynthesis and secretion of the microbial sulfated peptide RaxX and binding to the rice XA21 immune receptor. *Proceedings of the National Academy of Sciences, 116*(17), 8525–8534.

Lv, H., Li, J., Wu, Y., Garyali, S., & Wang, Y., (2016). Transporter and its engineering for secondary metabolites. *Applied Microbiology and Biotechnology., 100*(14), 6119–6130. https://doi.org/10.1007/s00253-016-7605-6.

Ma, J., Chen, J., Wang, M., Ren, Y., Wang, S., Lei, C., Cheng, Z., & Sodmergen, (2018). Disruption of OsSEC3A increases the content of salicylic acid and induces plant defense responses in rice. *Journal of Experimental Botany, 69*(5), 1051–1064.

Ma, X., Zhang, X., Liu, H., & Li, Z., (2020). Highly efficient DNA-free plant genome editing using virally delivered CRISPR-Cas9. *Nature Plants, 6*, 773–779.

Ma, X., Zhu, Q., Chen, Y., & Liu, Y. G., (2016). CRISPR/Cas9 platforms for genome editing in plants: Developments and applications. *Molecular Plant, 9*, 961–974.

Malnoy, M., Viola, R., Jung, M. H., Koo, O. J., Kim, S., Kim, J. S., Velasco, R., & Nagamangala, K. C., (2016). DNA-free genetically edited grapevine and apple protoplast using CRISPR/Cas9 ribonucleoproteins. *Frontiers in Plant Science, 7*, 1904. https://doi.org/10.3389/fpls.2016.01904.

Mao, Y., Botella, J. R., Liu, Y., & Zhu, J., (2019). Gene editing in plants: Progress and challenges. *National Science Review, 6*(3), 421–437.

McKenna, A., Findlay, G. M., Gagnon, J. A., Horwitz, M. S., Schier, A. F., & Shendure, J., (2016). Whole-organism lineage tracing by combinatorial and cumulative genome editing. *Science (New York, N.Y.), 353*(6298), aaf7907. https://doi.org/10.1126/science.aaf7907.

Mehta, D., Sturchler, A., Anjanappa, R. B., Zaidi, S. S., Hirsch-Hoffmann, M., Gruissem, W., & Vanderschuren, H., (2019). Linking CRISPR-Cas9 interference in cassava to the evolution of editing resistant geminiviruses. *Genome Biology, 20*, 80.

Mercx, S., Smargiasso, N., Chaumont, F., De Pauw, E., Boutry, M., & Navarre, C., (2017). Inactivation of the β(1,2)-xylosyl transferase and the α(1,3)-fucosyl transferase genes in *Nicotiana tabacum* BY-2 cells by a multiplex CRISPR/Cas9 strategy results in glycoproteins without plant-specific glycans. *Frontiers in Plant Science, 8*, 403. https://doi.org/10.3389/fpls.2017.00403.

Midha, S., Bansal, K., Sharma, S., Kumar, N., Patil, P. P., Chaudhry, V., & Patil, P. B., (2016). Genomic resource of rice seed associated bacteria. *Frontiers in Microbiology, 6*, 1551. https://doi.org/10.3389/fmicb.2015.01551.

Miles, L. A., Garippa, R. J., & Poirier, J. T., (2016). Design, execution, and analysis of pooled *in vitro* CRISPR/Cas9 screens. *The FEBS Journal, 283*(17), 3170–3180. https://doi.org/10.1111/febs.13770.

Monreal, C. M., DeRosa, M., Mallubhotla, S. C., Bindraban, P. S., & Dimkpa, C., (2016). Nanotechnologies for increasing the crop use efficiency of fertilizer-micronutrients. *Biology and Fertility of Soils, 52*(3), 423–437. https://doi.org/10.1007/s00374-015-1073-5.

Morgan, S. L., Mariano, N. C., Bermudez, A., Arruda, N. L., Wu, F., Luo, Y., Shankar, G., et al., (2017). Manipulation of nuclear architecture through CRISPR-mediated chromosomal looping. *Nature Communications, 8*, 15993. https://doi.org/10.1038/ncomms15993.

Nelles, D. A., Fang, M. Y., O'Connell, M. R., Xu, J. L., Markmiller, S. J., Doudna, J. A., & Yeo, G. W., (2016). Programmable RNA tracking in live cells with CRISPR/Cas9. *Cell, 165*(2), 488–496. https://doi.org/10.1016/j.cell.2016.02.054.

Netea, M. G., & Van, D. M. J. W., (2017). Trained immunity: An ancient way of remembering. *Cell Host & Microbe., 21*(3), 297–300. https://doi.org/10.1016/j.chom.2017.02.003.

Niu, B., Paulson, J. N., Zheng, X., & Kolter, R., (2017). Simplified and representative bacterial community of maize roots. *Proceedings of the National Academy of Sciences, 114*(12), E2450–E2459. https://doi.org/10.1073/pnas.1616148114.

Ortigosa, A., Gimenez-Ibanez, S., Leonhardt, N., & Solano, R., (2019). Design of a bacterial speck resistant tomato by CRISPR/ Cas9-mediated editing of SlJAZ2. *Plant Biotechnology Journal, 17*(3), 665–673.

Peng, A., Chen, S., Lei, T., Xu, L., He, Y., Wu, L., Yao, L., & Zou, X., (2017). Engineering canker-resistant plants through CRISPR/ Cas9-targeted editing of the susceptibility gene CsLOB1 promoter in citrus. *Plant Biotechnology Journal, 15*(12), 1509–1519.

Piatek, A., Ali, Z., Baazim, H., Li, L., Abulfaraj, A., Al-Shareef, S., Aouida, M., & Mahfouz, M. M., (2015). RNA-guided transcriptional regulation in planta via synthetic dCas9-based transcription factors. *Plant Biotechnology Journal, 13*(4), 578–589. https://doi.org/10.1111/pbi.12284.

Pieterse, C. M., Zamioudis, C., Berendsen, R. L., Weller, D. M., Van, W. S. C., & Bakker, P. A., (2014). Induced systemic resistance by beneficial microbes. *Annual Review of Phytopathology, 52*, 347–375. https://doi.org/10.1146/annurev-phyto-082712-102340.

Pinski, A., Betekhtin, A., Hupert-Kocurek, K., Mur, L., & Hasterok, R., (2019). Defining the genetic basis of plant- endophytic bacteria interactions. *International Journal of Molecular Sciences, 20*(8), 1947. https://doi.org/10.3390/ijms20081947.

Podolsky, I. A., Seppala, S., Lankiewicz, T. S., Brown, J. L., Swift, C. L., & O'Malley, M. A., (2019). Harnessing nature's anaerobes for biotechnology and bioprocessing. *Annual Review of Chemical and Biomolecular Engineering, 10*, 105–128. https://doi.org/10.1146/annurev-chembioeng-060718-030340.

Pramanik, D., Shelake, R. M., Kim, M. J., & Kim, J. Y., (2021). CRISPR-Mediated engineering across the central dogma in plant biology for basic research and crop improvement. *Molecular Plant, 14*(1), 127–150.

Price, A., Sampson, T., Ratner, H., Grakoui, A., & Weiss, D., (2015). Cas9-mediated targeting of viral RNA in eukaryotic cells. *Proceedings of the National Academy of Sciences of the United States of America, 112*(19), 6164–6169.

Pyne, M. E., Narcross, L., & Martin, V., (2019). Engineering plant secondary metabolism in microbial systems. *Plant Physiology., 179*(3), 844–861. https://doi.org/10.1104/pp.18.01291.

Pyott, D., Sheehan, E., & Molnar, A., (2016). Engineering of CRISPR/Cas9- mediated potyvirus resistance in transgene-free *Arabidopsis* plants. *Molecular Plant Pathology, 17*, 1276–1288.

Qi, L. S., Larson, M. H., Gilbert, L. A., Doudna, J. A., Weissman, J. S., Arkin, A. P., & Lim, W. A., (2013). Repurposing CRISPR as an RNA-guided platform for sequence-specific control of gene expression. *Cell, 152*(5), 1173–1183.

Qin, P., Parlak, M., Kuscu, C., Bandaria, J., Mir, M., Szlachta, K., Singh, R., et al., (2017). Live cell imaging of low-and non-repetitive chromosome loci using CRISPR-Cas9. *Nature Communications, 8*(1), 1–10. https://doi.org/10.1038/ncomms14725.

Qiu, Z., Egidi, E., Liu, H., Kaur, S., & Singh, B. K., (2019). New frontiers in agriculture productivity: Optimized microbial inoculants and in situ microbiome engineering. *Biotechnology Advances, 37*(6), 107371. https://doi.org/10.1016/j.biotechadv.2019.03.010.

Robaglia, C., & Caranta, C., (2006). Translation initiation factors: A weak link in plant RNA virus infection. *Trends in Plant Science, 11*(1), 40–45.

Roberts, B., Haupt, A., Tucker, A., Grancharova, T., Arakaki, J., Fuqua, M. A., Nelson, A., et al., (2017). Systematic gene tagging using CRISPR/Cas9 in human stem cells to illuminate cell organization. *Molecular Biology of the Cell, 28*(21), 2854–2874. https://doi.org/10.1091/mbc.E17-03-0209.

Rodriguez, P. A., Rothballer, M., Chowdhury, S. P., Nussbaumer, T., Gutjahr, C., & Falter-Braun, P., (2019). Systems biology of plant-microbiome interactions. *Molecular Plant, 12*(6), 804–821. https://doi.org/10.1016/j.molp.2019.05.006.

Rosenberg, E., & Zilber-Rosenberg, I., (2018). The hologenome concept of evolution after 10 years. *Microbiome, 6*(1), 1–14. https://doi.org/10.1186/s40168-018-0457-9.

Rybakova, D., Mancinelli, R., Wikstrom, M., Birch-Jensen, A. S., Postma, J., Ehlers, R. U., Goertz, S., & Berg, G., (2017). The structure of the *Brassica napus* seed microbiome is cultivar-dependent and affects the interactions of symbionts and pathogens. *Microbiome., 5*(1), 1–16. https://doi.org/10.1186/s40168-017-0310-6.

Sakuma, T., & Yamamoto, T., (2017). Magic wands of CRISPR-Lots of choices for gene knock-in. *Cell Biology and Toxicology, 33*(6), 501–505. https://doi.org/10.1007/s10565-017-9409-6.

Sanfacon, H., (2015). Plant translation factors and virus resistance. *Viruses, 7*(7), 3392–3419.

Savary, S., Willocquet, L., Pethybridge, S. J., Esker, P., McRoberts, N., & Nelson, A., (2019). The global burden of pathogens and pests on major food crops. *Nature Ecology & Evolution, 3*(3), 430–439. https://doi.org/10.1038/s41559-018-0793-y.

Shanmugam, K., Ramalingam, S., Venkataraman, G., & Hariharan, G. N., (2019). The CRISPR/Cas9 system for targeted genome engineering in free-living fungi: Advances and opportunities for lichenized fungi. *Frontiers in Microbiology, 10*, 62. https://doi.org/10.3389/fmicb.2019.00062.

Shi, T. Q., Liu, G. N., Ji, R. Y., Shi, K., Song, P., Ren, L. J., Huang, H., & Ji, X. J., (2017). CRISPR/Cas9-based genome editing of the filamentous fungi: The state of the art. *Applied Microbiology and Biotechnology, 101*(20), 7435–7443. https://doi.org/10.1007/s00253-017-8497-9.

Shmakov, S., Abudayyeh, O., Makarova, K., Wolf, Y., Gootenberg, J., Semenova, E., Minakhin, L., et al., (2015). Discovery and functional characterization of diverse class 2 CRISPR-Cas systems. *Molecular Cell, 60*(3), 385–397.

Spoel, S. H., & Dong, X., (2012). How do plants achieve immunity? Defense without specialized immune cells. *Nature Reviews Immunology, 12*(2), 89–100. https://doi.org/10.1038/nri3141.

Su, Z., Bernardo, A., Tian, B., Chen, H., Wang, S., Ma, H., Cai, S., et al., (2019). A deletion mutation in TaHRC confers Fhb1 resistance to *Fusarium* head blight in wheat. *Nature Genetics, 51*(7), 1099–1105.

Sun, J., Wang, Q., Jiang, Y., Wen, Z., Yang, L., Wu, J., & Yang, S., (2018). Genome editing and transcriptional repression in *Pseudomonas putida* KT2440 via the type II CRISPR system. *Microbial Cell Factories, 17*(1), 1–17. https://doi.org/10.1186/s12934-018-0887-x.

Sun, Y., Zhang, X., Wu, C., He, Y., Ma, Y., Hou, H., Guo, X., et al., (2016). Engineering herbicide-resistant rice plants through CRISPR/Cas9-mediated homologous recombination of acetolactate synthase. *Molecular Plant, 9*(4), 628–631.

Tashkandi, M., Ali, Z., Aljedaani, F., Shami, A., & Mahfouz, M. M., (2018). Engineering resistance against tomato yellow leaf curl virus via the CRISPR/Cas9 system in tomato. *Plant Signaling and Behavior, 13*(10), e1525996.

Thrall, P. H., Hochberg, M. E., Burdon, J. J., & Bever, J. D., (2007). Coevolution of symbiotic mutualists and parasites in a community context. *Trends in Ecology & Evolution, 22*(3), 120–126. https://doi.org/10.1016/j.tree.2006.11.007.

Tripathi, J., Ntui, V., Ron, M., Muiruri, S., Britt, A., & Tripathi, L., (2019). CRISPR/Cas9 editing of endogenous banana streak virus in the B genome of *Musa* spp. overcomes a major challenge in banana breeding. *Communications Biology, 2*, 46.

Tyagi, S., Kumar, R., Kumar, V., Won, S. Y., & Shuk, P., (2020). Engineering disease resistant plants through CRISPR-Cas9 technology. *GM Crops and Food, 12*(1), 125–144.

Upson, J. L., Zess, E. K., Bialas, A., Wu, C. H., & Kamoun, S., (2018). The coming of age of EvoMPMI: Evolutionary molecular plant-microbe interactions across multiple timescales. *Current Opinion in Plant Biology, 44*, 108–116. https://doi.org/10.1016/j.pbi.2018.03.003.

Van, E. H. P., Reuber, L., & Van, D. D. D., (2019). GM approaches to improve disease resistance in crops. *New Phytol., 225*, 70–86. https://doi.org/10.1111/nph.15967.

Waghunde, R. R., Shelake, R. M., Shinde, M. S., & Hayashi, H., (2017). Endophyte microbes: A weapon for plant health management. In: *Microorganisms for Green Revolution* (pp. 303–325). Springer, Singapore. https://doi.org/10.1007/978-981-10-6241-4_16.

Wang, A., & Krishnaswamy, S., (2012). Eukaryotic translation initiation factor 4E-mediated recessive resistance to plant viruses and its utility in crop improvement. *Molecular Plant Pathology, 13*(7), 795–803.

Wang, A., (2015). Dissecting the molecular network of virus-plant interactions: The complex roles of host factors. *Annual Review of Phytopathology, 53*, 45–66.

Wang, F., Wang, C., Liu, P., Lei, C., Hao, W., Gao, Y., Liu, Y. G., & Zhao, K., (2016). Enhanced rice blast resistance by CRISPR/cas9-targeted mutagenesis of the ERF transcription factor gene OsERF922. *PLoS One, 11*(4), e0154027.

Wang, L., Wang, L., Tan, Q., Fan, Q., Zhu, H., Hong, Z., Zhang, Z., & Duanmu, D., (2016). Efficient inactivation of symbiotic nitrogen fixation related genes in *Lotus japonicus* using CRISPR-Cas9. *Frontiers in Plant Science, 7*, 1333. https://doi.org/10.3389/fpls.2016.01333.

Wang, Y., Cheng, X., Shan, Q., Zhang, Y., Liu, J., Gao, C., & Qiu, J. L., (2014). Simultaneous editing of three homoeoalleles in hexaploid bread wheat confers heritable resistance to powdery mildew. *Nature Biotechnology, 32*, 947–951.

Weber, J., Valiante, V., Nodvig, C. S., Mattern, D. J., Slotkowski, R. A., Mortensen, U. H., & Brakhage, A. A., (2017). Functional reconstitution of a fungal natural product gene cluster by advanced genome editing. *ACS Synthetic Biology, 6*(1), 62–68. https://doi.org/10.1021/acssynbio.6b00203.

Woo, J. W., Kim, J., Kwon, S. I., Corvalán, C., Cho, S. W., Kim, H., Kim, S. G., et al., (2015). DNA-free genome editing in plants with preassembled CRISPR-Cas9 ribonucleoproteins. *Nature Biotechnology, 33*(11), 1162–1164.

Wu, J., Du, G., Chen, J., & Zhou, J., (2015). Enhancing flavonoid production by systematically tuning the central metabolic pathways based on a CRISPR interference system in *Escherichia coli*. *Scientific Reports, 5*(1), 1–14. https://doi.org/10.1038/srep13477.

Xie, K., Minkenberg, B., & Yang, Y., (2015). Boosting CRISPR/Cas9 multiplex editing capability with the endogenous tRNA-processing system. *Proceedings of the National Academy of Sciences of the United States of America, 112*(11), 3570–3575.

Xin, X. F., Kvitko, B., & He, S. Y., (2018). *Pseudomonas syringae*: What it takes to be a pathogen. *Nature reviews. Microbiology, 16*(5), 316–328. https://doi.org/10.1038/nrmicro.2018.17.

Xu, X., & Qi, L. S., (2019). A CRISPR-dCas toolbox for genetic engineering and synthetic biology. *Journal of Molecular Biology, 431*(1), 34–47. https://doi.org/10.1016/j.jmb.2018.06.037.

Yi, Y., Li, Z., Song, C., & Kuipers, O. P., (2018). Exploring plant-microbe interactions of the rhizobacteria *Bacillus subtilis* and *Bacillus mycoides* by use of the CRISPR-Cas9 system. *Environmental Microbiology, 20*(12), 4245–4260. https://doi.org/10.1111/1462-2920.14305.

Yin, K., Han, T., Xie, K., Zhao, J., Song, J., & Liu, Y., (2019). Engineer complete resistance to cotton leaf curl Multan virus by the CRISPR/Cas9 system in *Nicotiana benthamiana*. *Phytopathology Research., 1*, 9.

Zaidi, S. S. E. A., Mukhtar, M. S., & Mansoor, S., (2018). Genome editing: Targeting susceptibility genes for plant disease resistance. *Trends in Biotechnology, 36*(9), 898–906. https://doi.org/10.1016/j.tibtech.2018.04.005.

Zeilmaker, T., Ludwig, N. R., Elberse, J., Seidl, M. F., Berke, L., Van, D. A., Schuurink, R. C., et al., (2015). Downy mildew resistant 6 and DMR6-like oxygenase 1 are partially redundant but distinct suppressors of immunity in *Arabidopsis*. *Plant Journal, 81*(2), 210–222.

Zhang, T., Zhao, Y., Ye, J., Cao, X., Xu, C., Chen, B., An, H., et al., (2019). Establishing CRISPR/Cas13a immune system conferring RNA virus resistance in both dicot and monocot plants. *Plant Biotechnology Journal, 17*(7), 1185–1187.

Zhang, T., Zheng, Q., Yi, X., An, H., Zhao, Y., Ma, S., & Zhou, G., (2018). Establishing RNA virus resistance in plants by harnessing CRISPR immune system. *Plant Biotechnology Journal, 16*, 1415–1423.

Zhang, Y., Bai, Y., Wu, G., Zou, S., Chen, Y., Gao, C., & Tang, D., (2017). Simultaneous modification of three homoeologs of TaEDR1 by genome editing enhances powdery mildew resistance in wheat. *The Plant Journal, 91*(4), 714–724.

Zheng, K., Wang, Y., Li, N., Jiang, F. F., Wu, C. X., Liu, F., Chen, H. C., & Liu, Z. F., (2018). Highly efficient base editing in bacteria using a Cas9-cytidine deaminase fusion. *Communications Biology, 1*(1), 1–6. https://doi.org/10.1038/s42003-018-0035-5.

Zhou, J., Peng, Z., Long, J., Sosso, D., Liu, B., Eom, J. S., Huang, S., et al., (2015). Gene targeting by the TAL effector PthXo2 reveals cryptic resistance gene for bacterial blight of rice. *The Plant Journal, 82*(4), 632–643.

Zhou, Z., Tan, H., Li, Q., Chen, J., Gao, S., Wang, Y., Chen, W., & Zhang, L., (2018). CRISPR/Cas9-mediated efficient targeted mutagenesis of RAS in *Salvia miltiorrhiza*. *Phytochemistry, 148*, 63–70. https://doi.org/10.1016/j.phytochem.2018.01.015.

CHAPTER 14

Unraveling the Application of CRISPR/ Cas System in Natural Product Research

PROTHA BISWAS,[1] MIMOSA GHORAI,[1] SAMAPIKA NANDY,[1] DEVENDRA KUMAR PANDEY,[2] MAHIPAL S. SHEKHAWAT,[3] ARABINDA GHOSH,[4] POTSHANGBAM NONGDAM,[5] ABDEL RAHMAN AL-TAWAHA,[6] ERCAN BURSAL,[7] MUDDASARUL HODA,[8] JOGINDER SINGH,[9] and ABHIJIT DEY[1*]

[1]*Department of Life Sciences, Presidency University, 86/1 College Street, Kolkata – 700073, West Bengal, India*

[2]*Department of Biotechnology, Lovely Professional University, Punjab, India*

[3]*Department of Plant Biology and Biotechnology, K.M. Center for Post Graduate Studies, Lawspet, Pondicherry, Tamil Nadu, India*

[4]*Department of Botany, Guwahati University, Guwahati, Assam, India*

[5]*Department of Biotechnology, Manipur University, Canchipur, Imphal, Manipur, India*

[6]*Department of Biological Sciences, Al-Hussein Bin Talal University, Maan, Jordan*

[7]*Department of Biochemistry, Mus Alparslan University, Turkey*

[8]*Department of Biological Sciences, Aliah University, Kolkata, West Bengal, India*

[9]*Department of Microbiology, School of Bioengineering and Biosciences, Lovely Professional University, Phagwara – 144411, Punjab, India*

[*]*Corresponding author. E-mail: abhijit.dbs@presiuniv.ac.in*

CRISPR/Cas-Mediated Genome Editing in Plants, Deepu Pandita, Anu Pandita (Eds.)

ABSTRACT

Plants produce a wide range of natural products that protect them from herbivores and microbes, attract pollinators, and disperse seeds, among the ecological functions. Humans use natural products for a wide range of purposes beyond their obvious medical uses, including as flavoring agents, fragrances, cosmetics, and food. As demand for natural products increases and gene engineering systems are developed, researchers are attempting to modify the plant genome to modulate the biosynthesis pathway for a compound of interest or block the pathway for the synthesis of unwanted compounds. CRISPR/Cas9, the clustered regularly interspaced short palindromic repeats (CRISPR), has had tremendous success in genome editing. This system is highly efficient, easy to use, and accurate, so that it revolutionized genome editing in living organisms. The chapter highlights the CRISPR/Cas technology and its application in the study of natural products in microbes, algae, fungi, and even higher plants, as well as its limitations and future prospects.

14.1 INTRODUCTION

"Natural product" is an extremely broad range of compounds derived from biological sources that are organic chemical compounds in their natural state has a diverse range of chemical properties. A variety of natural products have been studied and used by humans throughout history for various diseases. Among the multitude of natural products that may be discovered and developed are pharmaceuticals, agrochemistry, cosmetics, and fine chemicals (McChesney et al., 2007). Many biotechnological interventions are being applied to modulate plant secondary metabolite pathways to increase the production of valuable compounds used in therapy, diet, and industry (Dey et al., 2017). In recent years, with the advent of new technologies like genetic engineering (GE), molecular genetics, gene editing, and screening, natural products have become once again attractive to pharmaceutical companies as potential leads (Pelzer et al., 2015).

The development of genome editing (GE) has led to new possibilities of accurately editing living organisms' genomes, and has revolutionized biological research as well. Gene editing is a technique that involves introducing mutations in DNA, such as deletions, insertions, or base substitutions, to target sequences to modify their structure (Manghwar et al., 2019). In the beginning, mega endonucleases, and homing endonucleases (HEs) were used to engineer DNA and the use of zinc finger nucleases (ZFNs) to edit plant genomes was found to be successful subsequently (Silva et al., 2011; Weinthal et al., 2010). Transcription activator-like effector nucleases (TALEN) is developed as a third tool (Zhang et al., 2013). Programmable endonucleases such as ZNFs and TALENS have been successful in the past, but there is hindrance in their success, such as programmability at the protein level and their ability to associate with DNA with specific modifications viz. gene knock-out by creating double-stranded breaks (DSBs) in DNA targets using the FokI protein-based endonuclease (Shukla et al., 2009; Gaj et al., 2013; Char et al., 2015; Agarwal et al., 2018). Researchers developed an efficient and dominant method for editing DNA after overcoming challenges such as inconvenient tools, lack of specificity, and time-consuming methods. A new generation of genome editing tools has been developed following the discovery of CRISPR (clustered regularly interspaced short palindromic repeats) in bacteria and CRISPR-associated (Cas) methods (Cong et al., 2013; Upadhyay et al., 2013). In the early 21st century, CRISPR/Cas technologies revolutionized genetic research, rendering it more democratized and enabling us to manipulate genomes with unprecedented precision (Knott and Doudna, 2018). With the use of CRISPR/Cas methods, it has become easier and more precise to determine the function of several DNA coding and noncoding sequences in models plants. Since CRISPR/Cas gene editing does not produce foreign DNA sequences in plants, an inherently hereditable, targeted mutation is produced as well as sufficient protection against transgenes (Ricroch et al., 2017). The CRISPR-Cas systems combine short RNAs and proteins to cleave specific DNA sequences. By undergoing integration of short guide RNAs (sgRNAs) into bacterial vectors, bacteria are able to efficiently and effectively express sgRNA through rapid, efficient, and effective expression of foreign DNA sequences. Using the protospacers, CRISPR/Cas can eliminate DNA sequences that match the CRISPR/Cas system (Rehman et al., 2021). To edit a genome with CRISPR/Cas9, Cas9 and sgRNA are first synthesized, and a complex Cas9-sgRNA is formed afterwards which is then by random collision finds PAM in the genome and binds to it and checks for spacer complementarity in flanking DNA. Thus, Cas9 binds to the target DNA, undergoes conformational changes, and then its RuvC-like domains and HNH domain trigger nuclease activity, resulting in a break in the double-stranded DNA (dsDNA) (Nishimasu et al., 2014; Sternberg et al., 2014). By triggering CRISPR/Cas9, DNA double-strand breaks are caused on the genome, which are repaired by cellular machinery in reverse order, either via nonhomologous end joining (NHEJ) or homology-directed repair (HDR) (Shalem et al., 2015).

There are many research articles on the application of the CRISPR/Cas system for the enhancement of industrially and pharmaceutically important natural products from different plant species. This chapter summarizes the CRISPR/Cas system as well as its methods of applications including microbes, fungi, algae, and higher plants for increasing production of natural products. A discussion is also given of the system's bottlenecks and its future prospects.

14.2 CRISPR/CAS APPLICATION IN DIFFERENT BIOLOGICAL SYSTEMS FOR ENHANCEMENT OF NATURAL PRODUCTS

14.2.1 APPLICATION OF CRISPR/CAS TOOL IN BACTERIA

The CRISPR/Cas technology has been applied to bacteria producing natural products, notably *Streptomyces*, *Myxobacteria*, *Bacillus*, *Pseudomonas*, and *Cyanobacteria* (Bibb et al., 1978). In order to manipulate genetic sequences in a directed way, PCR-targeting developed as one method widely used, but it has a low throughput and is time-consuming. CRISPR/Cas system allows genomes to be manipulated more efficiently than ever before. The CRISPR/Cas technique has also been successfully used on some Actinomycetes besides *Streptomycetes*, like *Corynebacteria*, *Actinoplanes*, and *Mycobacteria* (Choudhary et al., 2015; Cleto et al., 2016; Wolf et al., 2016; Jiang et al., 2017; Liu et al., 2017; Peng et al., 2017; Rock et al., 2017; Park et al., 2018). pCRISPomyces-2, pCRISPR-Cas9, pKCcas9dO and CRISPR/Cas9-CodA(sm) are the types of CRISPR/Cas genome editing systems rely on Sp(d)Cas9 in a one-vector approach and attach with Sp(d)Cas9 gene, sgRNA(s), and editing template if used together on a single plasmid in actinomycetes (Tong et al., 2015; Zeng et al., 2015; Cobb et al., 2015; Huang et al., 2015). The homing device in most cases is sgRNA, except for pCRISPomyces-1 (Cobb et al., 2015). Nonhomologous end-joining (NHEJ) repair has been used for introducing deletions. The CRISPRi system, made possible through a Cas9 variant that is catalytically dead (dCas9), was also used and successfully applied (Tong et al., 2015). Gram-positive bacteria such as *Bacillus* produce enzymes and molecules that are beneficial. A CRISPR/Cas9-based system for *B. subtilis* generates mutations with a 100% efficiency rate, as well as knockouts, knock-ins, and knockdowns of genes

(Westbrook et al., 2016). By combining multiple sgRNAs and the MazF gene, a counter-selectable marker, it was possible to simultaneously edit multiple genes at the same time. As a result of the SpCas9 gene's interaction with the lacA and thrC loci and HR, genome editing occurs using the sgRNA transcription cassette (Westbrook et al., 2016). Gram-negative *myxobacteria* are also known to produce a number of SMs, and the species *Myxococcus xanthus* has been used as a model plant for the use of gene editing techniques (Yang et al., 2017). Two plasmids were needed for the CRISPR/Cas9 system: the pSWU30 plasmid and the pBJ113 suicide plasmid. By interacting with plasmid, Cas9:sgRNA introduces double-strand breaks resulting in a double crossover, thereby facilitating high-fidelity in-frame deletion (Yang et al., 2017). Furthermore, controls of gene transcription were developed with a CRISPRi system with dCas9 induced by xylose. In another genetic editing toolkit that is based on nickase, CRISPR/SpCas9 was built for *Bacillus licheniformis* (Li et al., 2018). There are many species of *Pseudomonas* capable of producing natural products, including mupirocin, which is a topical antibiotic used to treat superficial wound infections (El-Sayed et al., 2003; Gross and Loper, 2009). Recently *Staphylococcus aureus* Cas9 (SaCas9), *Streptococcus thermophilus* CRISPR1 Cas9 (Sth1Cas9), and *Francisella tularensis* subsp. *novicida* U112 Cpf1 (fnCpf1) are found to be used as an alternative to *Streptococcus pyogenes* Cas9 (SpCas9) system. Other species such as *Streptomyces coelicolor*, *St. lividans*, *St. albus*, *St. formicae*, *St. rimosus* are also investigated for natural products using CRISPR-Cas9 genome editing tool (Yeo et al., 2019). *Streptococcus pasteurianus* dCas9 was used to set up a CRISPRi system for modulating gene transcription. Ptac promoter is required to transcribe dCas9 in the *Mycobacteria* attTn7 locus, while a Ptet promoter is needed to transcribe sgRNA (Tan et al., 2018). The natural products of *P. putida* and *P. aeruginosa* were also tested (Tong et al., 2019).

14.2.2 APPLICATION OF CRISPR/CAS TOOL IN BLUE GREEN ALGAE

A variety of products are produced by the autotrophic bacteria *cyanobacteria*, which are used in industry as a source of bioactive natural products (Dittmann et al., 2015). To introduce point mutations, markerless knockouts and knock-ins to three model organisms' *cyanobacteria Synechococcus*, *Synechocystis*, and *Anabaena*, a CRISPR/Cpf1 toolkit was developed (Ungerer and Pakrasi, 2016). CRISPR/Cas9-based genome editing also improved *Synechococcus elongatus* PCC 7942, which is based on two plasmid-based systems, SpCas9 and crRNA:tracrRNA transcript cassette (Li et al., 2016). A codon-optimized SpCas9 expression transiently expressed in the genome of *S. elongatus* UTEX 2973 allowed the system to function, but no DNA was altered. In the *Synechococcus sp.* strain PCC 7002 SgRNA was incorporated into the acsA gene and Cas9 was introduced into the NS1 gene via the CRISPRi system for gene transcription modulation (Gondon et al., 2016). The PCC 7120 strain of *Anabaena* sp. was also given a CRISPR system that is capable of efficiently regulating gene expression, carrying both sgRNAs and dCas9 (Higo et al., 2018).

14.2.3 APPLICATION OF CRISPR/CAS TOOL IN FUNGI

The fungi in nature are capable of producing many valuable natural products, such as organic acids, enzymes, polyunsaturated fatty acids, flavors, pigments, and particularly pharmacologically active compounds, such as antibiotics and tumor-fighting agents (Tong et al., 2019). The original SpCas9 coding sequence should be amended to add the nuclear localization signal (NLS) for fungi and it is necessary for the double strand cuts to be repaired by a highly active NHEJ pathway and specifically for the sgRNA transcript to be properly transcribed without the usual 5' caps and 3' poly A-tail that are typical of bacteria (Cho et al., 2017; tong et al., 2019). One single plasmid containing a CRISPR/SpCas9 genome editing toolkit integrated with codon-optimized SpCas9 genes to the 30-SV40 NLS sequence was introduced into *Aspergillus* (Nødvig et al., 2015). In this case, DSBs can be repaired using microhomology-mediated end joining (MMEJ) instead of NHEJ or HR (Zhang et al., 2016). The CRISPR/Cas9 genome editing toolkit currently includes four common fungal markers: hygR, AFUM_pyrG, AN_argB, and bleR. These markers have been successfully used in several species of *Aspergillus,* such as *A. brasiliensis*, *A. carbonarius*, *A. luchuensis*, *A. niger*, *A. nidulans* and *A. aculeatus* (Weber et al., 2017). Using this method, a single nucleotide insertion was successful in restabilizing the trypacidin synthesis pathway in *A. fumigatus* by inserting a new nucleotide insertion (Weber et al., 2017). *Ganoderma*, another fungus higher up, has also been set up with genome editing tools. An industrially viable natural product can also be produced using *Penicillium chrysogenum*, a filamentous fungus utilizing the CRISPR/Cas9 method (Qin et al., 2017). *Penicillium chrysogenum* is reported to have been directly delivered with preassembled CRISPR/Cas9 ribonucleoproteins (RNPs) of Cas9:sgRNA complexes (Pohl et al., 2016). Also, in Mucor circinelloides, Cas9:sgRNA complexes were used for CRISPR/Cas9-based production of bioactive natural products (Nagy et al., 2017).

14.2.4 APPLICATION OF CRISPR/CAS TOOL IN ALGAE

In addition to producing vitamins, carotenoids, and fatty acids, microalgae produce bioactive SMs, including antioxidants, antivirals, antibacterials, antifungals, anti-inflammatory agents, antitumor agents, and antimalarials (Michalak and Chojnacka, 2015; de Jesus Raposo and de Morais, 2015). Using CRISPR/Cas technology, microalgae genome editing is also possible. CRISPR/Cas9 genome editing system was created using a *Chlamydomonas reinhardtii* plasmid containing SpCas9 codon optimized for *Chlamydomonas* and the U6 promoter-controlled sgRNA. However, Cas9 and dCas9 were found to be toxic to *C. reinhardtii* (Jiang et al., 2014). An *in vitro* Cas9:sgRNA complex was created to reduce Cas9 toxicity and enhance editing efficiency (Shin et al., 2016). Researchers established an HR-based CRISPR-saCas9 system comprising two plasmids, one containing Cas9 from Staphylococcus aureus, and the other bearing sgRNA and editing templates for *C. reinhardtii* (Baek et al., 2016; Greiner et al., 2017). Using a genetic editing method using single-stranded DNA, an alternative delivery system for *C. reinhardtii* RNPs has been developed to replace plasmid-based delivery systems (Ferenczi et al., 2017).

14.2.5 APPLICATION OF CRISPR/CAS TOOL IN HIGHER PLANTS

Plant CRISPR/Cas applications have been largely preliminary proofs of concept studies. For plant genome editing, researchers used NLS-modified CRISPR/Cas9 enzymes in the NHEJ-mediated repair of DNA DSBs combined with human-coded or plant-coded codons. *Agrobacterium* was used to deliver the CRISPR/Cas9 components via the U6 promoter and the majority of the transcription of the sgRNA was controlled by the U6 promoter (Tong et al., 2019).

14.2.5.1 CAMELINA SATIVA

Among *Brassicales* members, *Camelina sativa* or flax sees has attracted attention for its distinct composition and fatty acid content including linolenic, oleic, linoleic, and eicosenoic acid, which offer a high industrial value in biofuel, personal care, cosmetics, and pharmaceuticals (Gugel and Falk, 2006). A genome editing method named CRISPR/Cas9 has been used for this plant in order to knock out fatty acid desaturase 2 (FAD2) genes and raise the concentration of oleic acid using CRISPR/Cas9-based genome editing tool (Morineau et al., 2017; Jiang et al., 2017). One single sgRNA altered the oil content of *C. sativa* by targeting three conserved homologous genes (Aznar-Moreno and Durrett, 2017).

14.2.5.2 PAPAVER SOMNIFERUM

Researchers successfully developed a gene knockout system, which operates through gene inactivation and DSB formation by NHEJ repair and Cas9, in the medical plant *Papaver somniferum* (opium poppy), which produces morphine. *Agrobacterium*-mediated transformation of leaves with synthetic plasmids that express sgRNA and human codon-optimized SpCas9 vector inactivated 40OMT2 through benzyl isoquinoline synthesis inhibition (Alagoz et al., 2016).

14.2.5.3 DIOSCOREA ZINGIBERENSIS

The genus *Dioscorea* sp. produces the hormone diosgenin, which has a huge variety of medicinal properties; it is also a precursor to numerous hormones. The pharmacological properties also include neuroprotection, cardiovascular, hyperlipidemia, anti-osteoporotic, anti-tumor, anti-inflammatory, anti-oxidant, anti-thrombosis, anti-allergic, and anthelmintic properties (Pandey et al., 2017; Zhang et al., 2018; Feng et al., 2018). In *Dioscorea zingiberensis*, the genome editing procedure was carried out by transforming with *A. tumefaciens* and using OsU3 and 35S promoters for Cas9 and sgRNA expression cassettes, respectively. Using an edited version of the farnesyl pyrophosphate synthase gene (Dzfps), a high mutation frequency was obtained (Feng et al., 2018).

14.2.5.4 DENDROBIUM OFFICINALE

Using *Agrobacterium*-mediated transformation of the perennial plant *Dendrobium officinale*, a member of the Orchidaceae family, researchers can genetically modify the SMs with 10–100% editing efficiency using CRISPR/Cas9 system (Kui et al., 2017).

14.2.5.5 SALAVIA MILTIORRHIZA

It is a well-recognized ancient herb that produces diterpene compounds that have medicinal importance. It is commonly used as a remedy for diabetes, cardiovascular illness, and strokes (Jia et al., 2019; Fan et al., 2019; Shi et al., 2019). The *Agrobacterium rhizogenes*-based CRISPR/Cas9 editing method was successfully applied to *S. tilicitiorhiza* in the knockout of SmCPS1 (diterpene synthase gene) (Li et al., 2017a). *S. miltiorrhiza* has also been knocked out using the CRISPR/Cas9 tool to knock out the rosmarinic acid synthase gene (SmRAS) associated with the biosynthesis of phenolic acids (Zhou et al., 2018). In a study to edit SmPAL1 in the *S. miltiorrhiza* phenylpropane metabolism pathway, software was used to design the CIRSPR/Cas9 target site for SmPAL1. In order to screen sgRNAs for a powerful screening tool, SmPAL1-g1, SmPAL1-g2, and SmPAL1-g3 of the CRISPR gene have been designed (Qiu et al., 2018).

14.2.5.6 NICOTIANA TABACUM

N. tabacum is an herbaceous perennial plant that produces terpenoids, alkaloids, flavonoids, and phenylpropanoids along with tobacco. A method for gene editing using CRISPR/Cas9 has been developed for knocking out and deleting genes in suspended *N. tabacum* cell suspensions (Mercx et al., 2016). CRISPR/Cas9 was used to knock out two genes in tobacco protoplasts (NtPDS and NtPDR6) (Gao et al., 2015). Inhibiting six glycosyltransferase genes through CRISPR/Cas9 produced *Nicotiana* recombinant proteins that lack core α-1,3-fucose and β-1,2-xylose (Jansing et al., 2019). Moreover, the CRISPR/Cas9 editing tool has enabled non-transgenic tobacco without nicotine to be developed (Schachtsiek and Stehle, 2019).

14.2.6 ISOLATION OF NATURAL PRODUCTS USING CRISPR/CAS: CHALLENGES AND SOLUTIONS (FIGURE 14.1)

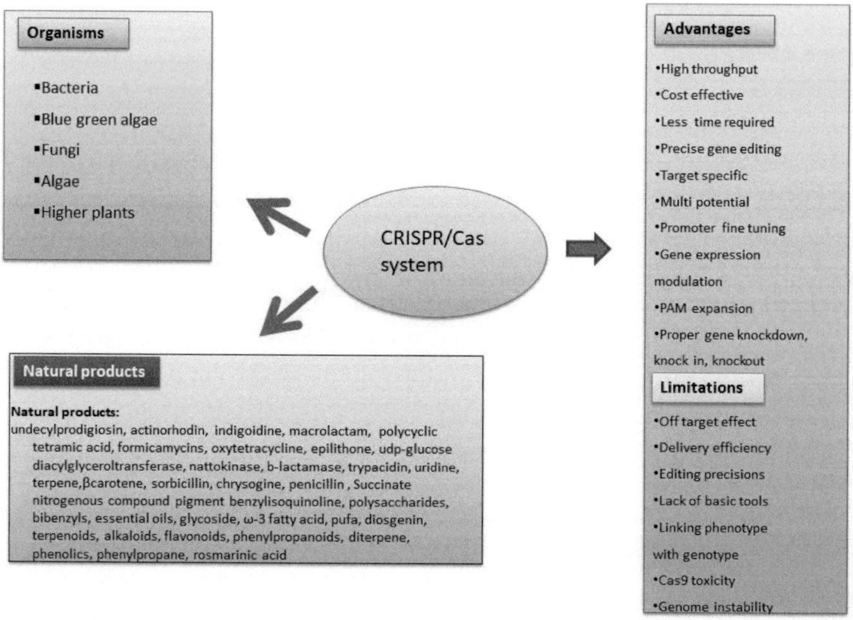

FIGURE 14.1 Natural product optimization in different organisms using CRISPR/Cas system with advantages and limitations.

14.2.6.1 OFF TARGET EFFECTS (OTES)

These are some of the most harmful side effects that can result from targeting a gene with the CRISPR toolkit. By optimizing sgRNAs so that on-target effects are maximized, off-target effects (OTEs) can be minimized. A synthetic switch can control Cas9 transcription and translation by minimizing DNA mismatches and secondary structures promoting the binding of the off-target SgRNAs. To optimize Cas9 specificity, coupled Cas9 nickases can be engineered using protein engineering specificity (Doench et al., 2016; Kimberland et al., 2018; Chen, 2019; Shen et al., 2019; Tong et al., 2019).

14.2.6.2 EDITING PRECISION

Plants and fungi, however, tend to have a mechanism where a DSB is formed via NHEJ, which generally produces random indels around the DSB that causes frameshift mutations. In order to overcome this limitation, a variety of strategies have succeeded in suppressing the native NHEJ pathway viz.

DNA ligase IV inhibition by Scr7 molecule, improvement of suppressors like resveratrol and L755507 for NHEJ or HDR repair pathways (Maruyama et al., 2015; Li et al., 2017b).

14.2.6.3 DELIVERY EFFICIENCY

The lack of suitable plasmid vector systems and ineffective transfection strategies prevent CRISPR/Cas9 from being used in many organisms. For CRISPR/Cas9 to work properly, either Cas9-expressing plasmids or Cas9:sgRNA complexes prepared *in vitro* before application must be utilized. Moreover, sgRNA complexed with Cas9 protein can also be delivered directly to the nucleus through carriers' nanoparticles (Zhang et al., 2017; Tong et al., 2019).

14.2.6.4 INSUFFICIENT BASIC SUPPORT

There are only a few genetic tools available to genetically manipulate most natural product producers, such as vectors, selection markers, characterized promoters, and general transformation rates. The aforementioned aspects inhibit the development and implementation of CRISPR/Cas (Tong et al., 2019).

14.2.6.5 LINKING GENOTYPE WITH PHENOTYPE

The use of high-throughput biosensors has demonstrated success in linking genotype to phenotype; however, the applications of these sensors are limited across many fields. The MAGIC (GRAMMY) technology can be used to document genotypes and phenotypes at a high throughput level using CRISPR technology in recent days (Tong et al., 2019; Lian et al., 2019).

14.2.6.6 CAS TOXICITY

In most bacteria and chlorophytes, high Cas9 expression is toxic, causing growth to be negatively impacted. A lot of evidence suggests that Cas9 can be transiently expressed to reduce its toxic effects by reducing the amount of Cas9 in the cells or using weak promoters or by using Cas9 variants, such as

Cas9D10A, to modify the size and frequency of gene edits for target genes. It is possible to reduce toxicity as well as OTEs by integrating anti-Cas9 components into various toolkits modulating the Cas9 activity (Wu et al., 2015; Dow et al., 2015; Shin et al., 2017; Ye et al., 2020).

14.2.6.7 THE GENOME'S INSTABILITY

In the case of cellular DSBs, they cause genomic instability leading to cell death if they aren't repaired. To maintain genomic stability, NHEJ, and HR are used for genome joining in eukaryotic genomes and bacterial genomes, respectively (Chayot et al., 2011). There is significant concern about genome instability resulting from CRISPR/Cas9 in a number of organisms (Hoff et al., 2018). CRISPR-BEST (CRISPR-base editing system), CRISPRi, and deaminase-based DNA base editors based on single-nucleotide resolution have been reported as efficient genome editing tools (Table 14.1) (Eid et al., 2018; Tong et al., 2020).

14.2.6.8 NATIVE INFLUENCES OF CRISPR

In the past, cell factories have been engineered using CRISPR/Cas, but this has often caused interference with cells that already contain CRISPR systems. Researchers found that *Streptomyces avermitilis* has a native type I-E CRISPR/Cas system that encounters foreign nucleic acids and crosses talk to the externally introduced type CRISPR/Cas system (Guo et al., 2011; Sangal et al., 2013; Qiu et al., 2016).

14.2.6.9 CRISPR-RESISTANCE RESULTING FROM UNIQUE FEATURES OF GENOMES

A high GC content in *Streptomycetes* genomes, as well as in related genera, makes genetic manipulation challenges. Due to the abundance of PAMs in SpCas9 (NGGs in SpCas9) at every location within the genome, the nonspecific binding may result in increased off-target effects due to editing. Cas nucleases such as Cas12a may also be limited by the TTTV PAM for fnCas12a, which is AT-rich (Tong et al., 2019).

TABLE 14.1 Application of CRISPR/Cas Tools for the Improvement of Natural Products in Different Organisms

Name of the Organism	Organism Type	Cas Type	Natural Products	Methods	References
Streptomyces lividans	Bacteria	SpCas9	Undecylprodigiosin, actinorhodin	Knockout, knockin, HR DSB repair,	Cobb et al. (2015); Zhang et al. (2017)
St. coelicolor	Bacteria	SpCas9	Actinorhodin, undecylprodigiosin	Knockout, HR DSB repair, NHEJ DSB repair	Zeng et al. (2015); Tong et al. (2015); Huang et al. (2015)
St. albus	Bacteria	SpCas9	Indigoidine macrolactam, polycyclic tetramic acid	Knockin, knockout, HR DSB repair	Cobb et al. (2015); Zhang et al. (2017)
St. formicae	Bacteria	SpCas9	Formicamycins	Knockout, HR DSB repair	Qin et al. (2017)
St. rimosus	Bacteria	SpCas9	Oxytetracycline	Knockout, HR DSB repair	Jia et al. (2017)
Myxococcus Xanthus	Bacteria	dCas9	Epilithone	Indel, HR DSB repair	Yang et al. (2017)
Bacillus subtilis	Bacteria	Cas9	UDP-glucose diacylglyceroltransferase	Knockin, knockout, HR DSB repair	Westbrook et al. (2016)
B. licheniformis	Bacteria	Cas9n	Nattokinase	Knockout, NHEJ DSB repair	Li et al. (2018)
Chlamydomonas reinhardtii	Algae	SpCas9	Pigment	Knockout, HR DSB repair	Jiang et al. (2014)
Synechococcus elongatus	Algae	SpCas9	Succinate	Knockin, HR DSB repair	Li et al. (2016)
Anabaena	Algae	dCas9	Nitrogenous compound	Knockdown	Higo et al. (2018)
Aspergillus aculeatus	Fungi	SpCas9	b-lactamase	Insertion, NHEJ DSB repair	Nødvig et al. (2015)
A. fumigatus	Fungi	SpCas9	Trypacidin	Insertion, MMEJ DSB repair	Zhang et al. (2016)
Penicillium chrysogenum	Fungi	SpCas9	Sorbicillin, chrysogine, penicillin	Knockin, knockout, HR DSB repair	Pohl et al. (2016)
Ganoderma. lingzhi	Fungi	SpCas9	Uridine	Indel, HR DSB repair	Qin et al. (2017)
Mucor circinelloides	Fungi	SpCas9	Terpene, βcarotene	Indel, NHEJ DSB repair	Nagy et al. (2017)

TABLE 14.1 *(Continued)*

Name of the Organism	Organism Type	Cas Type	Natural Products	Methods	References
Papaver somniferum	Higher plant	Cas9	Benzylisoquinoline	Knockout, NHEJ DSB repair	Alagoz et al. (2016)
Dendrobium officinale	Higher plant	Cas9	Polysaccharides, bibenzyls, essential oils, glycoside, etc.	Knockout, NHEJ DSB repair	Kui et al. (2017)
Camelina sativa	Higher plant	Cas9	ω-3 fatty acid, pufa	Knockout, NHEJ DSB repair	Morineau et al. (2017); Jiang et al. (2017)
Dioscorea zingiberensis	Higher plant	Cas9	Diosgenin	Knockout, NHEJ DSB repair	Feng et al. (2018)
Nicotiana tabacum	Higher plant	Cas9	Terpenoids, alkaloids, flavonoids, and phenylpropanoids	Knockout, NHEJ DSB repair	Mercx et al. (2016); Gao et al. (2015)
Salavia miltiorrhiza	Higher plant	Cas9	Diterpene, phenolics, phenylpropane, rosmarinic acid	Knockout, NHEJ DSB repair	Li et al. (2017); Zhou et al. (2018); Qiu et al. (2018)

14.3 CONCLUSION AND FUTURE PROSPECTS

In recent years, CRISPR/Cas systems have been widely accepted as an efficient means of improving agricultural crops through the insertion, deletion, substitution, fine-tuning, or replacement of genetic material. The use of CRISPR/Cas technologies in natural product research is still in its infancy, despite being widely discussed over the past few years. Technology based on CRISPR/Cas could be used to build new cell factories, to understand biosynthetic pathways, and to clone large DNA fragments, with the potential to create gene clusters. Furthermore, this tool is capable of creating large mutant libraries and antiviral breeding techniques in addition to the ability to modify gene expression. The most widely used gene knockout method is SpCas9, but CRISPRi and CRISPRa can be engineered to knock down and activate target genes as well (Tong et al., 2019). Comparing CRISPR/Cas with previously available technologies like ZFNs (zinc finger nucleases) and TALENs (transcription activator-like effector nucleases), CRISPR/Cas is more affordable, straightforward, and effective. Hairy root cultures based on *Agrobacterium rhizogenes* are a promising method for synthetic biology production of diverse medicinal compounds using CRISPR/Cas9 induced polyploidy.

In spite of the fact that nearly all organisms tested engineered with CRISPR/Cas9 functioned, the reports are largely limited to proof-of-concept findings. Since its inception, its effectiveness and scalability have increased substantially; however, further improvements are needed. Although CRISPR/Cas9 has limitations, other techniques are being developed to enhance its editing power. CRISPR/Cas9 needs to be improved to better detect and avoid off-target effects (OTEs) in order to increase its efficiency. The CRISPR/Cas9 system is too large for encapsulation in viral vectors, making a small-sized system essential for efficient GE in crops. New methods or strategies must be developed to restrain endogenous NHEJ or to harness the CRISPR system. Cas9 can only recognize NGG PAM sites, so multiple sites can be selected in order to broaden the scope of the application. Gene transfer using *Agrobacterium* and tissue culture is time-consuming and not feasible for all plants, so new technologies using RNP, virus, or nanoparticle-based systems need to take these into account for increasing the efficiency of the CRISPR/Cas system. The sequences of the whole genomes and mRNA of many medicinally and industrially important plants are still unknown. Numerous enzymes and genes control metabolic pathways, thus for SMs, single-gene modification is not as effective. Thus, using CRISPR/Cas, it is possible to edit multiple genes associated with a pathway to improve the biosynthesis of natural products by plants. During the next few years, we expect all of these areas to advance rapidly and fulfill the needs of pharmaceutical and industrial industries for natural products from plants.

KEYWORDS

- **CRISPR/Cas**
- **CRISPRi**
- **gene modification**
- **genome editing**
- **natural products**
- **non-homologous end joining**
- **off-target**

REFERENCES

Agarwal, A., Yadava, P., Kumar, K., Singh, I., Kaul, T., Pattanayak, A., & Agrawal, P. K., (2018). Insights into maize genome editing via CRISPR/Cas9. *Physiology and Molecular Biology of Plants, 24*(2), 175–183.

Alagoz, Y., Gurkok, T., Zhang, B., & Unver, T., (2016). Manipulating the biosynthesis of bioactive compound alkaloids for next-generation metabolic engineering in opium poppy using CRISPR-Cas 9 genome editing technology. *Scientific Reports, 6*(1), 1–9.

Aparicio, T., De Lorenzo, V., & Martínez-García, E., (2018). CRISPR/Cas9-based counterselection boosts recombineering efficiency in *Pseudomonas putida. Biotechnology Journal, 13*(5), 1700161.

Aznar-Moreno, J. A., & Durrett, T. P., (2017). Simultaneous targeting of multiple gene homeologs to alter seed oil production in *Camelina sativa. Plant and Cell Physiology, 58*(7), 1260–1267.

Baek, K., Kim, D. H., Jeong, J., Sim, S. J., Melis, A., Kim, J. S., Jin, E., & Bae, S., (2016). DNA-free two-gene knockout in *Chlamydomonas reinhardtii* via CRISPR-Cas9 ribonucleoproteins. *Sci Rep., 6*, 30620.

Bibb, M. J., Ward, J. M., & Hopwood, D. A., (1978). Transformation of plasmid DNA into *Streptomyces* at high frequency. *Nature, 274*(5669), 398–400.

Char, S. N., Neelakandan, A. K., Nahampun, H., Frame, B., Main, M., Spalding, M. H., Becraft, P. W., et al., (2017). An *Agrobacterium*-delivered CRISPR/Cas9 system for high-frequency targeted mutagenesis in maize. *Plant Biotechnology Journal, 15*(2), 257–268.

Chayot, R., Montagne, B., Mazel, D., & Ricchetti, M., (2010). An end-joining repair mechanism in *Escherichia coli. Proceedings of the National Academy of Sciences, 107*(5), 2141–2146.

Chen, S. J., (2019). Minimizing off-target effects in CRISPR-Cas9 genome editing. *Cell Biol. Toxicol., 35*, 399–401.

Cho, J. S., Choi, K. R., Prabowo, C. P. S., Shin, J. H., Yang, D., Jang, J., & Lee, S. Y., (2017). CRISPR/Cas9-coupled recombineering for metabolic engineering of *Corynebacterium glutamicum. Metabolic Engineering, 42*, 157–167.

Choudhary, E., Thakur, P., Pareek, M., & Agarwal, N., (2015). Gene silencing by CRISPR interference in mycobacteria. *Nature Communications, 6*(1), 1–11.

Cleto, S., Jensen, J. V., Wendisch, V. F., & Lu, T. K., (2016). *Corynebacterium glutamicum* metabolic engineering with CRISPR interference (CRISPRi). *ACS Synth. Biol., 5*, 375–385.

Cobb, R. E., Wang, Y., & Zhao, H., (2015). High-efficiency multiplex genome editing of *Streptomyces* species using an engineered CRISPR/Cas system. *ACS Synth. Biol., 4*, 723–728. doi. org/10.1021/sb50, 351.

Cong, L., Ran, F. A., Cox, D., Lin, S., Barretto, R., Habib, N., Hsu, P. D., et al., (2013). Multiplex genome engineering using CRISPR/Cas systems. *Science, 339*(6121), 819–823.

De Jesus, R. M. F., & De Morais, A. M. M. B, (2015). Microalgae for the prevention of cardiovascular disease and stroke. *Life Sciences, 125*, 32–41.

Dittmann, E., Gugger, M., Sivonen, K., & Fewer, D. P., (2015). Natural product biosynthetic diversity and comparative genomics of the *cyanobacteria*. *Trends in Microbiology, 23*(10), 642–652.

Doench, J. G., Fusi, N., Sullender, M., Hegde, M., Vaimberg, E. W., Donovan, K. F., Smith, I., et al., (2016). Optimized sgRNA design to maximize activity and minimize off-target effects of CRISPR-Cas9. *Nature Biotechnology, 34*(2), 184–191.

Dow, L. E., Fisher, J., O'rourke, K. P., Muley, A., Kastenhuber, E. R., Livshits, G., Tschaharganeh, D. F., et al., (2015). Inducible *in vivo* genome editing with CRISPR-Cas9. *Nature Biotechnology, 33*(4), 390–394.

Eid, A., Alshareef, S., & Mahfouz, M. M., (2018). CRISPR base editors: Genome editing without double-stranded breaks. *Biochemical Journal, 475*(11), 1955–1964.

El-Sayed, A. K., Hothersall, J., Cooper, S. M., Stephens, E., Simpson, T. J., & Thomas, C. M., (2003). Characterization of the mupirocin biosynthesis gene cluster from *Pseudomonas fluorescens* NCIMB 10586. *Chemistry & Biology, 10*(5), 419–430.

Feng, S., Song, W., Fu, R., Zhang, H., Xu, A., & Li, J., (2018). Application of the CRISPR/Cas9 system in *Dioscorea zingiberensis*. *Plant Cell, Tissue and Organ Culture (PCTOC), 135*(1), 133–141.

Ferenczi, A., Pyott, D. E., Xipnitou, A., & Molnar, A., (2017). Efficient targeted DNA editing and replacement in *Chlamydomonas reinhardtii* using Cpf1 ribonucleoproteins and single-stranded DNA. *Proceedings of the National Academy of Sciences, 114*(51), 13567–13572.

Gaj, T., Gersbach, C. A., & Barbas, III. C. F., (2013). ZFN, TALEN, and CRISPR/Cas-based methods for genome engineering. *Trends in Biotechnology, 31*(7), 397–405.

Gao, J., Wang, G., Ma, S., Xie, X., Wu, X., Zhang, X., Wu, Y., et al., (2015). CRISPR/Cas9-mediated targeted mutagenesis in *Nicotiana tabacum*. *Plant Molecular Biology, 87*(1, 2), 99–110.

Gordon, G. C., Korosh, T. C., Cameron, J. C., Markley, A. L., Begemann, M. B., & Pfleger, B. F., (2016). CRISPR interference as a titratable, trans-acting regulatory tool for metabolic engineering in the cyanobacterium *Synechococcus* sp. strain PCC 7002. *Metabolic Engineering, 38*, 170–179.

Greiner, A., Kelterborn, S., Evers, H., Kreimer, G., Sizova, I., & Hegemann, P., (2017). Targeting of photoreceptor genes in *Chlamydomonas reinhardtii* via zinc-finger nucleases and CRISPR/Cas9. *The Plant Cell, 29*(10), 2498–2518.

Gross, H., & Loper, J. E., (2009). Genomics of secondary metabolite production by *Pseudomonas* spp. *Natural Product Reports, 26*(11), 1408–1446.

Gugel, R. K., & Falk, K. C., (2006). Agronomic and seed quality evaluation of *Camelina sativa* in western Canada. *Canadian Journal of Plant Science, 86*(4), 1047–1058.

Guo, P., Cheng, Q., Xie, P., Fan, Y., Jiang, W., & Qin, Z., (2011). Characterization of the multiple CRISPR loci on *Streptomyces* linear plasmid pSHK1. *Acta Biochim. Biophys. Sin., 43*(8), 630–639.

Higo, A., Isu, A., Fukaya, Y., Ehira, S., & Hisabori, T., (2018). Application of CRISPR interference for metabolic engineering of the heterocyst-forming multicellular cyanobacterium *Anabaena sp.* PCC 7120. *Plant and Cell Physiology, 59*(1), 119–127.

Hoff, G., Bertrand, C., Piotrowski, E., Thibessard, A., & Leblond, P., (2018). Genome plasticity is governed by double strand break DNA repair in *Streptomyces*. *Scientific Reports, 8*(1), 1–11.

Huang, H., Zheng, G., Jiang, W., Hu, H., & Lu, Y., (2015). One-step high-efficiency CRISPR/Cas9-mediated genome editing in *Streptomyces*. *Acta Biochimica et Biophysica Sinica, 47*(4), 231–243.

Jansing, J., Sack, M., Augustine, S. M., Fischer, R., & Bortesi, L., (2019). CRISPR/Cas9-mediated knockout of six glycosyltransferase genes in *Nicotiana benthamiana* for the production of recombinant proteins lacking β-1, 2-xylose and core α-1, 3-fucose. *Plant Biotechnology Journal, 17*(2), 350–361.

Jia, Q., Zhu, R., Tian, Y., Chen, B., Li, R., Li, L., Wang, L., et al., (2019). *Salvia miltiorrhiza* in diabetes: A review of its pharmacology, phytochemistry, and safety. *Phytomedicine, 58*, 152871.

Jiang, W. Z., Henry, I. M., Lynagh, P. G., Comai, L., Cahoon, E. B., & Weeks, D. P., (2017). Significant enhancement of fatty acid composition in seeds of the allohexaploid, *Camelina sativa*, using CRISPR/Cas9 gene editing. *Plant Biotechnology Journal, 15*(5), 648–657.

Jiang, W., Brueggeman, A. J., Horken, K. M., Plucinak, T. M., & Weeks, D. P., (2014). Successful transient expression of Cas9 and single guide RNA genes in *Chlamydomonas reinhardtii*. *Eukaryotic Cell, 13*(11), 1465–1469.

Jiang, Y., Qian, F., Yang, J., Liu, Y., Dong, F., Xu, C., Sun, B., et al., (2017). CRISPR-Cpf1 assisted genome editing of *Corynebacterium glutamicum*. *Nature Communications, 8*(1), 1–11.

Kimberland, M. L., Hou, W., Alfonso-Pecchio, A., Wilson, S., Rao, Y., Zhang, S., & Lu, Q., (2018). Strategies for controlling CRISPR/Cas9 off-target effects and biological variations in mammalian genome editing experiments. *Journal of Biotechnology, 284*, 91–101.

Kui, L., Chen, H., Zhang, W., He, S., Xiong, Z., Zhang, Y., Yan, L., et al., (2017). Building a genetic manipulation toolbox for orchid biology: Identification of constitutive promoters and application of CRISPR/Cas9 in the orchid, *Dendrobium officinale*. *Frontiers in Plant Science, 7*, 2036.

Li, B., Cui, G., Shen, G., Zhan, Z., Huang, L., Chen, J., & Qi, X., (2017a). Targeted mutagenesis in the medicinal plant *Salvia miltiorrhiza*. *Sci. Rep., 7*, 43320.

Li, G., Zhang, X., Zhong, C., Mo, J., Quan, R., Yang, J., Liu, D. W., et al., (2017b). Small molecules enhance CRISPR/Cas9-mediated homology-directed genome editing in primary cells. *Sci. Rep., 7*, 8943.

Li, H., Shen, C. R., Huang, C. H., Sung, L. Y., Wu, M. Y., & Hu, Y. C., (2016). CRISPR-Cas9 for the genome engineering of *cyanobacteria* and succinate production. *Metabolic Engineering, 38*, 293–302.

Li, K., Cai, D., Wang, Z., He, Z., & Chen, S., (2018). Development of an efficient genome editing tool in *Bacillus licheniformis* using CRISPR-Cas9 nickase. *Applied and Environmental Microbiology, 84*(6), e02608–17.

Liu, J., Wang, Y., Lu, Y., Zheng, P., Sun, J., & Ma, Y., (2017). Development of a CRISPR/Cas9 genome editing toolbox for *Corynebacterium glutamicum*. *Microbial Cell Factories, 16*(1), 1–17.

Manghwar, H., Lindsey, K., Zhang, X., & Jin, S., (2019). CRISPR/Cas system: Recent advances and future prospects for genome editing. *Trends in Plant Science, 24*(12), 1102–1125.

Maruyama, T., Dougan, S. K., Truttmann, M. C., Bilate, A. M., Ingram, J. R., & Ploegh, H. L., (2015). Increasing the efficiency of precise genome editing with CRISPR-Cas9 by inhibition of nonhomologous end joining. *Nature Biotechnology, 33*(5), 538–542.

McChesney, J. D., Venkataraman, S. K., & Henri, J. T., (2007). Plant natural products: Back to the future or into extinction? *Phytochemistry, 68*(14), 2015–2022.

Mercx, S., Tollet, J., Magy, B., Navarre, C., & Boutry, M., (2016). Gene inactivation by CRISPR-Cas9 in *Nicotiana tabacum* BY-2 suspension cells. *Frontiers in Plant Science, 7*, 40.

Michalak, I., & Chojnacka, K., (2015). Algae as production systems of bioactive compounds. *Engineering in Life Sciences, 15*(2), 160–176.

Morineau, C., Bellec, Y., Tellier, F., Gissot, L., Kelemen, Z., Nogué, F., & Faure, J. D., (2017). Selective gene dosage by CRISPR-Cas9 genome editing in hexaploid *Camelina sativa*. *Plant Biotechnology Journal, 15*(6), 729–739.

Nagy, G., Szebenyi, C., Csernetics, Á., Vaz, A. G., Tóth, E. J., Vágvölgyi, C., & Papp, T., (2017). Development of a plasmid free CRISPR-Cas9 system for the genetic modification of *Mucor circinelloides*. *Scientific Reports, 7*(1), 1–10.

Nishimasu, H., Ran, F. A., Hsu, P. D., Konermann, S., Shehata, S. I., Dohmae, N., Ishitani, R., et al., (2014). Crystal structure of Cas9 in complex with guide RNA and target DNA. *Cell, 156*(5), 935–949.

Nødvig, C. S., Nielsen, J. B., Kogle, M. E., & Mortensen, U. H., (2015). A CRISPR-Cas9 system for genetic engineering of filamentous fungi. *PloS One, 10*(7), e0133085.

Pandey, D. K., Nazir, A., & Dey, A., (2017). Isolation and characterization of phosphate solubilizing bacteria from rhizosphere of *Dioscorea alata* stimulating growth and diosgenin production. *Proceedings of the National Academy of Sciences, India Section B: Biological Sciences, 87*(4), 1143–1152.

Park, J., Shin, H., Lee, S. M., Um, Y., & Woo, H. M., (2018). RNA-guided single/double gene repressions in *Corynebacterium glutamicum* using an efficient CRISPR interference and its application to industrial strain. *Microbial. Cell Factories, 17*(1), 1–10.

Pelzer, S., Vente, A., & Bechthold, A., (2005). Novel natural compounds obtained by genome-based screening and genetic engineering. *Current Opinion in Drug Discovery & Development, 8*(2), 228–238.

Peng, F., Wang, X., Sun, Y., Dong, G., Yang, Y., Liu, X., & Bai, Z., (2017). Efficient gene editing in *Corynebacterium glutamicum* using the CRISPR/Cas9 system. *Microbial Cell Factories, 16*(1), 1–13.

Pohl, C., Kiel, J. A., Driessen, A. J., Bovenberg, R. A., & Nygard, Y., (2016). CRISPR/Cas9 based genome editing of *Penicillium chrysogenum*. *ACS Synthetic Biology, 5*(7), 754–764.

Qin, H., Xiao, H., Zou, G., Zhou, Z., & Zhong, J. J., (2017). CRISPR-Cas9 assisted gene disruption in the higher fungus *Ganoderma* species. *Process Biochemistry, 56*, 57–61.

Qiu, J. R., Su, Y. K., Song, Z. Q., Fang, X. S., Li, J. Y., Zhang, J., & Wang, J. H., (2018). Directing construction of CRISPR/Cas9 vector of SmPAL1 in *Salvia miltiorrhiza* by target efficiency detection *in vitro*. *Zhongguo Zhong Yao za zhi= Zhongguo zhongyao zazhi= China Journal of Chinese Materia Medica, 43*(21), 4226–4230.

Qiu, Y., Wang, S., Chen, Z., Guo, Y., & Song, Y., (2016). An active type IE CRISPR-Cas system identified in *Streptomyces avermitilis*. *PloS One, 11*(2), e0149533.

Rehman, S., Rehman, I. U., Jan, B., Rashid, I., Reshi, Z. A., & Ganie, A. H., (2021). Genome editing: Applications for medicinal and aromatic plants. In: *Medicinal and Aromatic Plants* (pp. 119–144). Academic Press.

Ricroch, A., Clairand, P., & Harwood, W., (2017). Use of CRISPR systems in plant genome editing: Toward new opportunities in agriculture. *Emerging Topics in Life Sciences, 1*(2), 169–182.

Rock, J. M., Hopkins, F. F., Chavez, A., Diallo, M., Chase, M. R., Gerrick, E. R., Pritchard, J. R., et al., (2017). Programmable transcriptional repression in mycobacteria using an orthogonal CRISPR interference platform. *Nature Microbiology, 2*(4), 1–9.

Sangal, V., Fineran, P. C., & Hoskisson, P. A., (2013). Novel configurations of type I and II CRISPR–Cas systems in *Corynebacterium diphtheriae*. *Microbiology, 159*(Pt_10), 2118–2126.

Schachtsiek, J., & Stehle, F., (2019). Nicotine-free, nontransgenic tobacco (*Nicotiana tabacum L.*) edited by CRISPR-Cas9. *Plant Biotechnology Journal, 17*(12), 2228.

Selle, K., & Barrangou, R., (2015). Harnessing CRISPR–Cas systems for bacterial genome editing. *Trends in Microbiology, 23*(4), 225–232.

Shalem, O., Sanjana, N. E., & Zhang, F., (2015). High-throughput functional genomics using CRISPR–Cas9. *Nature Reviews Genetics, 16*(5), 299–311.

Shen, C. C., Hsu, M. N., Chang, C. W., Lin, M. W., Hwu, J. R., Tu, Y., & Hu, Y. C., (2019). Synthetic switch to minimize CRISPR off-target effects by self-restricting Cas9 transcription and translation. *Nucleic Acids Research, 47*(3), e13 (1–11).

Shin, J., Jiang, F., Liu, J. J., Bray, N. L., Rauch, B. J., Baik, S. H., Nogales, E., et al., (2017). Disabling Cas9 by an anti-CRISPR DNA mimic. *Science Advances, 3*(7), e1701620.

Shin, S. E., Lim, J. M., Koh, H. G., Kim, E. K., Kang, N. K., Jeon, S., Kwon, S., et al., (2016). CRISPR/Cas9-induced knockout and knock-in mutations in *Chlamydomonas reinhardtii*. *Scientific Reports, 6*(1), 1–15.

Silva, G., Poirot, L., Galetto, R., Smith, J., Montoya, G., Duchateau, P., & Pâques, F., (2011). Meganucleases and other tools for targeted genome engineering: Perspectives and challenges for gene therapy. *Current Gene Therapy, 11*(1), 11–27.

Sternberg, S. H., Redding, S., Jinek, M., Greene, E. C., & Doudna, J. A., (2014). DNA interrogation by the CRISPR RNA-guided endonuclease Cas9. *Nature, 507*(7490), 62–67.

Tan, S. Z., Reisch, C. R., & Prather, K. L., (2018). A robust CRISPR interference gene repression system in *Pseudomonas*. *Journal of Bacteriology, 200*(7), e00575–17.

Tong, Y., Charusanti, P., Zhang, L., Weber, T., & Lee, S. Y., (2015). CRISPR-Cas9 based engineering of actinomycetal genomes. *ACS Synth. Biol., 4*, 1020–1029.

Tong, Y., Weber, T., & Lee, S. Y., (2019). CRISPR/Cas-based genome engineering in natural product discovery. *Natural Product Reports, 36*(9), 1262–1280.

Tong, Y., Whitford, C. M., Blin, K., Jørgensen, T. S., Weber, T., & Lee, S. Y., (2020). CRISPR–Cas9, CRISPRi and CRISPR-BEST-mediated genetic manipulation in *Streptomycetes*. *Nature Protocols, 15*(8), 2470–2502.

Ungerer, J., & Pakrasi, H. B., (2016). Cpf1 is a versatile tool for CRISPR genome editing across diverse species of *Cyanobacteria*. *Sci. Rep., 6*, 39681.

Upadhyay, S. K., Kumar, J., Alok, A., & Tuli, R., (2013). RNA-guided genome editing for target gene mutations in wheat. *G3: Genes, Genomes, Genetics, 3*(12), 2233–2238.

Weber, J., Valiante, V., Nødvig, C. S., Mattern, D. J., Slotkowski, R. A., Mortensen, U. H., & Brakhage, A. A., (2017). Functional reconstitution of a fungal natural product gene cluster by advanced genome editing. *ACS Synthetic Biology, 6*(1), 62–68.

Weinthal, D., Tovkach, A., Zeevi, V., & Tzfira, T., (2010). Genome editing in plant cells by zinc finger nucleases. *Trends in Plant Science, 15*(6), 308–321.

Westbrook, A. W., Moo-Young, M., & Chou, C. P., (2016). Development of a CRISPR-Cas9 tool kit for comprehensive engineering of *Bacillus subtilis*. *Applied and Environmental Microbiology, 82*(16), 4876–4895.

Wolf, T., Gren, T., Thieme, E., Wibberg, D., Zemke, T., Pühler, A., & Kalinowski, J., (2016). Targeted genome editing in the rare actinomycete *Actinoplanes* sp. SE50/110 by using the CRISPR/Cas9 system. *Journal of Biotechnology, 231*, 122–128.

Wu, X., Scott, D. A., Kriz, A. J., Chiu, A. C., Hsu, P. D., Dadon, D. B., Cheng, A. W., et al., (2014). Genome-wide binding of the CRISPR endonuclease Cas9 in mammalian cells. *Nature Biotechnology, 32*(7), 670–676.

Yang, Y. J., Wang, Y., Li, Z. F., Gong, Y., Zhang, P., Hu, W. C., Sheng, D. H., & Li, Y. Z., (2017). Increasing on-target cleavage efficiency for CRISPR/Cas9-induced large fragment deletion in *Myxococcus xanthus*. *Microbial Cell Factories, 16*(1), 1–15.

Ye, S., Enghiad, B., Zhao, H., & Takano, E., (2020). Fine-tuning the regulation of Cas9 expression levels for efficient CRISPR-Cas9 mediated recombination in *Streptomyces*. *Journal of Industrial Microbiology and Biotechnology, 47*(4, 5), 413–423.

Yeo, W. L., Heng, E., Tan, L. L., Lim, Y. W., Lim, Y. H., Hoon, S., Zhao, H., et al., (2019). Characterization of Cas proteins for CRISPR-Cas editing in *Streptomycetes*. *Biotechnology and Bioengineering, 116*(9), 2330–2338.

Zeng, H., Wen, S., Xu, W., He, Z., Zhai, G., Liu, Y., Deng, Z., & Sun, Y., (2015). Highly efficient editing of the actinorhodin polyketide chain length factor gene in *Streptomyces coelicolor* M145 using CRISPR/Cas9-CodA (sm) combined system. *Applied Microbiology and Biotechnology, 99*(24), 10575–10585.

Zhang, C., Meng, X., Wei, X., & Lu, L., (2016). Highly efficient CRISPR mutagenesis by microhomology-mediated end joining in *Aspergillus fumigatus*. *Fungal Genetics and Biology, 86*, 47–57.

Zhang, L., Wang, P., Feng, Q., Wang, N., Chen, Z., Huang, Y., Zheng, W., & Jiang, X., (2017). Lipid nanoparticle-mediated efficient delivery of CRISPR/Cas9 for tumor therapy. *NPG Asia Materials, 9*(10), e441 (1–8).

Zhang, X., Jin, M., Tadesse, N., Dang, J., Zhou, T., Zhang, H., Wang, S., et al., (2018). *Dioscorea zingiberensis* CH Wright: An overview on its traditional use, phytochemistry, pharmacology, clinical applications, quality control, and toxicity. *Journal of Ethnopharmacology, 220*, 283–293.

Zhang, Y., Zhang, F., Li, X., Baller, J. A., Qi, Y., Starker, C. G., Bogdanove, A. J., & Voytas, D. F., (2013). Transcription activator-like effector nucleases enable efficient plant genome engineering. *Plant Physiology, 161*(1), 20–27.

Zhou, Z., Tan, H., Li, Q., Chen, J., Gao, S., Wang, Y., Chen, W., & Zhang, L., (2018). CRISPR/Cas9-mediated efficient targeted mutagenesis of RAS in *Salvia miltiorrhiza*. *Phytochemistry, 148*, 63–70.

CHAPTER 15

Unraveling the Promise of CRISPR/Cas Systems in Modulating Nutrients and Antinutrients in Fruits

PROTHA BISWAS,[1] MIMOSA GHORAI,[1] SAMAPIKA NANDY,[1] DEVENDRA KUMAR PANDEY,[2] MAHIPAL S. SHEKHAWAT,[3] ARABINDA GHOSH,[4] POTSHANGBAM NONGDAM,[5] ABDEL RAHMAN AL-TAWAHA,[6] ERCAN BURSAL,[7] MUDDASARUL HODA,[8] JOGINDER SINGH,[9] and ABHIJIT DEY[1]

[1]*Department of Life Sciences, Presidency University, 86/1 College Street, Kolkata – 700073, West Bengal, India*

[2]*Department of Biotechnology, Lovely Professional University, Punjab, India*

[3]*Department of Plant Biology and Biotechnology, K.M. Center for Post Graduate Studies, Lawspet, Pondicherry, Tamil Nadu, India*

[4]*Department of Botany, Guwahati University, Guwahati, Assam, India*

[5]*Department of Biotechnology, Manipur University, Canchipur, Imphal, Manipur, India*

[6]*Department of Biological Sciences, Al-Hussein Bin Talal University, Maan, Jordan*

[7]*Department of Biochemistry, Mus Alparslan University, Turkey*

[8]*Department of Biological Sciences, Aliah University, Kolkata, West Bengal, India*

[9]*Department of Microbiology, School of Bioengineering and Biosciences, Lovely Professional University, Phagwara – 144411, Punjab, India*

[*]*Corresponding author. E-mail: abhijit.dbs@presiuniv.ac.in*

CRISPR/Cas-Mediated Genome Editing in Plants, Deepu Pandita, Anu Pandita (Eds.)

ABSTRACT

Throughout human history, fruits have been consumed as a staple food that provides many nutrients, including vitamins, carbohydrates, minerals, and dietary fiber. In addition, fruits are rich in secondary metabolites (SMs), which have therapeutic potential and possess antioxidant properties, and can combat many diseases, such as heart disease, hypertension, etc. Bringing fruit plants into cultivation serves as a means to provide food or increase the product's value. Due to the labor and complications associated with conventional breeding, researchers are discovering cost-effective, efficient, and fast-targeted genome editing techniques. Plant science has seen an increase in attention in recent years due to the discovery of sequence-specific nucleases, such as CRISPR/Cas, and ribonucleoproteins (RNPs), which have greatly advanced the process of breeding since it is simple and highly effective. Our topic for this chapter is the CRISPR/Cas technology, its use in the fruit industry to optimize nutrients and antinutrients, bottlenecks, and what the future holds.

15.1 INTRODUCTION

In the majority of cultures around the world, fruits are a staple food that are essential sources of nutrients (Wang et al., 2019). Fruit quality consists of various nutritional elements such as sugars and vitamins and bioactive compounds, including anthocyanins, malate, and lycopene (Satpute and Jagdale, 2016). These compounds are natural components found in certain foods in limited quantities and prevent disease, such as cardiovascular diseases and cancers, by preventing the release of toxic compounds into the body (Kris-Etherton et al., 2002). Many efforts have been made to enhance the fruit's nutritional content and bioactive compounds so that its quality can be enhanced internally. With the development of CRISPR (clustered regularly interspaced short palindromic repeat) technology, the industry has opened up a new horizon for fruit crop improvement, and plant breeding has been transformed in the process.

In recent years, the rapid progress in high-throughput sequencing technology and biotechnology has allowed researchers to make significant progress on gene mapping in a short period of time (Rong and Golic, 2000; Gong and Golic, 2003). The field of functional genomics, gene therapy, and research into transgenic organisms has revolutionized many fields through the use of genome editing as an approach. Editing the genome refers to the process of making targeted alterations at the genetic level through the use of sequence-specific nucleases that bind to individual segments of DNA and cleave them (Zhang et al., 2014). To alter the sequence of target sequences, DNA mutations, such as insertions, deletions, or base substitutions, are introduced as part of the method (Manghwar et al., 2019). The first techniques used to engineer DNA within a genome were homing endonucleases (HEs) and mega endonucleases (Silva et al., 2011). Nucleases with utility-engineered components include zinc-finger nucleases (ZFNs), which possess a Cys2-His2 DNA-binding domain and it also contains a DNA cleaving domain or DSB domain of the FokI restriction endonuclease (Kim et al., 1996). In addition, there is another tool called Transcription Activator-Like Nucleases (TALEN) that is derived from *Xanthomonas* bacterium (Zhang et al., 2013). These proteins recognize specific nucleotides through a sequence of 33–35 amino acid repeat. In order to target specific DNA sequences, TALENs can be programmed to rearrange repeated amino acid recognition motif sequences (Zhang et al., 2014). Programmable endonucleases such as ZNFs and TALENS have demonstrated success in the past. While they are directly involved in DNA modification and can be programmed at the protein level, they interact directly with DNA (Shukla et al., 2009; Char et al., 2015; Agarwal et al., 2018). In recent years, researchers have developed methods for DNA editing that are efficient, more precise, and faster than these inconvenient, time-consuming tools. In bacteria, CRISPR, and associated Cas systems discovery have led to a new generation of genome editing tools called CRISPR-associated (Cas) (Cong et al., 2013; Upadhyay et al., 2013). The aim of this chapter is to discuss the CRISPR/Cas system, its methodology, limitations, and application in plants to optimize the nutrition and antinutrient content of fruits.

15.2 THE MECHANISM BEHIND CRISPR/CAS

CRISPR/Cas can be used to create a genetically modified plant that has a heritable, targeted mutation and ideal protection against transgenes caused by foreign DNA sequences. In order to determine the function of various DNA sequences, model plants have been effectively manipulated using simple and straightforward tools such as CRISPR/Cas (Ricroch et al., 2017). Based on principles drawn from adaptive immunity in prokaryotic type II cells, CRISPR/Cas is a modern development in genetic engineering (GE) (Barrangou et al., 2007). There is a 40% and 90% resistance to foreign DNA elements in bacteria and archaea that are mediated by CRISPR/CRISPR-associated systems (Cas) (Makarova et al., 2011).

Streptococcus pyogenes is an example of a CRISPR/Cas9 system that has been extensively studied. By recognizing and cleaving the external genetic material of bacteria and archaea related to their adaptive immunity systems, CRISPR-Cas9 functions naturally (Mojica et al., 2009; Thurtle-Schmidt and Lo, 2018). Several CRISPR/Cas types exist (I, II, and III), each of which uses different molecular mechanisms to break down nucleic acids. The Cas family proteins act as a complex for targeting crRNA-guided genes in type I and type III systems, while DNA is recognized and cleaved by type II, only one protein is involved. Originating from type II CRISPR/Cas9 genetic system, CRISPR/Cas9 confers immunity to bacteria against viral invasion and the spread of plasmids (Doudna and Charpentier, 2014). Proteins and short RNAs are combined to cleave DNA in CRISPR-Cas systems. By integrating short guide RNAs (sgRNAs) into their genomes via protospacers, the bacteria can express sgRNA efficiently, effectively, and quickly. A CRISPR/Cas system eliminates a sequence that matches protospacers with DNA (Rehman et al., 2021). To function, the CRISPR/Cas9 system requires a DNA target of 20 base pairs (guide RNA or gRNA) as well as a protospacer adjacent motif (PAM) in the DNA of the host that is a (5'-NGG-3' or 5'-NAG-3') trinucleotide (Jinek et al., 2012). A PAM sequence is close to the crRNA-targeted sequences on the invader's DNA, which plays a key role during type I and type II interference and adaptation stages (Doudna and Charpentier, 2014). Cas9 is used in conjunction with synthetic guide RNA (sgRNA) in this tool to modify DNA at a desired location in the genome (Hsu et al., 2014). SgRNA directs the activity of the Cas9 nuclease, leading to DSBs and mutations viz. insertions, deletions, and substitutions at the target sites (Xie et al., 2014). CRISPR-Cas9 gene editing begins with searching target sites by sgRNA and creating a DSB by endonuclease Cas9. The DNA is then repaired either by nonhomologous end-joinings (NHEJs) or by homology-directed repairs (HDRs), resulting in subsequent mutations (Hsu et al., 2014; Shalem et al., 2015). In HDR, sister chromatids are involved in a natural, endogenous process. A number of mutations can be created in the process of DNA repair by NHEJ, and this is a major cause of genome rearrangements (Knoll et al., 2014). There are two lobes on Endonuclease Cas9, namely, the globular REC (recognition) lobe and the NUC (nuclease) lobe (Nishimasu et al., 2014). Two nuclease domains are involved in Cas9 proteins: the HNH nuclease domain breaks the complementary DNA strand, while the RuvC-like domain sets up the opposite pairing (Figure 15.1) (Jiang and Doudna, 2017).

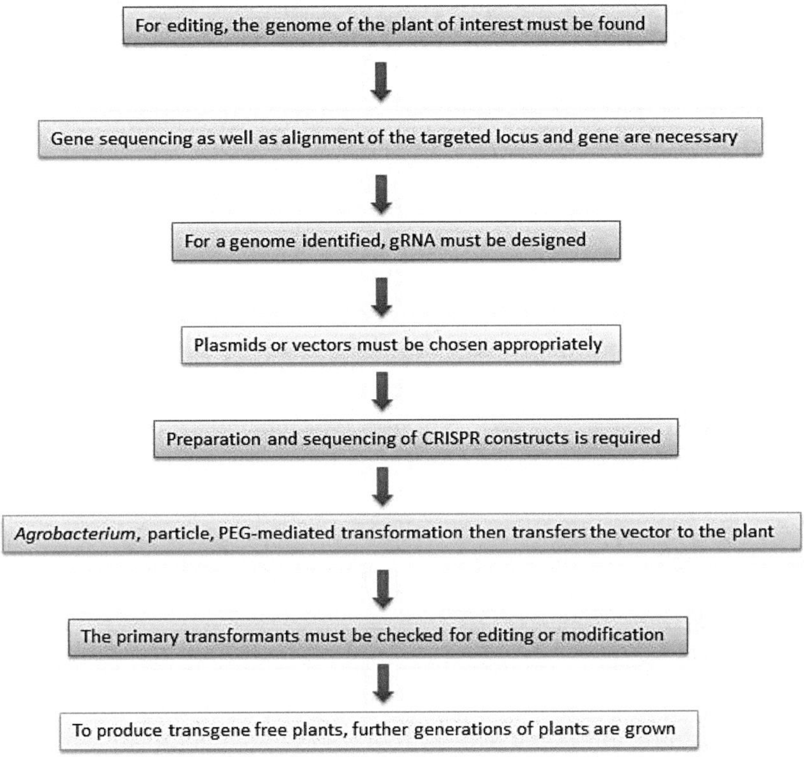

FIGURE 15.1 Method for editing genomes in plants with CRISPR/Cas system mediated by *agrobacterium*.

15.3 ENHANCING NUTRIENT AND ANTINUTRIENT CONTENT OF VARIOUS FRUIT PLANTS BY CRISPR/CAS TOOLS

The nutritional and health benefits of fruit are numerous, including sugar, fiber, vitamins, and minerals, which is why fruit is an important part of our diet. The banana, breadfruit, and date fruit are staple foods in cultures across Asia, Africa, and South America. Despite the importance of fruit crops as a source of nutrition, the difficult breeding process for fruit crops makes improvement of their nutritional quality relatively difficult, so they have received little attention for a long time. Therefore, transgenesis or genome engineering is the best approach for them. The field of CRISPR/Cas9-mediated genome editing in fruit crops has just emerged and few reports have been published to date. It has been reported that CRISPR/Cas can improve fruit production, increase locule number, improve size, color, and also be used to generate resistance to some viral and bacterial diseases as well as causing parthenogenesis; however, little is known about CRISPR/Cas-based fruit nutrition or antinutrient content optimization. In subsections, we detail some reports about CRISPR/Cas technology's application to fruits for optimizing their nutrients and antinutrients.

15.3.1 CRISPR/CAS APPLICATION IN TOMATO

The tomato fruit in its various forms is consumed widely, making it an important crop in the food industry. There have been numerous publications about CRISPR-Cas9 usage in tomato. They describe domestication of tomatoes, resistance to biotic stresses, resistance to abiotic stresses, and improvements in quality of tomato fruit. Fruit's internal attributes, such as sugars and vitamins, as well as bioactive elements, such as lycopene, anthocyanins, and malates are an essential part of determining fruit's quality. A geminivirus replicated on a tomato genome and was able to modify it with high precision and frequency by using CRISPR/Cas9. The promoter that controls the anthocyanin gene in tomatoes, causing ectopic expression of the anthocyanin gene in the tomato tissues, was introduced upstream from that gene (Čermák et al., 2015). The tomato contains higher concentrations of GABA (gamma-aminobutyric acid), a nonproteinogenic amino acid that has hypotensive effects, making it one of the most popular vegetables in the world. With CRISPR/Cas9 technology, it was possible to increase the GABA content in tomato by deleting the autoinhibitory domains of glutamate decarboxylase (GAD) (the key enzyme for GABA biosynthesis) genes SlGAD2 and SlGAD3 (Nonaka et al., 2017). By targeting phytoene desaturase gene and GABA synthesis controlling genes with pYLCRISPR/Cas9 system and single-site guide RNAs, tomato fruit accumulated 19 times more GABA than that with standard genome editing techniques (Li et al., 2018). Lycopene, a bioactive component found in tomatoes, is effective in treating chronic diseases, such as cancer and heart disease, while reducing the risk of diabetes and heart disease (Li and Xu, 2014). A gene controlling tomato malate content was identified through CRISPR-Cas9 technology as aluminum-activated malate transporter 9 (ALMT9) genes (Ye et al., 2017). Several CRISPR-Cas9 genome editing methods have been used in four stress-tolerant wild tomato accessions to achieve desired traits of vitamin C biosynthesis (SlGGP1 encodes a vitamin C-biosynthetic enzyme). Two independent CRISPR lines with mutations in the SlGGP1 uORF greatly increased the ascorbic acid content of the foliar leaves, showing that CRISPR-Cas9 can greatly impact tomato ascorbic acid levels (Li et al., 2018). The fruit lycopene accumulation of *Solanum pimpinellifolium* was improved by 500% using CRISPR-Cas9 genome engineering, when modifying the CycB gene, comparable with wide *S. lycopersicum* (Zsögön et al., 2018). Wild tomatoes are known to produce atroviolacea (ATV) and anthocyanin fruit (Aft), which increase the amount of anthocyanin inside tomato fruits. In the absence of multiple anthocyanin-related genes, SlAN2-like mutant tomatoes produce a lower level of anthocyanins than wild-type tomatoes. Therefore, SlAN2-like is responsible for the Aft phenotype (Yan et al., 2020). On tomato (Indigo Rose), the same experiment was performed using CRISPR/Cas, which again demonstrated that SlAN2 regulates anthocyanin biosynthesis in tomato plant vegetative tissues (Zhi et al., 2020). Using CRISPR/Cas9 to alter the SlHY5 gene in purple tomato, or 'Indigo Rose,' scientists have shown that there are other transcription factors that are independent of Hy5 that contribute to anthocyanin biosynthesis (Qiu et al., 2019). Based on the research with the commercial tomato, we discovered that a QTL (quantitative trait locus) on chromosome 4 influences levels of PHEVs (phenylalanine-derived volatiles), such as 2-phenylethanol, phenylacetaldehyde, and 1-nitro-2-phenylethane. FLORAL4 was discovered to be a gene that is involved in PHEV locus regulation, allowing a finer map to be created. The fruit of tomato trees accumulated PHEV largely due to FLORAL4, as revealed by reverse genetics (CRISPR-Cas9) (Tikunov et al., 2020). Sl-LIP8 mutants were found to have a significantly lower level of fruit short chain FA-VOCs (fatty acid-derived volatile organic compounds) that was reduced by CRISPR/Cas9 gene editing, suggesting that this enzyme is necessary for this pathway (Li et al., 2020).

15.3.2 CRISPR/CAS APPLICATION IN BANANA

Banana crops are ranked fourth on the global food crop ranking behind rice, wheat, and maize, and their consumption is second to that of mangos. The most economically valuable tropical fruit, bananas are grown in over 130 countries around the globe (Perrier et al., 2011; Dash and Rai, 2016). Traditional breeding of bananas only results in sterility since they have a triploid genome; so, gene editing might be an alternative path to genetic modification (GM) (Dash and Rai, 2016). The application of biotechnological approaches based on innovative biotechnology, such as GE, can therefore be considered a useful option when cultivating bananas. Researchers have used *Agrobacterium* and CRISPR/Cas9 to edit the genome of banana using CRISPR/Cas9, which successfully mutated two phytoene desaturase RAS-PDS genes that enabled them to make bananas rich in carotenoids in banana cv. Rasthali (Kaur et al., 2018). Phytoene desaturase was also successfully modified using a CRISPR/Cas9 system with a system for processing glycine tRNA in Cavendish bananas. A recent report states that a β-carotenoid-rich bananas were grown with CRISPR/Cas9 gene editing to create bananas enriched with lycopene epsilon-cyclase (LCYε) (Kaur et al., 2020). CRISPR/Cas9 is being used in India to create genetically enhanced banana cultivars to improve banana fruit nutrition, which is lacking in popular, high-yielding cultivars.

15.3.3 CRISPR/CAS APPLICATION IN GRAPES

In a study, researchers used the CRISPR/Cas9 system for editing the genome and targeting specific genes in Chardonnay grape plants and suspension cells. To control the biosynthesis of tartaric acid (TA) in grape, two single guide RNAs (sgRNA) targeting contiguous sites of the gene encoded L-idonate dehydrogenase (IdnDH) were developed. The sgRNA1/Cas9 expressed by transgenic plants and the regenerated plants cell mass expressed indel mutations with 100% mutation frequency at the target site, as well as 100% mutation frequency in the transgenic cells (Ren et al., 2016). Through the CRISPR/Cas9 tool, grapes were mutated by combining a synthetic sgRNA expression construct targeting the phytoene desaturase (VvPDS) gene with an embryonic Calli expression construct (Nakajima et al., 2017).

15.3.4 CRISPR/CAS APPLICATION IN KIWIFRUIT

Kiwifruits (Actinidia sp.) are an important fruit crop both economically and nutritionally, as they contain high levels of Vitamin C, as well as fibers, minerals, and health-promoting compounds. Using a sgRNA/Cas9-based cloning technique, researchers targeted the kiwifruit phytoene desaturase gene (AcPDS) by synthesizing two gRNA-containing primers. This study found that the (PTG)/Cas9 system had superior efficiency to the CRISPR/Cas9 system for kiwifruit genome editing, which provides valuable insight for other plants to use the CRISPR/Cas9 system in the future (Wang et al., 2018).

15.3.5 CRISPR/CAS APPLICATION IN WATERMELON

In the Cucurbitaceae family, *Citrullus lanatus* (with 2n=2×=22) is one of the most commonly eaten fruits in the world. Increasing the sweetness of watermelon fruit flesh is a goal of modern breeding. To better understand how sugars are accumulated in fruit flesh, the CRISPR/Cas9 method was used to generate mutant lines of mutant *C. lanatus* that are lacking ClAGA2. Compared with their wild-type counterparts, the mutants showed markedly lower concentrations of soluble sugars, glucose, fructose, and sucrose, but they had higher concentrations of raffinose. These findings suggest that ClAGA2 could have played an early role in the selection of sweet watermelons by facilitating glucose, fructose, and sucrose metabolisms in fruit flesh (Guo et al., 2019). There are novel sugar transporters for vacuole found in watermelon fruit (ClVST1) that accumulate sucrose in the fruit phloem. With the CRISPR/Cas9 system and associated sgRNA application, watermelon sugar content and total biomass decreases when ClVST1 is knocked out while sugar content increases when ClVST1 is overexpressed (Red et al., 2020). Watermelon DNA can also be precisely edited with CRISPR-Cas9 using the target gene *ClPDS,* phytoene desaturase. A 100% genome editing efficiency was achieved using the CRISPR/Cas9 system, where elements are inserted or deleted as needed at expected locations (Tian et al., 2017).

15.3.6 CRISPR/CAS APPLICATION IN CITRUS

In sweet orange, with the CRISPR/Cas9 genome editing system, sgRNA was used to identify the symbiotic pathway for carotenoid biosynthesis with successful mutation of phytoene desaturase encoded by CsPDS (Table 15.1) (Jia and Wang, 2014).

15.4 CRISPR/CAS SYSTEM LIMITATIONS

CRISPR/Cas9 is a very advantageous technology and has many applications, but it does have some limitations. CRISPR/Cas9 system development attempts have resulted in mutated plant varieties of interest, but there are few pressing obstacles that could hinder its usefulness. Because of a lack of basic molecular resources, including promoters, scorable markers, vectors, and transformation protocols, CRISPR/Cas genome editing is still not feasible in a variety of higher organisms, including plants that produce valuable natural

TABLE 15.1 Nutrient and Antinutrient Regulation in Fruits Using CRISPR/Cas

Fruit	Targeted Gene	Gene Function	References
Tomato	*GABA-TP1*, *GABA-TP2*, *GABA-TP3*, *CAT9*, *SSADH*, *SlGAD2*, and *SlGAD3*	Increased GABA content	Li et al. (2018); Nonaka et al. (2017)
	ANT1, *ANT2*, *HY5*	Anthocyanin biosynthesis, decreased content	Čermák et al. (2015); Yan et al. (2020); Zhi et al. (2020); Qiu et al. (2019)
	CycB	Increased lycopene content	Zsögön et al. (2018)
	FLORAL4	Regulates phenylalanine-derived volatiles	Tikunov et al. (2020)
	LIP8	Regulates short-chain fatty-acid-derived volatile organic compounds	Li et al. (2020)
Banana	*MaRAS-PDS1,2*	Increased carotenoid content	Kaur et al. (2018)
	MaLCYepsilon	Fortification of β-carotene	Kaur et al. (2020)
Watermelon	*ClAGA2*, *ClVST1*	Sugar accumulation	Guo et al. (2019); Red et al. (2020)
	ClPDS	Carotenoid synthesis	Tian et al. (2017)
Grapes	*IdnDH*	Tartaric acid biosynthesis	Ren et al. (2016)
	VvPDS	Carotenoid synthesis	Nakajima et al. (2017)
Kiwifruit	*AcPDS*	Carotenoid synthesis	Wang et al. (2018)
Sweet orange	*CsPDS*	Carotenoid synthesis	Jia and Wang (2014)

products (Tong et al., 2019). There is a major limitation of this technology in that mutations can be introduced at loci similar to the target sites but not identical. Identifying these mutations is difficult and requires scanning for mutations at sites similar to gRNA targets in the genome. In the case of a non-homologous end join, multiple mutations can occur simultaneously at the point of cleavage; for each mutation to be determined, genome sequencing will be required (Sharma et al., 2017). As a consequence of cleaving identical or homologous sequences of DNA, the CRISPR/Cas9 system can cause mutations in unintended sites, known as off-target mutations causing death of transformation of cells. CRISPR/Cas9 has some major concerns in regard to gene therapy because they typically cause 50% of off-target effects (OTEs) (Zhang et al., 2015). Various variants of Cas9 have also been developed that have reduced the amount of editing errors while preserving efficiency. Choosing a target site that has the fewest off-target sites is crucial to minimize OTEs, as is ensuring that the complementary sequences of the gRNA and the complementary sequences of the gRNA match. CasOT can effectively reduce the off target effect by searching for it, maintaining the dose, and replacing the Cas9 with nickase (Hsu et al., 2013; Ding et al., 2013; Zhang et al., 2014). PAM-dependent cleavage by CRISPR/Cas9 limits the number of DNA targets that can be targeted. In order to produce gRNAs, the RGR gene's transcribed mRNA was synthesized with ribozyme sequences at either end (Gao and Zhao, 2014). The CRISPR/Cas9 system has poor delivery or transfection efficiency, which limits its use. The lipid nanoparticles encapsulated in plasmid (DNA) deliver Cas9 and sgRNA via plasmids for successful transfection of Cas9 and sgRNA (Zhang et al., 2017).

15.5 CONCLUSION AND FUTURE PROSPECTS

Plant-based foods, medicines, and other products are in high demand due to the rise in the number of people. It is crucial that we develop new plant varieties that are high yielding and able to withstand environmental changes, even though the world's food requirements are growing. Fruit crops are essential for human nutrition, but they have received little attention due to the difficulty of breeding them. Many crops have been engineered due to the achievement of CRISPR/Cas9 gene editing in recent years to exhibit desirable characteristics, including resistance to biotic and abiotic stresses, higher yields, improved texture, and superior nutritional value, and the acceleration of domestication processes. However, there has been surprisingly little reporting of experiments to use CRISPR/Cas to increase fruit content in terms of nutrients and antinutrients currently. It is possible to enhance anthocyanin biosynthesis by modifying several genes involved in phenolic synthesis. Through manipulation of biosynthetic pathways, it is possible to get rid of undesirable properties in fruits while improving their nutritional value. Since most important fruit plants still have undetermined genomes today, genome editing is very difficult to modify different biosynthetic pathways in fruit plants. The CRISPR/Cas technology allows scientists to create precise, efficient, and targeted mutagenesis to increase fruit yield, nutritional value, and yield efficiency and opens a new era in plant research.

Plants cultivated for horticulture fruit lack standardized protocols for cell regeneration and complete tissue culture strategies required for the CRISPR/Cas system. To overcome the difficulty of using complete tissue culture approaches to grow fruit plants in horticulture, new systems of transformation have been developed. CRISPR/Cas system will become more effective and cost-effective with the use of protoplast transfection, nanotechnology, and viral vectors. Several genes are often regulated simultaneously in plant engineering, and the CRISPR/Cas system can edit multiple genes either by associating many short RNAs into one CRISPR vector or by using endogenous RNA processing systems to produce thousands of genes. As CRISPR technology is optimized, fruit plant species can be improved more quickly by editing individual genes with greater precision and creating mutations in short amounts of time.

Developing new tools for gene editing, deploying CRISPR/Cas technology, and easier methods for detecting heritable mutants will contribute to increased commercialization of gene editing efforts in fruits plants. The future of CRISPR/Cas technology depends on researchers coming up with strategies to strengthen it and overcome its bottlenecks. We hope with this gene-engineering application, food security can be increased and nutritional deficiencies of the ever-increasing global population, one of the biggest challenges of modern civilization, can be confronted.

KEYWORDS

- **antinutrients**
- **CRISPR/Cas**
- **fruit plants**
- **gene editing**
- **tartaric acid**
- **transcription activator-like nucleases**
- **zinc-finger nucleases**

REFERENCES

Agarwal, A., Yadava, P., Kumar, K., Singh, I., Kaul, T., Pattanayak, A., & Agrawal, P. K., (2018). Insights into maize genome editing via CRISPR/Cas9. *Physiology and Molecular Biology of Plants, 24*(2), 175–183.

Banga, S. S., & Boyd, J. B., (1992). Oligonucleotide-directed site-specific mutagenesis in *Drosophila melanogaster*. *Proceedings of the National Academy of Sciences, 89*(5), 1735–1739.

Barrangou, R., Birmingham, A., Wiemann, S., Beijersbergen, R. L., Hornung, V., & Smith, A. V. B., (2015). Advances in CRISPR-Cas9 genome engineering: Lessons learned from RNA interference. *Nucleic Acids Research, 43*(7), 3407–3419.

Čermák, T., Baltes, N. J., Čegan, R., Zhang, Y., & Voytas, D. F., (2015). High-frequency, precise modification of the tomato genome. *Genome Biology, 16*(1), 232.

Char, S. N., Neelakandan, A. K., Nahampun, H., Frame, B., Main, M., Spalding, M. H., Becraft, P. W., et al., (2017). An *Agrobacterium*-delivered CRISPR/Cas9 system for high-frequency targeted mutagenesis in maize. *Plant Biotechnology Journal, 15*(2), 257–268.

Cong, L., Ran, F. A., Cox, D., Lin, S., Barretto, R., Habib, N., Hsu, P. D., et al., (2013). Multiplex genome engineering using CRISPR/Cas systems. *Science, 339*(6121), 819–823.

Dash, P. K., & Rai, R., (2016). Translating the "banana genome" to delineate stress resistance, dwarfing, parthenocarpy and mechanisms of fruit ripening. *Frontiers in Plant Science, 7*, 1543.

Ding, Q., Regan, S. N., Xia, Y., Oostrom, L. A., Cowan, C. A., & Musunuru, K., (2013). Enhanced efficiency of human pluripotent stem cell genome editing through replacing TALENs with CRISPRs. *Cell Stem Cell, 12*(4), 393.

Doudna, J. A., & Charpentier, E., (2014). The new frontier of genome engineering with CRISPR-Cas9. *Science, 346*(6213).

Gao, Y., & Zhao, Y., (2014). Self-processing of ribozyme-flanked RNAs into guide RNAs *in vitro* and *in vivo* for CRISPR-mediated genome editing. *Journal of Integrative Plant Biology, 56*(4), 343–349.

Gong, W. J., & Golic, K. G., (2003). Ends-out, or replacement, gene targeting in *Drosophila*. *Proceedings of the National Academy of Sciences, 100*(5), 2556–2561.

Guo, S., Zhao, S., Sun, H., Wang, X., Wu, S., Lin, T., Ren, Y., et al., (2019). Resequencing of 414 cultivated and wild watermelon accessions identifies selection for fruit quality traits. *Nature Genetics, 51*(11), 1616–1623.

Hsu, P. D., Lander, E. S., & Zhang, F., (2014). Development and applications of CRISPR-Cas9 for genome engineering. *Cell, 157*(6), 1262–1278.

Hsu, P. D., Scott, D. A., Weinstein, J. A., Ran, F. A., Konermann, S., Agarwala, V., Li, Y., et al., (2013). DNA targeting specificity of RNA-guided Cas9 nucleases. *Nature Biotechnology, 31*(9), 827–832.

Jia, H., & Wang, N., (2014). Targeted genome editing of sweet orange using Cas9/sgRNA. *PloS One, 9*(4), e93806.

Jiang, F., & Doudna, J. A., (2017). CRISPR–Cas9 structures and mechanisms. *Annual Review of Biophysics, 46*, 505–529.

Jinek, M., Chylinski, K., Fonfara, I., Hauer, M., Doudna, J. A., & Charpentier, E., (2012). A programmable dual-RNA–guided DNA endonuclease in adaptive bacterial immunity. *Science, 337*(6096), 816–821.

Kaur, N., Alok, A., Shivani, Kumar, P., Kaur, N., Awasthi, P., Chaturvedi, S., et al., (2020). CRISPR/Cas9 directed editing of lycopene epsilon-cyclase modulates metabolic flux for β-carotene biosynthesis in banana fruit. *Metab. Eng., 59*, 76–86.

Kim, Y. G., Cha, J., & Chandrasegaran, S., (1996). Hybrid restriction enzymes: Zinc finger fusions to FokI cleavage domain. *Proceedings of the National Academy of Sciences, 93*(3), 1156–1160.

Klap, C., Yeshayahou, E., Bolger, A. M., Arazi, T., Gupta, S. K., Shabtai, S., Usadel, B., et al., (2017). Tomato facultative parthenocarpy results from Sl AGAMOUS-LIKE 6 loss of function. *Plant Biotechnology Journal, 15*(5), 634–647.

Knoll, A., Fauser, F., & Puchta, H., (2014). DNA recombination in somatic plant cells: Mechanisms and evolutionary consequences. *Chromosome Research, 22*(2), 191–201.

Kris-Etherton, P. M., Hecker, K. D., Bonanome, A., Coval, S. M., Binkoski, A. E., Hilpert, K. F., Griel, A. E., & Etherton, T. D., (2002). Bioactive compounds in foods: Their role in the prevention of cardiovascular disease and cancer. *The American Journal of Medicine, 113*(9), 71–88.

Li, R., Li, R., Li, X., Fu, D., Zhu, B., Tian, H., Luo, Y., & Zhu, H., (2018). Multiplexed CRISPR/Cas9-mediated metabolic engineering of γ-aminobutyric acid levels in *Solanum lycopersicum*. *Plant Biotechnol. J., 16*(2), 415–427.

Li, T., Yang, X., Yu, Y., Si, X., Zhai, X., Zhang, H., Dong, W., et al., (2018). Domestication of wild tomato is accelerated by genome editing. *Nature Biotechnology, 36*(12), 1160–1163.

Li, X., & Xu, J., (2014). Meta-analysis of the association between dietary lycopene intake and ovarian cancer risk in postmenopausal women. *Scientific Reports, 4*(1), 1–9.

Li, X., Tieman, D., Liu, Z., Chen, K., & Klee, H. J., (2020). Identification of a lipase gene with a role in tomato fruit short-chain fatty acid-derived flavor volatiles by genome-wide association. *The Plant Journal, 104*(3), 631–644.

Makarova, K. S., Haft, D. H., Barrangou, R., Brouns, S. J., Charpentier, E., Horvath, P., Moineau, S., et al., (2011). Evolution and classification of the CRISPR–Cas systems. *Nature Reviews Microbiology, 9*(6), 467–477.

Manghwar, H., Lindsey, K., Zhang, X., & Jin, S., (2019). CRISPR/Cas system: Recent advances and future prospects for genome editing. *Trends in Plant Science, 24*(12), 1102–1125.

Mojica, F. J., Díez-Villaseñor, C., García-Martínez, J., & Almendros, C., (2009). Short motif sequences determine the targets of the prokaryotic CRISPR defense system. *Microbiology, 155*(3), 733–740.

Naim, F., Dugdale, B., Kleidon, J., Brinin, A., Shand, K., Waterhouse, P., & Dale, J., (2018). Gene editing the phytoene desaturase alleles of cavendish banana using CRISPR/Cas9. *Transgenic Research*, 27.

Nakajima, I., Ban, Y., Azuma, A., Onoue, N., Moriguchi, T., Yamamoto, T., Toki, S., & Endo, M., (2017). CRISPR/Cas9-mediated targeted mutagenesis in grape. *PLoS One, 12*(5), e0177966.

Nishimasu, H., Ran, F. A., Hsu, P. D., Konermann, S., Shehata, S. I., Dohmae, N., Ishitani, R., et al., (2014). Crystal structure of Cas9 in complex with guide RNA and target DNA. *Cell, 156*(5), 935–949.

Nonaka, S., Arai, C., Takayama, M., Matsukura, C., & Ezura, H., (2017). Efficient increase of ɣ-aminobutyric acid (GABA) content in tomato fruits by targeted mutagenesis. *Scientific Reports, 7*(1), 7057.

Perrier, X., De Langhe, E., Donohue, M., Lentfer, C., Vrydaghs, L., Bakry, F., Carreel, F., et al., (2011). Multidisciplinary perspectives on banana (*Musa* spp.) domestication. *Proc. Natl. Acad. Sci. U S A, 108*(28), 11311–11318.

Qiu, Z., Wang, H., Li, D., Yu, B., Hui, Q., Yan, S., Huang, Z., Cui, X., & Cao, B., (2019). Identification of candidate HY5-dependent and-independent regulators of anthocyanin biosynthesis in tomato. *Plant and Cell Physiology, 60*(3), 643–656.

Rehman, S., Rehman, I. U., Jan, B., Rashid, I., Reshi, Z. A., & Ganie, A. H., (2021). Genome editing: Applications for medicinal and aromatic plants. In: *Medicinal and Aromatic Plants* (pp. 119–144). Academic Press.

Ren, C., Liu, X., Zhang, Z., Wang, Y., Duan, W., Li, S., & Liang, Z., (2016). CRISPR/ Cas9-mediated efficient targeted mutagenesis in chardonnay (*Vitis vinifera* L.). *Scientific Reports, 6*(1), 1–9.

Ren, Y., Sun, H., Zong, M., Guo, S., Ren, Z., Zhao, J., Li, Met al., (2020). Localization shift of a sugar transporter contributes to phloem unloading in sweet watermelons. *New Phytologist, 227*(6), 1858–1871.

Ricroch, A., Clairand, P., & Harwood, W., (2017). Use of CRISPR systems in plant genome editing: Toward new opportunities in agriculture. *Emerging Topics in Life Sciences, 1*(2), 169–182.

Rodríguez-Leal, D., Lemmon, Z. H., Man, J., Bartlett, M. E., & Lippman, Z. B., (2017). Engineering quantitative trait variation for crop improvement by genome editing. *Cell, 171*(2), 470–480.

Rong, Y. S., & Golic, K. G., (2000). Gene targeting by homologous recombination in *Drosophila*. *Science, 288*(5473), 2013–2018.

Satpute, M. R., & Jagdale, S. M., (2016). Color, size, volume, shape and texture feature extraction techniques for fruits: A review. *Int. Res. J. Eng. Technol, 3*, 703–708.

Shalem, O., Sanjana, N. E., & Zhang, F., (2015). High-throughput functional genomics using CRISPR–Cas9. *Nature Reviews Genetics, 16*(5), 299–311.

Sharma, S., Kaur, R., & Singh, A., (2017). Recent advances in CRISPR/Cas mediated genome editing for crop improvement. *Plant Biotechnology Reports, 11*(4), 193–207.

Shu, P., Li, Z., Min, D., Zhang, X., Ai, W., Li, J., Zhou, J., et al., (2020). CRISPR/Cas9-mediated SlMYC2 mutagenesis adverse to tomato plant growth and MeJA-induced fruit resistance to *Botrytis cinerea*. *Journal of Agricultural and Food Chemistry, 68*(20), 5529–5538.

Shukla, V. K., Doyon, Y., Miller, J. C., DeKelver, R. C., Moehle, E. A., Worden, S. E., Mitchell, J. C., et al., (2009). Precise genome modification in the crop species *Zea mays* using zinc-finger nucleases. *Nature, 459*(7245), 437–441.

Silva, G., Poirot, L., Galetto, R., Smith, J., Montoya, G., Duchateau, P., & Pâques, F., (2011). Meganucleases and other tools for targeted genome engineering: Perspectives and challenges for gene therapy. *Current Gene Therapy, 11*(1), 11–27.

Thurtle-Schmidt, D. M., & Lo, T. W., (2018). Molecular biology at the cutting edge: A review on CRISPR/CAS9 gene editing for undergraduates. *Biochemistry and Molecular Biology Education, 46*(2), 195–205.

Tian, S., Jiang, L., Gao, Q., Zhang, J., Zong, M., Zhang, H., Ren, Y., et al., (2017). Efficient CRISPR/Cas9-based gene knockout in watermelon. *Plant Cell Reports, 36*(3), 399–406.

Tikunov, Y. M., Roohanitaziani, R., Meijer-Dekens, F., Molthoff, J., Paulo, J., Finkers, R., Capel, I., et al., (2020). The genetic and functional analysis of flavor in commercial tomato:

The FLORAL4 gene underlies a QTL for floral aroma volatiles in tomato fruit. *The Plant Journal, 103*(3), 1189–1204.

Tong, Y., Weber, T., & Lee, S. Y., (2019). CRISPR/Cas-based genome engineering in natural product discovery. *Natural Product Reports, 36*(9), 1262–1280.

Upadhyay, S. K., Kumar, J., Alok, A., & Tuli, R., (2013). RNA-guided genome editing for target gene mutations in wheat. *G3: Genes, Genomes, Genetics, 3*(12), 2233–2238.

Wang, T., Zhang, H., & Zhu, H., (2019). CRISPR technology is revolutionizing the improvement of tomato and other fruit crops. *Horticulture Research, 6*(1), 1–13.

Wang, Z., Wang, S., Li, D., Zhang, Q., Li, L., Zhong, C., Liu, Y., & Huang, H., (2018). Optimized paired-sgRNA/Cas9 cloning and expression cassette triggers high-efficiency multiplex genome editing in kiwifruit. *Plant Biotechnology Journal, 16*(8), 1424–1433.

Xie, K., Zhang, J., & Yang, Y., (2014). Genome-wide prediction of highly specific guide RNA spacers for CRISPR–Cas9-mediated genome editing in model plants and major crops. *Molecular Plant, 7*(5), 923–926.

Xu, J., Hua, K., & Lang, Z., (2019). Genome editing for horticultural crop improvement. *Horticulture Research, 6*(1), 1–16.

Yan, S., Chen, N., Huang, Z., Li, D., Zhi, J., Yu, B., Liu, X., et al., (2020). Anthocyanin fruit encodes an R2R3-MYB transcription factor, SlAN2-like, activating the transcription of SlMYBATV to fine-tune anthocyanin content in tomato fruit. *New Phytologist, 225*(5), 2048–2063.

Ye, J., Wang, X., Hu, T., Zhang, F., Wang, B., Li, C., Yang, T., et al., (2017). An InDel in the promoter of Al-ACTIVATED MALATE TRANSPORTER9 selected during tomato domestication determines fruit malate contents and aluminum tolerance. *The Plant Cell, 29*(9), 2249–2268.

Zhang, J., Guo, S., Ji, G., Zhao, H., Sun, H., Ren, Y., Tian, Set al., (2020). A unique chromosome translocation disrupting ClWIP1 leads to gynoecy in watermelon. *The Plant Journal, 101*(2), 265–277.

Zhang, L., Wang, P., Feng, Q., Wang, N., Chen, Z., Huang, Y., Zheng, W., & Jiang, X., (2017). Lipid nanoparticle-mediated efficient delivery of CRISPR/Cas9 for tumor therapy. *NPG Asia Materials, 9*(10), e441 (1–8).

Zhi, J., Liu, X., Li, D., Huang, Y., Yan, S., Cao, B., & Qiu, Z., (2020). CRISPR/Cas9-mediated SlAN2 mutants reveal various regulatory models of anthocyanin biosynthesis in tomato plant. *Plant Cell Reports, 39*(6), 799–809.

Zsögön, A., Čermák, T., Naves, E. R., Notini, M. M., Edel, K. H., Weinl, S., Freschi, L., et al., (2018). De novo domestication of wild tomato using genome editing. *Nature Biotechnology, 36*(12), 1211–1216.

CHAPTER 16

CRISPR/Cas-Based Strategies for Resistance Against Plant Viruses

SAURABH PANDEY[1] and SUNIDHI MISHRA[2]

[1]Department of Agriculture, Guru Nanak Dev University, Amritsar, Punjab – 143005, India

[2]Department of Horticulture, College of Agriculture, Korea, Indira Gandhi Agriculture University, Raipur – 497335, Chhattisgarh, India

[*]Corresponding author. E-mail: E-mails: saurabh_agri@gndu.ac.in; pandey.saurabh784@gmail.com

ABSTRACT

Plant viruses invading crops emerged as a significant concern for crop production worldwide. Modern biotechnological interventions to engineer crop resistance against viruses have recently become an effective strategy. Among these biotechnological tools, genome editing based on clustered regularly interspaced short palindromic repeats (CRISPR)/CRISPR-associated system (Cas) offers attractive DNA/RNA editing vectors for virus resistance in plants. These tools provide tremendous flexibility in the virus or host gene editing based upon RNA or DNA genome. The specificity of genome editing proves this to be an excellent protection measure. This chapter will explore different CRISPR/Cas tools, their applications against viral genes, and host susceptibility factors. We will also discuss the challenges and opportunities of applying these tools in tandem with antiviral breeding to control plant viruses.

CRISPR/Cas-Mediated Genome Editing in Plants, Deepu Pandita, Anu Pandita (Eds.)

16.1 INTRODUCTION

Plants are affected by an array of pathogens, including fungi, bacteria, fungus, viruses, and nematodes, leading to substantial global economic losses. Plant viruses contribute around half of the plant disease worldwide, with an estimated loss of US $30 billion annually (Nicaise, 2014; Sastry and Zitter, 2014). To date, more than 1,500 virus species have been discovered and, based upon their genome, classified into 26 different families (https://talk.ictvonline.org/taxonomy/). Viruses are obligate intracellular parasites with a small number of proteins that rely upon the host to complete their infection cycle. This makes their control difficult because drugs targeting the virus will ultimately kill the host cells. Therefore, there is a dire need for molecular and breeding interventions to prevent and control plant viruses to generate virus-resistant crops. In the last century, virus resistance/tolerance was developed through molecular breeding that discovered and transferred resistance gene/quantitative trait loci (QTLs) to the popular cultivars.

Genome editing is the group of techniques that offer alteration of the target genome at the nucleotide level. Among them, CRISPR (clustered regularly interspaced short palindromic repeats)/CRISPR-associated (Cas) systems are the widely adopted genome editing technique initially working in bacterial immunity against infecting organisms' nucleic acids (Sood et al., 2021). This technique is unique in its ability to edit the genome at particular nucleotide sequence precisely. Molecular understanding of this system led to the evolution of techniques to edit DNA and RNA sequences in different organisms, including plants and plant viruses. CRISPR system consists of an engineered guide RNA (gRNA) with its endonuclease Cas effector for genome editing. These Cas effectors differ in their structure, sequence, and function and form the basis of two distinct CRISPR/Cas system classes (Makarova et al., 2015, 2017a, b). Class 1 comprises a multi-protein effector complex and includes type I, III, and IV, whereas class 2 contains a single effector protein and includes type II, V, and VI (Makarova et al., 2017a, b; Shmakov et al., 2017). The single effector protein class is most widely adopted for genetic manipulation at the nucleic acid level (Hsu et al., 2013; Garcia-Doval and Jinek, 2017; Ji et al., 2019). Type II and V Cas effectors are used as DNA specific, whereas type VI is used RNA precise editing. Recently, the application of CRISPR/Cas-mediated genome editing for plant virus resistance has been reported by various groups (Ma et al., 2016; Hadidi et al., 2016; Mahas and Mahfouz, 2018; Langner et al., 2018; Li et al., 2019; Li and Wang, 2019; Chen et al., 2019; Zhang et al., 2019b). These methods follow two approaches to control the virus. Either they interfere with the viral genome to inhibit replication and infection or target host susceptibility genes deployed by viruses to complete these processes.

16.2 POTENTIAL OF CRISPR AS A TOOL TO ENGINEER DISEASE RESISTANCE

Genome editing technologies provide specific editing, thereby reducing the risk of off-target mutations. With the sequencing of most crop genomes, these techniques can be applied to precisely edit target gene alleles in complex genomes of polyploid crops such as strawberry, wheat, potato, rapeseed, and cotton (Wilson et al., 2019; Wang et al., 2014, 2018; Johansen et al., 2019; Braatz et al., 2017). Among genome editing techniques, the CRISPR/Cas system, with its amenability for multiplexing, straightforward nature of use with relatively low cost, made it a choice for editing (Thompson et al., 2018). CRISPR/Cas system specificity lies in 20 nucleotide long replaceable sequences in sgRNA complementary to the target sequence and acts as a target recognition sequence for Cas effector. This feature permits easy manipulation of sgRNAs according to our need by consecutive polymerase chain reactions (Li et al., 2013).

The most common Cas effector is Cas9 isolated from *Streptococcus pyogenes* increases its stringency by NGG sequence requirement adjacent to sgRNA recognition site, which is called a PAM (Doudna and Charpentier, 2014). This PAM sequence changes according to different CRISPR systems (Leenay and Beisel, 2017). With the genome sequence available for most crops, designing specific sgRNA targeting specific genes of interest responsible for particular traits such as disease resistance is possible. The efficiency and specificity of sg RNA depend on its secondary structure and number of mismatches outside the seed region (10–12 nucleotide region near PAM) (Kumlehn et al., 2018; Jiang and Doudna, 2017). These mismatches generate one drawback of the early CRISPR system, i.e., OTEs that were reduced by using mutated Cas effector proteins engineered for nicking activity instead of double standard breaks (DSBs). For example, the D10A mutation in Cas 9 effector converted it to Cas9 nickase (nCas9), producing two different nicks at opposite strands (Schiml et al., 2016). This paired binding and nicking activity of nCas9 increases the process specificity (Doudna and Charpentier, 2014). Further recent advances reduced the OTEs more precisely, for example, SpCas9-HF1, HypaCas9, and HiFi Cas9 (Kleinstiver et al., 2016;

Chen et al., 2017; Vakulskas et al., 2018). Advanced bioinformatics software for designing guide RNA (gRNA) was developed that helped reduce the OTEs (Khatodia et al., 2016). Nicking through the Cas effector causes DSB that is ultimately repaired by two pathways: (i) error-prone non-homologous end joining (NHEJ); and (ii) template-based Homology-directed repair (HDR). Theoretically, the CRISPR system will work until the availability of the last target sites recognized by the sgRNA/Cas complex, which is beneficial in multiple editing of the same site in the target genome. Template-based repair is often a method of choice because it enables the introduction of the desired sequence. However, it requires the delivery of repair templates along with the CRISPR system. Repair templates could be provided with the help of viral vectors, which increases the possibility of HDR (Baltes et al., 2014; Cermak et al., 2015). HDR mediated target base editing was successfully implemented with the help of improved Cas effector, for example, dead Cas9 (dCas9) and nCas9 in tomato, rice, and cotton crops (Komor et al., 2016; Li et al., 2017; Shimatani et al., 2017; Qin et al., 2020; Molla et al., 2020). Induced point mutations through nCas9 can create herbicide-resistant rice crops (Shimatani et al., 2017). Besides that, prime editing can be used for genome editing, which uses a repair template for sequence repair (Anzalone et al., 2019). In prime editing, Cas9 nickase fused with reverse transcriptase (RT) to precisely edit with the help of gRNA (Marzec and Hensel, 2020). Prime editing was successfully implemented in rice and wheat plants (Lin et al., 2020). Both these editing methods are amenable to virus resistance breeding. Also, CRISPR enables multiple mutations simultaneously for a specific gene/allele or for multiple genes/alleles, which help target essential genes of viruses at the same time with reduced OTEs (Hahn and Nekrasov, 2019; Langner et al., 2018) and with designing tools can be essentially escaped.

16.3 APPROACHES TO ENGINEER PATHOGEN RESISTANCE

Plant-pathogen interaction is a complex phenomenon governed by the arms race between them. Genome editing technologies enable us to mutate the pathogen genome or host susceptibility factors to respond to pathogen resistance. Also, these techniques will elucidate the molecular mechanism of host plant susceptibility and pathogen evolution. In subsections, will discuss plant resistance engineering approaches coming under transgene free and transgene-based regulation.

16.3.1 TRANSGENE FREE APPROACH BY CRISPR/CAS-INDUCED MUTATIONS

Conventionally CRISPR based transgenic plants are identified by PCR, and subsequent certification requires sequencing for knowing the exact sites of desired and off-site mutations (Kim and Kim, 2016). Genetically modified organisms (GMOs) regulation in the United States of America does not apply for CRISPR-based crops, but in Europe, these crops fall under the GMO regulations (Waltz, 2018; Sprink et al., 2016). The former system is based on product-based legislation, whereas the latter is based on the process. The counter-argument of the process-based evaluation is that it is impossible to trace the actual process involved in generating a small amount of mutation, whether natural, genome-edited, or chemically induced (Urnov et al., 2018). With sequence-specific nucleases, induced mutations in the target organism can be inherited independently of construct integration. This suggests in sexually propagated crops CRISPR-based transgenic allows segregation of mutated loci from the construct integration site and thus produces transgene-free crops. Transient expression of CRISPR-based constructs can be used in vegetatively propagated crops for the segregation of mutant loci from the construct (Zhang et al., 2016a; Nadakuduti et al., 2018). Scientists have proposed another way of bypassing the GMOs regulation by deploying ribonucleoprotein (RNP) complexes carrying only Cas9 and sgRNA through biolistics (Kim and Kim, 2016; Metje-Sprink et al., 2019). Practical application of this technique is not yet reported and needs substantial experimentation. However, some of the DNA-free systems have tried in maize, *Chlamydomonas reinhardtii,* wheat, petunia, grapevine, apple, lettuce, tobacco, rice, and Arabidopsis paved the way for its future GMOs regulation-free application (Svitashev et al., 2016; Baek et al., 2016; Liang et al., 2017; Subburaj et al., 2016; Malnoy et al., 2016; Woo et al., 2015). Some of these potential approaches in terms of plant-pathogen interaction are listed in this section.

16.3.1.1 TARGETING HOST SUSCEPTIBILITY FACTOR

Plant pathogen often requires host proteins and enzymes for the completion of their life cycle. Therefore, a simple knockout event through CRISPR could dramatically improve the plant resistance (Zaidi et al., 2018). Non-functional alleles of the MLO locus govern mildew resistance in many crops.

It was shown in the case of tomato crops when MLO locus was mutated plant became resistant against the powdery mildew disease (Nekrasov et al., 2017). Genome editing techniques enabled multiple allele mutation of MLO in the wheat crop for providing resistance against *Blumeria graminis f. sp. tritici* (Wang et al., 2014). Another down mildew resistance locus, DNR6, was a susceptibility factor in Arabidopsis and potato (van Damme et al., 2008; Sun et al., 2016). Knock out of DMR6 in tomato proposed to provide multiple pathogen resistance (de Toledo Thomazella et al., 2016). Similarly, SWEET gene mutation and promoters in rice provided resistance against *Xanthomonas oryzae pv. oryzae* (Zhou et al., 2015; Oliva et al., 2019; Xu et al., 2019). Also, an induced mutation in the ERF922 allele in rice provided resistance against blast fungus *Magnaporthe oryzae* (Wang et al., 2016a). Likewise, in plant-virus interaction, eukaryotic initiation factors were targeted for virus resistance. For example, a mutation in the eIF4E gene of cucumber provided resistance against the *Cucumber vein yellowing virus* (CVYV) (Chandrasekaran et al., 2016). Therefore, different host susceptibility factors are the potent target for resistance against fungus, bacterial, and virus pathogens. However, the involvement of these genes in essential plant developmental processes makes them critical and needs further evaluation for yield related trade-offs in different crop models.

16.3.1.2 ALTERATION IN THE PROMOTER REGION OF SUSCEPTIBILITY/RESISTANCE GENES

Deleting, inserting, and rearranging cis-elements in the promoter region of resistance/susceptibility genes is another promising strategy for plant pathogen resistance. Pathogens often evolve effector molecules to evade the plant defense, which works through cis-element binding of susceptibility genes. These regions can be deleted to provide resistance without affecting their integrity in plant development. For example, deletion of cis-element of rice OsSWEET14 gene increased plant resistance against *Xanthomonas oryzae* by targeting AvrXa7 effector (Li et al., 2012). Similarly, other SWEET gene promoters in rice, such as *OsSWEET11* and *OsSWEET13,* were targeted through genome editing techniques (Jiang et al., 2013; Oliva et al., 2019). Alteration of PthA4 effector-binding sites at the promoter region of *CsLOB1* increases the resistance against *Xanthomonas citri* in Duncan grapefruit (Jia et al., 2016). Placing cis elements through HDR via a repair template will ensure common elements' proper spacing and functioning on gene expression. Also, introducing the effector binding site as a new element in the promoter region could trigger stress-related expressions. For example, in natural conditions, Bs3 gene expression in pepper activated by avrBS3 effector from *Xanthomonas campestris* (Rivas and Genin, 2011; Boch et al., 2014). Therefore, activation of these effector binding sites could be an excellent strategy against biotrophic plant pathogens. These small changes (up to 20 bp) could bypass the GMOs regulations as this much gain or loss of nucleotides occurs in nature spontaneously (ZKBS (German Biosafety Commission), 2012).

16.3.1.3 ALTERATION IN THE CODING REGION OF DEFENSE GENES

Editing the coding sequence of the host defense gene through HDR is also a potent strategy by replacing the triplets without changing the amino acid sequences in the open reading frame. This can change the mRNA structure in such a way that it can evade pathogens. Pathogen often deploys small RNAs to target plant defense genes in the phenomenon called cross-kingdom RNA interference (RNAi) by hijacking its RNAi machinery (Pandey et al., 2020b). Some of the targets by this approach were identified in tomato were defense signaling component MAPKKK4, in Arabidopsis AtFEI2, AtPMR6, and AtWRKY7 against *Botrytis cinerea* interaction (Weiberg et al., 2013; Wang et al., 2017). Moreover, *Verticillium dahliae* utilizes this approach as identified in the Arabidopsis AGO1 gene important in RNAi (Wang et al., 2016b). With the enormous dataset available for sRNA during host-pathogen interaction, this approach might reveal many more potential targets for plant defense with almost zero fitness cost expectation.

16.3.1.4 ALTERATION IN AMINO ACID SEQUENCES OF TARGETED PLANT PROTEINS

Pathogen recognizes certain amino acids in plant proteins to establish the infection employing effector proteins. Such proteins cleaved by the effector proteases, such as HopB1 from *Pseudomonas syringae*, cleaves the signaling receptor BAK1, or AvrPphB targeting other components such as BIK1 and PBL1 PBL2, PBL6, and PBS1 (Li et al., 2016; Zhang et al., 2010). These plant amino acid recognition sites could further be explored for plant defense because a single amino acid change in the target site could result in cleavage protection from pathogen proteases. The double mutation in BIK1 caused a dominant-negative effect on flg22-induced plant defense, suggesting these residues' functional importance (Zhang et al., 2010). Therefore, the impact of target amino acid changes should be thoroughly checked before making any changes in the amino acid sequences of the host protein. The application of this approach to enhance disease resistance was shown for wheat R-gene where substituting two amino acids at the binding site increases the hypersensitive response against powdery mildew (Stirnweis et al., 2014). Induction of post-translational modification by bacterial effectors was also reported, for example, bacterial effectors induced RIN4 hyper phosphorylation to reduce the RIN4-ROC1 protein interaction, thus increasing susceptibility (Rodriguez et al., 2016). These sites can be replaced by the help of genome editing to minimize the effector-triggered host susceptibility. Therefore, base editing with modern editing tools offers plant biologists an exciting effector targeted amino acid change to enhance plant defense.

These four approaches offer race-specific resistance, which can be enhanced by introducing mutation stacking in the host for evolving pathogen and increasing resistance durability.

16.3.2 TRANSGENIC APPROACHES TO ENHANCE DISEASE RESISTANCE

Heterologous genes transfer is traditionally the method of choice for generating resistance plants. CRISPR-based genome editing transfers the desired DNA fragment via HDR in empty spaces existing in high-expressing defense genes under pathogen attack. This precision could relax the regulatory prohibition of transferred resistance genes by reducing possible side effects on host gene integration. Flanking regions are known in this type of integration, enabling successful events determination by PCR that eliminates the need for selectable marker uses. A selectable marker is a primary reason for hesitancy and public concern in the adaptability of the transgenic crop. Initial transgenic crops were heavily under scrutiny due to random integration of foreign genes that could lead to significant side effects and require selectable markers such as antibiotic resistance genes (Gepts, 2002). Inserting foreign DNA requires the regulatory norms to be implemented as its unknown behavior in the recipient plant species with high cost (Waltz, 2018). In the following, we have listed CRISPR-mediated transgenic approaches for plant resistance against pathogen.

16.3.2.1 GAIN OF FUNCTION APPROACH

Resistance breeding in plants could potentially be adopted in susceptible species by HDR-mediated transfer of resistance genes in the actively expressed regions of chromatin. These insertions can be successfully and precisely implemented through the CRISPR/Cas system and result in the target species gain of function. Further co-regulatory modules of different resistance genes enhance the resistance while decreasing the dysregulation and fitness cost (Karasov et al., 2017). Resistance genes consist of a nucleotide-binding site with leucine-rich repeat (NLR) and require helper and sensor NLR to activate and recognize pathogen effectors. This co-regulation ensures the activation of resistance genes only during pathogen attacks (Wu et al., 2017). However, a limited number of resistance genes were discovered in crops and requires further exploration of helper NLRs interacting with them. These resistance genes transfer through genome editing provide a fascinating approach to develop pathogen resistance (Wang et al., 2019). The durability of dominant resistance is often breakable as they provide race-specific resistance, which dilutes with pathogen evolution through mutation shown by various reports (Dangl et al., 2013; Pandolfi et al., 2017; Kourelis and van der Hoorn, 2018). Therefore, resistance gene stacking/pyramiding with the help of CRISPR/Cas-mediated HDR can facilitate durable resistance against constantly evolving pathogens (Pandolfi et al., 2017).

Pathogen recognition receptors (PRRs) are molecules recognizing a broader pathogen group to provide durable resistance. Application of PRRs expression to confer broad-spectrum resistance has been shown between rice and banana, Arabidopsis, and wheat and Arabidopsis and tomato (Tripathi et al., 2014; Schoonbeek et al., 2015; Lacombe et al., 2010). Conservation of immune signaling pathways between different groups of plants facilitates the plant immunity activation in this approach (Holton et al., 2015). Using this approach, plants can detect evolved and a new pathogen spectrum to initiate the defense responses against them, thus reducing the yield loss (Karasov et al., 2017). Also, stress-induced priming changes the chromatin states for allowing higher defense gene expression in further exposure (Karasov et al., 2017). These permissive chromatin regions could be utilized for heterologous expression of resistance and PRRs genes. Potential candidates for a gain of function approach include signaling components, antimicrobial peptides, transcription factors, antimicrobial metabolite, and structural genes (Pandey et al., 2018, 2019, 2020a).

16.3.2.2 LOSS OF FUNCTION/KNOCKDOWN APPROACH

Loss of function or knockdown approaches is potent approaches for conferring plant disease resistance. The primary mechanism of these approaches is RNAi which operates in eukaryotic cells. Knockdown is dependent upon small RNA molecules (sRNAs). In plant-pathogen interaction, transfer of sRNAs between host and pathogen is observed, which is responsible for cross-kingdom RNAi (Hua et al., 2018; Pandey et al., 2020b). Here pathogen-derived sRNA crosses the host barrier to target their defense genes (Weiberg et al., 2013). Similarly, host plant-derived sRNAs target different pathogens and their virulence factors (Zhang et al., 2011, 2016b; Shen et al., 2014). In plants, RNAi is present naturally, suggesting its ability to target pathogenic plant RNAs by introducing artificial miRNA with pathogen genome homology. These targets include viral replicases and viral coat proteins (CPs) (Dong and Ronald, 2019). Here CRISPR/Cas system, through constitutive expression of sgRNAs against the virus genome, could transform the host immune system (Ji et al., 2015; Zhang et al., 2018b; Mushtaq et al., 2020). Host plant cells infection by virus triggers the system and degrade pathogen genome thus provide the resistance. Specific designing tools for reducing off-targets in hosts should be considered. With the advancements of CRISPR/Cas proteins, the choice of system depends upon the target, for example, Cas9 and Cas13 effectors could target DNA and RNA viruses, respectively (Wolter and Puchta, 2018; Ali et al., 2018). Another advantage of the system is that it is host cell-specific, with the controlled processing of targeted nucleic acid that reduces the potential off-targets (Smith et al., 2017). Also, CRISPR/Cas can affect the virus evolution, as shown in the case of geminiviruses-infected cassava plants (Rybicki, 2019; Mehta et al., 2019). Therefore, CRISPR/Cas appears to be a suitable method for controlling virus pathogens with consideration of yield and growth-related trade-offs.

16.3.3 VIRUS RESISTANCE BY TARGETING VIRAL GENOME

16.3.3.1 DNA VIRUSES

Two DNA virus families, Caulimoviridae and Geminiviridae, are causing immense plant damage with 485 single-stranded and 85 double-stranded DNA (dsDNA) containing virus members. Genome editing technologies emerged as a preferential tool for targeting DNA virus genomes directly. For example, zinc-finger nucleases (ZFN) and transcription activator-like effector nucleases (TALENS)-mediated targeting of viral replication region were reported in tobacco curly shoot virus (TbCSV), and tomato yellow leaf curl China virus (TYLCCNV) (Cheng et al., 2015; Chen et al., 2014). The emergence of the CRISPR/Cas system for precise editing shifted to focus on antiviral manipulation through genome editing. DNA interference through sgRNAs targeting replication-associated region and intergenic region (IR) of virus genome showed effective viral resistance against cotton leaf curl Multan virus (CLCuMuV), bean yellow dwarf virus (BeYDV), and beet severe curly top virus (BSCTV) in transgenic Arabidopsis and *Nicotiana benthamiana* plants (Baltes et al., 2015; Ji et al., 2015; Yin et al., 2019). Similarly, targeting CP and replicase gene showed lower accumulation of virus particles for tomato yellow leaf curl virus (TYLCV) in transgenic *N. benthamiana* and tobacco plants (Tashkandi et al., 2018). Another approach adopted for the targeting of the viral genome is the designed sgRNAs against stem-loop sequence present within the IR, which has shown more excellent resistance in comparison to CP or replicase targeting sgRNAs in geminiviruses Merremia mosaic virus (MeMV), different strains of TYLCV, and cotton leaf curl Kokhran virus (CLCuKoV) (Ali et al., 2016). Similarly, single sgRNA targeting conserved IR region of TYLCV origin of replication confers broad-spectrum resistance against plant geminiviruses such as MeMV and beet curly top virus (BCTV) (Ali et al., 2015). All these reports suggest the intergenic region is a conserved and best possible target for sgRNA design to provide DNA virus resistance. Duplexed sgRNA approach targeting CP and movement protein (MP) also increased resistance against banana streak virus (BSV) and wheat dwarf virus (WDV), respectively (Kis et al., 2019; Tripathi et al., 2019). Similar approaches could be utilized for controlling the cotton leaf curl virus (CLCuV) (Uniyal et al., 2019). These successful examples of CRISPR/Cas mediated editing potentially be adopted against devastating banana pathogen banana bunchy top virus (BBTV) (Rao et al., 2017). Transgenic plants in Arabidopsis expressing sgRNA targeting CP region with Cas9 protein conferred resistance against cauliflower mosaic virus (CaMV) (Liu et al., 2018).

16.3.3.2 RNA VIRUSES

The Discovery of RNA targeting effector proteins paved the path for engineering resistance against RNA viruses. Examples of such effectors were Cas13, discovered from *Leptotrichia shahii* (LshCas13a), and Cas9, found from *Francisella novicida* (FnCas9) (Shmakov et al., 2015; Price et al., 2015). LshCas13a was the first discovered Cas effector protein programmable for RNA targeting (Abudayyeh et al., 2016). With the combination of protospacers, this effector can be deployed for mRNA targeting, as shown in bacteria (Abudayyeh et al., 2017). These effectors proved to be a new tool targeting RNA viruses. Application of this type of virus genome engineering was first shown for mammalian cells against hepatitis C virus (HCV), where FnCas9 effector protein along with RNA-targeting guide RNA (rgRNA) reduced the viral titer by inhibiting protein production (Price et al., 2015). Further, FnCas9 was codon-optimized for plant uses with designed sgRNA against tobacco mosaic virus (TMV) in Arabidopsis and cucumber mosaic virus (CMV) in *N. benthamiana* transgenic plants that provided viral resistance (Zhang et al., 2018). Similarly, LshCas13a was reprogrammed to target ssRNA plant viruses against turnip mosaic virus (TuMV) in transgenic Arabidopsis and *N. benthamiana* plants (Aman et al., 2018a, b). Also, utilization of this system was done for enhanced resistance against rice stripe mosaic virus (RSMV), TMV, and Southern rice black-streaked dwarf virus (SRBSDV) in transgenic rice and tobacco plants (Zhang et al., 2019a). In potato virus Y (PVY), different viral essential genes, such as P3, CP, and Nib were targeted by the Cas13 effector to inhibit the viral infection cycle (Zhan et al., 2019). Newer Cas13 were identified with prokaryotic and eukaryotic nucleotide-binding domains (HEPN) from *Ruminococcus flavefaciens* and named Cas13d (CasRx). This module of CRISPR/Cas13 is responsible for CRISPR RNA maturation and cleavage with RNAse activity (Konermann et al., 2018). This variant of Cas13 was better than other variants viz. Cas13a, Cas13b, etc., as shown by targeting TuMV CP, HC-Pro, and GFP genomic region (Mahas et al., 2019). Recently this variant was used to target SARS-CoV-2 (novel coronavirus) in mammalian cells (Abbott et al., 2020).

16.3.4 VIRUS RESISTANCE BY TARGETING HOST FACTORS

The most crucial host gene against the virus is the resistance gene; however, they are often dominant and linked with undesired traits that lead to a trade-off between crop resistance and yield and quality-related traits. Similarly, host susceptibility factors are essential for plant growth and development. Such as *ssi2* mutant of Arabidopsis accumulates defense hormone salicylic acid (SA), which provides resistance against CMV but are defective in growth and development (Sekine et al., 2004). Therefore, scientists were prompted to discover novel host factors that can be used for plant viral resistance without affecting its growth or development-related traits. In this section, we will discuss such host factors discovered by the scientists against the Geminiviridae and Potyviridae family of viruses for the antiviral breeding of crops.

16.3.4.1 HOST FACTORS NEGATIVELY REGULATING PLANT DEFENSE

Some plant genes negatively regulate the plant defense signaling pathways. For example, rgs-CaM represses the RDR6 transcription and promotes degradation of SGS3 protein by autophagy pathway, which is an essential component of RNA silencing machinery, thus inhibits RNA silencing (Li et al., 2018). Similarly, a mutation in homeodomain leucine zipper protein 1 (HAT1), HAT2, and HAT3 promoted the accumulation of jasmonic acid (JA) and SA, thus providing defense against CMV infection (Zou et al., 2016).

16.3.4.2 HOST FACTORS INVOLVED IN THE VIRUS LIFE CYCLE

The virus genome is minimal, encodes fewer proteins, thus hijacking host proteins to complete life cycle-related processes such as replication, translation, and movement. For example, replication of tobamoviruses requires tobamovirus multiplication 1 (TOM1) and related homologs from the host (Yamanaka et al., 2000). Similarly, eukaryotic translation factors (eEF1A and eEF4) were hijacked in the case of potyvirus translation, whereas plasma-membrane-associated cation binding protein-1 (PCaP1) and vesical transfer protein Sec24a were essential for potyvirus movement (Wang, 2015).

16.3.4.3 HOST-FACTORS MODIFYING VIRAL PROTEINS

Virus proteins sometimes require proteins for post-translation modification to interact and infect plant cells effectively. For example, shaggy-related kinases from host phosphorylate the geminivirus C4 protein, thus causing the disease (Lozano-Durán et al., 2011). Similarly, the CK2 protein-mediated phosphorylation of CP protein of potato virus A (PVA) is required for its interaction with CP-interacting protein (CPIP) and HSP90 to complete replication and increased CP deposition (Lõhmus et al., 2017).

16.3.4.4 HOST FACTORS RELATED TO OTHER CELLULAR PROCESSES

Plant cellular processes are often crucial for plant metabolism; however, some components are beneficial for virus behaviors. For example, BRN2, Bearskin2B are important secondary cell wall synthesis factors, whereas phenylpropanoid metabolism factors (4-Coumarate, CoA ligase1, and 4CL1) helps geminivirus infection through phloem tissue (Lozano-Durán et al., 2011).

Identification of these host factors will help generate genetic resources by using the CRISPR/Cas system for virus resistance. The use of viral vectors could enhance the efficiency of delivering sgRNA and Cas protein because of their higher expression during the infection with the virus. Considering this, two virus-induced vectors for gene editing were developed with trans-activated pV86 and pC86 promoters of BSCTV to reduce OTEs (Ji et al., 2018). Their application in transient and stable transformation in *N. benthamiana* and Arabidopsis plants could inhibit the BSCTV infection without any off-targets. Further, a single-step transgene-free delivery system for CRISPR/Cas from Sonchus yellow net virus (SYNV) was also developed to save time and money for virus resistance programs (Ma et al., 2020). Also, this system is from a negative-stranded RNA virus that could not integrate into the host genome. We need more viral vector systems for optimized delivery and targeting of the host genome precisely for different monocot and dicot plants that can be engineered for virus resistance.

16.4 RISK ASSOCIATED WITH CRISPR/CAS STRATEGY

CRISPR/Cas system emerged as a potential and attractive tool for genome editing to control plant viruses. However, every new technique has some considerations and challenges for its use. In the case of host editing, recombination, and mutation of an endogenous gene could inactivate their function, while viral genome mutation leads to their possible evolution. It was demonstrated that genetic recombination, mutation, and reassortment of viral genes could lead to the evolution of DNA and RNA viruses (Mehta et al., 2019), as shown in the African cassava mosaic virus. Similarly, viral escape events after genome editing in viral ORFs raised the concern of more viral variants using more sgRNAs targeting them (Ali et al., 2016; Liu et al., 2018). Though, the combination of sgRNA delays the resistance breakdown compared to single sgRNA targeting (Liu et al., 2018). Similarly, related

OTEs should also be considered for potential risk in future application of CRISPR/Cas system along with siRNA generation against dsRNA domains present in the sgRNA (Zhang et al., 2018b).

16.5 FUTURE PERSPECTIVES

Plant viruses emerged as a threat to global food security. Their control is constantly challenged by continuous evolution and mutation in their genome. Therefore, there is a dire need to devise strategies to control them. Molecular breeding of resistance genes in popular cultivars was the method of choice for virus control in the last century. However, these kinds of resistance break down with the virus evolution. The advent of CRISPR/Cas as a genome-editing tool expedited genetic resource generation for virus resistance. This evolving technique has precision for editing with various Cas effectors specifically targeting DNA and RNA viruses. Similarly, exploration of recessive resistance genes and host susceptibility factors in plants with understanding genome editing tools offers a repertoire of genetic resources that can be utilized for viral resistance breeding. Compared to editing the viral genome, host susceptibility factors targeting provides a new avenue for engineering resistant plants against viruses without the risk of virus evolution and escape. Also, virus dissemination could be prevented through these host factors targeting insect vectors. A newly developed virus-derived vector could enhance the efficiency of sgRNA and Cas effector protein delivery for efficient editing of the host genome. Moreover, multiplexing of different sgRNAs with Cas proteins could build the durable resistance against the plant viruses offered through multiplex CRISPR platforms. This technology is in the initial years of its application for plant virus resistance. It needs further consideration for their experiment complexity, off-targets, virus escape and evolution, and public acceptance for practical field-level application.

KEYWORDS

- **CRISPR-Cas9**
- **disease resistance**
- **plant resistance**
- **plant stress**
- **plant viruses**
- **virus genome editing**

REFERENCES

Abbott, T., Dhamdhere, G., Liu, Y., Lin, X., Goudy, L., Zeng, L., et al., (2020). Development of CRISPR as an antiviral strategy to combat SARS-CoV-2 and influenza. *Cell, 181*, 865–876. doi: 10.1016/j.cell.2020.04.020.

Abudayyeh, O., Gootenberg, J., Essletzbichler, P., Han, S., Joung, J., Belanto, J., et al., (2017). RNA targeting with CRISPR/Cas13. *Nature, 550*, 280–284.

Abudayyeh, O., Gootenberg, J., Konermann, S., Joung, J., Slaymaker, I., Cox, D., et al., (2016). C2c2 is a single-component programmable RNA-guided RNA-targeting CRISPR effector. *Science, 353*, aaf5573. doi: 10.1126/science.aaf 5573.

Ali, Z., Abulfaraj, A., Idris, A., Ali, S., Tashkandi, M., & Mahfouz, M., (2015). CRISPR/Cas9-mediated viral interference in plants. *Genome Biol., 16*, 238.

Ali, Z., Ali, S., Tashkandi, M., Zaidi, S. S., & Mahfouz, M., (2016). CRISPR/Cas9-mediated immunity to geminiviruses: Differential interference and evasion. *Sci. Rep., 6*, 26912.

Ali, Z., Mahas, A., & Mahfouz, M., (2018). CRISPR/Cas13 as a tool for RNA interference. *Trends Plant Sci., 23*, 374–378.

Aman, R., Ali, Z., Butt, H., Mahas, A., Aljedaani, F., Khan, M. Z., et al., (2018a). RNA virus interference via CRISPR/Cas13a system in plants. *Genome Biol., 19*, 1.

Aman, R., Mahas, A., Butt, H., Aljedaani, F., & Mahfouz, M., (2018b). Engineering RNA virus interference via the CRISPR/Cas13 machinery in *Arabidopsis*. *Viruses, 10*, 732. doi: 10.3390/v10120732.

Anzalone, A. V., Randolph, P. B., Davis, J. R., Sousa, A. A., Koblan, L. W., Levy, J. M., Chen, P. J., et al., (2019). Search-and-replace genome editing without double-strand breaks or donor DNA. *Nature, 576*, 149–157.

Baek, K., Kim, D. H., Jeong, J., Sim, S. J., Melis, A., Kim, J. S., Jin, E., & Bae, S., (2016). DNA-free two-gene knockout in *Chlamydomonas reinhardtii* via CRISPR/Cas9 ribonucleoproteins. *Sci. Rep., 6*, 30620.

Baltes, N. J., Gil-Humanes, J., Cermak, T., Atkins, P. A., & Voytas, D. F., (2014). DNA replicons for plant genome engineering. *Plant Cell, 26*, 151–163.

Baltes, N., Hummel, A., Konecna, E., Cegan, R., Bruns, A., Bisaro, D., et al., (2015). Conferring resistance to geminiviruses with the CRISPR/Cas prokaryotic immune system. *Nat. Plants, 1*, 15145.

Boch, J., Bonas, U., & Lahaye, T., (2014). TAL effectors–pathogen strategies and plant resistance engineering. *New Phytol., 204*, 823–832.

Braatz, J., Harloff, H. J., Mascher, M., Stein, N., Himmelbach, A., & Jung, C., (2017). CRISPRCas9 targeted mutagenesis leads to simultaneous modification of different homoeologous gene copies in polyploid oilseed rape (*Brassica napus*). *Plant Physiol., 174*, 935–942.

Cermak, T., Baltes, N. J., Cegan, R., Zhang, Y., & Voytas, D. F., (2015). High-frequency, precise modification of the tomato genome. *Genome Biol., 16*, 232.

Chandrasekaran, J., Brumin, M., Wolf, D., Leibman, D., Klap, C., Pearlsman, M., Sherman, A., et al., (2016). Development of broad virus resistance in non-transgenic cucumber using CRISPR/Cas9 technology. *Mol. Plant Pathol., 17*, 1140–1153.

Chen, J. S., Dagdas, Y. S., Kleinstiver, B. P., Welch, M. M., Sousa, A. A., Harrington, L. B., Sternberg, S. H., et al., (2017). Enhanced proofreading governs CRISPRCas9 targeting accuracy. *Nature, 550*, 407–410.

Chen, K., Wang, Y., Zhang, R., Zhang, H., & Gao, C., (2019). CRISPR/Cas genome editing and precision plant breeding in agriculture. *Annu. Rev. Plant Biol., 70*, 667–697. doi: 10.1146/annurev-arplant-050718-100049.

Chen, W., Qian, Y., Wu, X., Sun, Y., Wu, X., & Cheng, X., (2014). Inhibiting replication of *Begomoviruses* using artificial zinc finger nucleases that target viral-conserved nucleotide motif. *Virus Genes., 48*, 494–501. doi: 10.1007/s11262-014-1041-4.

Cheng, X., Li, F., Cai, J., Chen, W., Zhao, N., Sun, Y., et al., (2015). Artificial TALE as a convenient protein platform for engineering broad-spectrum resistance to *Begomoviruses*. *Viruses, 7*, 4772–4782. doi: 10.3390/v7082843.

Dangl, J. L., Horvath, D. M., & Staskawicz, B. J., (2013). Pivoting the plant immune system from dissection to deployment. *Science, 341*, 746–751.

De Toledo, T. D. P., Brail, Q., Dahlbeck, D., & Staskawicz, B. J., (2016). *CRISPR/Cas9 Mediated Mutagenesis of a DMR6 Ortholog in Tomato Confers Broad-Spectrum Disease Resistance*. bioRxiv. https://doi.org/10.1101/064824.

Dong, O. X., & Ronald, P. C., (2019). Genetic engineering for disease resistance in plants: Recent progress and future perspectives. *Plant Physiol., 180*, 26–38.

Doudna, J. A., & Charpentier, E., (2014). Genome editing. The new frontier of genome engineering with CRISPR/Cas9. *Science, 346*, 1258096.

Garcia-Doval, C., & Jinek, M., (2017). Molecular architectures and mechanisms of class 2 CRISPR-associated nucleases. *Curr. Opin. Struct. Biol., 47*, 157–166. doi: 10.1016/j.sbi.2017.10.015.

Gepts, P., (2002). A comparison between crop domestication, classical plant breeding, and genetic engineering. *Crop Sci., 42*, 1780–1790.

Hadidi, A., Flores, R., Candresse, T., & Barba, M., (2016). Next-generation sequencing and genome editing in plant virology. *Front. Microbiol., 7*, 1325. doi: 10.3389/fmicb.2016.01325.

Hahn, F., & Nekrasov, V., (2019). CRISPR/Cas precision: Do we need to worry about off targeting in plants? *Plant Cell Rep., 38*, 437–441.

Holton, N., Nekrasov, V., Ronald, P. C., & Zipfel, C., (2015). The phylogenetically-related pattern recognition receptors EFR and XA21 recruit similar immune signaling components in monocots and dicots. *PLoS Pathog., 11*, e1004602.

Hsu, P., Scott, D., Weinstein, J., Ran, F., Konermann, S., Agarwala, V., et al., (2013). DNA targeting specificity of RNA-guided Cas9 nucleases. *Nat. Biotechnol., 31*, 827–832. doi: 10.1038/nbt.2647.

https://talk.ictvonline.org/taxonomy/

Hua, C., Zhao, J. H., & Guo, H. S., (2018). Trans kingdom RNA silencing in plant-fungal pathogen interactions. *Mol. Plant, 11*, 235–244.

Ji, X., Si, X., Zhang, Y., Zhang, H., Zhang, F., & Gao, C., (2018). Conferring DNA virus resistance with high specificity in plants using virus-inducible genome-editing system. *Genome Biol., 19*, 197.

Ji, X., Wang, D., & Gao, C., (2019). CRISPR editing-mediated antiviral immunity: A versatile source of resistance to combat plant virus infections. *Sci. China Life Sci., 62*, 1246–1249. doi: 10.1007/s11427-019-9722-2.

Ji, X., Zhang, H., Zhang, Y., Wang, Y., & Gao, C., (2015). Establishing a CRISPR/Cas-like immune system conferring DNA virus resistance in plants. *Nat. Plants, 1*, 15144.

Jia, H., Orbovic, V., Jones, J. B., & Wang, N., (2016). Modification of the PthA4 effector binding elements in type I CsLOB1 promoter using Cas9/ sgRNA to produce transgenic

Duncan grapefruit alleviating XccDeltapthA4:dCsLOB1.3 infection. *Plant Biotechnol. J., 14,* 1291–1301.

Jiang, F., & Doudna, J. A., (2017). CRISPR/Cas9 structures and mechanisms. *Annu. Rev. Biophys., 46,* 505–529.

Jiang, W., Zhou, H., Bi, H., Fromm, M., Yang, B., & Weeks, D. P., (2013). Demonstration of CRISPR/Cas9/sgRNA-mediated targeted gene modification in *Arabidopsis*, tobacco, sorghum and rice. *Nucleic Acids Res., 41*, e188.

Johansen, I. E., Liu, Y., Jørgensen, B., Bennett, E. P., Andreasson, E., Nielsen, K. L., Blennow, A., & Petersen, B. L., (2019). High efficacy full allelic CRISPR/Cas9 gene editing in tetraploid potato. *Sci. Rep., 9,* 17715.

Karasov, T. L., Chae, E., Herman, J. J., & Bergelson, J., (2017). Mechanisms to mitigate the trade-off between growth and defense. *Plant Cell, 29*, 666–680.

Khatodia, S., Bhatotia, K., Passricha, N., Khurana, S. M. P., & Tuteja, N., (2016). The CRISPR/Cas genome-editing tool: Application in improvement of crops. *Front. Plant Sci., 7*, 506.

Kim, J., & Kim, J. S., (2016). Bypassing GMO regulations with CRISPR gene editing. *Nat. Biotechnol., 34*, 1014–1015.

Kis, A., Hamar, É., Tholt, G., Bán, R., & Havelda, Z., (2019). Creating highly efficient resistance against wheat dwarf virus in barley by employing CRISPR/Cas9 system. *Plant Biotechnol. J., 17*, 1004–1006. doi: 10.1111/pbi.13077.

Kleinstiver, B. P., Pattanayak, V., Prew, M. S., Tsai, S. Q., Nguyen, N. T., Zheng, Z., & Joung, J. K., (2016). High-fidelity CRISPR/Cas9 nucleases with no detectable genome-wide off-target effects. *Nature, 529*, 490–495.

Komor, A. C., Kim, Y. B., Packer, M. S., Zuris, J. A., & Liu, D. R., (2016). Programmable editing of a target base in genomic DNA without double stranded DNA cleavage. *Nature, 533*, 420–424.

Konermann, S., Lotfy, P., Brideau, N., Oki, J., Shokhirev, M., & Hsu, P., (2018). Transcriptome engineering with RNA-targeting type VI-D CRISPR effectors. *Cell, 173*, 665–676.e14.

Kourelis, J., & Van, D. H. R. A. L., (2018). Defended to the nines: 25 Years of resistance gene cloning identifies nine mechanisms for R protein function. *Plant Cell, 30*, 285–299.

Kumlehn, J., Pietralla, J., Hensel, G., Pacher, M., & Puchta, H., (2018). The CRISPR/Cas revolution continues: From efficient gene editing for crop breeding to plant synthetic biology. *J. Integr. Plant Biol., 60*, 1127–1153.

Lacombe, S., Rougon-Cardoso, A., Sherwood, E., Peeters, N., Dahlbeck, D., Van, E. H. P., Smoker, M., Rallapalli, G., Thomma, B. P., Staskawicz, B., et al., (2010). Interfamily transfer of a plant pattern recognition receptor confers broad-spectrum bacterial resistance. *Nat. Biotechnol., 28*, 365–369.

Langner, T., Kamoun, S., & Belhaj, K., (2018). CRISPR crops: Plant genome editing toward disease resistance. *Annu. Rev. Phytopathol., 56*, 479–512. doi: 10.1146/annurev-phyto-080417-050158.

Leenay, R. T., & Beisel, C. L., (2017). Deciphering, communicating, and engineering the CRISPR PAM. *J. Mol. Biol., 429,* 177–191.

Li, F., & Wang, A., (2019). RNA-targeted antiviral immunity: More than just RNA silencing. *Trends Microbiol., 27,* 792–805. doi: 10.1016/j.tim.2019.05.007.

Li, F., Liu, W., & Zhou, X., (2019). Pivoting plant immunity from theory to the field. *Sci. China Life Sci., 62*, 1539–1542. doi: 10.1007/s11427-019-1565-1.

Li, F., Yang, X., Bisaro, D., & Zhou, X., (2018). The bC1 protein of geminivirus betasatellite complexes: A target and repressor of host defenses. *Mol. Plant., 11,* 1424–1426. doi: 10.1016/j.molp.2018.10.007.

Li, J. F., Norville, J. E., Aach, J., McCormack, M., Zhang, D., Bush, J., Church, G. M., & Sheen, J., (2013). Multiplex and homologous recombination-mediated genome editing in *Arabidopsis* and *Nicotiana benthamiana* using guide RNA and Cas9. *Nat. Biotechnol., 31*, 688–691.

Li, J., Sun, Y., Du, J., Zhao, Y., & Xia, L., (2017). Generation of targeted point mutations in rice by a modified CRISPR/Cas9 system. *Mol. Plant, 10*, 526–529.

Li, L., Kim, P., Yu, L., Cai, G., Chen, S., Alfano, J. R., & Zhou, J. M., (2016). Activation-dependent destruction of a co-receptor by a *Pseudomonas syringae* effector dampens plant immunity. *Cell Host Microbe., 20*, 504–514.

Li, T., Liu, B., Spalding, M. H., Weeks, D. P., & Yang, B., (2012). High-efficiency TALEN-based gene editing produces disease-resistant rice. *Nat. Biotechnol., 30*, 390–392.

Liang, Z., Chen, K., Li, T., Zhang, Y., Wang, Y., Zhao, Q., Liu, J., Zhang, H., Liu, C., Ran, Y., et al., (2017). Efficient DNA-free genome editing of bread wheat using CRISPR/Cas9 ribonucleoprotein complexes. *Nat. Commun., 8*, 14261.

Lin, Q., Zong, Y., Xue, C., Wang, S., Jin, S., Zhu, Z., Wang, Y., Anzalone, A. V., Raguram, A., Doman, J. L., et al., (2020). Prime genome editing in rice and wheat. *Nat. Biotechnol., 38*, 582–585.

Liu, H., Soyars, C., Li, J., Fei, Q., He, G., Peterson, B., et al., (2018). CRISPR/Cas9- mediated resistance to cauliflower mosaic virus. *Plant Direct., 2*, e00047. doi: 10.1002/pld3.47.

Lõhmus, A., Hafrén, A., & Mäkinen, K., (2017). Coat protein regulation by CK2, CPIP, HSP70, and CHIP is required for potato virus a replication and coat protein accumulation. *J. Virol., 91*, e01316.

Lozano-Durán, R., Rosas-Díaz, T., Luna, A. P., & Bejarano, E. R., (2011). Identification of host genes involved in geminivirus infection using a reverse genetics approach. *PLoS One, 6*, e22383. doi: 10.1371/journal.pone.0022383.

Ma, X., Zhang, X., Liu, H., & Li, Z., (2020). Highly efficient DNA-free plant genome editing using virally delivered CRISPR/Cas9. *Nat. Plants, 6*, 773–779. doi: 10.1038/s41477-020-0704-5.

Ma, X., Zhu, Q., Chen, Y., & Liu, Y., (2016). CRISPR/Cas9 platforms for genome editing in plants: Developments and applications. *Mol. Plant, 9*, 961–974. doi: 10.1016/j.molp.2016.04.009.

Mahas, A., & Mahfouz, M., (2018). Engineering virus resistance via CRISPR/Cas systems. *Curr. Opin. Virol., 32*, 1–8. doi: 10.1016/j.coviro.2018.06.002.

Mahas, A., Aman, R., & Mahfouz, M., (2019). CRISPR/Cas13d mediates robust RNA virus interference in plants. *Genome Biol., 20*, 263.

Makarova, K. S., Zhang, F., & Koonin, E. V., (2017a). SnapShot: Class 1 CRISPRCas systems. *Cell, 168*, 946–946.e1.

Makarova, K. S., Zhang, F., & Koonin, E. V., (2017b). SnapShot: Class 2 CRISPRCas systems. *Cell, 168*, 328–328.e1. http://dx.doi.org/10.1016/j.cell.2017.02.018.

Makarova, K., Wolf, Y., Alkhnbashi, O., Costa, F., Shah, S., Saunders, S., et al., (2015). An updated evolutionary classification of CRISPR/Cas systems. *Nat. Rev. Microbiol., 13*, 722–736.

Malnoy, M., Viola, R., Jung, M. H., Koo, O. J., Kim, S., Kim, J. S., Velasco, R., & Nagamangala, K. C., (2016). DNA-free genetically edited grapevine and apple protoplast using CRISPR/Cas9 ribonucleoproteins. *Front. Plant Sci., 7*, 1904.

Marzec, M., & Hensel, G., (2020). Prime editing: Game changer for modifying plant genomes. *Trends Plant Sci., 25,* 722–724.

Mehta, D., Sturchler, A., Anjanappa, R. B., Zaidi, S. S., Hirsch-Hoffmann, M., Gruissem, W., & Vanderschuren, H., (2019). Linking CRISPR/Cas9 interference in cassava to the evolution of editing-resistant geminiviruses. *Genome Biol., 20*, 80.

Metje-Sprink, J., Menz, J., Modrzejewski, D., & Sprink, T., (2019). DNA-free genome editing: Past, present and future. *Front. Plant Sci., 9*, 1957.

Molla, K. A., Shih, J., & Yang, Y., (2020). Single nucleotide editing for zebra3 and wsl5 phenotypes in rice using CRISPR/Cas9-mediated adenine base editors. *aBiotech, 1*, 106–118.

Mushtaq, M., Mukhtar, S., Sakina, A., Dar, A. A., Bhat, R., Deshmukh, R., Molla, K., Kundoo, A. A., & Dar, M. S., (2020). Tweaking genome-editing approaches for virus interference in crop plants. *Plant Physiol. Biochem., 147*, 242–250.

Nadakuduti, S. S., Buell, C. R., Voyta, D. F., Starker, C. G., & Douches, D. S., (2018). Genome editing for crop improvement - applications in clonally propagated polyploids with a focus on potato (*Solanum tuberosum* L.). *Front. Plant Sci., 9*, 1607.

Nekrasov, V., Wang, C., Win, J., Lanz, C., Weigel, D., & Kamoun, S., (2017). Rapid generation of a transgene-free powdery mildew resistant tomato by genome deletion. *Sci. Rep., 7*, 482.

Nicaise, V., (2014). Crop immunity against viruses: Outcomes and future challenges. *Front. Plant Sci., 5,* 660. doi: 10.3389/fpls.2014.00660.

Oliva, R., Ji, C., Atienza-Grande, G., HuguetTapia, J. C., Perez-Quintero, A., Li, T., Eom, J. S., Li, C., Nguyen, H., Liu, B., et al., (2019). Broad spectrum resistance to bacterial blight in rice using genome editing. *Nat. Biotechnol., 37*, 1344–1350.

Pandey, S., Muthamilarasan, M., Sharma, N., Chaudhry, V., Dulani, P., Shweta, S., et al., (2019). Characterization of DEAD-box family of RNA helicases in tomato provides insights into their roles in biotic and abiotic stresses. *Environ. Exp. Bot., 158*, 107–116.

Pandey, S., Prasad, A., Sharma, N., & Prasad, M., (2020a). Linking the plant stress responses with RNA helicases. *Plant Sci., 299*, 110607. https://doi.org/10.1016/j.plantsci.2020.110607.

Pandey, S., Sahu, P., Kulshreshtha, R., & Prasad, M., (2018). Role of host transcription factors in modulating defense response during plant virus interaction. In: Patil, B., (ed.), *Genes, Genetics and Transgenics for Virus Resistance in Plants* (pp. 25–54). Caister Academic Press, UK.

Pandey, S., Sharma, N., & Prasad, M., (2020b). Role of RNA-interacting proteins in modulating plant-microbe interactions. *Adv. Genet., 105*, 67–94 https://doi.org/10.1016/bs.adgen.2019.12.001.

Pandolfi, V., Neto, J. R. C. F., Da Silva, M. D., Amorim, L. L. B., Wanderley-Nogueira, A. C., De Oliveira, S. R. L., Kido, E. A., et al., (2017). Resistance (R) genes: Applications and prospects for plant biotechnology and breeding. *Curr. Protein Pept. Sci., 18*, 323–334.

Price, A., Sampson, T., Ratner, H., Grakoui, A., & Weiss, D., (2015). Cas9- mediated targeting of viral RNA in eukaryotic cells. *Proc. Natl. Acad. Sci. U. S. A., 112*, 6164–6169. doi: 10.1073/pnas.1422340112.

Qin, L., Li, J., Wang, Q., Xu, Z., Sun, L., Alariqi, M., Manghwar, H., Wang, G., Li, B., Ding, X., et al., (2020). High-efficient and precise base editing of C$G to T$A in the allotetraploid cotton (*Gossypium hirsutum*) genome using a modified CRISPR/Cas9 system. *Plant Biotechnol. J., 18*, 45–56.

Rao, X., Wu, Z., Zhou, L., Sun, J., & Li, H., (2017). Genetic diversity of banana bunchy top virus isolates from China. *Acta. Virol., 61*, 217–222. doi: 10.4149/av_2017_02_13.

Rivas, S., & Genin, S., (2011). A plethora of virulence strategies hidden behind nuclear targeting of microbial effectors. *Front. Plant Sci., 2*, 104.

Rodriguez, E., El Ghoul, H., Mundy, J., & Petersen, M., (2016). Making sense of plant autoimmunity and' negative regulators. *FEBS J., 283*, 1385–1391.

Rybicki, E. P., (2019). CRISPR/Cas9 strikes out in cassava. *Nat. Biotechnol., 37*, 727–728.

Sastry, S. K., & Zitter, T. A., (2014). Management of virus and viroid diseases of crops in the tropics. In: *Plant Virus and Viroid Diseases in the Tropics: Epidemiology and Management* (Vol. 2, pp. 149–480). (Dordrecht: Springer). doi:10.1007/978-94-007-7820-7_2.

Schiml, S., Fauser, F., & Puchta, H., (2016). Repair of adjacent single-strand breaks is often accompanied by the formation of tandem sequence duplications in plant genomes. *Proc. Natl. Acad. Sci. U S A., 113*, 7266–7271.

Schoonbeek, H. J., Wang, H. H., Stefanato, F. L., Craze, M., Bowden, S., Wallington, E., Zipfel, C., & Ridout, C. J., (2015). Arabidopsis EF-Tu receptor enhances bacterial disease resistance in transgenic wheat. *New Phytol., 206*, 606–613.

Sekine, K., Nandi, A., Ishihara, T., Hase, S., Ikegami, M., Shah, J., et al., (2004). Enhanced resistance to cucumber mosaic virus in the *Arabidopsis thaliana* ssi2 mutant is mediated via an SA-independent mechanism. *Mol. Plant Microbe Interact., 17*, 623–632. doi: 10.1094/mpmi.2004.17.6.623.

Shen, D., Suhrkamp, I., Wang, Y., Liu, S., Menkhaus, J., Verreet, J. A., Fan, L., & Cai, D., (2014). Identification and characterization of microRNAs in oilseed rape (*Brassica napus*) responsive to infection with the pathogenic fungus *Verticillium longisporum* using brassica AA (*Brassica rapa*) and CC (*Brassica oleracea*) as reference genomes. *New Phytol., 204*, 577–594.

Shimatani, Z., Kashojiya, S., Takayama, M., Terada, R., Arazoe, T., Ishii, H., Teramura, H., Yamamoto, T., Komatsu, H., Miura, K., et al., (2017). Targeted base editing in rice and tomato using a CRISPR/Cas9 cytidine deaminase fusion. *Nat. Biotechnol., 35*, 441–443.

Shmakov, S., Abudayyeh, O., Makarova, K., Wolf, Y., Gootenberg, J., Semenova, E., et al., (2015). Discovery and functional characterization of diverse class 2 CRISPR/Cas systems. *Mol. Cell, 60,* 385–397. doi: 10.1016/j.molcel.2015.10.008.

Shmakov, S., Smargon, A., Scott, D., Cox, D., Pyzocha, N., Yan, W., et al., (2017). Diversity and evolution of class 2 CRISPR/Cas systems. *Nat. Rev. Microbiol., 15*, 169–182.

Smith, I., Greenside, P. G., Natoli, T., Lahr, D. L., Wadden, D., Tirosh, I., Narayan, R., et al., (2017). Evaluation of RNAi and CRISPR technologies by large-scale gene expression profiling in the connectivity map. *PLoS Biol., 15*, e2003213.

Sood, P., Pandey, S., & Prasad, M., (2021). Biotechnological advances and advanced mutation breeding techniques for crop plants. In: Kumar, P., & Kumar, T. A., (eds.), *Crop Improvement Biotechnological Advances* (pp. 139–165). CRC Press doi.org/10.1201/9781003099079.

Sprink, T., Eriksson, D., Schiemann, J., & Hartung, F., (2016). Regulatory hurdles for genome editing: Process- vs. product-based approaches in different regulatory contexts. *Plant Cell Rep., 35*, 1493–1506.

Stirnweis, D., Milani, S. D., Jordan, T., Keller, B., & Brunner, S., (2014). Substitutions of two amino acids in the nucleotide-binding site domain of a resistance protein enhance the hypersensitive response and enlarge the PM3F resistance spectrum in wheat. *Mol. Plant Microbe Interact., 27*, 265–276.

Subburaj, S., Chung, S. J., Lee, C., Ryu, S. M., Kim, D. H., Kim, J. S., Bae, S., & Lee, G. J., (2016). Site-directed mutagenesis in *Petunia x hybrid* protoplast system using direct delivery of purified recombinant Cas9 ribonucleoproteins. *Plant Cell Rep., 35*, 1535–1544.

Sun, K., Wolters, A. M., Vossen, J. H., Rouwet, M. E., Loonen, A. E., Jacobsen, E., Visser, R. G., & Bai, Y., (2016). Silencing of six susceptibility genes results in potato late blight resistance. *Transgenic Res., 25*, 731–742.

Svitashev, S., Schwartz, C., Lenderts, B., Young, J. K., & Mark, C. A., (2016). Genome editing in maize directed by CRISPR/Cas9 ribonucleoprotein complexes. *Nat. Commun., 7*, 13274.

Tashkandi, M., Ali, Z., Aljedaani, F., Shami, A., & Mahfouz, M. M., (2018). Engineering resistance against tomato yellow leaf curl virus via the CRISPR/Cas9 system in tomato. *Plant Signal. Behav., 13,* e1525996. doi: 10.1080/15592324.2018.1525996.

Thompson, D. B., Aboulhouda, S., Hysolli, E., Smith, C. J., Wang, S., Castanon, O., & Church, G. M., (2018). The future of multiplexed eukaryotic genome engineering. *ACS Chem. Biol., 13*, 313–325.

Tripathi, J. N., Lorenzen, J., Bahar, O., Ronald, P., & Tripathi, L., (2014). Transgenic expression of the rice Xa21 pattern-recognition receptor in banana (Musa sp.) confers resistance to *Xanthomonas campestris* pv. musacearum. *Plant Biotechnol. J., 12*, 663–673.

Tripathi, J., Ntui, V., Ron, M., Muiruri, S., Britt, A., & Tripathi, L., (2019). CRISPR/Cas9 editing of endogenous banana streak virus in the B genome of *Musa spp.* overcomes a major challenge in banana breeding. *Commun. Biol., 2*, 46.

Uniyal, A. P., Yadav, S. K., & Kumar, V., (2019). The CRISPR/Cas9, genome editing approach: A promising tool for drafting defense strategy against *Begomoviruses* including cotton leaf curl viruses. *J. Plant Biochem. B., 28*, 121–132. doi: 10.1007/s13562-019-00491-6.

Urnov, F. D., Ronald, P. C., & Carroll, D., (2018). A call for science-based review of the European court's decision on gene-edited crops. *Nat. Biotechnol., 36*, 800–802.

Vakulskas, C. A., Dever, D. P., Rettig, G. R., Turk, R., Jacobi, A. M., Collingwood, M. A., Bode, N. M., McNeill, M. S., Yan, S., Camarena, J., et al., (2018). A high-fidelity Cas9 mutant delivered as a ribonucleoprotein complex enables efficient gene editing in human hematopoietic stem and progenitor cells. *Nat. Med., 24*, 1216–1224.

Van, D. M., Huibers, R. P., Elberse, J., & Van, D. A. G., (2008). Arabidopsis DMR6 encodes a putative 2OG-Fe(II) oxygenase that is defense-associated but required for susceptibility to downy mildew. *Plant J., 54*, 785–793.

Waltz, E., (2018). With a free pass, CRISPR-edited plants reach market in record time. *Nat. Biotechnol., 36,* 6, 7.

Wang, A., (2015). Dissecting the molecular network of virus-plant interactions: The complex roles of host factors. *Annu. Rev. Phytopathol., 53*, 45–66. doi:10.1146/annurev-phyto-080614-120001.

Wang, F., Wang, C., Liu, P., Lei, C., Hao, W., Gao, Y., Liu, Y. G., & Zhao, K., (2016a). Enhanced rice blast resistance by CRISPR/Cas9-Targeted mutagenesis of the ERF transcription factor gene OsERF922. *PLoS One, 11*, e0154027.

Wang, L., Zhao, L., Zhang, X., Zhang, Q., Jia, Y., Wang, G., Li, S., Tian, D., Li, W. H., & Yang, S., (2019). Large-scale identification and functional analysis of NLR genes in blast resistance in the Tetep rice genome sequence. *Proc. Natl. Acad. Sci. U S A, 116*, 18479–18487.

Wang, M., Weiberg, A., Dellota, E. Jr., Yamane, D., & Jin, H., (2017). Botrytis small RNA Bc-siR37 suppresses plant defense genes by cross-kingdom RNAi. *RNA Biol., 14,* 421–428.

Wang, M., Weiberg, A., Lin, F. M., Thomma, B. P., Huang, H. D., & Jin, H., (2016b). Bidirectional cross-kingdom RNAi and fungal uptake of external RNAs confer plant protection. *Nat. Plants, 2,* 16151.

Wang, P., Zhang, J., Sun, L., Ma, Y., Xu, J., Liang, S., Deng, J., Tan, J., Zhang, Q., Tu, L., et al., (2018). High efficient multisites genome editing in allotetraploid cotton (*Gossypium hirsutum*) using CRISPR/Cas9 system. *Plant Biotechnol. J., 16,* 137–150.

Wang, Y., Cheng, X., Shan, Q., Zhang, Y., Liu, J., Gao, C., & Qiu, J. L., (2014). Simultaneous editing of three homoeoalleles in hexaploid bread wheat confers heritable resistance to powdery mildew. *Nat. Biotechnol., 32*, 947–951.

Weiberg, A., Wang, M., Lin, F. M., Zhao, H., Zhang, Z., Kaloshian, I., Huang, H. D., & Jin, H., (2013). Fungal small RNAs suppress plant immunity by hijacking host RNA interference pathways. *Science, 342*, 118–123.

Wilson, F. M., Harrison, K., Armitage, A. D., Simkin, A. J., & Harrison, R. J., (2019). CRISPR/Cas9- mediated mutagenesis of phytoene desaturase in diploid and octoploid strawberry. *Plant Methods, 15*, 45.

Wolter, F., & Puchta, H., (2018). The CRISPR/Cas revolution reaches the RNA world: Cas13, a new Swiss army knife for plant biologists. *Plant J., 94*, 767–775.

Woo, J. W., Kim, J., Kwon, S. I., Corvalan, C., Cho, S. W., Kim, H., Kim, S. G., et al., (2015). DNA-free genome editing in plants with preassembled CRISPR/Cas9 ribonucleoproteins. *Nat. Biotechnol., 33*, 1162–1164.

Wu, C. H., Abd-El-Haliem, A., Bozkurt, T. O., Belhaj, K., Terauchi, R., Vossen, J. H., & Kamoun, S., (2017). NLR network mediates immunity to diverse plant pathogens. *Proc. Natl. Acad. Sci. U S A., 114*, 8113–8118.

Xu, Z., Xu, X., Gong, Q., Li, Z., Li, Y., Wang, S., Yang, Y., Ma, W., Liu, L., Zhu, B., et al., (2019). Engineering broad-spectrum bacterial blight resistance by simultaneously disrupting variable TALE-binding elements of multiple susceptibility genes in rice. *Mol. Plant, 12,* 1434–1446.

Yin, K., Han, T., Xie, K., Zhao, J., Song, J., & Liu, Y., (2019). Engineer complete resistance to cotton leaf curl multan virus by the CRISPR/Cas9 system in *Nicotiana benthamiana. Phytopathol. Res., 1,* 9.

Zaidi, S. S., Mukhtar, M. S., & Mansoor, S., (2018). Genome editing: Targeting susceptibility genes for plant disease resistance. *Trends Biotechnol., 36*, 898–906.

Zhan, X., Zhang, F., Zhong, Z., Chen, R., Wang, Y., Chang, L., et al., (2019). Generation of virus-resistant potato plants by RNA genome targeting. *Plant Biotechnol. J., 17*, 1814–1822. doi: 10.1111/pbi.13102.

Zhang, J., Li, W., Xiang, T., Liu, Z., Laluk, K., Ding, X., Zou, Y., Gao, M., Zhang, X., Chen, S., et al., (2010). Receptor-like cytoplasmic kinases integrate signaling from multiple plant immune receptors and are targeted by a *Pseudomonas syringae* effector. *Cell Host Microbe., 7,* 290–301.

Zhang, Q., Xing, H. L., Wang, Z. P., Zhang, H. Y., Yang, F., Wang, X. C., et al., (2018a). Potential high-frequency off-target mutagenesis induced by CRISPR/Cas9 in *Arabidopsis* and its prevention. *Plant Mol. Biol., 96*, 44–456.

Zhang, T., Zhao, Y. L., Zhao, J. H., Wang, S., Jin, Y., Chen, Z. Q., Fang, Y. Y., et al., (2016b). Cotton plants export microRNAs to inhibit virulence gene expression in a fungal pathogen. *Nat. Plants, 2*, 16153.

Zhang, T., Zhao, Y., Ye, J., Cao, X., Xu, C., Chen, B., et al., (2019a). Establishing CRISPR/Cas13a immune system conferring RNA virus resistance in both dicot and monocot plants. *Plant Biotechnol. J., 17*, 1185–1187. doi: 10.1111/pbi. 13095.

Zhang, T., Zheng, Q., Yi, X., An, H., Zhao, Y., Ma, S., & Zhou, G., (2018b). Establishing RNA virus resistance in plants by harnessing CRISPR immune system. *Plant Biotechnol. J., 16*, 1415–1423. doi: 10.1111/pbi.12881.

Zhang, W., Gao, S., Zhou, X., Chellappan, P., Chen, Z., Zhou, X., Zhang, X., et al., (2011). Bacteria-responsive microRNAs regulate plant innate immunity by modulating plant hormone networks. *Plant Mol. Biol., 75*, 93–105.

Zhang, Y., Liang, Z., Zong, Y., Wang, Y., Liu, J., Chen, K., Qiu, J. L., & Gao, C., (2016a). Efficient and transgene-free genome editing in wheat through transient expression of CRISPR/Cas9 DNA or RNA. *Nat. Commun., 7*, 12617.

Zhang, Y., Malzahn, A., Sretenovic, S., & Qi, Y., (2019b). The emerging and uncultivated potential of CRISPR technology in plant science. *Nat. Plants, 5*, 778–794. doi: 10.1038/s41477-019-0461-5.

Zhou, J., Peng, Z., Long, J., Sosso, D., Liu, B., Eom, J. S., Huang, S., Liu, S., Vera, C. C., Frommer, W. B., et al., (2015). Gene targeting by the TAL effector PthXo2 reveals cryptic resistance gene for bacterial blight of rice. *Plant J., 82*, 632–643.

ZKBS (German Biosafety Commission), (2012). *Position Statement of the ZKBS on New Plant Breeding Techniques*. http://www.bvl.bund.de/DE/06_Gentechnik/04_Fachmeldungen/2013/2013_08_01_Fa_neue_Techniken_Pflanzenzuechtung.html?nn=1471850 (accessed on 04 September 2022).

Zou, L., Deng, X., Han, X., Tan, W., Zhu, L., Xi, D., et al., (2016). Role of transcription factor HAT1 in modulating *Arabidopsis thaliana* response to cucumber mosaic virus. *Plant Cell Physiol., 57*, 1879–1889.

CHAPTER 17

Regulatory Status of CRISPR/Cas Genome-Edited Crops

HEERAH EIJAZ[1] and FAYAZ AHMAD DAR[2*]

[1]*Department of Bioresources, University of Kashmir, Srinagar – 190006, Jammu and Kashmir, India*

[2]*Department of Bioresources, Amar Singh College Campus, Cluster University, Srinagar – 190008, Jammu and Kashmir, India*

[*]*Corresponding author. E-mail: dfayaz5u@gmail.com*

ABSTRACT

Genome editing through CRISPR/Cas technology is transforming the way crop improvement is carried out, because of its efficiency over other gene-editing techniques. Besides, it also allows accurate and stable adjustments to the plant's gene function. CRISPR/Cas technology-based gene modifications offer several advantages due to its simplicity, high specificity, and ability to multiplex successfully thus providing better options for future crop breeding programs. Genome editing is a process that enables plants to produce specific genetic variants that are unrecognizable from the natural variants. This technology is widely used in agriculture and environmental science. It is generally believed that new varieties with targeted mutations generated by making use of genome editing systems should not be regulated under the regulations of genetically modified plants (GMP). However, regulations on the use of genome-edited crops in different countries across the globe vary significantly and are mainly based on either a specific process for creating novel crop variety or the composition of the final product created after genetic modification (GM). In this chapter, we will be discussing CRISPR

CRISPR/Cas-Mediated Genome Editing in Plants, Deepu Pandita, Anu Pandita (Eds.)

technology, its applications, regulatory status in different countries, and biosafety concerns arising from its use keeping in view the present scenario.

17.1 INTRODUCTION

In biological research, CRISPR is quickly becoming an important tool, because of its two major features. Primarily, CRISPR sequences are found in more than 40% of bacteria and 90% of archaea (Mojica et al., 2000). Second, CRISPR elements are located near CRISPR-associated genes, which are well-conserved regions (Jansen et al., 2002). CRISPR sequences along with the Cas9 enzyme system together form the basis of technology called CRISPR-Cas9 technology. Therefore, CRISPR/Cas-based system is a powerful genome-editing tool specifically meant for addressing the changes in the target sequence of the genome.

The full form of CRISPR is clustered regularly interspaced short palindromic repeats. The acronym Cas stands for CRISPR-associated system, and it was initially discovered in bacteria as part of their virus defense system. The Cas system consists of various CRISPR-associated proteins, for instance, Cas3, Cas6, Cas8, Cas9, and so on (Beus et al., 2020). These Cas proteins form the basis for the classification of the CRISPR/Cas system into three types, namely Type-1, Type-2, and Type-3. Type-1 system forms a complex which recognizes the foreign DNA and starts DNA degradation, by associating with multiple Cas proteins. Type-2 systems depend on the association of both crRNA and tracrRNA (a short RNA code that works in tandem with the repeat sequence) with a single multifunctional protein Cas9 and thus are used in genome editing. Whereas, Type-3 system also relies on the association with multiple Cas proteins, however, instead of attacking foreign DNA, they attack foreign RNA, allowing them to attack RNA-based viruses (Makarova and Koonin, 2015). Besides, endonucleases are essential components of the CRISPR/Cas9 system for Cas9 to cleave DNA, therefore being the most prominent and the most commonly used one, useful in describing the basic principle of the Cas9 system. CRISPR/Cas9 is a genome-editing tool that is based on two important components one is a DNA cutting enzyme (Cas9, a nuclease), and the second is a guiding piece of genetic material (guide RNA or gRNA) that determines its position in the genome.

This procedure involves cutting a DNA sequence and attaching a Cas9 enzyme to it. The aim is to insert or remove a specific strand of DNA (Purnhagen et al., 2018). It was discovered that the system works well in animals, humans, plants as well as in bacteria and fungi. In bacteria, for example, the CRISPR/Cas9 system confers acquired resistance to viral infection. In crop and farm animals, it is used to knock out the undesirable genes, and in medicines, this technique is used to knock out the gene that causes disease or correct naturally occurring mutations that result in disease (for instance, to knock out the cancer gene). Furthermore, the CRISPR/Cas9 system has been successfully used for a variety of crop species, including citrus fruits, sorghum, soybean, maize, rice, potato, tomato, poplar, and wheat (Liu et al., 2021). In the case of animal genetics, CRISPR/Cas9 has been used to reduce the chicken egg white protein allergenicity (Oishi et al., 2016).

17.2 APPLICATIONS OF CRISPR/CAS9 TECHNOLOGY

CRISPR/Cas9 technology has been applied in a variety of species of plants, animals, fungi, and mammals to manipulate genes in different ways like altering the nucleotide sequence, changing the gene expression, formation of new high-yielding cultivars, enhancement in abiotic and biotic stress tolerance and crops having high nutritional value (Figure 17.1).

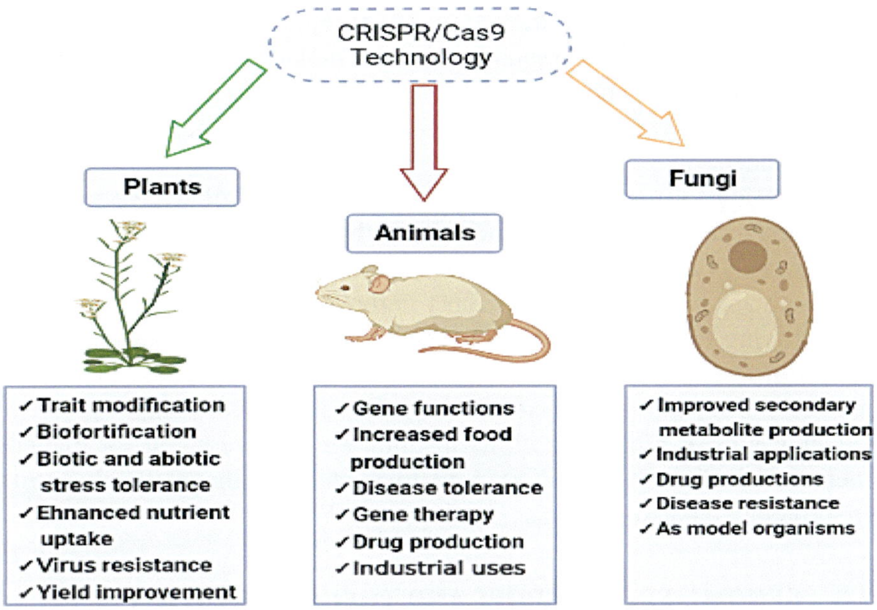

FIGURE 17.1 Some of the major applications of CRISPR/Cas9 technology in plants, animals, and fungi.

Biotech oil extracted from *Camelina sativa* seeds using CRISPR/Cas9 technology, which has improved fatty acid content, making it more favorable to human health, more resistant to oxidation, and more suitable for the manufacture of biofuels (Jiang et al., 2017). Similarly, in *Arabidopsis* CRISPR-induced mutation of the OST2 gene, a new allele that confers salt resistance in plants has been obtained (Osakabe et al., 2010). Besides, rice has been successfully modified to increase its resistance against *Magnaporthe oryzae*. The alteration in the ethylene dependent pathway was caused by targeting OsERF922 gene mutation using CRISPR technology (Liu et al., 2012; Wang et al., 2016a). The effectiveness of the CRISPR/Cas9 system in offering intervention against geminiviruses has been discovered. Different transformation techniques, such as a single-guide RNA/Cas9 (sgRNA/Cas9), have been employed to degrade/suppress the curly top virus genome in *N. benthamiana* (Wang et al., 2016a; Baltes et al., 2015). CRISPR/Cas9 technology has also been used in knocking in and knocking out of genes. For instance, knocking out the Eif4E gene that encodes the eukaryotic translation initiation factor essential for translation of viruses in *Cucumis sativus* has been achieved using CRISPR/Cas9 technology (Kamburova et al., 2017). Knocking out gene confers resistance to different types of viruses like papaya ringspot mosaic virus-W (PRSV-W), cucumber vein yellowing virus (CVYV) and zucchini yellow mosaic virus (ZYMV) (Fang and Tyler, 2016; Chandrasekaran et al., 2016). Similarly, CRISPR/Cas9 was utilized to knock out the gene encoding granule bound starch synthase (GBSS) in potato plants in just one cycle of transfection, resulting in the production of potato plants that synthesize amylopectin starch, a highly desired commercial feature (Andersson et al., 2017).

By altering the pathogenicity gene (Avr4/6) in an oomycete pathogen of Soybean called *Phytophthora sojae*, the CRISPR/Cas9 system has proven to be an effective approach for gene editing. In addition, it has been used to obtain plant resistance against herbicides (Svitashev et al., 2015). By modifying the ALS2 gene in maize, researchers were able to create a chlorsulfuron-resistant mutant corn plant (Kamburova et al., 2017). CRISPR/Cas9 technology has also been utilized to develop novel plants capable of producing human proteins such as insulin, which is required for diabetic patients, and albumin, which is used to treat hemorrhage, shock, burns, hypoproteinemia, and cirrhosis (Hastings and Wolf, 1992).

17.3 GENOME-EDITED CROPS AND THEIR REGULATION

GEEN (genome editing with engineered nucleases) has become a popular approach for studying gene function and creating genetic variants utilizing site-directed genomic modifications. GEEN delivers economic benefits when applied to commercially important plant species and thus becomes an efficient means of generating valuable novel phenotypes and traits (Wolt et al., 2016).

17.4 TECHNIQUES FOR GENOME EDITING

Oligonucleotide mediated mutagenesis (OMM) was the first method of genome editing in higher plants. It utilized chemically generated oligonucleotides with base substitution or addition mediated by endogenous DNA repair enzymes to cause site-specific gene targeting (Beetham et al., 1999). In contrast to the GEEN technique, OMM does not transport the nuclease to the site of action. Site-directed nucleases (SDNs), which were engineered to allow very selective targeting to any gene of interest, have been exploited in recent discoveries and breakthroughs in genome editing. Engineered mega nucleases (EMNs), also known as homing nucleases or LAGLIDADG endonucleases, transcriptional activators like effective nucleases (TALENs), zinc finger nucleases (ZFNs), and CRISPR in conjugation with Cas9 endonuclease (CRISPR/Cas9) have all been used for genome editing in higher plants (Gaj et al., 2013; Osakabe and Osakabe, 2015). Gene targeting using EMNs, ZFNs, and TALENS is achieved using protein-based DNA recognition domains, whereas CRISPR uses a single guide RNA (sgRNA) to instruct the Cas9 nuclease (Curtin et al., 2012; Jones, 2015).

17.5 GENOME ALTERATION OF ECONOMICALLY IMPORTANT PLANTS

CRISPR/Cas9 could be used to improve a variety of crop plants by creating crops with enhanced yield (Zhang et al., 2016; Xu et al., 2016), improved disease tolerance (Wang et al., 2016b), resistant to abiotic stresses (Shi et al., 2016; Haque et al., 2018) and better nutritional quality (Li et al., 2018). Woo et al. (2015) used the technique to manufacture tiny insertions and deletions that are equivalent to natural gene variants, which were then repaired using either a non-homologous end joining (NHEJ) or a donor template-based homology-directed repair (HDR) mechanism. Most CRISPR/Cas9-based gene-editing research has focused on the NHEJ pathway because it is the most common method in plants (Hilscher et al., 2016). The NHEJ mechanism aids in the creation of transgenic free plants that are not subject to regulatory inspection (Ishii and Araki, 2017); as they do not contain foreign components seen in GMOs, and the mutations caused are comparable to those seen in spontaneous random mutagenesis. As a result, once null segregants are obtained, plants created using NHEJ-based editing are currently unregulated. Compared to traditional breeding, NHEJ allows for faster crop breeding (Wieczorek and Wright, 2012; Waltz, 2016a; Sander and Jeschke, 2016). Exogenously given homologous template DNA sequences are also used in HDR to accurately repair DNA DSBs (Kim and Kim, 2014).

The genome editing of monocot and dicot crop species has been done using a single guide RNA (sgRNA) system, which is a simplified version of the type 2 CRISPR/Cas9 system from *Streptococcus pyogenes* (Jiang et al., 2013). *Agrobacterium tumefaciens* transferred genes producing a non-functional mutant green fluorescence protein (GFP) and Cas9/sgRNA to sorghum and *Arabidopsis*. Rice protoplast cells were used to create constructs that target the promoter region of the OsSWEET14 and OsSWEET11 bacterial blight susceptibility genes (Wolt et al., 2016).

17.6 BIOSAFETY CONCERNS ABOUT GENOME-EDITED CROP PLANTS

Researchers continue to debate the biosafety and socio-economic implications of genome editing in plants. They are most frequently associated with target gene site selection, guide RNA (gRNA) design, off-target mutation, and vector transformation systems. However, an off-target mutation poses a significant risk (Liang et al., 2018; Pineda et al., 2019). Several techniques were employed by the researchers, among which every technique has its limitations, therefore requiring the attention of researchers in the community of scientists (Liang et al., 2018; Murovec et al., 2018). Although the technology of delivering *in-vitro* pre-assembled CRISPR/Cas9 ribonucleoproteins (RNPs) has been employed in numerous crops, it has some limitations, including limited stability and a high level of technical demand, which must be addressed (Malnoy et al., 2016; Subburaj et al., 2016; Murovec et al., 2018). Delivering *in-vitro* pre-assembled CRISPR/Cas9 RNPs, protein engineering, tailored gRNA, the controlled activity of Cas9 protein and modifying genetic circuits of CRISPR to execute on pre-defined logic have all been used to reduce off-target mutations (Malnoy et al., 2016; Svitashev et al., 2016; Zhang et al., 2016; Liang et al., 2018).

Another issue with CRISPR/Cas9 technology is that the Cas9 protein may trigger an immunological response in mice when given *via* adeno-associated virus, posing a risk of immunogenic side effects (Chew et al.,

2016). The specificity of Cas9 in targeting a small number of sites is exacerbated by the necessity of the PAM sequence (Spencer and Zhang, 2017; El-Mounadi et al., 2020). The protein engineering helped in identifying the mutation in Cas9 that altered the PAM recognition sequence and enhanced its integrity and reliability to other motifs (Kleinstiver et al., 2015; Leenay and Beisel, 2017). Cas9 and gRNA design have been improved further using Fok1 fusions, paired nicking, and shortened gRNA (Wyvekens et al., 2015). Moreover, Cas9 variants, Cas9 homologs derived from bacterial strains and unique Cas9 proteins such as Cpf1 can be used to overcome the drawbacks (Nakade et al., 2017; Pineda et al., 2019; El-Mounadi et al., 2020; Fiaz et al., 2021).

The lack of understanding of the ideas and uses of genome editing is a cause of social concern. Transgenic plants, genome-edited plants, and genetically engineered plants all have distinct characteristics. Plants that have had their genomes modified may or may not be transgenic. A genome-edited plant will be termed non-transgenic if the transgene containing the CRISPR/Cas9 gene is eliminated using the gene segregation technique. As a result, teaching and raising public understanding regarding genome editing concepts can prevent and correct misconceptions (Garcia Ruiz et al., 2018; Eckerstorfer et al., 2019; El-Mounadi et al., 2020).

17.7 ETHICS OF THE CRISPR GENE EDITING SYSTEM

Genetic studies and research have changed dramatically as a result of genome editing technology. Among those genome editing technologies, CRISPR/Cas9 technology of genome editing stands out due to its various advantages like high accuracy, cost-effectiveness, efficiency, easy handling, and less time to modify the genetic background of plants, animals, and even humans (de Lecuona et al., 2017). However, despite the CRISPR/Cas9 system is widely used for the manipulation of genes, holds several ethical, moral, social, political, and scientific concerns, which needs to be addressed by the researchers of the scientific community to identify the problems, elucidate the concepts, and promote the discussions among all the stakeholders such as societies, universities, science, and technology organizations to prepare an ethical and legal framework for its wider acceptance (Lyon, 2017).

To address these concerns, precautionary measures have been recommended based on the damage that has been caused to the environment and health. Although agriculture is the only source of food and nutrition for all humans on this planet, it necessitates unique agricultural strategies for long-term food production. Similarly, genome editing techniques are playing an important role in developing plants and crops that can withstand both abiotic and biotic stresses. However, the ethical along with social and ecological concerns for CRISPR/Cas9 system in present day agriculture is still debatable. The ethical implications around the CRISPR genome-editing system have grown beyond the system's limitations, necessitating intra- and inter-governmental attention to address the challenges. In the industrialized world, however, there is a societal debate on how to guide and regulate the use of new technologies (Bartkowski et al., 2018; Fiaz et al., 2021).

17.8 REGULATORY BACKGROUND

The term genetically modified refers to plants whose genome structure has been altered *via* targeted mutagenesis and which previously did not exist (Wang et al., 2016c; Duensing et al., 2018; Friedrichs et al., 2019). Gene editing, on the other hand, refers to DNA changes in targeted plants or organisms that are similar to those that occur naturally as a result of deletions, substitutions, and insertions in traditional breeding approaches (Nature Plants Editorial, 2018). The Cartagena Protocol Biosafety set the foundation for regulating the release and international trade of genetically modified organisms (GMOs). However, there have been significant differences in the production, use, and regulation of genetically engineered plants. While some countries refuse to consume them and have banned their production, others freely cultivate and eat them (Garcia Ruiz et al., 2018; El-Mounadi et al., 2020).

There are two frameworks for regulating genome-edited crops. Primarily, regulating the process of creating genome-edited crops and secondly, regulating the final product features (Eckerstorfer et al., 2019; Van Vu et al., 2019). There are different regulations related to genome-edited crops in different countries (Table 17.1). Many countries have not formulated biosafety related regulations for genome-edited crops or for deregulating these crops (El-Mounadi et al., 2020), but several nations have published guidelines (Eckerstorfer et al., 2019). Access to the market for genetically engineered plant products, social acceptance and biosafety concerns are all emerging issues that researchers and government institutions are working to address, researchers, and government institutions are collaborating to provide benefits to people all over the world (Eckerstorfer et al., 2019). Using genome editing, it is possible to create plants that are genetically diverse and are similar to those created using conventional breeding methods (Barman et al., 2019). The genome-edited plants could be free of biosafety regulations dictated by conventional transgenic plants formulated on the transgene clean plant model (Van Vu et al., 2019). Global regulatory principles for genetically modified crops have an impact on genome editing in crops around the world, and these policies differ by country (Fiaz et al., 2021).

17.8.1 CANADA REGULATION POLICIES FOR GENOME EDITED CROPS

Canada has stayed dedicated in its commitment to scientific ideas that were established 25 years ago by the domestic regulatory framework for plants. According to Canadian regulations, any gene-editing technique which leads to the development of a new product must undergo further regulatory monitoring for allergenicity, toxicity, and impact on non-target organisms (Smyth, 2017). In collaboration with Health Canada, the Canadian Food Inspection Agency (CFIA) provides recommendations on determining novelty based on significant new qualities without taking into account the process of their formation (Schuttelaar, 2015; Gao, 2018). In this context, the first commercially manufactured herbicide-tolerant canola variety was licensed for cultivation by the CFIA and Health Canada (Jones, 2015).

TABLE 17.1 Genome Editing-based Regulations in Some Countries Across the World

Sl. No.	Name of the Country	Regulation Process	Method	Year
1.	India	GMO definition encompasses genome editing. Discussion is still going on.	Process-based	1989
2.	New Zealand	Genome-edited crops are regulated as GMO	Process-based	1996
3.	Australia	If no foreign DNA, then not regulated as GMO	Process-based	2000
4.	Israel	If no foreign DNA, then not regulated as GMO	Process-based	2000
5.	Pakistan	Absence of proper regulation	Process-based	2005
6.	Canada	Not regulated unless trait identified as novel	Product-based	2015
7.	China	No formal regulation	Case by case	2016
8.	Brazil	If no foreign DNA, then not regulated as GMO	Case by case	2018
9.	European Union	Genome-edited crops are regulated as GMOs	Process-based	2018
10.	United States	Mostly non-transgenic plants not regulated	Product-based	2018
11.	Japan	If no foreign DNA, then not regulated as GMO	Product-based	2019

17.8.2 UNITED STATES REGULATION POLICIES FOR GENOME-EDITED CROPS

In the United States, the regulatory environment is vastly different. The main bodies in charge of evaluating genetically modified crops are the United States Department of Agriculture (USDA), the Food and Drug Administration (FDA) and the Environmental Protection Agency (EPA). As a result, policies are developed and implemented within a well-coordinated framework. The USDA found in March 2018 that genome editing is identical to traditional breeding in some circumstances and does not require regulatory oversight under American regulations. The engineered mushroom that was discovered to be resistant to browning and waxy corn that contained starch composed entirely of amylopectin was among the first GE crops certified for commercial production using the CRISPR/Cas9 technology (Waltz, 2016b). In addition, the US adopted a form of a hybrid system that includes plants having modified characters *via* specific regulations (Agapito-Tenfen et al., 2018).

17.8.3 CHINA REGULATION POLICIES FOR GENOME-EDITED CROPS

Chinese researchers have published the largest number of genomes editing papers and invested heavily in genome editing research (Cohen and Desai, 2019). However, there are currently no formal regulations in place (Gao et al., 2018). In China, the foreign policies of the government based on genome editing are closely monitored. As of 2001, the State Council of China released the "Regulation on Administration of Agricultural Genetically Modified Organism Safety," which established GMOs are organisms, animals, plants, and animal-derived products with their genes altered by genome manipulation technology used in agricultural production and development. As a result, genome-edited products will be subject to GMO regulations (Gao et al., 2018). Caixa Gao, a senior CRISPR/Cas researcher, estimated that transferring a cultivar from the lab to the farm would take six months (Cohen, 2019).

17.8.4 NEW ZEALAND REGULATION POLICIES FOR GENOME-EDITED CROPS

New Zealand enacted the hazardous substances and new organisms (HSNO) act in 1996 to regulate process-based GMOs. By definition, a GMO is the result of any genes/genome manipulated through *in-vitro* technologies. The HSNO regulatory framework is therefore central to regulating a large number of genomes editing technologies (Fritsche et al., 2018). Under HSNO Act – Section 26, organisms or products which have been modified with no foreign genes, and which function similarly to chemically induced mutagenesis, will not be subject to regulations that govern GMOs (Kershen, 2015; Fritsche et al., 2018). In this case, the High Court reversed the interpretation. In response, the act was updated to clarify that all mutation techniques after 1996 result in GMOs. Plants whose genomes have been edited were regulated in New Zealand since then as GMOs (Kershen, 2015).

17.8.5 JAPAN REGULATION POLICIES FOR GENOME-EDITED CROPS

In Japan, the authorities are still debating the regulatory framework on GMOs. The Japanese Ministry of Health, Labor, and Welfare (MHLW) recently declared that foods that are not derived from transgenic approach are not considered genetically modified organisms (GMOs). They are unregulated as long as the genetic engineering (GE) induced DNA double-strand break (DSB) involves either a base pair deletion or a one-to-multiple base pair insertion (El-Mounadi et al., 2020).

17.8.6 EUROPEAN UNION (EU) REGULATION POLICIES FOR GENOME-EDITED CROPS

European Union (EU) opposed the cultivation and consumption of GMOs (Waltz, 2016b). The European Court of Justice (ECJ) declared in July 2018 that gene-edited crops must follow the same strict restrictions as traditionally genetically modified (GM) organisms (Metje-Sprink et al., 2018). Only mutagenesis procedures that have been traditionally used in a wide range of applications and have a long track record of safety are exempted from this criterion, according to ECJ (El-Mounadi et al., 2020; Fiaz et al., 2021).

17.8.7 INDIA'S REGULATION POLICIES FOR GENOME-EDITED CROPS

In 1989, the Indian government passed laws regulating the use of GMOs and their products, aside from genome editing technology (Fiaz et al., 2021).

India's Department of Biotechnology (2020) devised genome editing guidelines based on the categorization of plants in regulatory groups according to the types of genome editing. Group I plant have a few SDN-2 template-based base-pair modifications. Group II plants have a single or few base pair alterations or modifications based on SDN-1 or ODM, while group III plants have a lot of DNA alterations and foreign DNA insertions (Menz et al., 2020). GM foods, according to India's Food Safety and Standard Authorities are foods or elements obtained using a biotechnological process from genetically manipulated organisms (Chimata and Bharti, 2019). So far, there have been no new regulations governing genome editing technologies. Instead, they remain under existing regulatory frameworks (Friedrichs et al., 2019; Fiaz et al., 2021).

17.8.8 AUSTRALIAN REGULATION POLICIES FOR GENOME-EDITED CROPS

Introducing the gene technology (GT) act in 2000, Australia defined GMOs as organisms created by gene modifications or modification of genetic material. The regulatory development influenced the genome editing technologies utility, therefore producing transgenic plants that have no risk to the environment and human health (Mallapaty, 2019; Tchetvertakov, 2019). Regulations governing GT were enacted in 2001. According to Schedule-I of these standards organisms originating from the DNA transfer/exchange are not deemed GMO. Organisms transformed utilizing CRISPR/Cas9 and other unguided site-directed nuclease activity (SDN) repairs were excluded from being regulated as GMO. As a result, GMO control extends to ODM, SDN-2 and SDN-3 (Mallapaty, 2019; Thygesen, 2019).

17.8.9 PAKISTAN REGULATION POLICIES FOR GENOME-EDITED CROPS

The Biosafety Rules and the guidelines for agricultural biotechnology were drafted in 2005 following the Environment Protection Act. However, they have been in legislative limbo since 2010, following the 18th Amendment in 2010. The country's potential to benefit from the adoption of crops with high-stress tolerance and high-yielding cultivars has been hampered by the lack of a robust regulatory framework. Although, Pakistan generally follows on process-based approach in the case of genetically engineered based crops (Ishii and Araki, 2017).

17.8.10 ISRAEL REGULATION POLICIES FOR GENOME-EDITED CROPS

In March 2017, Israel formulated that plants whose genomes have been edited are not considered GMO, therefore reconfirming the statement that was proposed in 2016. In the same way, National Committee for Transgenic Plants has taken a decision not to include minor genomic sequence changes like nucleotide deletions, additions, and so on under such regulations in the case of genome-edited plants (Menz et al., 2020).

17.9 CONCLUSIONS AND FUTURE PERSPECTIVE

Genome editing techniques, such as CRISPR/Cas9, as well as innovative plant breeding strategies, have enormous potential for improving crop output, disease resistance, quality, and nutritional traits. This technology has contributed significantly in resolving the issues of food deficit currently prevailing in the world. In the USA, based on innovative plant products and an amicable regulatory environment, genome-edited plants and their products having significant market value are extensively investigated to become available for human consumption (Synthego, 2020). Nowadays, one of the main concerns about CRISPR-edited plant products is their acceptance among the producers and consumers of such products (Bartkowski et al., 2018; Wang et al., 2019). Many people are of the view that the final products obtained from gene editing having quality traits should be considered for food safety assessment and the method should be overlooked (Bain et al., 2020). Globally, there are diverse recognitions regarding the acceptance and rejection of gene-edited products, due to their altering regulatory requirements. In this context, it has been proposed by the scientific community globally to adopt uniform regulatory measures applicable to all the nations to recognize the worth of gene-editing technology apart from its research objective (Bhatta and Malla, 2020).

KEYWORDS

- **biosafety**
- **CRISPR/Cas9 technology**
- **double standard breaks**
- **ethical status**
- **genome-edited crops**
- **regulatory framework**

REFERENCES

Agapito-Tenfen, S. Z., Okoli, A. S., Bernstein, M. J., Wikmark, O. G., & Myhr, A. I., (2018). Revisiting risk governance of GM plants: The need to consider new and emerging gene-editing techniques. *Front. Plant Sci., 9,* 1874. doi: 10.3389/fpls.2018.01874.

Andersson, M., Turesson, H., Nicolia, A., Fält, A. S., Samuelsson, M., & Hofvander, P., (2017). Efficient targeted multiallelic mutagenesis in tetraploid potato (*Solanum tuberosum*) by transient CRISPR-Cas9 expression in protoplasts. *Plant Cell Rep., 36*(1), 117–128. doi: 10.1007/s00299-016-2062-3.

Bain, C., Lindberg, S., & Selfa, T., (2020). Emerging sociotechnical imaginaries for gene-edited crops for foods in the United States: Implications for governance. *Agric. Hum. Values, 37*(2), 265–279. doi: 10.1007/s10460-019-09980-9.

Baltes, N. J., Hummel, A. W., Konecna, E., Cegan, R., Bruns, A. N., Bisaro, D. M., & Voytas, D. F., (2015). Conferring resistance to *Geminiviruses* with the CRISPR-Cas prokaryotic immune system. *Nat. Plants, 1*(10), article 15145. doi: 10.1038/nplants.2015.145.

Barman, H. N., Sheng, Z., Fiaz, S., Zhong, M., Wu, Y., Cai, Y., Wang, W., et al., (2019). Generation of a new thermo-sensitive genic male sterile rice line by targeted mutagenesis of TMS5 gene through Crispr/Cas9 System. *B.M.C. Plant Biol., 19*(1), 109. doi: 10.1186/s12870-019-1715-0.

Bartkowski, B., Theesfeld, I., Pirscher, F., & Timaeus, J., (2018). Snipping around for food: Economic, ethical and policy implications of CRISPR/Cas genome editing. *Geoforum, 96*, 172–180. doi: 10.1016/j.geoforum.2018.07.017.

Beetham, P. R., Kipp, P. B., Sawycky, X. L., Arntzen, C. J., & May, G. D., (1999). A tool for functional plant genomics: Chimeric RNA/DNA oligonucleotides cause *in vivo* gene-specific mutations. *Proc. Natl. Acad. Sci. U. S. A, 96*(15), 8774–8778. doi: 10.1073/pnas.96.15.8774.

Beus, N. D., Dubois, J. L., Loo, R. V., Keller, H., Krens, F., Piotrowski, S., Rettenmaier, N., et al., (2020). W*hat is the CRISPR/Cas System and How is it Used in the COSMOS Project?* http://www.cosmos-h2020.eu/ (accessed on 04 September 2022).

Bhatta, B. P., & Malla, S., (2020). Improving horticultural crops via CRISPR/Cas9: Current successes and prospects: Review. *Plants (Basel), 9*(10), 1360. doi: 10.3390/plants9101360.

Chandrasekaran, J., Brumin, M., Wolf, D., Leibman, D., Klap, C., Pearlsman, M., Sherman, A., et al., (2016). Development of broad virus resistance in non-transgenic cucumber using CRISPR/Cas9 technology. *Mol. Plant Pathol., 17*(7), 1140–1153. doi: 10.1111/mpp.12375.

Chew, W. L., Tabebordbar, M., Cheng, J. K., Mali, P., Wu, E. Y., Ng, A. H., Zhu, K., et al., (2016). A multifunctional AAV-CRISPR-Cas9 and its host response. *Nat. Methods, 13*(10), 868–874. doi: 10.1038/nmeth.3993.

Chimata, M. K., & Bharti, G., (2019). Regulation of genome-edited technologies in India. *Transgenic Res., 28*(Suppl 2), 175–181. doi: 10.1007/s11248–019–00148-z.

Cohen, J., & Desai, N., (2019). With its CRISPR revolution, China has become a world leader in genome editing. *Science.* https://www.sciencemag.org/news/2019/08/its-crispr-revolution-chinabecomes-world-leader-genome-editing; doi: 10.1126/science.aay9689 (accessed on 04 September 2022).

Cohen, J., (2019). To feed its 1.4 billion, china bets big on genome editing of crops. *Science.* https://www.sciencemag.org/news/2019/07/feedits-14-billion-china-bets-big-genome-editing-crops; doi: 10.1126/science.aay8951 (accessed on 04 September 2022).

Curtin, S. J., Voytas, D. F., & Stupar, R. M., (2012). Genome engineering of crops with designer nucleases. *Plant Genome, 5*(2), 42–50. doi: 10.3835/plantgenome2012.06.0008.

De Lecuona, I., Casado, M., Marfany, G., Lopez, B. M. L., & Escarrabill, M., (2017). Gene editing in humans: Towards a global and inclusive debate for responsible research. *Yale J. Biol. Med., 90*(4), 673–681.

Duensing, N., Sprink, T., Parrott, W. A., Fedorova, M., Lema, M. A., Wolt, J. D., & Bartsch, D., (2018). Novel features and considerations for ERA and regulation of crops produced by genome editing. *Front. Bioeng. Biotechnol., 6,* 79. doi: 10.3389/fbioe.2018.00079.

Eckerstorfer, M. F., Engelhard, M., Heissenberger, A., Simon, S., & Teichmann, H., (2019). Plants developed by new genetic modification techniques—Comparison of existing regulatory frameworks in the EU and non-EU countries. *Front. Bioeng. Biotechnol., 7,* 26. doi: 10.3389/fbioe.2019.00026.

El-Mounadi, K., Morales-Floriano, M. L., & Garcia-Ruiz, H., (2020). Principles, applications and biosafety of plant genome editing using CRISPR-Cas9. *Front. Plant Sci., 11,* 56. doi: 10.3389/fpls.2020.00056.

Fang, Y., & Tyler, B. M., (2016). Efficient disruption and replacement of an effector gene in the oomycete *Phytophthora sojae* using CRISPR/Cas9. *Mol. Plant Pathol., 17*(1), 127–139. doi: 10.1111/mpp.12318.

Fiaz, S., Khan, S. A., Noor, M. A., Ali, H., Ali, N., Alharthi, B., Qayyum, A., & Nadeem, F., (2021). *CRISPR/Cas9 Regulations in Plant Science, 2021.* doi: 10.1016/B978-0-12-821910-2.00035-7.

Friedrichs, S., Takasu, Y., Kearns, P., Dagallier, B., Oshima, R., Schofield, J., et al., (2019). Meeting report of the O.E.C.D. conference on 'genome editing: Applications in agriculture-implications for health, environment and regulation.' *Transgen. Res., 28,* 419–463. doi: 10.1007/s11248-019-00154.

Fritsche, S., Poovaiah, C., MacRae, E., & Thorlby, G., (2018). A New Zealand perspective on the application and regulation of gene editing. *Front. Plant Sci., 9,* 1323. doi: 10.3389/fpls.2018.01323.

Gaj, T., Gersbach, C. A., & Barbas, C. F., (2013). ZFN, TALEN, and CRISPR/Cas-based methods for genome engineering. *Trends Biotechnol., 31*(7), 397–405. doi: 10.1016/j.tibtech.2013.04.004.

Gao, W., Xu, W. T., Huang, K. L., Guo, M. Z., & Luo, Y. B., (2018). Risk analysis for genome editing-derived food safety in China. *Food Control, 84,* 128–137. doi: 10.1016/j.foodcont.2017.07.032.

Garcia, R. M. T., Knapp, A. N., & Garcia-Ruiz, H., (2018). Profile of genetically modified plants authorized in Mexico. *G.M. Crops Food, 9*(3), 152–168. doi: 10.1080/21645698.2018.1507601.

Haque, E., Taniguchi, H., Hassan, M. M., Bhowmik, P., Karim, M. R., Śmiech, M., Zhao, K., et al., (2018). Application of CRISPR/Cas9 genome editing technology for the improvement of crops cultivated in tropical climates: Recent progress, prospects, and challenges. *Front. Plant Sci., 9,* 617. doi: 10.3389/fpls.2018.00617.

Hastings, G. E., & Wolf, P. G., (1992). The therapeutic use of albumin. *Arch. Fam. Med., 1*(2), 281–287.

Hilscher, J., Bürstmayr, H., & Stoger, E., (2016). Targeted modification of plant genomes for precision crop breeding. *Biotechnol. J., 12.* 1600173.

Ishii, T., & Araki, M. A., (2017). Future scenario of the global regulatory landscape regarding genome-edited crops. *G.M. Crops Food, 8*(1), 44–56. doi: 10.1080/21645698.2016.1261787.

Jansen, R., Embden, J. D., Gaastra, W., & Schouls, L. M., (2002). Identification of genes that are associated with DNA repeats in prokaryotes. *Mol. Microbiol., 43*(6), 1565–1575.us. doi: 10.1046/j.1365-2958.2002.02839.x.

Jiang, W. Z., Henry, I. M., Lynagh, P. G., Comai, L., Cahoon, E. B., & Weeks, D. P., (2017). Significant enhancement of fatty acid composition in seeds of the allohexaploid, *Camelina sativa*, using CRISPR/Cas9 gene editing. *Plant Biotechnol. J., 15*(5), 648–657. doi: 10.1111/pbi.12663.

Jiang, W., Zhou, H., Bi, H., Fromm, M., Yang, B., & Weeks, D. P., (2013). Demonstration of CRISPR/Cas9/sgRNA-mediated targeted gene modification in *Arabidopsis*, tobacco, sorghum and rice. *Nucleic Acids Res., 41*(20), e188. doi: 10.1093/nar/gkt780.

Jones, H. D., (2015). Regulatory uncertainty over genome editing. *Nat. Plants, 1,* 14011. doi: 10.1038/nplants.2014.11.

Kamburova, V. S., Nikitina, E. V., Shermatov, S. E., Buriev, Z. T., Kumpatla, S. P., Emani, C., & Abdurakhmonov, I. Y., (2017). Genome editing in plants: An overview of tools and applications. *Int. J. Agron., 2017,* 15, article ID 7315351. doi: 10.1155/2017/7315351.

Kershen, D. L., (2015). Sustainability council of New Zealand trust v. The environmental protection authority: Gene editing technologies and the law. *G.M. Crops Food, 6*(4), 216–222. doi: 10.1080/21645698.2015.1122859.

Kim, H., & Kim, J. S., (2014). A guide to genome engineering with programmable nucleases. *Nat. Rev. Genet., 15*(5), 321–334. doi: 10.1038/nrg3686.

Kleinstiver, B. P., Prew, M. S., Tsai, S. Q., Topkar, V. V., Nguyen, N. T., Zheng, Z., Gonzales, A. P., et al., (2015). Engineered CRISPR-Cas9 nucleases with altered PAM specificities. *Nature, 523,* 481–485. doi: 10.1038/nature14592.

Leenay, R. T., & Beisel, C. L., (2017). Deciphering, communicating, and engineering the CRISPR PAM. *J. Mol. Biol., 429*(2), 177–191. doi: 10.1016/j.jmb.2016.11.024.

Li, X., Wang, Y., Chen, S., Tian, H., Fu, D., Zhu, B., Luo, Y., & Zhu, H., (2018). Lycopene is enriched in tomato fruit by CRISPR/Cas9-mediated multiplex genome editing. *Front. Plant Sci., 9,* 559. doi: 10.3389/fpls.2018.00559.

Liang, Z., Chen, K., Zhang, Y., Liu, J., Yin, K., Qiu, J. L., & Gao, C., (2018). Genome editing of bread wheat using biolistic delivery of CRISPR/Cas9 *in vitro* transcripts or ribonucleoproteins. *Nat. Protoc., 13*(3), 413–430. doi: 10.1038/nprot.2017.145.

Liu, D., Chen, X., Liu, J., Ye, J., & Guo, Z., (2012). The rice ERF transcription factor OsERF922 negatively regulates resistance to *Magnaporthe oryzae* and salt tolerance. *J. Exp. Bot., 63*(10), 3899–3911. doi: 10.1093/jxb/ers079.

Liu, Q., Yang, F., Zhang, J., Liu, H., Rahman, S., Islam, S., Ma, W., & She, M., (2021). Application of CRISPR/Cas9 in crop quality improvement. *Int. J. Mol. Sci., 22*(8), 4206. https: doi: 10.3390/ijms22084206.

Lyon, J., (2017). Bioethics panels open door slightly to germline gene editing. *JAMA, 318*, 1639–1640. doi: 10.1001/jama.2017.13962.

Makarova, K. S., & Koonin, E. V., (2015). Annotation and classification of CRISPR-Cas systems. *Methods Mol. Biol., 1311,* 47–75. doi: 10.1007/978-1-4939-2687-9_4.

Mallapaty, S., (2019). *Australian Gene-Editing Rules Adopt 'Middle Ground.'* Nature. http://www.nature.com/articles/d41586-019-01282-8.

Malnoy, M., Viola, R., Jung, M. H., Koo, O. J., Kim, S., Kim, J. S., Velasco, R., & Nagamangala, K. C., (2016). DNA-free genetically edited grapevine and apple protoplast using CRISPR/Cas9 ribonucleoproteins. *Front. Plant Sci., 7,* 1904. doi: 10.3389/fpls.2016.01904.

Menz, J., Modrzejewski, D., Hartung, F., Wilhelm, R., & Sprink, T., (2020). Genome edited crops touch the market: A view on the global development and regulatory environment. *Front. Plant Sci., 11*, 586027. doi: 10.3389/fpls.2020.586027.

Metje-Sprink, J., Menz, J., Modrzejewski, D., & Sprink, T., (2018). DNA-free genome editing: Past, present and future. *Front. Plant Sci., 9,* 1957. doi: 10.3389/fpls.2018.01957.

Mojica, F. J., Díez-Villaseñor, C., Soria, E., & Juez, G., (2000). Biological significance of a family of regularly spaced repeats in the genomes of archaea, bacteria and mitochondria. *Mol. Microbiol., 36*(1), 244–246. doi: 10.1046/j.1365-2958.2000.01838.x.

Murovec, J., Guček, K., Bohanec, B., Avbelj, M., & Jerala, R., (2018). DNA-free genome editing of *Brassica oleracea* and *B. rapa* protoplasts using CRISPR-Cas9 ribonucleoprotein complexes. *Front. Plant Sci., 9,* 1594. doi: 10.3389/fpls.2018.01594.

Nakade, S., Yamamoto, T., & Sakuma, T., (2017). Cas9, Cpf1 and C2c1/2/3- what's next? *Bioengineered, 8*(3), 265–273. doi: 10.1080/21655979.2017.1282018.

Nature Plants Editorial, (2018). A CRISPR definition of genetic modification. *Nat. Plants, 4*(5), 233. doi: 10.1038/s41477-018-0158-1.

Oishi, I., Yoshii, K., Miyahara, D., Kagami, H., & Tagami, T., (2016). Targeted mutagenesis in chicken using CRISPR/Cas9 system. *Sci. Rep., 6*, 23980. doi: 10.1038/srep23980.

Osakabe, K., Osakabe, Y., & Toki, S., (2010). Site-directed mutagenesis in *Arabidopsis* using custom-designed zinc finger nucleases. *Proc. Natl. Acad. Sci. U. S. A, 107*(26), 12034–12039. doi: 10.1073/pnas.1000234107.

Osakabe, Y., & Osakabe, K., (2015). Genome editing with engineered nucleases in plants. *Plant Cell Physiol., 56*(3), 389–400. doi: 10.1093/pcp/pcu170.

Pineda, M., Lear, A., Collins, J. P., & Kiani, S., (2019). Safe CRISPR: Challenges and possible solutions. *Trends Biotechnol., 37*(4), 389–401. doi: 10.1016/j.tibtech.2018.09.010.

Purnhagen, K. P., Kok, E., Kleter, G., Schebesta, H., Visser, R. G. F., & Wesseler, J., (2018). EU court casts new plant breeding techniques into regulatory limbo. *Nat. Biotechnol., 36*(9), 799, 800. https://www.nature.com/articles/nbt.4251. doi: 10.1038/nbt.4251 (accessed on 04 September 2022).

Sander, J., & Jeschke, M., (2016). CRISPR-Cas advanced plant breeding. *Crop Insights, 26,* 18.

Schuttelaar, (2015). *The Regulatory Status of New Breeding Techniques in Countries Outside the European Union*. The Hague (NL): Schuttelaar & Partners. Available online at: https://www.nbtplatform.org/background-documents/rep-regulatory-status-of-nbts-oustide-the-eu-june-2015.pdf (accessed on 24 October 2022).

Shi, J., Gao, H., Wang, H., Lafitte, H. R., Archibald, R. L., Yang, M., Hakimi, S. M., et al., (2016). ARGOS. 8 variants generated by CRISPR-Cas9 improve maize grain yield under field drought stress conditions. *Plant Biotechnol. J., 15,* 207–216.

Smyth, S. J., (2017). Canadian regulatory perspectives on genome engineered crops. *G.M. Crops Food, 8*(1), 35–43. doi: 10.1080/21645698.2016.1257468.

Spencer, J. M., & Zhang, X., (2017). Deep mutational scanning of *S. pyogenes* Cas9 reveals important functional domains. *Sci. Rep., 7*(1), 16836. doi: 10.1038/s41598-017-17081-y.

Subburaj, S., Chung, S. J., Lee, C., Ryu, S. M., Kim, D. H., Kim, J. S., Bae, S., & Lee, G. J., (2016). Site-directed mutagenesis in *Petunia x hybrida* protoplast system using direct delivery of purified recombinant Cas9 ribonucleoproteins. *Plant Cell Rep., 35*(7), 1535–1544. doi: 10.1007/s00299-016-1937-7.

Svitashev, S., Schwartz, C., Lenderts, B., Young, J. K., & Mark, C. A., (2016). Genome editing in maize directed by CRISPR-Cas9 ribonucleoprotein complexes. *Nat. Commun., 7,* 13274. doi: 10.1038/ncomms13274.

Svitashev, S., Young, J. K., Schwartz, C., Gao, H., Falco, S. C., & Cigan, A. M., (2015). Targeted mutagenesis, precise gene editing, and site specific gene insertion in maize using Cas9 and guide RNA. *Plant Physiol., 169*(2), 931–945. doi: 10.1104/pp.15.00793.

Synthego. (2020). *CRISPR in Agriculture: An Era of Food Evolution.* https://www.synthego.com/blog/crispr-agriculture-foods#crops (accessed on 04 September 2022).

Tchetvertakov, G., (2019). *Australia Approves Cutting-Edge CRISPR Gene Editing Technology*. https://smallcaps.com.au/australia-approves-cutting-edge-crispr-gene-editing-technology/ (accessed on 24 October 2022).

Thygesen, P., (2019). Clarifying the regulation of genome editing in Australia: Situation for genetically modified organisms. *Transgen. Res., 28*(Suppl 2), 151–159. doi: 10.1007/s11248-019-00151-4.

Van, V. T., Sung, Y. W., Kim, J., Doan, D. T. H., Tran, M. T., & Kim, J. Y., (2019). Challenges and perspectives in homology-directed gene targeting in monocot plants. *Rice (N.Y.), 12*(1), 95. doi: 10.1186/s12284-019-0355-1.

Waltz, E., (2016a). CRISPR-edited crops free to enter market, skip regulation. *Nat. Biotechnol., 34*(6), 582. doi: 10.1038/nbt0616-582.

Waltz, E., (2016b). Gene-edited CRISPR mushroom escapes US regulation. *Nature, 532*(7599), 293. doi: 10.1038/nature.2016.19754.

Wang, F., Wang, C., Liu, P., Lei, C., Hao, W., Gao, Y., Liu, Y. G., & Zhao, K., (2016a). Enhanced rice blast resistance by CRISPR/Cas9-targeted mutagenesis of the ERF transcription factor gene OsERF922. *PLOS One, 11*(4), e0154027. doi: 10.1371/journal.pone.0154027.

Wang, G. P., Yu, X. D., Sun, Y. W., Jones, H. D., & Xia, L. Q., (2016c). Generation of marker-and/or backbone-free transgenic wheat plants via agrobacterium mediated transformation. *Front. Plant Sci., 7,* 1324. doi: 10.3389/fpls.2016.01324.

Wang, H., La Russa, M., & Qi, L. S., (2016b). CRISPR/Cas9 in genome editing and beyond. *Annu. Rev. Biochem., 85,* 227–264. doi: 10.1146/annurev-biochem-060815-014607.

Wang, T., Zhang, H., & Zhu, H., (2019). CRISPR technology is revolutionizing the improvement of tomato and other fruit crops. *Hortic. Res., 6*, 77. doi: 10.1038/s41438-019-0159-x, 6, 77.

Wieczorek, A., & Wright, M. G., (2012). History of agricultural biotechnology: How crop development has evolved. *Natl. Educ. Knowl., 3,* 9.

Wolt, J. D., Wang, K., & Yang, B., (2016). The regulatory status of genome-edited crops. *Plant Biotechnol. J., 14*(2), 510–518. doi: 10.1111/pbi.12444.

Woo, J. W., Kim, J., Kwon, S. I., Corvalán, C., Cho, S. W., Kim, H., Kim, S. G., et al., (2015). DNA-free genome editing in plants with preassembled CRISPR-Cas9 ribonucleoproteins. *Nat. Biotechnol., 33*(11), 1162–1164. doi: 10.1038/nbt.3389.

Wyvekens, N., Topkar, V. V., Khayter, C., Joung, J. K., & Tsai, S. Q., (2015). Dimeric Crispr RNA-guided FOKI-dCas9 nucleases directed by truncated gRNAs for highly specific genome editing. *Hum. Gene Ther., 26*, 425–431. doi: 10.1089/hum.2015.084.

Xu, X., Tao, Y., Gao, X., Zhang, L., Li, X., Zou, W., Ruan, K., et al., (2016). A CRISPR-based approach for targeted DNA demethylation. *Cell Discov., 2,* 16009. doi: 10.1038/celldisc.2016.9.

Zhang, Y., Liang, Z., Zong, Y., Wang, Y., Liu, J., Chen, K., Qiu, J. L., & Gao, C., (2016). Efficient and transgene-free genome editing in wheat through transient expression of CRISPR/Cas9 DNA or RNA. *Nat. Commun., 7,* 12617. doi: 10.1038/ncomms12617.

Index

B

C

D

F

N

O

S

T

U

V

W

X

Y

Z